Probability and Its Applications

T0237965

Published in association with the Applied Probability Trust

Editors: S. Asmussen, J. Gani, P. Jagers, T.G. Kurtz

Probability and Its Applications

Azencott et al.: Series of Irregular Observations. Forecasting and Model
Building. 1986

Bass: Diffusions and Elliptic Operators. 1997

Bass: Probabilistic Techniques in Analysis. 1995

Berglund/Gentz: Noise-Induced Phenomena in Slow-Fast Dynamical Systems:
A Sample-Paths Approach. 2006

Biagini/Hu/Øksendal/Zhang: Stochastic Calculus for Fractional Brownian Motion
and Applications. 2008

Chen: Eigenvalues, Inequalities and Ergodic Theory. 2005

Chen/Goldstein/Shao: Normal Approximation by Stein's Method. 2011

Costa/Fragoso/Marques: Discrete-Time Markov Jump Linear Systems. 2005

Daley/Vere-Jones: An Introduction to the Theory of Point Processes I: Elementary
Theory and Methods. 2nd ed. 2003, corr. 2nd printing 2005

Daley/Vere-Jones: An Introduction to the Theory of Point Processes II: General
Theory and Structure. 2nd ed. 2008

de la Peña/Gine: Decoupling: From Dependence to Independence, Randomly
Stopped Processes, U-Statistics and Processes, Martingales and Beyond. 1999

de la Peña/Lai/Shao: Self-Normalized Processes. 2009

Del Moral: Feynman-Kac Formulae. Genealogical and Interacting Particle
Systems with Applications. 2004

Durrett: Probability Models for DNA Sequence Evolution. 2002, 2nd ed. 2008

Ethier: The Doctrine of Chances. 2010

Feng: The Poisson–Dirichlet Distribution and Related Topics. 2010

Galambos/Simonelli: Bonferroni-Type Inequalities with Equations. 1996

Gani (ed.): The Craft of Probabilistic Modelling. A Collection of Personal
Accounts. 1986

Gawarecki/Mandrekar: Stochastic Differential Equations in Infinite Dimensions.
2011

Gut: Stopped RandomWalks. Limit Theorems and Applications. 1987

Guyon: Random Fields on a Network. Modeling, Statistics and Applications. 1995

Kallenberg: Foundations of Modern Probability. 1997, 2nd ed. 2002

Kallenberg: Probabilistic Symmetries and Invariance Principles. 2005

Last/Brandt: Marked Point Processes on the Real Line. 1995

Li: Measure-Valued Branching Markov Processes. 2011

Molchanov: Theory of Random Sets. 2005

Nualart: The Malliavin Calculus and Related Topics, 1995, 2nd ed. 2006

Rachev/Rueschendorf: Mass Transportation Problems. Volume I: Theory and
Volume II: Applications. 1998

Resnick: Extreme Values, Regular Variation and Point Processes. 1987

Schmidli: Stochastic Control in Insurance. 2008

Schneider/Weil: Stochastic and Integral Geometry. 2008

Serfozo: Basics of Applied Stochastic Processes. 2009

Shedler: Regeneration and Networks of Queues. 1986

Silvestrov: Limit Theorems for Randomly Stopped Stochastic Processes. 2004

Thorisson: Coupling, Stationarity and Regeneration. 2000

Zenghu Li

Measure-Valued Branching Markov Processes

 Springer

Zenghu Li
School of Mathematical Sciences
Beijing Normal University
No. 19 Xinjie Kouwai Street
Haidian District, Beijing 100875
People's Republic of China
lizh@bnu.edu.cn

Series Editors:

Søren Asmussen
Department of Mathematical Sciences
Aarhus University
Ny Munkegade
8000 Aarhus C
Denmark

Peter Jagers
Mathematical Statistics
Chalmers University of Technology
and University of Gothenburg
412 96 Göteborg
Sweden
jagers@chalmers.se

Joe Gani
Centre for Mathematics and its Applications
Mathematical Sciences Institute
Australian National University
Canberra, ACT 0200
Australia
gani@maths.anu.edu.au

Thomas G. Kurtz
Department of Mathematics
University of Wisconsin - Madison
480 Lincoln Drive
Madison, WI 53706-1388
USA
kurtz@math.wisc.edu

ISSN 1431-7028
ISBN 978-3-642-26620-1   ISBN 978-3-642-15004-3 (eBook)
DOI 10.1007/978-3-642-15004-3
Springer Heidelberg Dordrecht London New York

Mathematics Subject Classification (2010): 60J68, 60J80, 60G57, 60J70, 60J85, 60J35, 60J40

Cover design: VTEX, Vilnius

Printed on acid-free paper

Springer is part of Springer Science+Business Media (www.springer.com)

Preface

The books by Athreya and Ney (1972), Harris (1963) and Jagers (1975) contain a lot about finite-dimensional branching processes and their applications. Measure-valued branching processes with abstract underlying spaces were constructed in Watanabe (1968), who showed those processes arose as high-density limits of branching particle systems. The connection of measure-valued branching processes with stochastic evolution equations was investigated in Dawson (1975). A special class of measure-valued branching processes are known as Dawson–Watanabe superprocesses, which have been undergoing rapid development thanks to the contributions of a great number of researchers. The developments have been stimulated from different subjects including classical branching processes, interacting particle systems, stochastic partial differential equations and nonlinear partial differential equations. The study of superprocesses leads to better understanding of results in those subjects as well. We refer the reader to Dawson (1992, 1993), Dynkin (1994, 2002), Etheridge (2000), Le Gall (1999) and Perkins (1995, 2002) for detailed treatments of different aspects of the developments in the past decades. Branching processes give the mathematical modeling for populations evolving randomly in isolated environments. A useful and realistic modification of the branching model is the addition of immigration from outside sources. From the viewpoint of applications, branching models allowing immigration are clearly of great importance and physical appeal; see, e.g., Athreya and Ney (1972). This modification is also familiar in the setting of measure-valued processes; see, e.g., Dawson (1993), Dawson and Ivanoff (1978) and Dynkin (1991a).

The main purpose of this book is to give a compact and rigorous treatment of the basic theory of measure-valued branching processes and immigration processes. In the first part of the book, we give an analytic construction of Dawson–Watanabe superprocesses with general branching mechanisms. The spatial motions of those processes can be general Borel right processes in Lusin topological spaces. We show that the superprocesses arise as high-density limits of branching particle systems, giving the intuitive interpretations of the former. Under natural assumptions, it is shown that the superprocesses have Borel right realizations. From the general model, we use transformations to derive the existence and regularity of several different forms of the superprocesses including those in spaces of tempered measures, multitype models, age-structured models and time-inhomogeneous models. This unified treatment of the different models simplifies their constructions and gives useful perspectives for their properties. When the underlying space shrinks to a single point, the superprocess reduces to a one-dimensional continuous-state branching process. We discuss briefly extinction probabilities and limit theorems related to the latter. The theory of the one-dimensional processes requires much less prerequisite knowledge and is helpful for the reader in developing their intuitions for superprocesses. Under Feller type assumptions, several martingale problems for superprocesses are formulated and their equivalence are established. The martingale measures induced by those martingale problems are not necessarily orthogonal, but they are still worthy. To make the book essentially self-contained, overlaps of the first part with Dawson (1993) and Dynkin (1994) cannot be avoided completely, but we have made them as little as possible.

In the second part of the book we investigate the immigration structures associated with measure-valued branching processes. For that purpose, we first give some characterizations of entrance laws for those processes. We define immigration processes in an axiomatic way using skew convolution semigroups as in Li (1995/6). It is then proved that the skew convolution semigroups associated with a given measure-valued branching process are in one-to-one correspondence with its infinitely divisible probability entrance laws. The immigration superprocess has regularities similar to those of the Dawson–Watanabe superprocess if the corresponding probability entrance law is closable. Instead of establishing the results by repeating the techniques in the first part, we concentrate on the genuinely new or different aspects of the immigration processes and develop the theory on the bases of the processes without immigration. In this way, we hope to give the book a more compact and unified form.

The concept of skew convolution semigroups can actually be introduced in an abstract setting. Roughly speaking, such a semigroup gives the law of evolution of a system with branching structure under the perturbation of random extra forces. The immigration process is only a special case of this formulation. There is another special case investigated by Bogachev and Röckner (1995) and Bogachev et al. (1996), who formulated Ornstein–Uhlenbeck type processes on Hilbert spaces using generalized Mehler semigroups. Skew convolution semigroups were also used in Dawson and Li (2006) to study the affine Markov processes introduced in mathematical finance. In the last part of the book, we discuss briefly characterizations of generalized Mehler semigroups and properties of the corresponding Ornstein–Uhlenbeck type processes. We also show that a typical class of those processes arise as fluctuation limits of immigration superprocesses.

The main theory of Dawson–Watanabe superprocesses and immigration superprocesses is developed for general branching mechanisms that are not necessarily decomposable into local and non-local parts. Most of the results were obtained before only for specific classes of branching mechanisms. The emphasis here is the basic structures and regularities, rather than intensive properties of specific models. The setting of Borel right processes we have chosen is very convenient for the development of the theory. The title of the book stresses the applications of techniques from the theory of general Markov processes. Our main references for those are Ethier and Kurtz (1986) and Sharpe (1988). In the appendix we give a summary of the basic concepts and results that are frequently used. We hope the summary will help the reader in a quick start of the main parts of the book. In the last section of each chapter, comments on the history and recent development are given. This book can be used as a reference of the basics of Dawson–Watanabe superprocesses and immigration superprocesses. It can also be used in a course for graduate students specialized in probability and stochastic processes.

I would like to express my sincere thanks to Professor Zikun Wang for his advice and encouragement given to me for many years. I am deeply grateful to Professor Mufa Chen for his enormous help in my work. My special thanks are given to Professors Donald A. Dawson, Eugene B. Dynkin and Tokuzo Shiga, from whom I learned the theory of measure-valued processes. I have also benefited from stimulating discussions on this subject with many other experts including Professors Patrick J. Fitzsimmons, Klaus Fleischmann, Luis G. Gorostiza, Zhiming Ma, Hao Wang, Shinzo Watanabe, Jie Xiong and Xiaowen Zhou. I thank Professors Marco Fuhrman, Michael Röckner, Byron Schmuland, Wei Sun and Fengyu Wang for their advice on generalized Mehler semigroups. I am very grateful to Professors Peter Jagers, Thomas G. Kurtz, Jean-François Le Gall and Renming Song for valuable comments on earlier versions of this book. I want to thank Professors Wenming Hong, Yanxia Ren, Yongjin Wang, Kainan Xiang and Mei Zhang for helpful discussions. The materials in this book have been used for graduate courses in Beijing Normal University. I am indebted to my colleagues and students here, who provide a very pleasant research environment. In particular, I thank Congzao Dong, Hui He, Chunhua Ma, Rugang Ma, Li Wang and Xu Yang for reading the manuscript carefully and pointing out numerous typos and errors. I would like to express sincere gratitude to Dr. Marina Reizakis, the PIA series editor at Springer, for her advice and help. I want to thank the Natural Science Foundation and the Ministry of Education of China, who have supported my research in the past years. Finally I thank my wife and my son for their continuing moral support.

Beijing, China   *Zenghu Li*
                                                                                                                         May 18, 2010

Contents

Chapter 1
Random Measures on Metric Spaces

In this chapter, we discuss the basic properties of Laplace functionals of random measures, which provide an important tool in the study of measure-valued processes. In particular, we give some characterizations of the convergence of random measures in terms of their Laplace functionals. Based on these results, a general representation for the distributions of infinitely divisible random measures is established. We also give some characterizations of continuous functions on the positive half line with Lévy–Khintchine type representations. The reader is referred to Kallenberg (1975) for more complete discussions of random measures and Laplace functionals.

1.1 Borel Measures

Given a class $\mathscr{G}$ of functions on or subsets of some space E, let $\sigma(\mathscr{G})$ denote the σ-algebra on E generated by $\mathscr{G}$. If $\mathscr{F}$ is a class of functions, we define the classes $\mathrm{b}\mathscr{F} = \{f \in \mathscr{F} : f \text{ is bounded}\}$ and $\mathrm{p}\mathscr{F} = \{f \in \mathscr{F} : f \text{ is positive}\}$. Let $\mathbb{R}$ denote the real line and let $\mathbb{R}_+ = [0, \infty)$ denote the positive half line.

For a topological space E, let $\mathscr{B}(E)$ denote the σ-algebra on E generated by the class of open sets, which is referred to as the *Borel σ-algebra*. A real function defined on E is called a *Borel function* if it is measurable with respect to $\mathscr{B}(E)$. We also use $\mathscr{B}(E)$ to denote the set of Borel functions on E. Let $B(E) = \mathrm{b}\mathscr{B}(E)$ denote the Banach space of bounded Borel functions on E endowed with the supremum/uniform norm $\|\cdot\|$. For any $a \geq 0$, let $B_a(E)$ be the set of functions $f \in B(E)$ satisfying $\|f\| \leq a$. Let $C(E)$ denote the space of bounded continuous real functions on E. We use the superscript "+" to denote the subsets of positive elements of the function spaces, and the superscript "++" is used to denote those of positive elements bounded away from zero, e.g., $B(E)^+, C(E)^{++}$. If a metric d is specified on E, we denote by $C_u(E) := C_u(E, d)$ the subset of $C(E)$ of d-uniformly continuous real functions. If E is a locally compact space, then $C_0(E)$ denotes the space

Z. Li, *Measure-Valued Branching Markov Processes*,
Probability and Its Applications, DOI 10.1007/978-3-642-15004-3_1,
© Springer-Verlag Berlin Heidelberg 2011

of functions in $C(E)$ vanishing at infinity. Therefore $C_0(E) = C(E)$ when E is compact.

A *Borel measure*, or simply a *measure*, on some topological space E means a measure on the space $(E, \mathscr{B}(E))$. We write $\mu(f)$ or $\langle \mu, f \rangle$ for the integral of a function f with respect to a measure μ if the integral exists. The unit measure concentrated at a point $x \in E$ is denoted by δ_x. A measure μ on E is said to be *purely atomic* if it has the decomposition $\mu = \sum_i a_i \delta_{x_i}$ for countable families $\{a_i\} \subset [0, \infty)$ and $\{x_i\} \subset E$. We say μ is a *diffuse measure* if it does not charge any singleton.

Theorem 1.1 *Suppose that (E, d) is a metric space and A is a non-empty subset of E. For $x \in E$ let $d(x, A) = \inf\{d(x, y) : y \in A\}$. Then we have*

$$|d(x, A) - d(y, A)| \le d(x, y), \qquad x, y \in E. \tag{1.1}$$

In particular, $x \mapsto d(x, A)$ is a uniformly continuous function on E.

Proof. For any $x, y \in E$ and $z \in A$ we have

$$d(x, A) - d(y, z) \le d(x, z) - d(y, z) \le d(x, y).$$

Then we take the supremum over $z \in A$ in both sides to get

$$d(x, A) - d(y, A) \le d(x, y).$$

By the symmetry of $d(\cdot, \cdot)$ we have

$$d(y, A) - d(x, A) \le d(x, y).$$

Combining the two preceding inequalities gives (1.1). $\square$

Corollary 1.2 *For any metric space (E, d), we have $\sigma(C_u(E)^+) = \mathscr{B}(E)$.*

Proof. Since a continuous function is measurable, we have $\sigma(C_u(E)^+) \subset \mathscr{B}(E)$. Given a proper open subset $G \subset E$, let $f_n(x) = (1 \wedge d(x, G^c))^{1/n}$ for $x \in E$ and $n \ge 1$. By Theorem 1.1 we have $\{f_n\} \subset C_u(E)^+$. It is easy to see $f_n \to 1_G$ as $n \to \infty$, implying $G \in \sigma(C_u(E)^+)$. But we have $E \in \sigma(C_u(E)^+)$ clearly, so $\sigma(C_u(E)^+)$ contains all open subsets of E. That implies $\mathscr{B}(E) \subset \sigma(C_u(E)^+)$. $\square$

For a topological space E, let $M(E)$ denote the space of finite Borel measures on E and let $P(E)$ be the subset of $M(E)$ consisting of probability measures. We say a sequence $\{\mu_n\} \subset M(E)$ *converges weakly* to $\mu \in M(E)$ and write $\lim_{n \to \infty} \mu_n = \mu$ or $\mu_n \to \mu$ if $\lim_{n \to \infty} \mu_n(f) = \mu(f)$ for every $f \in C(E)$. The weak convergence is a topological concept. For functions $f_1, \ldots, f_k \in C(E)$ and open sets $G_1, \ldots, G_k \subset \mathbb{R}$ let

$$U(f_1, \ldots, f_k; G_1, \ldots, G_k) = \{\nu \in M(E) : \nu(f_i) \in G_i, 1 \le i \le k\}. \tag{1.2}$$

It is easy to show that the family $\mathscr{U}_0$ of sets $U(f_1, \ldots, f_k; G_1, \ldots, G_k)$ obtained by varying $k \geq 1$ and $\{(f_1, G_1), \ldots, (f_k, G_k)\}$ satisfies the axioms of a base for a topology of $M(E)$, which is called the *topology of weak convergence*. It is clear that $\lim_{n \to \infty} \mu_n = \mu$ if and only if μ_n converges to μ in this topology.

Proposition 1.3 *Let (E, d) be a metric space and let $\mathscr{G}$ be a set of functions on E which is closed under bounded pointwise convergence. Then*

(1) $C_u(E) \subset \mathscr{G}$ *implies* $B(E) \subset \mathscr{G}$;
(2) $C_u(E)^{++} \subset \mathscr{G}$ *implies* $B(E)^+ \subset \mathscr{G}$.

Proof. Since $C_u(E)$ is a vector space which contains 1_E and is closed under multiplication, the first assertion follows from Proposition A.2 and Corollary 1.2. Under the condition of the second assertion, we have

$$C_u(E) \subset \{f : e^f \in C_u(E)^{++}\} \subset \{f : e^f \in \mathscr{G}\}.$$

Then the first assertion implies $B(E) \subset \{f : e^f \in \mathscr{G}\}$. In particular, for any $h \in B(E)^{++}$ we have $\log h \in B(E) \subset \{f : e^f \in \mathscr{G}\}$ and hence $h = e^{\log h} \in \mathscr{G}$. That proves $B(E)^{++} \subset \mathscr{G}$, implying the second assertion since $\mathscr{G}$ is closed under bounded pointwise convergence. $\qquad\square$

Corollary 1.4 *Let (E, d) be a metric space and let $\mu, \nu \in M(E)$. If $\mu(f) = \nu(f)$ for every $f \in C_u(E)$, we have $\mu = \nu$.*

Proof. Let $\mathscr{G}$ be the family of functions $f \in B(E)$ such that $\mu(f) = \nu(f)$. Then $C_u(E) \subset \mathscr{G}$. By dominated convergence it is easy to show that $\mathscr{G}$ is closed under bounded pointwise convergence. Consequently, we have $B(E) \subset \mathscr{G}$ by Proposition 1.3. $\qquad\square$

Corollary 1.5 *Let (E, d) be a metric space. Then for every $f \in B(E)$ the mapping $\mu \mapsto \mu(f)$ from $M(E)$ to $\mathbb{R}$ is Borel measurable.*

Proof. Let $\mathscr{G}$ be the family of functions $f \in B(E)$ such that $\mu \mapsto \mu(f)$ is Borel measurable. Then $\mathscr{G}$ is closed under bounded pointwise convergence. For any $f \in C_u(E)$ the mapping $\mu \mapsto \mu(f)$ is continuous and hence Borel measurable. In other words, we have $C_u(E) \subset \mathscr{G}$. Then Proposition 1.3 implies $B(E) \subset \mathscr{G}$. $\qquad\square$

Theorem 1.6 *Suppose that (E, d) is a metric space. For any $\mu \in M(E)$ and any sequence $\{\mu_n\} \subset M(E)$ the following statements are equivalent:*

(1) $\lim_{n \to \infty} \mu_n = \mu$;
(2) $\lim_{n \to \infty} \mu_n(f) = \mu(f)$ *for every* $f \in C_u(E)$;
(3) $\lim_{n \to \infty} \mu_n(1) = \mu(1)$ *and* $\limsup_{n \to \infty} \mu_n(C) \leq \mu(C)$ *for every closed set* $C \subset E$;
(4) $\lim_{n \to \infty} \mu_n(1) = \mu(1)$ *and* $\liminf_{n \to \infty} \mu_n(G) \geq \mu(G)$ *for every open set* $G \subset E$;
(5) $\lim_{n \to \infty} \mu_n(B) = \mu(B)$ *for every* $B \in \mathscr{B}(E)$ *with* $\mu(\partial B) = 0$, *where* ∂B *is the boundary of* B.

Proof. The results are obvious if $\mu(1) = 0$. If $\mu(1) > 0$, then there is an index $n_0 \geq 1$ so that $\mu_n(1) > 0$ for all $n \geq n_0$. Let $\hat{\mu} = \mu(1)^{-1}\mu \in P(E)$ and let $\hat{\mu}_n = \mu_n(1)^{-1}\mu_n \in P(E)$ for $n \geq n_0$. It is easy to see that $\lim_{n \to \infty} \mu_n = \mu$ if and only if $\lim_{n \to \infty} \mu_n(1) = \mu(1)$ and $\lim_{n \to \infty} \hat{\mu}_n = \hat{\mu}$. Then the theorem follows from the results in the special case of probability measures; see, e.g., Ethier and Kurtz (1986, p.108) and Parthasarathy (1967, pp.41–42). $\qquad\square$

Theorem 1.7 *Suppose that E is a Borel subspace of a metrizable topological space F. Let $\mu \in M(E)$ and let $\{\mu_n\} \subset M(E)$ be a sequence. Let ν and ν_n denote respectively the extensions of μ and μ_n to F such that $\nu(F \setminus E) = \nu_n(F \setminus E) = 0$. Then $\lim_{n \to \infty} \nu_n = \nu$ in $M(F)$ if and only if $\lim_{n \to \infty} \mu_n = \mu$ in $M(E)$.*

Proof. Since the restriction of a bounded continuous function is also a bounded continuous function, $\lim_{n \to \infty} \mu_n = \mu$ in $M(E)$ implies $\lim_{n \to \infty} \nu_n = \nu$ in $M(F)$. For the converse, suppose that $\lim_{n \to \infty} \nu_n = \nu$ in $M(F)$. Then we have

$$\lim_{n \to \infty} \mu_n(E) = \lim_{n \to \infty} \nu_n(F) = \nu(F) = \mu(E).$$

For any closed subset C of E, there is a closed subset D of F such that $C = D \cap E$. It follows that

$$\limsup_{n \to \infty} \mu_n(C) = \limsup_{n \to \infty} \nu_n(D) \leq \nu(D) = \mu(C).$$

Then $\lim_{n \to \infty} \mu_n = \mu$ in $M(E)$ by Theorem 1.6. $\qquad\square$

If E is a separable metric space, its topology can be defined by a totally bounded metric $(x, y) \mapsto d(x, y)$. Indeed, E is homeomorphic to a subset of the countable product space $[0, 1]^\infty$ furnished with the product metric; see, e.g., Kelley (1955, p.125). Then the set of uniformly continuous functions $C_u(E)$ endowed with the supremum norm $\| \cdot \|$ is a separable Banach space; see, e.g., Parthasarathy (1967, p.43).

Theorem 1.8 *If E is a separable metric space, then $M(E)$ is separable.*

Proof. Let Q be a countable dense subset of $[0, \infty)$ and F a countable dense subset of E. We claim that the countable set

$$M_1 := \left\{ \sum_{i=1}^{n} \alpha_i \delta_{x_i} : x_1, \dots, x_n \in F; \alpha_1, \dots, \alpha_n \in Q; n \geq 1 \right\}$$

is dense in $M(E)$. To see that we first fix a totally bounded metric d on E compatible with its topology. Then for each integer $n \geq 1$ the space E has a finite covering $\{B_{n,i} : i = 1, \dots, p_n\}$ consisting of open balls of radius $1/n$. Take $x_{n,i} \in B_{n,i} \cap F$ for $i = 1, \dots, p_n$. Let $A_{n,1} = B_{n,1}$ and let $A_{n,i} = B_{n,i} \setminus (B_{n,1} \cup \cdots \cup B_{n,i-1})$ for $i = 2, \dots, p_n$. Given $\mu \in M(E)$ we take $\alpha_{n,i} \in Q$ so that $|\alpha_{n,i} - \mu(A_{n,i})| \leq 1/np_n$ and define $\mu_n = \sum_{i=1}^{p_n} \alpha_{n,i} \delta_{x_{n,i}} \in M_1$. Then for any $f \in C_u(E)$ we have

$$\begin{aligned}
|\mu_n(f) - \mu(f)| &\le \sum_{i=1}^{p_n} |\alpha_{n,i} f(x_{n,i}) - \mu(f 1_{A_{n,i}})| \\
&\le \sum_{i=1}^{p_n} |\alpha_{n,i} - \mu(A_{n,i})||f(x_{n,i})| \\
&\quad + \sum_{i=1}^{p_n} |\mu(A_{n,i}) f(x_{n,i}) - \mu(f 1_{A_{n,i}})| \\
&\le \|f\| \sum_{i=1}^{p_n} \frac{1}{n p_n} + \sum_{i=1}^{p_n} \sup_{y \in A_{n,i}} |f(x_{n,i}) - f(y)| \mu(A_{n,i}) \\
&\le \frac{\|f\|}{n} + \sup_{d(x,y) \le 2/n} |f(x) - f(y)| \mu(E).
\end{aligned}$$

By the uniform continuity of $f \in C_u(E)$, the right-hand side of the above inequality tends to zero as $n \to \infty$. Then M_1 is dense in $M(E)$. $\qquad\square$

Theorem 1.9 *Suppose that E is a separable metric space and d is a totally bounded metric on E for its topology. Let $S(E, d) = \{f_0, f_1, f_2, \ldots\}$ be a dense sequence in $C_u(E)$ with $f_0 \equiv 1$. Then $\mu_n \to \mu$ in $M(E)$ if and only if $\mu_n(f_i) \to \mu(f_i)$ for every $f_i \in S(E, d)$.*

Proof. It is clear that $\mu_n \to \mu$ in $M(E)$ implies $\mu_n(f_i) \to \mu(f_i)$ for every $f_i \in S(E, d)$. Conversely, suppose that $\mu_n(f_i) \to \mu(f_i)$ for every $f_i \in S(E, d)$. For $f \in C_u(E)$ and $f_i \in S(E, d)$ we have

$$\begin{aligned}
|\mu_n(f) - \mu(f)| &\le \mu_n(|f - f_i|) + \mu(|f - f_i|) + |\mu_n(f_i) - \mu(f_i)| \\
&\le \|f - f_i\|[\mu_n(1) + \mu(1)] + |\mu_n(f_i) - \mu(f_i)|.
\end{aligned}$$

Since there is a sequence $\{f_{k_i}\} \subset S(E, d)$ satisfying $\|f_{k_i} - f\| \to 0$, it is easy to conclude $|\mu_n(f) - \mu(f)| \to 0$. Then $\mu_n \to \mu$ by Theorem 1.6. $\qquad\square$

Corollary 1.10 *Suppose that E is a separable metric space and d is a totally bounded metric on E for its topology. Let $S_1(E, d) = \{h_0, h_1, h_2, \ldots\}$ be a dense sequence in $\{f \in C_u(E)^+ : \|f\| \le 1\}$ with $h_0 \equiv 1$. Then $\mu_n \to \mu$ in $M(E)$ if and only if $\mu_n(h_i) \to \mu(h_i)$ for every $h_i \in S_1(E, d)$.*

Proof. If $\mu_n \to \mu$ in $M(E)$, we clearly have $\mu_n(h_i) \to \mu(h_i)$ for every $h_i \in S_1(E, d)$. Conversely, suppose that $\mu_n(h_i) \to \mu(h_i)$ for every $h_i \in S_1(E, d)$. Then $\mu_n(f) \to \mu(f)$ for every $f \in \mathscr{D}$, where $\mathscr{D} = \{a h_i + b h_j : h_i, h_j \in S_1(E, d)$ and a, b are rationals$\}$ is a countable dense subset of $C_u(E)$. Then we have $\mu_n \to \mu$ in $M(E)$ by Theorem 1.9. $\qquad\square$

Given a separable metric space E, we fix a totally bounded metric d compatible with its topology and let $S_1(E, d)$ be as in Corollary 1.10. Then a metric ρ on $M(E)$ is defined by

$$\rho(\mu, \nu) = \sum_{i=0}^{\infty} \frac{1}{2^i} (1 \wedge |\mu(h_i) - \nu(h_i)|), \qquad \mu, \nu \in M(E). \tag{1.3}$$

This metric is compatible with the weak convergence topology of $M(E)$. In other words, we have $\mu_n \to \mu$ in $M(E)$ if and only if $\rho(\mu_n, \mu) \to 0$. The countable family $\mathcal{U}_1$ of sets

$$U(h_0, h_1, \ldots, h_k; (a_0, b_0), (a_1, b_1), \ldots, (a_k, b_k))$$

obtained by varying the integer $k \geq 1$, the functions $h_i \in S_1(E, d)$ and the pairs of rationals $a_i < b_i$ is a countable base of the topology of $M(E)$.

Theorem 1.11 *For a separable metric space E we have $\mathscr{B}(M(E)) = \sigma(\{\mu \mapsto \mu(f) : f \in C(E)^+\}) = \sigma(\{\mu \mapsto \mu(f) : f \in C(E)\})$.*

Proof. It is easy to see that $\mathscr{B}_0 := \sigma(\{\mu \mapsto \mu(f) : f \in C(E)^+\})$ contains the countable family $\mathcal{U}_1$. Since every open subset of $M(E)$ is the union of some elements of this family, all those open subsets belong to $\mathscr{B}_0$ and hence $\mathscr{B}(M(E)) \subset \mathscr{B}_0 \subset \sigma(\{\mu \mapsto \mu(f) : f \in C(E)\})$. On the other hand, for any $f \in C(E)$ the mapping $\mu \mapsto \mu(f)$ is continuous on $M(E)$. Then we have $\sigma(\{\mu \mapsto \mu(f) : f \in C(E)\}) \subset \mathscr{B}(M(E))$. $\qquad\square$

Corollary 1.12 *If E is a separable metric space, then $\mathscr{B}(M(E)) = \sigma(\{\mu \mapsto \mu(f) : f \in B(E)\}) = \sigma(\{\mu \mapsto \mu(A) : A \in \mathscr{B}(E)\})$.*

Proof. Let $\mathscr{B}_1 = \sigma(\{\mu \mapsto \mu(f) : f \in B(E)\})$ and $\mathscr{B}_2 = \sigma(\{\mu \mapsto \mu(A) : A \in \mathscr{B}(E)\})$. Then $\mathscr{B}_1 \supset \mathscr{B}_2$ obviously. By Theorem 1.11 it is easy to see $\mathscr{B}(M(E)) \subset \mathscr{B}_1$. Then Corollary 1.5 implies $\mathscr{B}(M(E)) = \mathscr{B}_1$. For a simple function $f \in B(E)$, the mapping $\mu \mapsto \mu(f)$ is clearly measurable with respect to $\mathscr{B}_2$. By an approximation argument one sees $\mu \mapsto \mu(f)$ is measurable with respect to $\mathscr{B}_2$ for an arbitrary $f \in B(E)$. Then $\mathscr{B}_2 \supset \mathscr{B}_1$. $\qquad\square$

Theorem 1.13 *Suppose that $(F, \mathscr{F})$ is a general measurable space and E is a separable metric space. For $A \in \mathscr{B}(E)$ and $\mu \in M(E)$ write $l_A(\mu) = \mu(A)$. Then ψ is a measurable map from $(F, \mathscr{F})$ to $(M(E), \mathscr{B}(M(E)))$ if and only if for every $A \in \mathscr{B}(E)$ the composition $l_A \circ \psi$ is a measurable real function on $(F, \mathscr{F})$.*

Proof. Suppose that ψ is a measurable map from $(F, \mathscr{F})$ to $(M(E), \mathscr{B}(M(E)))$. By Corollary 1.12 the real function l_A on $M(E)$ is Borel for every $A \in \mathscr{B}(E)$. Then the composition $l_A \circ \psi$ is a measurable function on $(F, \mathscr{F})$. Conversely, suppose for every $A \in \mathscr{B}(E)$ the composition $l_A \circ \psi$ is a measurable function on $(F, \mathscr{F})$. Then for $B \in \mathscr{B}(\mathbb{R})$ we have $\psi^{-1}(l_A^{-1}(B)) \in \mathscr{F}$, so $\psi^{-1}(\{l_A^{-1}(B) : A \in \mathscr{B}(E), B \in \mathscr{B}(\mathbb{R})\}) \subset \mathscr{F}$. It follows that

$$\begin{aligned}
\mathscr{F} &\supset \sigma(\psi^{-1}(\{l_A^{-1}(B) : A \in \mathscr{B}(E), B \in \mathscr{B}(\mathbb{R})\})) \\
&= \psi^{-1}(\sigma(\{l_A^{-1}(B) : A \in \mathscr{B}(E), B \in \mathscr{B}(\mathbb{R})\})) \\
&= \psi^{-1}(\sigma(\{l_A : A \in \mathscr{B}(E)\})) = \psi^{-1}(\mathscr{B}(M(E))).
\end{aligned}$$

Then ψ is a measurable map from $(F, \mathscr{F})$ to $(M(E), \mathscr{B}(M(E)))$. $\qquad\square$

Theorem 1.14 *If E is a compact metric space, then $M(E)$ is a locally compact separable and metrizable space. Moreover, for any $b \geq 0$ the set $M_b := \{\mu \in M(E) : \mu(E) \leq b\}$ is compact.*

Proof. Since E is a compact metric space, it is separable. Then $M(E)$ is separable by Theorem 1.8. Let d be a metric on E for its topology and let $S_1(E,d)$ be as in Corollary 1.10. The topology of $M(E)$ can be defined by the metric ρ given by (1.3). For any $\mu \in M(E)$ let $T(\mu) = (\mu(h_0), \mu(h_1), \mu(h_2), \ldots)$. It is easy to see that T is a homeomorphism between $M(E)$ and a subset of the countable product space $\mathbb{R}_+^\infty$. Observe that $T(M_b) \subset [0,b]^\infty \subset \mathbb{R}_+^\infty$. We claim that $T(M_b)$ is closed in $[0,b]^\infty$. To see this, suppose that $\{\mu_n\} \subset M_b$ and $T(\mu_n) \to (\alpha_0, \alpha_1, \alpha_2, \ldots)$ in $[0,b]^\infty$. We need to show $(\alpha_0, \alpha_1, \alpha_2, \ldots) \in T(M_b)$. For $f \in C_u(E)^+$ satisfying $\|f\| \leq 1$ let $\{h_{i_k}\} \subset S_1(E,d)$ be a sequence so that $\|h_{i_k} - f\| \to 0$ as $k \to \infty$. For $n \geq m \geq 1$ we have

$$|\mu_n(f) - \mu_m(f)| \leq \|f - h_{i_k}\|[\mu_n(1) + \mu_m(1)] + |\mu_n(h_{i_k}) - \mu_m(h_{i_k})|$$

and hence

$$\limsup_{m,n\to\infty} |\mu_n(f) - \mu_m(f)| \leq 2\alpha_0 \|f - h_{i_k}\|.$$

Then letting $k \to \infty$ gives

$$\limsup_{m,n\to\infty} |\mu_n(f) - \mu_m(f)| = 0.$$

By linearity, the above relation holds for all $f \in C_u(E)$, so the limit $\lambda(f) = \lim_{n\to\infty} \mu_n(f)$ exists for each $f \in C_u(E)$. Clearly, $f \mapsto \lambda(f)$ is a positive linear functional on $C_u(E)$. By the Riesz representation theorem, there exists $\mu \in M(E)$ so that $\mu(f) = \lambda(f)$ for every $f \in C_u(E)$. In particular, $\mu(h_i) = \lambda(h_i) = \alpha_i$ for all $i \geq 0$. It follows that $\mu(1) = \alpha_0 = \lim_{n\to\infty} \mu_n(1) \leq b$ and hence $\mu \in M_b$. That shows $(\alpha_0, \alpha_1, \alpha_2, \ldots) = T(\mu) \in T(M_b)$. Then $T(M_b)$ is a closed subset of $[0,b]^\infty$. Since $[0,b]^\infty$ is compact, so is $T(M_b)$. It follows that M_b is compact and $M(E)$ locally compact. $\square$

Corollary 1.15 *Let E be a compact metric space and let $\bar{M}(E) := M(E) \cup \{\Delta\}$ be the one-point compactification of $M(E)$. Then $\mu_n \to \Delta$ if and only if $\mu_n(E) \to \infty$.*

Proof. It is easy to see that $\{\bar{M}(E) \setminus M_b : b \geq 0\}$ is a local base at Δ. Then the assertion is evident. $\square$

A metrizable space E is called a *Lusin topological space* if it is homeomorphic to a Borel subset of a compact metric space. Such a space is clearly separable. A measurable space $(F, \mathscr{F})$ is called a *Lusin measurable space* if it is measurably isomorphic to $(E, \mathscr{B}(E))$ with E being a Lusin topological space.

Theorem 1.16 *If E is a Lusin topological space, then $M(E)$ is a Lusin topological space.*

Proof. Since E is a Lusin topological space, we may embed it into some compact metric space F as a Borel subset. Theorem 1.7 implies that $M(E)$ is homeomorphic to $\{\mu \in M(F) : \mu(F \setminus E) = 0\}$. By Corollary 1.12 the mapping $\mu \mapsto \mu(F \setminus E)$ is $\mathscr{B}(M(F))$-measurable. Then $\{\mu \in M(F) : \mu(F \setminus E) = 0\}$ is a Borel subset of the locally compact separable and metrizable space $M(F)$, which is an open subset of its one-point compactification $\bar{M}(F) := M(F) \cup \{\Delta\}$. Therefore $M(E)$ is homeomorphic to a Borel subset of the compact metrizable space $\bar{M}(F)$.   □

Example 1.1 Let $[0,1]^\infty$ be the countable product of the unit interval furnished with the product metric q. Suppose that (E,d) is a separable metric space with the dense sequence $F := \{x_1, x_2, \ldots\}$. For any $x \in E$ write

$$g(x) = (1 \wedge d(x,x_1), 1 \wedge d(x,x_2), \ldots).$$

Then g is a homeomorphism between E and $g(E) \subset [0,1]^\infty$. This homeomorphism induces a totally bounded metric on E compatible with its original topology. For $\varepsilon > 0$ let

$$g(F)^\varepsilon = \{y \in [0,1]^\infty : q(y, g(x_i)) < \varepsilon \text{ for some } i \geq 1\}.$$

Clearly, each $g(F)^\varepsilon$ is an open set in the compact metric space $[0,1]^\infty$. If (E,d) is complete in addition, then $g(E) = \cap_{n=1}^\infty g(F)^{1/n}$. Consequently, a complete separable metric space is a Lusin topological space.

1.2 Laplace Functionals

In this section, we assume E is a Lusin topological space. Recall that $M(E)$ is the space of finite measures on E equipped with the topology of weak convergence. Given a finite measure Q on $M(E)$, we define the *Laplace functional* L_Q of Q by

$$L_Q(f) = \int_{M(E)} e^{-\nu(f)} Q(\mathrm{d}\nu), \qquad f \in B(E)^+. \tag{1.4}$$

Theorem 1.17 *A finite measure on $M(E)$ is uniquely determined by the restriction of its Laplace functional to $C(E)^+$.*

Proof. Suppose that Q_1 and Q_2 are finite measures on $M(E)$ and $L_{Q_1}(f) = L_{Q_2}(f)$ for all $f \in C(E)^+$. Let $\mathscr{K} = \{\nu \mapsto e^{-\nu(f)} : f \in C(E)^+\}$ and let $\mathscr{L} = \{F \in B(M(E)) : Q_1(F) = Q_2(F)\}$. Then $\mathscr{K}$ is closed under multiplication and $\mathscr{L}$ is a monotone vector space containing $\mathscr{K}$. By Theorem 1.11 it is easy to show $\sigma(\mathscr{K}) = \mathscr{B}(M(E))$. Then Proposition A.1 implies $\mathscr{L} \supset \mathrm{b}\sigma(\mathscr{K}) = B(M(E))$. That proves the desired result.   □

Theorem 1.18 *Let $Q_1, Q_2, \ldots$ and Q be finite measures on $M(E)$. If $Q_n \to Q$ weakly, then $L_{Q_n}(f) \to L_Q(f)$ for $f \in C(E)^+$. Conversely, if $L_{Q_n}(f) \to L_Q(f)$ for all $f \in C(E)^{++} \cup \{0\}$, then $Q_n \to Q$ weakly.*

Proof. If $Q_n \to Q$ weakly, we have $\lim_{n\to\infty} L_{Q_n}(f) = L_Q(f)$ for $f \in C(E)^+$ clearly. Now assume $\lim_{n\to\infty} L_{Q_n}(f) = L_Q(f)$ for all $f \in C(E)^{++} \cup \{0\}$. Let F be a compact metric space so that E is embedded into F as a Borel subset. Then $M(F)$ is a locally compact separable and metrizable space. Let $\bar{M}(F) = M(F) \cup \{\Delta\}$ be its one-point compactification, which is a compact metrizable space by Theorem 1.14. We identify $M(E)$ with the Borel subset of $M(F)$ consisting of measures supported by E and regard $Q_1, Q_2, \ldots$ and Q as finite measures on $\bar{M}(F)$. Then

$$\lim_{n\to\infty} Q_n(\bar{M}(F)) = \lim_{n\to\infty} Q_n(M(E)) = \lim_{n\to\infty} L_{Q_n}(0) = L_Q(0), \qquad (1.5)$$

and hence $\{Q_n\} \subset M(\bar{M}(F))$ is a bounded sequence. By another application of Theorem 1.14 we conclude that $\{Q_n\}$ is relatively compact. Let $\{Q_{n_k}\} \subset \{Q_n\}$ be a subsequence that converges to some $\bar{Q} \in M(\bar{M}(F))$. By (1.5) we have

$$\bar{Q}(\bar{M}(F)) = \lim_{k\to\infty} Q_{n_k}(\bar{M}(F)) = L_Q(0). \qquad (1.6)$$

Moreover, for any $\bar{f} \in C(F)^{++}$,

$$\int_{\bar{M}(F)} e^{-\nu(\bar{f})} \bar{Q}(d\nu) = \lim_{k\to\infty} \int_{\bar{M}(F)} e^{-\nu(\bar{f})} Q_{n_k}(d\nu) = L_Q(f), \qquad (1.7)$$

where $f = \bar{f}|_E$ denotes the restriction of $\bar{f}$ to E and $e^{-\Delta(\bar{f})} = 0$ by convention. By letting $f \to 0+$ in (1.7) we find $\bar{Q}(M(F)) = L_Q(0)$, so $\bar{Q}$ is supported by $M(F)$. From (1.7) we have

$$\int_{M(F)} e^{-\nu(\bar{f})} \bar{Q}(d\nu) = L_Q(f) = \int_{M(E)} e^{-\nu(f)} Q(d\nu).$$

Then the uniqueness of Laplace functionals implies $\bar{Q}$ is supported by $M(E)$ and its restriction to $M(E)$ coincides with Q. By Theorem 1.7 we have $\lim_{n\to\infty} Q_{n_k} = Q$ weakly on $M(E)$. In the same way, one shows that every convergent subsequence of $\{Q_n\}$ has the same limit Q. Thus $\lim_{n\to\infty} Q_n = Q$ weakly on $M(E)$. $\qquad \square$

A Lusin topological space E with the Borel σ-algebra is isomorphic to a compact metric space with the Borel σ-algebra. Indeed, a complete separable metric space is at most of the cardinality of the continuum and two Borel subsets of complete separable metric spaces are isomorphic if and only if they have the same cardinality; see, e.g., Parthasarathy (1967, pp.8–14). Consequently, $(E, \mathcal{B}(E))$ is in fact isomorphic to a compact subset of the real line with its Borel σ-algebra. Then we can and do introduce a metric r into E so that (E, r) becomes a compact metric space while the Borel σ-algebra induced by r coincides with $\mathcal{B}(E)$. Let $S_2(E, r)$ be a dense sequence in $C(E, r)^{++}$ including all strictly positive rationals and let $\bar{S}_2(E, r) = S_2(E, r) \cup \{0\}$.

Proposition 1.19 *Suppose that L is a functional on $\bar{S}_2(E, r)$ and there is a sequence $\{f_n\} \subset S_2(E, r)$ such that $\lim_{n\to\infty} f_n = 0$ in bounded pointwise conver-*

gence and $\lim_{n\to\infty} L(f_n) = L(0)$. *If there is a sequence of finite measures* $\{Q_n\}$ *on* $M(E)$ *such that*

$$\lim_{n\to\infty} L_{Q_n}(f) = L(f), \qquad f \in \bar{S}_2(E,r), \tag{1.8}$$

then there is a finite measure Q *on* $M(E)$ *such that* $L_Q(f) = L(f)$ *for every* $f \in \bar{S}_2(E,r)$ *and* $\lim_{n\to\infty} Q_n = Q$ *weakly on* $M(E,r)$.

Proof. This is a modification of the proof of Theorem 1.18. By Theorem 1.14, the space $M(E,r)$ is locally compact, separable and metrizable. Let $\bar{M}(E,r) = M(E,r) \cup \{\Delta\}$ be its one-point compactification. Then (1.8) implies that $\{Q_n\}$ is a bounded sequence of measures on $\bar{M}(E,r)$. By Theorem 1.14 the sequence is relatively compact in $M(\bar{M}(E,r))$. Choose any subsequence $\{Q_{n_k}\} \subset \{Q_n\}$ that converges to a finite measure $Q \in M(\bar{M}(E,r))$. Then

$$Q(\bar{M}(E,r)) = \lim_{k\to\infty} Q_{n_k}(\bar{M}(E,r)) = \lim_{k\to\infty} L_{Q_{n_k}}(0) = L(0). \tag{1.9}$$

By (1.8) for any $f \in S_2(E,r)$ we have

$$\int_{\bar{M}(E,r)} e^{-\nu(f)} Q(d\nu) = \lim_{k\to\infty} \int_{\bar{M}(E,r)} e^{-\nu(f)} Q_{n_k}(d\nu) = L(f), \tag{1.10}$$

where $e^{-\Delta(f)} = 0$ by convention. It follows that

$$Q(M(E,r)) = \lim_{n\to\infty} \int_{\bar{M}(E,r)} e^{-\nu(f_n)} Q(d\nu) = \lim_{n\to\infty} L(f_n) = L(0). \tag{1.11}$$

In view of (1.9) and (1.11) we have $Q(\{\Delta\}) = 0$, so (1.10) implies $L_Q(f) = L(f)$ for $f \in S_2(E,r)$. By Theorem 1.7 we have $\lim_{k\to\infty} Q_{n_k} = Q$ weakly on $M(E,r)$. In the same way, if $\{Q'_{n_k}\} \subset \{Q_n\}$ is another subsequence converging to a finite measure Q' on $\bar{M}(E,r)$, then $Q'(\{\Delta\}) = 0$ and $L_{Q'}(f) = L(f)$ for $f \in S_2(E,r)$. Consequently,

$$\int_{M(E)} e^{-\nu(f)} Q(d\nu) = \int_{M(E)} e^{-\nu(f)} Q'(d\nu)$$

first for $f \in S_2(E,r)$ and then for all $f \in C(E,r)^+$ by dominated convergence, so $Q = Q'$ by Theorem 1.17. Therefore we must have $\lim_{n\to\infty} Q_n = Q$ weakly on $M(E,r)$. □

Theorem 1.20 *Let* $\{Q_n\}$ *be a sequence of finite measures on* $M(E)$ *and let* L *be a functional on* $B(E)^+$ *continuous with respect to bounded pointwise convergence. If* $\lim_{n\to\infty} L_{Q_n}(f) = L(f)$ *for all* $f \in B(E)^+$, *then there is a finite measure* Q *on* $M(E)$ *such that* $L = L_Q$ *and* $\lim_{n\to\infty} Q_n = Q$ *by weak convergence.*

Proof. By Proposition 1.19, there is a finite measure Q on $M(E)$ such that $L_Q(f) = L(f)$ for all $f \in \bar{S}_2(E,r)$ and $\lim_{n\to\infty} Q_n = Q$ weakly on $M(E,r)$. Let

$\mathcal{G} = \{f \in B(E)^+ : L_Q(f) = L(f)\}$. Then $\bar{S}_2(E, r) \subset \mathcal{G}$. Since both $f \mapsto L(f)$ and $f \mapsto L_Q(f)$ are continuous in bounded pointwise convergence and $\bar{S}_2(E, r)$ is dense in $C(E, r)^+$, we have $C(E, r)^+ \subset \mathcal{G}$, so Proposition 1.3 implies $B(E)^+ \subset \mathcal{G}$. That is, $L_Q(f) = L(f)$ for all $f \in B(E)^+$. It then follows that $\lim_{n \to \infty} L_{Q_n}(f) = L_Q(f)$ for all $f \in B(E)^+$. By Theorem 1.18, we have $\lim_{n \to \infty} Q_n = Q$ weakly on $M(E)$. □

Corollary 1.21 *Let $\{Q_n\}$ be a sequence of finite measures on $M(E)$. If $L_{Q_n}(f) \to L(f)$ uniformly in $f \in B_a(E)^+$ for each $a \geq 0$, then there is a finite measure Q on $M(E)$ such that $L = L_Q$ and $\lim_{n \to \infty} Q_n = Q$ by weak convergence.*

Theorem 1.22 *Suppose that $(F, \mathcal{F})$ is a measurable space and to each $z \in F$ there corresponds a finite measure $Q_z(d\nu)$ on $M(E)$. If $z \mapsto L_{Q_z}(f)$ is $\mathcal{F}$-measurable for every $f \in C(E)^+$, then $Q_z(d\nu)$ is a kernel from $(F, \mathcal{F})$ to $(M(E), \mathcal{B}(M(E)))$.*

Proof. Let $\mathcal{L}$ denote the set of functions $G \in B(M(E))$ so that $z \mapsto Q_z(G)$ is $\mathcal{F}$-measurable. Then $\mathcal{L} \supset \mathcal{K} := \{\nu \mapsto e^{-\nu(f)} : f \in C(E)^+\}$. By Proposition A.1 and Theorem 1.11 we have $\mathcal{L} \supset b\sigma(\mathcal{K}) = B(M(E))$. Then $Q_z(d\nu)$ is a kernel from $(F, \mathcal{F})$ to $(M(E), \mathcal{B}(M(E)))$. □

Let $M(E)^\circ = M(E) \setminus \{0\}$, where 0 is the null measure. We often use a variation of the Laplace functional in dealing with σ-finite measures on $M(E)^\circ$. A typical case is considered in the following:

Theorem 1.23 *Let Q_1 and Q_2 be two σ-finite measures on $M(E)^\circ$. If for every $f \in C(E)^+$,*

$$\int_{M(E)^\circ} (1 - e^{-\nu(f)}) Q_1(d\nu) = \int_{M(E)^\circ} (1 - e^{-\nu(f)}) Q_2(d\nu) \qquad (1.12)$$

and the value is finite, then we have $Q_1 = Q_2$.

Proof. By setting $Q_1(\{0\}) = Q_2(\{0\}) = 0$ we extend Q_1 and Q_2 to σ-finite measures on $M(E)$. Taking the difference of (1.12) for f and $f + 1$ we obtain

$$\int_{M(E)} e^{-\nu(f)} (1 - e^{-\nu(1)}) Q_1(d\nu) = \int_{M(E)} e^{-\nu(f)} (1 - e^{-\nu(1)}) Q_2(d\nu).$$

Then the result of Theorem 1.17 implies that

$$(1 - e^{-\nu(1)}) Q_1(d\nu) = (1 - e^{-\nu(1)}) Q_2(d\nu)$$

as finite measures on $M(E)$. Since $1 - e^{-\nu(1)}$ is strictly positive on $M(E)^\circ$, it follows that $Q_1 = Q_2$ as σ-finite measures on $M(E)^\circ$. □

For any integer $m \geq 1$, we can also consider the Laplace functionals of finite measures on the product space $M(E)^m$. The results proved above can be modified obviously to the multi-dimensional setting. In particular, we have the following:

Theorem 1.24 *Let $Q_1, Q_2, \ldots$ and Q be finite measures on $M(E)^m$. Then $Q_n \to Q$ weakly if and only if*

$$\lim_{n \to \infty} \int_{M(E)^m} \exp \left\{ - \sum_{i=1}^{m} \nu_i(f_i) \right\} Q_n(\mathrm{d}\nu_1, \ldots, \mathrm{d}\nu_m)$$

$$= \int_{M(E)^m} \exp \left\{ - \sum_{i=1}^{m} \nu_i(f_i) \right\} Q(\mathrm{d}\nu_1, \ldots, \mathrm{d}\nu_m)$$

for all $\{f_1, \ldots, f_m\} \subset C(E)^+$.

Suppose that $h \in p\mathscr{B}(E)$ is a strictly positive function and let $M_h(E)$ be the space of Borel measures μ on E satisfying $\mu(h) < \infty$, which is sometimes referred to as the space of *tempered measures*. A topology on $M_h(E)$ can be defined by the convention:

$$\mu_n \to \mu \text{ in } M_h(E) \text{ if and only if } \mu_n(hf) \to \mu(hf) \text{ for all } f \in C(E).$$

In particular, if h is a strictly positive continuous function on E, we have

$$\mu_n \to \mu \text{ in } M_h(E) \text{ if and only if } \mu_n(f) \to \mu(f) \text{ for all } f \in C_h(E),$$

where $C_h(E)$ is the set of continuous functions f on E such that $|f| \le \text{const} \cdot h$. A random variable X taking values in $M_h(E)$ is also called a *random measure* on E. Let $B_h(E)$ be the set of functions $f \in \mathscr{B}(E)$ satisfying $|f| \le \text{const} \cdot h$. Given a finite measure Q on $M_h(E)$, we define the *Laplace functional L_Q of Q* by

$$L_Q(f) = \int_{M_h(E)} \mathrm{e}^{-\nu(f)} Q(\mathrm{d}\nu), \qquad f \in B_h(E)^+. \tag{1.13}$$

This is a generalization of (1.4). By increasing limits we can easily extend the Laplace functional to all functions $f \in B(E)^+$, or even to all $f \in p\mathscr{B}(E)$, with the convention $\mathrm{e}^{-\infty} = 0$. We shall make those extensions whenever they are needed. The *Laplace functional* of a random measure X taking values in $M_h(E)$ means the Laplace functional of its distribution on $M_h(E)$. It is easy to see that the mapping $\mu(\mathrm{d}x) \mapsto h(x)\mu(\mathrm{d}x)$ defines a homeomorphism between $M_h(E)$ and $M(E)$. Then the results proved above can also be modified to the space $M_h(E)$.

If $E = \{a_1, \ldots, a_d\}$ is a finite set containing d elements, the mapping $\mu \mapsto (\mu(\{a_1\}), \ldots, \mu(\{a_d\}))$ gives a homeomorphism between $M(E)$ and $\mathbb{R}_+^d$. Then the results for $M(E)$ can be restated for the space $\mathbb{R}_+^d$. In particular, we define the *Laplace transform* of a finite measure G on $\mathbb{R}_+^d$ by

$$L_G(\lambda) = \int_{\mathbb{R}_+^d} \mathrm{e}^{-(\lambda, u)} G(\mathrm{d}u), \qquad \lambda \in \mathbb{R}_+^d, \tag{1.14}$$

where $(\cdot, \cdot)$ denotes the Euclidean inner product on $\mathbb{R}^d$. This is essentially a special form of the Laplace functional defined by (1.4).

1.3 Poisson Random Measures

Suppose that E is a Lusin topological space. Let $h \in p\mathscr{B}(E)$ be a strictly positive function and let $\lambda \in M_h(E)$. A random measure X on E taking values from $M_h(E)$ is called a *Poisson random measure* with *intensity* λ provided:

(1) for each $B \in \mathscr{B}(E)$ with $\lambda(B) < \infty$, the random variable $X(B)$ has the Poisson distribution with parameter $\lambda(B)$, that is,

$$\mathbf{P}\{X(B) = n\} = \frac{\lambda(B)^n}{n!}\mathrm{e}^{-\lambda(B)}, \qquad n = 0, 1, 2, \ldots;$$

(2) if $B_1, \ldots, B_n \in \mathscr{B}(E)$ are disjoint and $\lambda(B_i) < \infty$ for each $i = 1, \ldots, n$, then $X(B_1), \ldots, X(B_n)$ are mutually independent random variables.

Theorem 1.25 *A random measure X on E is Poissonian with intensity $\lambda \in M_h(E)$ if and only if its Laplace functional is given by*

$$\mathbf{E}\exp\{-X(f)\} = \exp\left\{-\int_E (1 - \mathrm{e}^{-f(x)})\lambda(\mathrm{d}x)\right\}, \quad f \in B_h(E)^+. \tag{1.15}$$

Proof. Suppose that X is a Poisson random measure on E with intensity λ. Let $B_1, \ldots, B_n \in \mathscr{B}(E)$ be disjoint sets satisfying $\lambda(B_i) < \infty$ for each $i = 1, \ldots, n$. For any constants $\alpha_1, \ldots, \alpha_n \geq 0$ we can use the above two properties to see

$$\mathbf{E}\exp\left\{-\sum_{i=1}^n \alpha_i X(B_i)\right\} = \exp\left\{-\sum_{i=1}^n (1 - \mathrm{e}^{-\alpha_i})\lambda(B_i)\right\}. \tag{1.16}$$

Then we get (1.15) by approximating $f \in B_h(E)^+$ by simple functions and using dominated convergence. Conversely, if the Laplace functional of X is given by (1.15), we may apply the equality to the simple function $f = \sum_{i=1}^n \alpha_i 1_{B_i}$ to get (1.16). Then X satisfies the above two properties in the definition of a Poisson random measure on E with intensity λ. $\square$

Corollary 1.26 *Suppose that X_1 and X_2 are independent Poisson random measures on E with intensities λ_1 and $\lambda_2 \in M_h(E)$, respectively. Then $X_1 + X_2$ is a Poisson random measure on E with intensity $\lambda_1 + \lambda_2$.*

Theorem 1.27 *For any $\lambda \in M_h(E)$, there exists a Poisson random measure with intensity λ.*

Proof. We assume $\lambda \neq 0$ to avoid triviality. Let $\{E_1, E_2, \ldots\} \subset \mathscr{B}(E)$ be a sequence of disjoint sets so that $E = \cup_{i=1}^\infty E_i$ and $0 < \lambda(E_i) < \infty$. For each $i \geq 1$ let η_i be a Poisson random variable with parameter $\lambda(E_i)$ and let $\{\xi_{i1}, \xi_{i2}, \ldots\}$ be a sequence of random variables on E with identical distribution $\lambda(E_i)^{-1}\lambda|_{E_i}$, where $\lambda|_{E_i}$ denotes the restriction of λ to E_i. Suppose that $\{\eta_i, \xi_{ij} : i, j = 1, 2, \ldots\}$ are mutually independent. Then we can define a σ-finite random measure on E by $X := \sum_{i=1}^\infty \sum_{j=1}^{\eta_i} \delta_{\xi_{ij}}$. For $f \in B_h(E)^+$ we have

$$\mathbf{E}\exp\{-X(f)\} = \mathbf{E}\exp\left\{-\sum_{i=1}^{\infty}\sum_{j=1}^{\eta_i} f(\xi_{ij})\right\}$$

$$= \prod_{i=1}^{\infty}\sum_{n=0}^{\infty} e^{-\lambda(E_i)}\frac{\lambda(E_i)^n}{n!}\mathbf{E}\left[\exp\left\{-\sum_{j=1}^{n} f(\xi_{ij})\right\}\right]$$

$$= \prod_{i=1}^{\infty}\sum_{n=0}^{\infty} e^{-\lambda(E_i)}\frac{1}{n!}\left[\int_{E_i} e^{-f(x)}\lambda(\mathrm{d}x)\right]^n$$

$$= \prod_{i=1}^{\infty}\exp\left\{-\lambda(E_i) + \int_{E_i} e^{-f(x)}\lambda(\mathrm{d}x)\right\}$$

$$= \exp\left\{-\int_E \left(1 - e^{-f(x)}\right)\lambda(\mathrm{d}x)\right\}.$$

Thus X is a Poisson random measure on E with intensity λ.   □

Proposition 1.28 *Suppose that X is a Poisson random measure on E with intensity $\lambda \in M_h(E)$. Let $\tilde{X} = X - \lambda$. Then for $f, g \in B_h(E)^+$ we have:*

(1) $\mathbf{E}[X(g)e^{-X(f)}] = \lambda(ge^{-f})\mathbf{E}[e^{-X(f)}];$
(2) $\mathbf{E}[X(g)^2 e^{-X(f)}] = [\lambda(g^2 e^{-f}) + \lambda(ge^{-f})^2]\mathbf{E}[e^{-X(f)}];$
(3) $\mathbf{E}[\tilde{X}(f)^4] = \lambda(f^4) + 3\lambda(f^2)^2.$

Proof. For any $\theta \geq 0$ we may apply (1.15) to the function $x \mapsto f(x) + \theta g(x)$ to get

$$\mathbf{E}\left[\exp\{-X(f + \theta g)\}\right] = \exp\left\{-\int_E \left(1 - e^{-f(x)-\theta g(x)}\right)\lambda(\mathrm{d}x)\right\}.$$

By differentiating both sides with respect to $\theta \geq 0$ at zero we get (1). The other two results can be obtained in similar ways.   □

Theorem 1.29 *Suppose that λ is a finite measure on E and μ is a probability measure on $(0, \infty)$ satisfying*

$$\int_0^{\infty} (1 \wedge u)\mu(\mathrm{d}u) < \infty.$$

Then there is a probability measure Q on $M(E)$ with Laplace functional given by

$$L_Q(f) = \exp\left\{-\int_E \lambda(\mathrm{d}x)\int_0^{\infty} \left(1 - e^{-uf(x)}\right)\mu(\mathrm{d}u)\right\}. \qquad (1.17)$$

Proof. We assume $\lambda \neq 0$ to avoid triviality. Let η be a Poisson random variable with parameter $\lambda(E)$ and let $\{\xi_1, \xi_2, \ldots\}$ be a sequence of random variables on E identically distributed according to $\lambda(E)^{-1}\lambda$. In addition, let $\{\theta_1, \theta_2, \ldots\}$ be a sequence of random variables with identical distribution μ. Suppose that $\{\eta, \xi_j, \theta_j : j = 1, 2, \ldots\}$ are mutually independent. Then a finite random measure on E is defined by $X = \sum_{j=1}^{\eta} \theta_j \delta_{\xi_j}$. As in the proof of Theorem 1.27 it is simple to see

$$\mathbf{E}\exp\{-X(f)\} = \exp\left\{-\int_E \lambda(\mathrm{d}x)\int_0^\infty \left(1 - \mathrm{e}^{-uf(x)}\right)\mu(\mathrm{d}u)\right\}.$$

for $f \in B(E)^+$. Then X has distribution Q on $M(E)$ given by (1.17). $\square$

A random measure X on E with distribution Q given by (1.17) is called a *compound Poisson random measure* with *intensity* λ and *height distribution* μ. The intuitive meanings of the parameters are clear from the construction of the random measure given in the above proof.

1.4 Infinitely Divisible Random Measures

Let E be a Lusin topological space. For probability measures Q_1 and Q_2 on $M(E)$, the product $Q_1 \times Q_2$ is a probability measure on $M(E)^2$. The image of $Q_1 \times Q_2$ under the mapping $(\mu_1, \mu_2) \mapsto \mu_1 + \mu_2$ is called the *convolution* of Q_1 and Q_2 and is denoted by $Q_1 * Q_2$, which is a probability measure on $M(E)$. According to the definition, for any $F \in b\mathscr{B}(M(E))$ we have

$$\int_{M(E)} F(\mu)(Q_1 * Q_2)(\mathrm{d}\mu)$$
$$= \int_{M(E)^2} F(\mu_1 + \mu_2)Q_1(\mathrm{d}\mu_1)Q_2(\mathrm{d}\mu_2). \tag{1.18}$$

Clearly, if X_1 and X_2 are independent random measures on E with distributions Q_1 and Q_2 on $M(E)$, respectively, the random measure $X_1 + X_2$ has distribution $Q_1 * Q_2$. It is easy to show that

$$L_{Q_1*Q_2}(f) = L_{Q_1}(f)L_{Q_2}(f), \qquad f \in B(E)^+. \tag{1.19}$$

Let $Q^{*0} = \delta_0$ and define $Q^{*n} = Q^{*(n-1)} * Q$ inductively for integers $n \geq 1$. We say a probability distribution Q on $M(E)$ is *infinitely divisible* if for each integer $n \geq 1$, there is a probability Q_n such that $Q = Q_n^{*n}$. In this case, we call Q_n the *n-th root* of Q. A random measure X on E is said to be *infinitely divisible* if its distribution on $M(E)$ is infinitely divisible.

Example 1.2 Poisson random measures and compound Poisson random measures are infinitely divisible. The n-th roots of their distributions can be obtained by replacing the intensity λ with λ/n.

The main purpose of this section is to give a characterization for the class of infinitely divisible probability measures on $M(E)$. Let $\mathscr{I}(E)$ denote the convex cone of all functionals U on $B(E)^+$ with the representation

$$U(f) = \lambda(f) + \int_{M(E)^\circ} \left(1 - \mathrm{e}^{-\nu(f)}\right)L(\mathrm{d}\nu), \qquad f \in B(E)^+, \tag{1.20}$$

where $\lambda \in M(E)$ and $(1 \wedge \nu(1))L(\mathrm{d}\nu)$ is a finite measure on $M(E)^\circ$. Let $S_2(E,r)$ be defined as in Section 1.2.

Proposition 1.30 *The measures λ and L in (1.20) are uniquely determined by the functional $U \in \mathscr{I}(E)$.*

Proof. Suppose that U can also be represented by (1.20) with (λ, L) replaced by (γ, G). For any constant $\theta \geq 0$, we can evaluate $U(f + \theta) - U(\theta)$ with the two representations and get

$$\gamma(f) + \int_{M(E)^\circ} \left(1 - e^{-\nu(f)}\right)e^{-\nu(\theta)}G(\mathrm{d}\nu)$$
$$= \lambda(f) + \int_{M(E)^\circ} \left(1 - e^{-\nu(f)}\right)e^{-\nu(\theta)}L(\mathrm{d}\nu).$$

Letting $\theta \to \infty$ gives $\gamma(f) = \lambda(f)$, and hence $\gamma = \lambda$. Then the above equality and Theorem 1.23 imply $e^{-\nu(\theta)}G(\mathrm{d}\nu) = e^{-\nu(\theta)}L(\mathrm{d}\nu)$, and so $G(\mathrm{d}\nu) = L(\mathrm{d}\nu)$. $\qquad\square$

Proposition 1.31 *Suppose that J is a functional on $S_2(E,r)$ and there is a sequence $\{f_n\} \subset S_2(E,r)$ such that $\lim_{n\to\infty} f_n = 0$ in bounded pointwise convergence and $\lim_{n\to\infty} J(f_n) = 0$. If there is a sequence $\{U_n\} \subset \mathscr{I}(E)$ so that*

$$\lim_{n\to\infty} U_n(f) = J(f), \qquad f \in S_2(E,r), \tag{1.21}$$

then J is the restriction to $S_2(E,r)$ of a functional $U \in \mathscr{I}(E)$.

Proof. Recall that $P(E)$ denotes the space of probability measures on E. For $\nu \in M(E)^\circ$ let $|\nu| = \nu(E)$ and $\hat{\nu} = |\nu|^{-1}\nu$. The mapping $F : \nu \mapsto (|\nu|, \hat{\nu})$ is clearly an homeomorphism between $M(E)^\circ$ and the product space $(0,\infty) \times P(E)$. For $0 \leq u \leq \infty$, $\pi \in P(E)$ and $f \in B(E)^+$ let

$$\xi(u, \pi, f) = \begin{cases} (1 - e^{-u})^{-1}(1 - e^{-u\pi(f)}) & \text{if } 0 < u < \infty, \\ \pi(f) & \text{if } u = 0, \\ 1 & \text{if } u = \infty. \end{cases} \tag{1.22}$$

Suppose that $U_n \in \mathscr{I}(E)$ is given by (1.21) with (λ, L) replaced by (λ_n, L_n). Let $\pi_n \in P(E)$ be such that $\lambda_n = |\lambda_n|\pi_n$ and let $H_n(\mathrm{d}u, \mathrm{d}\pi)$ be the image of $L_n(\mathrm{d}\nu)$ under the mapping F. Define the finite measure $G_n(\mathrm{d}u, \mathrm{d}\pi)$ on $[0,\infty] \times P(E)$ by $G_n(\{(0, \pi_n)\}) = |\lambda_n|$, $G_n((\{0,\infty\} \times P(E)) \setminus \{(0, \pi_n)\}) = 0$ and $G_n(\mathrm{d}u, \mathrm{d}\pi) = (1 - e^{-u})H_n(\mathrm{d}u, \mathrm{d}\pi)$ for $0 < u < \infty$ and $\pi \in P(E)$. Then we have

$$U_n(f) = \int_{[0,\infty]} \int_{P(E)} \xi(u, \pi, f)G_n(\mathrm{d}u, \mathrm{d}\pi), \qquad f \in B(E)^+.$$

By (1.21) it is evident that $\{G_n\}$ is a bounded sequence in $M([0,\infty] \times P(E))$. Let (E, r) be the compact metric space as described in Section 1.2. Then Theorem 1.14 implies that $\{G_n\}$ viewed as a sequence of measures on $[0,\infty] \times P(E, r)$ is relatively compact. Take any subsequence $\{G_{n_k}\} \subset \{G_n\}$ such that $\lim_{k\to\infty} G_{n_k} = G$

weakly for a finite measure G on $[0, \infty] \times P(E, r)$. For any $f \in S_2(E, r)$ the mapping $(u, \pi) \mapsto \xi(u, \pi, f)$ is clearly continuous on $[0, \infty] \times P(E, r)$. By (1.21) we have

$$J(f) = \int_{[0,\infty]} \int_{P(E)} \xi(u, \pi, f) G(\mathrm{d}u, \mathrm{d}\pi), \qquad f \in S_2(E, r).$$

Observe also that $\lim_{n \to \infty} J(f_n) = 0$ implies $G(\{\infty\} \times P(E)) = 0$. Then the desired conclusion follows by a change of the integration variable. $\qquad \square$

Theorem 1.32 *Suppose that U is a functional on $B(E)^+$ continuous with respect to bounded pointwise convergence. If there is a sequence $\{U_n\} \subset \mathscr{I}(E)$ such that $U(f) = \lim_{n \to \infty} U_n(f)$ for all $f \in B(E)^+$, then $U \in \mathscr{I}(E)$.*

Proof. This is similar to the proof of Theorem 1.20 with an application of Proposition 1.31. $\qquad \square$

Corollary 1.33 *Suppose that $\{U_n\} \subset \mathscr{I}(E)$ and U is a functional on $B(E)^+$. If $U_n(f) \to U(f)$ uniformly in $f \in B_a(E)^+$ for each $a \geq 0$, then $U \in \mathscr{I}(E)$.*

Corollary 1.34 *Suppose that $\{U_n\} \subset \mathscr{I}(E)$ and there is a measure $\pi \in M(E)$ so that $U_n(f) \leq \pi(f)$ for all $n \geq 1$ and $f \in B(E)^+$. If $U(f) = \lim_{n \to \infty} U_n(f)$ for all $f \in B(E)^+$, then we have $U \in \mathscr{I}(E)$.*

Proof. Let U_n be given by (1.20) with (λ, L) replaced by (λ_n, L_n). For $f \in B(E)^+$ we can use monotone convergence to see

$$\lambda_n(f) + \int_{M(E)^\circ} \nu(f) L_n(\mathrm{d}\nu) = \lim_{\theta \to 0+} \theta^{-1} U_n(\theta f) \leq \pi(f).$$

Consequently, for any f and $g \in B(E)^+$ we have

$$|U_n(f) - U_n(g)| \leq \lambda_n(|f - g|) + \int_{M(E)^\circ} \nu(|f - g|) L_n(\mathrm{d}\nu)$$
$$\leq \pi(|f - g|),$$

and hence $|U(f) - U(g)| \leq \pi(|f - g|)$. Then U is continuous in bounded pointwise convergence and so $U \in \mathscr{I}(E)$ by Theorem 1.32. $\qquad \square$

Suppose that $U \in \mathscr{I}(E)$ has the representation (1.20). Let $N(\mathrm{d}\nu)$ be a Poisson random measure on $M(E)^\circ$ with intensity measure L. By Theorem 1.25 it is easy to show that

$$X := \lambda + \int_{M(E)^\circ} \nu N(\mathrm{d}\nu) \tag{1.23}$$

defines a random measure on E with $-\log L_X = U$. Let Q be the distribution of X on $M(E)$. By the same reasoning, for any integer $n \geq 1$ there is a probability

measure Q_n on $M(E)$ satisfying $-\log L_{Q_n} = n^{-1}U$, implying $Q_n^{*n} = Q$. Therefore Q is infinitely divisible. We write $Q = I(\lambda, L)$ if Q is an infinitely divisible
probability measure on $M(E)$ with $U = -\log L_Q \in \mathscr{I}(E)$ represented by (1.20).
The construction (1.23) for the corresponding random measure is called a *cluster
representation*.

Theorem 1.35 *The equation* $U = -\log L_Q$ *establishes a one-to-one correspondence between functionals* $U \in \mathscr{I}(E)$ *and infinitely divisible probability measures*
Q *on* $M(E)$.

Proof. By the above comments we only need to show if Q is an infinitely divisible
probability measure on $M(E)$, then $U := -\log L_Q \in \mathscr{I}(E)$. For $n \geq 1$ let Q_n be
the n-th root of Q. Then

$$U(f) = \lim_{n\to\infty} n[1 - e^{-n^{-1}U(f)}] = \lim_{n\to\infty} \int_{M(E)^\circ} \left(1 - e^{-\nu(f)}\right) n Q_n(\mathrm{d}\nu)$$

and the convergence is uniform in $f \in B_a(E)^+$ for each $a \geq 0$. By Corollary 1.33
we have $U \in \mathscr{I}(E)$. $\square$

Theorem 1.36 *Suppose that* $V : f \mapsto v(\cdot, f)$ *is an operator on* $B(E)^+$ *so that*
$v(x, \cdot) \in \mathscr{I}(E)$ *for all* $x \in E$. *Then we have the representation*

$$v(x, f) = \lambda(x, f) + \int_{M(E)^\circ} \left(1 - e^{-\nu(f)}\right) L(x, \mathrm{d}\nu), \quad f \in B(E)^+, \quad (1.24)$$

where $\lambda(x, \mathrm{d}y)$ *is a bounded kernel on* E *and* $(1 \wedge \nu(1))L(x, \mathrm{d}\nu)$ *is a bounded
kernel from* E *to* $M(E)^\circ$.

Proof. Under the assumption, for any fixed $x \in E$ we have the representation
(1.24), where $\lambda(x, \cdot) \in M(E)$ and $(1 \wedge \nu(1))L(x, \mathrm{d}\nu)$ is a finite measure on $M(E)^\circ$.
For every $f \in B(E)^+$,

$$x \mapsto w(x, f) := 2v(x, f + 1) - v(x, f + 2) - v(x, f)$$

is a bounded Borel function on E. Observe also that

$$w(x, f) = \int_{M(E)^\circ} e^{-\nu(f)} \left(1 - e^{-\nu(1)}\right)^2 L(x, \mathrm{d}\nu).$$

By Theorem 1.22 we see that $(1 - e^{-\nu(1)})^2 L(x, \mathrm{d}\nu)$ is a bounded kernel from E to
$M(E)^\circ$. In view of (1.24),

$$x \mapsto \int_{M(E)^\circ} \left(1 - e^{-\nu(1)}\right) L(x, \mathrm{d}\nu)$$

is a bounded function on E. It follows that $(1 \wedge \nu(1))L(x, \mathrm{d}\nu)$ is a bounded kernel
from E to $M(E)^\circ$. By another application of the relation (1.24) one sees that $x \mapsto$

$\lambda(x, f)$ is a bounded Borel function on E for every $f \in B(E)^+$. Then $\lambda(x, dy)$ is a bounded kernel on E. $\qquad \square$

Theorem 1.37 *If $U \in \mathscr{I}(E)$ and if $V : f \mapsto v(\cdot, f)$ is an operator on $B(E)^+$ such that $v(x, \cdot) \in \mathscr{I}(E)$ for all $x \in E$, then $U \circ V \in \mathscr{I}(E)$.*

Proof. By Theorem 1.36 it is easy to see that for any $\mu \in M(E)$ the functional $f \mapsto \mu(Vf)$ belongs to $\mathscr{I}(E)$. Then there is an infinitely divisible probability measure $Q(\mu, \cdot)$ on $M(E)$ satisfying $- \log L_{Q(\mu, \cdot)}(f) = \mu(Vf)$. By Theorem 1.22 we see that $Q(\mu, d\nu)$ is a probability kernel on $M(E)$. Let G be the infinitely divisible probability measure on $M(E)$ with $- \log L_G = U$ and define

$$Q(d\nu) = \int_{M(E)} G(d\mu) Q(\mu, d\nu), \qquad \nu \in M(E).$$

It is not hard to show that $- \log L_Q = U \circ V$. By the same reasoning, for each integer $n \geq 1$ there is a probability measure Q_n such that $- \log L_{Q_n} = n^{-1} U \circ V$. Then $Q = Q_n^{*n}$ and hence Q is infinitely divisible. By Theorem 1.35 we conclude that $U \circ V \in \mathscr{I}(E)$. $\qquad \square$

The following result gives a useful method for the calculation of the moments of infinitely divisible probability measures on $M(E)$.

Proposition 1.38 *Let $U \in \mathscr{I}(E)$ be given by (1.20). Then for any $f \in B(E)^+$ we have*

$$\frac{d}{d\theta} U(\theta f) = \lambda(f) + \int_{M(E)^\circ} \nu(f) e^{-\theta\nu(f)} L(d\nu) \tag{1.25}$$

and

$$\frac{d^n}{d\theta^n} U(\theta f) = (-1)^{n-1} \int_{M(E)^\circ} \nu(f)^n e^{-\theta\nu(f)} L(d\nu) \tag{1.26}$$

for $0 < \theta < \infty$ and $n = 2, 3, \ldots$.

Proof. For any $n \geq 1$ the function $z \mapsto z^n e^{-z}$ achieves its maximal value on $[0, \infty)$ at $z = n$. It follows that

$$\nu(f)^n e^{-\theta\nu(f)} \leq n^n \theta_0^{-n} e^{-n}, \qquad \theta \geq \theta_0 > 0, \nu \in M(E).$$

Fix $\theta_0 > 0$ and $f \in B(E)^+$ and let $F_n(\nu) = n^n \theta_0^{-n} e^{-n} \wedge \nu(f)^n$. It is easy to see that $L(F_n) < \infty$ and

$$\nu(f)^n e^{-\theta\nu(f)} \leq F_n(\nu), \qquad \theta \geq \theta_0, \nu \in M(E).$$

Then we have (1.25) and (1.26) by dominated convergence. $\qquad \square$

To close this section we give a characterization of infinitely divisible probability measures on the half line $[0, \infty)$. Write $\psi \in \mathscr{I}$ if $\lambda \mapsto \psi(\lambda)$ is a positive function on $[0, \infty)$ with the representation

$$\psi(\lambda) = \beta\lambda + \int_0^\infty (1 - e^{-\lambda u}) l(du), \tag{1.27}$$

where $\beta \geq 0$ and $(1 \wedge u) l(du)$ is a finite measure on $(0, \infty)$. As a consequence of Theorem 1.35 we have the following:

Theorem 1.39 *The relation* $\psi = -\log L_\mu$ *establishes a one-to-one correspondence between the functions* $\psi \in \mathscr{I}$ *and infinitely divisible probability measures* μ *on* $[0, \infty)$.

Example 1.3 Let $b > 0$ and $\alpha > 0$. The *Gamma distribution* γ on $[0, \infty)$ with parameters (b, α) is defined by

$$\gamma(B) = \frac{\alpha^b}{\Gamma(b)} \int_B x^{b-1} e^{-\alpha x} dx, \qquad B \in \mathscr{B}([0, \infty)).$$

This reduces to the *exponential distribution* when $b = 1$. The Gamma distribution has Laplace transform

$$L_\gamma(\lambda) = \left(\frac{\alpha}{\alpha + \lambda}\right)^b, \qquad \lambda \geq 0.$$

It is easily seen that γ is infinitely divisible and its n-th root is the Gamma distribution with parameters $(b/n, \alpha)$.

Example 1.4 For $c > 0$ and $0 < \alpha < 1$ the function $\lambda \mapsto c\lambda^\alpha$ admits the representation (1.27). Indeed, it is simple to show

$$\lambda^\alpha = \frac{\alpha}{\Gamma(1 - \alpha)} \int_0^\infty (1 - e^{-\lambda u}) \frac{du}{u^{1+\alpha}}, \qquad \lambda \geq 0. \tag{1.28}$$

The infinitely divisible probability measure ν on $[0, \infty)$ satisfying $-\log L_\nu(\lambda) = c\lambda^\alpha$ is known as the *one-sided stable distribution* with index $0 < \alpha < 1$. This distribution does not charge zero and is absolutely continuous with respect to the Lebesgue measure on $(0, \infty)$ with continuous density. For $\alpha = 1/2$ it has density

$$q(x) := \frac{c}{2\sqrt{\pi}} x^{-3/2} e^{-c^2/4x}, \qquad x > 0.$$

For a general index the density can be given using an infinite series; see, e.g., Sato (1999, p.88).

1.5 Lévy–Khintchine Type Representations

In this section, we present some criteria for continuous functions on $[0, \infty)$ to have Lévy–Khintchine type representations. The results are useful in the study of high-density limits of discrete branching processes. For an interval $T \subset \mathbb{R}$ let $\mathscr{C}(T)$ denote the set of continuous (not necessarily bounded) functions on T. For any $c \geq 0$ we define the *difference operator* Δ_c by

$$\Delta_c f(\lambda) = f(\lambda + c) - f(\lambda), \qquad \lambda, \lambda + c \in T, f \in \mathscr{C}(T).$$

Let Δ_c^0 be the identity and define $\Delta_c^n = \Delta_c^{n-1}\Delta_c$ for $n \geq 1$ inductively. Then we have

$$\Delta_c^m f(\lambda) = (-1)^m \sum_{i=0}^m \binom{m}{i}(-1)^i f(\lambda + ic). \tag{1.29}$$

We call $\theta \in \mathscr{C}([0, \infty))$ a *completely monotone function* if it satisfies

$$(-1)^i \Delta_c^i \theta(\lambda) \geq 0, \qquad \lambda \geq 0, c \geq 0, i = 0, 1, 2, \dots. \tag{1.30}$$

The *Bernstein polynomials* of a function $f \in \mathscr{C}([0, 1])$ are given by

$$B_{f,m}(s) = \sum_{i=0}^m \binom{m}{i} \Delta_{1/m}^i f(0)s^i, \qquad 0 \leq s \leq 1, m = 1, 2, \dots.$$

It is well-known that

$$B_{f,m}(s) \to f(s), \qquad s \in [0, 1], \tag{1.31}$$

uniformly as $m \to \infty$; see, e.g., Feller (1971, p.222).

Theorem 1.40 *A function $\theta \in \mathscr{C}([0, \infty))$ is the Laplace transform of a finite measure G on $[0, \infty)$ if and only if it is completely monotone.*

Proof. If $\theta \in \mathscr{C}([0, \infty))$ is the Laplace transform of a finite measure on $[0, \infty)$ it is clearly a completely monotone function. Conversely, suppose that (1.30) holds. For fixed $a > 0$, we let $\gamma_a(s) = \theta(a - as)$ for $0 \leq s \leq 1$. The complete monotonicity of θ implies

$$\Delta_{1/m}^i \gamma_a(0) \geq 0, \qquad i = 0, 1, \dots, m.$$

Then the Bernstein polynomial $B_{\gamma_a,m}(s)$ has positive coefficients, so $B_{\gamma_a,m}(e^{-\lambda/a})$ is the Laplace transform of a finite measure $G_{a,m}$ on $[0, \infty)$. By Theorem 1.20,

$$\theta(\lambda) = \lim_{a \to \infty} \lim_{m \to \infty} B_{\gamma_a,m}(e^{-\lambda/a}), \qquad \lambda \geq 0,$$

is the Laplace transform of a finite measure on $[0, \infty)$. $\qquad\qquad\square$

Now let us consider a general Lévy–Khintchine type representation for continuous functions on $[0, \infty)$. For $u \geq 0$ and $\lambda \geq 0$ let

$$\xi_n(u, \lambda) = e^{-\lambda u} - 1 - (1 + u^n)^{-1} \sum_{i=1}^{n-1} \frac{(-\lambda u)^i}{i!}, \qquad n = 1, 2, \ldots.$$

We are interested in functions $\phi \in \mathscr{C}([0, \infty))$ with the representation

$$\phi(\lambda) = \sum_{i=0}^{n-1} a_i \lambda^i + \int_0^\infty \xi_n(u, \lambda)(1 - e^{-u})^{-n} G(du), \quad \lambda \geq 0, \qquad (1.32)$$

where $n \geq 1$ is an integer, $\{a_0, \ldots, a_{n-1}\}$ is a set of constants and $G(du)$ is a finite measure on $[0, \infty)$. The value at $u = 0$ of the integrand in (1.32) is defined by continuity as $(-\lambda)^n/n!$.

Lemma 1.41 *A function $\eta \in \mathscr{C}([0, \infty))$ is a polynomial of degree less than $n \geq 1$ if and only if $\Delta_c^n \eta(0) = 0$ for all $c \geq 0$.*

Proof. If $\eta \in \mathscr{C}([0, \infty))$ is a polynomial of degree less than $n \geq 1$, one sees easily $\Delta_c^n \eta(0) = 0$ for all $c \geq 0$. For the converse, suppose that $\eta \in \mathscr{C}([0, \infty))$ and $\Delta_c^n \eta(0) = 0$ for all $c \geq 0$. Fix $a > 0$ and let $\eta_a(s) = \eta(as)$ for $0 \leq s \leq 1$. Since

$$\Delta_c^n \eta_a(0) = 0, \quad 0 \leq c \leq n^{-1},$$

the polynomials of η_a have degree less than n, that is,

$$B_{\eta_a, m}(s) = \sum_{i=0}^{n-1} b_i^{(m)} s^i, \quad m = n, n+1, \ldots.$$

The coefficients $b_i^{(m)}$ here can be represented as linear combinations of

$$B_{\eta_a, m}(1/n), B_{\eta_a, m}(2/n), \ldots, B_{\eta_a, m}(n/n).$$

By (1.31) the limits

$$\lim_{m \to \infty} b_i^{(m)} = b_i, \quad i = 0, 1, \ldots, n-1$$

exist and hence $\eta_a(s) = \sum_{i=0}^{n-1} b_i s^i$ for $0 \leq s \leq 1$. Setting $a_i = a^{-1} b_i$ we get

$$\eta(s) = \sum_{i=0}^{n-1} a_i s^i, \quad 0 \leq s \leq a.$$

Clearly, this formula holds in fact for all $s \geq 0$. $\qquad \square$

Theorem 1.42 *A function $\phi \in \mathscr{C}([0, \infty))$ has the representation (1.32) if and only if for every $c \geq 0$ the function*

$$\theta_c(\lambda) := (-1)^n \Delta_c^n \phi(\lambda), \qquad \lambda \geq 0 \tag{1.33}$$

is the Laplace transform of a finite measure on $[0, \infty)$.

Proof. Suppose that ϕ is given by (1.32). Using (1.29) it is simple to see

$$\theta_c(\lambda) = \int_0^\infty e^{-\lambda u}(1 - e^{-cu})^n(1 - e^{-u})^{-n}G(du),$$

where the integrand is defined as c^n at $u = 0$ by continuity. Thus θ_c is the Laplace transform of a finite measure on $[0, \infty)$. Conversely, assume θ_c is the Laplace transform of a finite measure G_c on $[0, \infty)$, that is,

$$\theta_c(\lambda) = \int_0^\infty e^{-\lambda u}G_c(du), \qquad \lambda \geq 0. \tag{1.34}$$

From (1.29) and the relation

$$(-1)^n\Delta_1^n\theta_c(\lambda) = \Delta_1^n\Delta_c^n\phi(\lambda) = \Delta_c^n\Delta_1^n\phi(\lambda) = (-1)^n\Delta_c^n\theta_1(\lambda)$$

it follows that

$$\int_0^\infty e^{-\lambda u}(1 - e^{-u})^nG_c(du) = \int_0^\infty e^{-\lambda u}(1 - e^{-cu})^nG(du),$$

where $G = G_1$. Therefore

$$G_c(du) = (1 - e^{-cu})^n(1 - e^{-u})^{-n}G(du), \qquad 0 < u < \infty \tag{1.35}$$

by the uniqueness of the Laplace transform. Let

$$\phi_0(\lambda) = \int_0^\infty \xi_n(u, \lambda)(1 - e^{-u})^{-n}G(du), \qquad \lambda \geq 0. \tag{1.36}$$

The function $\eta(\lambda) := \phi(\lambda) - \phi_0(\lambda)$ is continuous and

$$
\begin{aligned}
(-1)^n\Delta_c^{n+1}\eta(\lambda) &= (-1)^n\Delta_c[\Delta_c^n\phi(\lambda) - \Delta_c^n\phi_0(\lambda)] \\
&= \Delta_c\Big[\theta_c(\lambda) - \int_0^\infty e^{-\lambda u}(1 - e^{-cu})^n(1 - e^{-u})^{-n}G(du)\Big] \\
&= \Delta_c[G_c(\{0\}) - c^nG(\{0\})] = 0,
\end{aligned}
$$

where we used (1.35) for the third equality. By Lemma 1.41 the function η is a polynomial of degree less than $n + 1$, say $\eta(\lambda) = \sum_{i=0}^n a_i\lambda^i$. By (1.33) and (1.36), we have

$$n!a_n = \Delta_1^n\eta(\lambda) = (-1)^n\Big[\theta_1(\lambda) - \int_0^\infty e^{-\lambda u}G(du)\Big] = 0.$$

Then ϕ has the representation (1.32). $\qquad\square$

Based on the above two theorems we can give canonical representations for the limit functions of some sequences involving probability generating functions. Let $\{\alpha_k\}$ be a sequence of positive numbers and let $\{g_k\}$ be a sequence of probability generating functions, that is,

$$g_k(z) = \sum_{i=0}^{\infty} p_{ki} z^i, \qquad |z| \le 1,$$

where $p_{ki} \ge 0$ and $\sum_{i=0}^{\infty} p_{ki} = 1$. We first consider the sequence of functions $\{\psi_k\}$ defined by

$$\psi_k(\lambda) = \alpha_k[1 - g_k(1 - \lambda/k)], \qquad 0 \le \lambda \le k. \tag{1.37}$$

Theorem 1.43 *If the sequence $\{\psi_k\}$ defined by (1.37) converges to some $\psi \in \mathscr{C}([0, \infty))$, then the limit function belongs to the class $\mathscr{I}$ defined by (1.27).*

Proof. For any $c, \lambda \ge 0$ and sufficiently large $k \ge 1$ we have

$$\Delta_c \psi_k(\lambda) = -\alpha_k \Delta_c g_k(1 - \cdot/k)(\lambda).$$

Since for each integer $i \ge 1$ the i-th derivative $g_k^{(i)}$ is a power series with positive coefficients, so is $\Delta_c^m g_k^{(i)}$ for any $m \ge 1$. In particular, we have

$$(-1)^i \frac{\mathrm{d}^i}{\mathrm{d}\lambda^i} \Delta_c \psi_k(\lambda) = -k^{-i} \alpha_k \Delta_c g_k^{(i)}(1 - \cdot/k)(\lambda) \ge 0.$$

By the mean-value theorem, one sees inductively $(-1)^i \Delta_h^i \Delta_c \psi_k(\lambda) \ge 0$. Letting $k \to \infty$ we obtain $(-1)^i \Delta_h^i \Delta_c \psi(\lambda) \ge 0$. Then $\Delta_c \psi(\lambda)$ is a completely monotone function of $\lambda \ge 0$, so by Theorem 1.40 it is the Laplace transform of a finite measure on $[0, \infty)$. Since $\psi(0) = \lim_{k\to\infty} \psi_k(0) = 0$, by Theorem 1.42 there is a finite measure F on $[0, \infty)$ so that

$$\psi(\lambda) = \int_0^\infty \left(1 - \mathrm{e}^{-\lambda u}\right)(1 - \mathrm{e}^{-u})^{-1} F(\mathrm{d}u),$$

where the value of the integrand at $u = 0$ is defined as λ by continuity. Then (1.27) follows with $\beta = F(\{0\})$ and $l(\mathrm{d}u) = (1 - \mathrm{e}^{-u})^{-1} F(\mathrm{d}u)$ for $u > 0$. $\qquad \square$

Example 1.5 Suppose that g is a probability generating function so that $\beta := g'(1-) < \infty$. Let $\alpha_k = k$ and $g_k(z) = g(z)$. Then the sequence $\psi_k(\lambda)$ defined by (1.37) converges to $\beta\lambda$ as $k \to \infty$.

Example 1.6 For any $0 < \alpha \le 1$ the function $\psi(\lambda) = \lambda^\alpha$ has the representation (1.27). For $\alpha = 1$ that is trivial, and for $0 < \alpha < 1$ that follows from (1.28). Let $\psi_k(\lambda)$ be defined by (1.37) with $\alpha_k = k^\alpha$ and $g_k(z) = 1 - (1 - z)^\alpha$. Then $\psi_k(\lambda) = \lambda^\alpha$ for $0 \le \lambda \le k$.

In the study of limit theorems of branching models, we shall also need to consider the limit of another function sequence defined as follows. Let $\{\alpha_k\}$ and $\{g_k\}$ be given as above and let

$$\phi_k(\lambda) = \alpha_k[g_k(1 - \lambda/k) - (1 - \lambda/k)], \qquad 0 \leq \lambda \leq k. \qquad (1.38)$$

Theorem 1.44 *If the sequence $\{\phi_k\}$ defined by (1.38) converges to some $\phi \in \mathscr{C}([0,\infty))$, then the limit function has the representation*

$$\phi(\lambda) = a\lambda + c\lambda^2 + \int_0^\infty \left(e^{-\lambda u} - 1 + \frac{\lambda u}{1 + u^2}\right) m(du), \qquad (1.39)$$

where $c \geq 0$ and a are constants, and $m(du)$ is a σ-finite measure on $(0,\infty)$ satisfying

$$\int_0^\infty (1 \wedge u^2) m(du) < \infty. \qquad (1.40)$$

Proof. Since $\phi(0) = \lim_{k\to\infty} \phi_k(0) = 0$, arguing as in the proof of Theorem 1.43 we see that ϕ has the representation (1.32) with $n = 2$ and $a_0 = 0$, which can be rewritten into the equivalent form (1.39). □

Now let us consider a special case of the function $\phi \in \mathscr{C}([0,\infty))$ given by (1.39). Observe that if the measure $m(du)$ satisfies the integrability condition

$$\int_0^\infty (u \wedge u^2)\, m(du) < \infty, \qquad (1.41)$$

we have

$$\phi(\lambda) = b\lambda + c\lambda^2 + \int_0^\infty \left(e^{-\lambda u} - 1 + \lambda u\right) m(du), \qquad (1.42)$$

where

$$b = a - \int_0^\infty \frac{u^3}{1 + u^2} m(du).$$

Proposition 1.45 *The function $\phi \in \mathscr{C}([0,\infty))$ with the representation (1.39) is locally Lipschitz if and only if (1.41) holds.*

Proof. For computational convenience we first rewrite (1.39) as

$$\phi(\lambda) = b_1\lambda + c\lambda^2 + \int_0^\infty \left(e^{-\lambda u} - 1 + \lambda u 1_{\{u \leq 1\}}\right) m(du), \qquad (1.43)$$

where

$$b_1 = a + \int_0^\infty \left(\frac{u}{1 + u^2} - u 1_{\{u \leq 1\}}\right) m(du).$$

By applying dominated convergence to (1.43), for each $\lambda > 0$ we have

$$\phi'(\lambda) = b_1 + 2c\lambda + \int_{(0,1]} u\left(1 - e^{-\lambda u}\right) m(du) - \int_{(1,\infty)} u e^{-\lambda u} m(du).$$

Then we use monotone convergence to the two integrals to get

$$\phi'(0+) = b_1 - \int_{(1,\infty)} um(du).$$

If ϕ is locally Lipschitz, we have $\phi'(0+) > -\infty$ and the integral on the right-hand side is finite. This together with (1.40) implies (1.41). Conversely, if (1.41) holds, then ϕ' is bounded on each bounded interval and so ϕ is locally Lipschitz.   □

Corollary 1.46 *If the sequence $\{\phi_k\}$ defined by (1.38) is uniformly Lipschitz on each bounded interval and $\phi_k(\lambda) \to \phi(\lambda)$ for all $\lambda \geq 0$ as $k \to \infty$, then the limit function has the representation (1.42).*

Example 1.7 Suppose that g is a probability generating function so that $g'(1-) = 1$ and $c := g''(1-)/2 < \infty$. Let $\alpha_k = k^2$ and $g_k(z) = g(z)$. By Taylor's expansion it is easy to show that the sequence $\phi_k(\lambda)$ defined by (1.38) converges to $c\lambda^2$ as $k \to \infty$.

Example 1.8 For $0 < \alpha < 1$ the function $\phi(\lambda) = -\lambda^\alpha$ has the representation (1.39). That follows from (1.28) as we notice

$$\int_0^\infty \left(\frac{u}{1+u^2}\right)\frac{du}{u^{1+\alpha}} = \int_0^\infty \left(\frac{1}{1+u^2}\right)\frac{du}{u^\alpha} < \infty.$$

The function is the limit of the sequence $\phi_k(\lambda)$ defined by (1.38) with $\alpha_k = k^\alpha$ and $g_k(z) = 1 - (1-z)^\alpha$.

Example 1.9 For any $1 \leq \alpha \leq 2$ the function $\phi(\lambda) = \lambda^\alpha$ can be represented in the form of (1.42). In particular, for $1 < \alpha < 2$ we have

$$\lambda^\alpha = \frac{\alpha(\alpha - 1)}{\Gamma(2 - \alpha)} \int_0^\infty \left(e^{-\lambda u} - 1 + \lambda u\right)\frac{du}{u^{1+\alpha}}, \qquad \lambda \geq 0.$$

Let $\phi_k(\lambda)$ be defined by (1.38) with $\alpha_k = \alpha k^\alpha$ and $g_k(z) = z + \alpha^{-1}(1-z)^\alpha$. Then $\phi_k(\lambda) = \lambda^\alpha$ for $0 \leq \lambda \leq k$.

1.6 Notes and Comments

For the theory of convergence of probability measures on metric spaces we refer to Billingsley (1999), Ethier and Kurtz (1986) and Parthasarathy (1967). The standard reference of random measures is Kallenberg (1975).

A slightly different form of Proposition 1.19 was given in Dynkin (1989a). Theorem 1.20 can be found in Dynkin (1989b, 1991a). See Dynkin (1989b) for another proof of Theorem 1.23. Some earlier forms of Theorem 1.32 were given in Silverstein (1969) and Watanabe (1968). Theorem 1.40 is a modification of a theorem of Bernstein; see, e.g., Feller (1971, p.439) or Berg et al. (1984, p.135). The proof of Theorem 1.42 is from Li (1991).

The Laplace transform provides an important tool for the study of probability measures on the half line $\mathbb{R}_+$. To characterize a probability measure μ on the real line $\mathbb{R}$ one usually uses its *characteristic function* defined by

$$\hat{\mu}(t) := \int_{\mathbb{R}} e^{itx} \mu(\mathrm{d}x), \qquad t \in \mathbb{R}.$$

It is well-known that the probability measure is infinitely divisible if and only if its characteristic function is given by the *Lévy–Khintchine formula*:

$$\hat{\mu}(t) = \exp\left\{ iat - ct^2 + \int_{\mathbb{R}\setminus\{0\}} \left(e^{ity} - 1 - ity\mathbf{1}_{\{|y|\leq 1\}} \right) L(\mathrm{d}y) \right\}, \quad (1.44)$$

where $c \geq 0$ and a are constants and $L(\mathrm{d}y)$ is a σ-finite (Lévy) measure on $\mathbb{R} \setminus \{0\}$ such that

$$\int_{\mathbb{R}\setminus\{0\}} (1 \wedge y^2) L(\mathrm{d}y) < \infty.$$

For the infinitely divisible probability measure μ given by (1.44) we have

$$\int_{\mathbb{R}} |x| \mu(\mathrm{d}x) < \infty$$

if and only

$$\int_{\mathbb{R}\setminus\{0\}} (|y| \wedge y^2) L(\mathrm{d}y) < \infty;$$

see, e.g., Sato (1999, p.163). In this case, we can rewrite the Lévy–Khintchine representation into

$$\hat{\mu}(t) = \exp\left\{ ibt - ct^2 + \int_{\mathbb{R}\setminus\{0\}} \left(e^{ity} - 1 - ity \right) L(\mathrm{d}y) \right\}, \qquad (1.45)$$

where

$$b = \int_{\mathbb{R}} x \mu(\mathrm{d}x).$$

In particular, a *one-sided stable distribution* with index $1 < \alpha < 2$ is obtained by taking $b = c = 0$ and $L(\mathrm{d}y) = hy^{-1-\alpha}1_{\{y>0\}}\mathrm{d}y$ for some constant $h > 0$ in (1.45).

Chapter 2
Measure-Valued Branching Processes

A measure-valued process describes the evolution of a population that evolves according to the law of chance. In this chapter we provide some basic characterizations and constructions for measure-valued branching processes. In particular, we establish a one-to-one correspondence between those processes and cumulant semigroups. Some results for nonlinear integral evolution equations are proved, which lead to an analytic construction of a class of measure-valued branching processes, the so-called Dawson–Watanabe superprocesses. We shall construct the superprocesses for admissible killing densities and general branching mechanisms that are not necessarily decomposable into local and non-local parts. A number of moment formulas for the superprocesses are also given.

2.1 Definitions and Basic Properties

Suppose that E is a Lusin topological space and $(Q_t)_{t\geq 0}$ is a conservative transition semigroup on $M(E)$. We say $(Q_t)_{t\geq 0}$ satisfies the *branching property* provided

$$Q_t(\mu_1 + \mu_2, \cdot) = Q_t(\mu_1, \cdot) * Q_t(\mu_2, \cdot), \qquad t \geq 0,\ \mu_1, \mu_2 \in M(E). \quad (2.1)$$

Given the transition semigroup $(Q_t)_{t\geq 0}$, for $t \geq 0$ and $f \in B(E)^+$ let

$$V_t f(x) = -\log \int_{M(E)} e^{-\nu(f)} Q_t(\delta_x, d\nu), \qquad x \in E. \quad (2.2)$$

We say $(Q_t)_{t\geq 0}$ satisfies the *regular branching property* if for every $t \geq 0$ and $f \in B(E)^+$ the function $V_t f$ belongs to $B(E)^+$ and

$$\int_{M(E)} e^{-\nu(f)} Q_t(\mu, d\nu) = \exp\{-\mu(V_t f)\}, \qquad \mu \in M(E). \quad (2.3)$$

Z. Li, *Measure-Valued Branching Markov Processes*,
Probability and Its Applications, DOI 10.1007/978-3-642-15004-3_2,
© Springer-Verlag Berlin Heidelberg 2011

Clearly, $(Q_t)_{t\geq 0}$ has the branching property (2.1) if it satisfies the regular branching property (2.3).

Theorem 2.1 *If $(Q_t)_{t\geq 0}$ satisfies the branching property (2.1), then for any probability measures N_1 and N_2 on $M(E)$ we have*

$$(N_1 * N_2)Q_t = (N_1 Q_t) * (N_2 Q_t), \qquad t \geq 0. \qquad (2.4)$$

Proof. For any $t \geq 0$ and $f \in B(E)^+$,

$$
\begin{aligned}
&\int_{M(E)} e^{-\nu(f)} (N_1 * N_2) Q_t(d\nu) \\
&= \int_{M(E)} (N_1 * N_2)(d\mu) \int_{M(E)} e^{-\nu(f)} Q_t(\mu, d\nu) \\
&= \int_{M(E)^2} N_1(d\mu_1) N_2(d\mu_2) \int_{M(E)} e^{-\nu(f)} Q_t(\mu_1 + \mu_2, d\nu) \\
&= \int_{M(E)^2} N_1(d\mu_1) N_2(d\mu_2) \int_{M(E)^2} e^{-\nu_1(f)-\nu_2(f)} Q_t(\mu_1, d\nu_1) Q_t(\mu_2, d\nu_2) \\
&= \int_{M(E)} e^{-\nu_1(f)} (N_1 Q_t)(d\nu_1) \int_{M(E)} e^{-\nu_2(f)} (N_2 Q_t)(d\nu_2).
\end{aligned}
$$

Then (2.4) follows by the uniqueness of the Laplace functional. $\square$

Proposition 2.2 *If $(Q_t)_{t\geq 0}$ satisfies the branching property (2.1) and K is an infinitely divisible probability measure on $M(E)$, then KQ_t is an infinitely divisible probability measure on $M(E)$ for any $t \geq 0$.*

Proof. For $n \geq 1$ let K_n be the n-th root of K. By applying (2.4) inductively we have $(K_n Q_t)^{*n} = (K_n^{*n})Q_t = KQ_t$. Then KQ_t is infinitely divisible. $\square$

Suppose that T is an interval on the real line and $(\mathscr{F}_t)_{t\in T}$ is a filtration. A Markov process $\{(X_t, \mathscr{F}_t) : t \in T\}$ in $M(E)$ with transition semigroup $(Q_t)_{t\geq 0}$ satisfying the branching property (2.1) is called a *measure-valued branching process* (MB-process). In particular, we call $\{(X_t, \mathscr{F}_t) : t \in T\}$ a *regular MB-process* if $(Q_t)_{t\geq 0}$ satisfies the regular branching property defined by (2.2) and (2.3).

Theorem 2.3 *Suppose that $\{(X_t, \mathscr{F}_t) : t \in T\}$ and $\{(Y_t, \mathscr{G}_t) : t \in T\}$ are two independent MB-processes with transition semigroup $(Q_t)_{t\geq 0}$. Let $Z_t = X_t + Y_t$ and $\mathscr{H}_t = \sigma(\mathscr{F}_t \cup \mathscr{G}_t)$. Then $\{(Z_t, \mathscr{H}_t) : t \in T\}$ is also an MB-process with transition semigroup $(Q_t)_{t\geq 0}$.*

Proof. Let $r \leq t \in T$ and suppose $F \in b\mathscr{F}_r$ and $G \in b\mathscr{G}_r$. For any $f \in B(E)^+$ we use the independence of $\{(X_t, \mathscr{F}_t) : t \in T\}$ and $\{(Y_t, \mathscr{G}_t) : t \in T\}$ and the branching property (2.1) to see that

$$
\begin{aligned}
&\mathbf{E}\Big[FG \exp\{-Z_t(f)\}\Big] \\
&= \mathbf{E}\Big[F \exp\{-X_t(f)\}\Big] \mathbf{E}\Big[G \exp\{-Y_t(f)\}\Big]
\end{aligned}
$$

$$= \mathbf{E}\left[F \int_{M(E)} \mathrm{e}^{-\nu(f)} Q_{t-r}(X_r, \mathrm{d}\nu)\right]\mathbf{E}\left[G \int_{M(E)} \mathrm{e}^{-\nu(f)} Q_{t-r}(Y_r, \mathrm{d}\nu)\right]$$
$$= \mathbf{E}\left[FG \int_{M(E)} \mathrm{e}^{-\nu(f)} Q_{t-r}(Z_r, \mathrm{d}\nu)\right].$$

Then Proposition A.1 implies

$$\mathbf{E}\Big[H \exp\{-Z_t(f)\}\Big] = \mathbf{E}\Big[H \int_{M(E)} \mathrm{e}^{-\nu(f)} Q_{t-r}(Z_r, \mathrm{d}\nu)\Big]$$

for any $H \in \mathrm{b}\mathscr{H}_r$. That gives the desired result.   □

Recall that $\mathscr{I}(E)$ denotes the convex cone of functionals on $B(E)^+$ with the representation (1.20). Let $(V_t)_{t\geq 0}$ be a family of operators on $B(E)^+$ and let $v_t(x, f) = V_t f(x)$. We call $(V_t)_{t\geq 0}$ a *cumulant semigroup* provided:

(1) $v_t(x, \cdot) \in \mathscr{I}(E)$ for all $t \geq 0$ and $x \in E$;
(2) $V_r V_t = V_{r+t}$ for every $r, t \geq 0$.

By Theorem 1.36, if $(V_t)_{t\geq 0}$ is a cumulant semigroup, each operator V_t has the canonical representation

$$V_t f(x) = \lambda_t(x, f) + \int_{M(E)^\circ} \left(1 - \mathrm{e}^{-\nu(f)}\right) L_t(x, \mathrm{d}\nu), \quad f \in B(E)^+, \quad (2.5)$$

where $\lambda_t(x, \mathrm{d}y)$ is a bounded kernel on E and $(1 \wedge \nu(1))L_t(x, \mathrm{d}\nu)$ is a bounded kernel from E to $M(E)^\circ$.

Theorem 2.4 *The relation* (2.3) *establishes a one-to-one correspondence between cumulant semigroups* $(V_t)_{t\geq 0}$ *on* $B(E)^+$ *and transition semigroups* $(Q_t)_{t\geq 0}$ *on* $M(E)$ *satisfying the regular branching property.*

Proof. Suppose that $(V_t)_{t\geq 0}$ is a cumulant semigroup. By Theorem 1.35 we see that (2.3) defines an infinitely divisible probability measure $Q_t(\mu, \cdot)$ on $M(E)$. From $V_r V_t = V_{r+t}$ we have $Q_r Q_t = Q_{r+t}$. That is, $(Q_t)_{t\geq 0}$ is a transition semigroup on $M(E)$. Conversely, suppose that $(Q_t)_{t\geq 0}$ is a transition semigroup on $M(E)$ satisfying the regular branching property. Then $Q_t(\mu, \cdot)$ is an infinitely divisible probability measure on $M(E)$. This is true in particular for $\mu = \delta_x$, and so $V_t f(x)$ has the representation (2.5) by Theorems 1.35 and 1.36. The semigroup property of $(V_t)_{t\geq 0}$ follows from that of $(Q_t)_{t\geq 0}$.   □

Example 2.1 Let $M_a(E)$ and $M_d(E)$ denote respectively the subset of $M(E)$ of purely atomic measures and that of diffuse measures. Then each $\mu \in M(E)$ has the unique decomposition $\mu = \mu_a + \mu_d$ for $\mu_a \in M_a(E)$ and $\mu_d \in M_d(E)$. The mappings $\mu \mapsto \mu_a$ and $\mu \mapsto \mu_d$ are measurable; see Kallenberg (1975, pp.10–11). Take two distinct real constants c_a and c_d and let $Q_t(\mu, \cdot)$ be the unit mass concentrated at $\mathrm{e}^{c_a t}\mu_a + \mathrm{e}^{c_d t}\mu_d$. Then $(Q_t)_{t\geq 0}$ satisfies the branching property (2.1), but it is not regular in the sense of (2.3).

Theorem 2.5 *Suppose that E is a compact metric space. If $(V_t)_{t\geq 0}$ is a cumulant semigroup on E preserving $C(E)^{++}$ and $V_t f(x) \to f(x)$ pointwise as $t \to 0$ for every $f \in C(E)^{++}$, then (2.3) defines a Feller semigroup $(Q_t)_{t\geq 0}$ on $M(E)$. Conversely, if $(Q_t)_{t\geq 0}$ is a Feller semigroup having the branching property (2.1), then it satisfies the regular branching property (2.3) with cumulant semigroup $(V_t)_{t\geq 0}$ preserving $C(E)^{++}$ and $V_t f(x) \to f(x)$ pointwise as $t \to 0$ for every $f \in C(E)^{++}$.*

Proof. If $(V_t)_{t\geq 0}$ is a cumulant semigroup on E that preserves $C(E)^{++}$ and $V_t f(x) \to f(x)$ pointwise as $t \to 0$ for every $f \in C(E)^{++}$, it is simple to see that (2.3) defines a Feller semigroup on $M(E)$. For the converse, suppose that $(Q_t)_{t\geq 0}$ is a Feller semigroup on $M(E)$ having the branching property (2.1). Given $f \in B(E)^+$ we define $V_t f(x)$ by (2.2). For any $f \in C(E)^{++}$ we clearly have $V_t f \in C(E)^+$. If $\mu = \sum_{i=1}^{n}(p_i/q_i)\delta_{x_i}$ for $x_i \in E$ and integers p_i and $q_i \geq 1$, we have (2.3) by easy calculations based on (2.1). By an approximating argument, the equality holds for all $\mu \in M(E)$ and $f \in C(E)^{++}$. The extension from $f \in C(E)^{++}$ to $f \in B(E)^+$ is immediate by Proposition 1.3. Then $(Q_t)_{t\geq 0}$ satisfies the regular branching property. If there exists $f \in C(E)^{++}$ so that $V_t f \notin C(E)^{++}$, the compactness of E assures the existence of a point $x_0 \in E$ satisfying $V_t f(x_0) = 0$, so the function $\mu \mapsto \exp\{-\mu(V_t f)\}$ does not belong to $C_0(M(E))$, yielding a contradiction. Then $(V_t)_{t\geq 0}$ preserves $C(E)^{++}$. Since $Q_t F(\mu) \to F(\mu)$ pointwise as $t \to 0$ for every $F \in C_0(M(E))$, we have $V_t f(x) \to f(x)$ pointwise as $t \to 0$ for every $f \in C(E)^{++}$. $\square$

In the rest of the book, we will only consider regular MB-processes and will omit the adjective "regular". Given the transition semigroup $(Q_t)_{t\geq 0}$ of an MB-process in $M(E)$, we use $(Q_t^\circ)_{t\geq 0}$ to denote its restriction to $M(E)^\circ$.

Proposition 2.6 *Suppose that $(Q_t)_{t\geq 0}$ is defined by (2.3) with $(V_t)_{t\geq 0}$ given by (2.5). If $N = I(\eta, H)$ is an infinitely divisible probability measure on $M(E)$, then $NQ_t = I(\eta_t, H_t)$ is infinitely divisible for every $t \geq 0$, where*

$$\eta_t = \int_E \eta(dy)\lambda_t(y,\cdot) \quad and \quad H_t = \int_E \eta(dy)L_t(y,\cdot) + HQ_t^\circ. \qquad (2.6)$$

Proof. We first note that NQ_t is infinitely divisible by Proposition 2.2. For $t \geq 0$ and $f \in B(E)^+$ we have

$$-\log \int_{M(E)} e^{-\nu(f)} NQ_t(d\nu)$$

$$= \eta(V_t f) + \int_{M(E)^\circ} \left(1 - e^{-\nu(V_t f)}\right) H(d\nu)$$

$$= \int_E \eta(dy)\lambda_t(y,f) + \int_E \eta(dy)\int_{M(E)^\circ}\left(1 - e^{-\nu(f)}\right)L_t(y,d\nu)$$

$$\quad + \int_{M(E)^\circ}\left(1 - e^{-\nu(f)}\right)HQ_t^\circ(d\nu).$$

Then $NQ_t = I(\eta_t, H_t)$ with (η_t, H_t) given by (2.6). $\square$

Corollary 2.7 *Suppose that* $(Q_t)_{t\geq 0}$ *is defined by* (2.3) *with* $(V_t)_{t\geq 0}$ *given by* (2.5). *Then for any* $t \geq r \geq 0$ *and* $x \in E$ *we have*

$$\lambda_{r+t}(x, \cdot) = \int_E \lambda_r(x, \mathrm{d}y)\lambda_t(y, \cdot) \tag{2.7}$$

and

$$L_{r+t}(x, \cdot) = \int_E \lambda_r(x, \mathrm{d}y)L_t(y, \cdot) + \int_{M(E)^\circ} L_r(x, \mathrm{d}\mu)Q_t^\circ(\mu, \cdot). \tag{2.8}$$

Proof. This follows by applying Proposition 2.6 to the infinitely divisible probability measure $Q_r(\delta_x, \cdot)$ on $M(E)$. □

Suppose that $(Q_t)_{t\geq 0}$ is the transition semigroup of an MB-process defined by (2.3). In general, the corresponding cumulant semigroup $(V_t)_{t\geq 0}$ has the representation (2.5). Let E° be the set of points $x \in E$ such that $\lambda_t(x, E) = 0$ for all $t > 0$. Then $x \in E^\circ$ if and only if

$$V_t f(x) = \int_{M(E)^\circ} \left(1 - \mathrm{e}^{-\nu(f)}\right) L_t(x, \mathrm{d}\nu), \quad t > 0, f \in B(E)^+. \tag{2.9}$$

In view of (2.8), we have the following:

Proposition 2.8 *For any* $x \in E^\circ$ *the family of* σ-*finite measures* $\{L_t(x, \cdot) : t > 0\}$ *on* $M(E)^\circ$ *constitute an entrance law for the restricted semigroup* $(Q_t^\circ)_{t\geq 0}$.

2.2 Integral Evolution Equations

Let E be a Lusin topological space. Suppose that $\xi = (\Omega, \mathscr{F}, \mathscr{F}_t, \xi_t, \mathbf{P}_x)$ is a Borel right process in E with transition semigroup $(P_t)_{t\geq 0}$. Let $\{K(t) : t \geq 0\}$ be a continuous additive functional of ξ which is *admissible* in the sense that each $\omega \mapsto K_t(\omega)$ is measurable with respect to the σ-algebra $\mathscr{F}^0 := \sigma(\{\xi_t : t \geq 0\})$ and

$$k(t) := \sup_{x \in E} \mathbf{P}_x\big[K(t)\big] \to 0, \quad t \to 0. \tag{2.10}$$

For any $\beta \in B(E)$ we write

$$K_t(\beta) = \int_0^t \beta(\xi_s)K(\mathrm{d}s), \quad t \geq 0.$$

Let $\mathrm{b}\mathscr{E}(K)$ denote the set of functions $\beta \in B(E)$ so that $t \mapsto \mathrm{e}^{-K_t(\beta)}$ is a locally bounded stochastic process. Recall that $\|\cdot\|$ denotes the supremum norm of functions on E.

Proposition 2.9 *Let* $f \in B(E)$ *and* $b, \beta \in b\mathscr{E}(K)$. *If the two locally bounded functions* $h, u \in \mathscr{B}([0, \infty) \times E)$ *satisfy*

$$u(t, x) = \mathbf{P}_x\left[e^{-K_t(\beta)}f(\xi_t)\right] + \mathbf{P}_x\left[\int_0^t e^{-K_s(\beta)}h(t - s, \xi_s)K(ds)\right], \quad (2.11)$$

they also satisfy

$$u(t, x) = \mathbf{P}_x\left[e^{-K_t(b)}f(\xi_t)\right] + \mathbf{P}_x\left\{\int_0^t e^{-K_s(b)}h(t - s, \xi_s)K(ds)\right\}$$

$$- \mathbf{P}_x\left\{\int_0^t e^{-K_s(b)}[\beta(\xi_s) - b(\xi_s)]u(t - s, \xi_s)K(ds)\right\}. \quad (2.12)$$

Proof. Let $K_t^r(\beta) = K_t(\beta) - K_r(\beta)$ for $t \geq r \geq 0$. Since $s \mapsto \mathscr{F}_s$ is a right continuous filtration, the process

$$s \mapsto \mathbf{P}_x[e^{-K_t^s(\beta)}f(\xi_t)|\mathscr{F}_s] = e^{K_s(\beta)}\mathbf{P}_x[e^{-K_t(\beta)}f(\xi_t)|\mathscr{F}_s]$$

is a.s. right continuous. Let $g = \beta - b \in b\mathscr{E}(K)$. By the Markov property of ξ,

$$\mathbf{P}_x\left\{\int_0^t g(\xi_s)e^{-K_s(b)}\mathbf{P}_{\xi_s}\left[e^{-K_{t-s}(\beta)}f(\xi_{t-s})\right]K(ds)\right\}$$

$$= \mathbf{P}_x\left\{\int_0^t g(\xi_s)e^{-K_s(b)}\mathbf{P}_x\left[e^{-K_t^s(\beta)}f(\xi_t)|\mathscr{F}_s\right]K(ds)\right\}$$

$$= \lim_{n\to\infty}\mathbf{P}_x\left\{\sum_{i=1}^n\int_{(i-1)t/n}^{it/n}g(\xi_s)e^{-K_s(b)}\mathbf{P}_x\left[e^{-K_t^{it/n}(\beta)}f(\xi_t)|\mathscr{F}_{it/n}\right]K(ds)\right\}$$

$$= \lim_{n\to\infty}\sum_{i=1}^n\mathbf{P}_x\left\{\mathbf{P}_x\left[\int_{(i-1)t/n}^{it/n}g(\xi_s)e^{-K_s(b)}e^{-K_t^{it/n}(\beta)}f(\xi_t)K(ds)\Big|\mathscr{F}_{it/n}\right]\right\}$$

$$= \lim_{n\to\infty}\sum_{i=1}^n\mathbf{P}_x\left\{\int_{(i-1)t/n}^{it/n}g(\xi_s)e^{-K_s(b)}e^{-K_t^{it/n}(\beta)}f(\xi_t)K(ds)\right\}$$

$$= \mathbf{P}_x\left\{\int_0^t g(\xi_s)e^{-K_s(b)}e^{-K_t^s(\beta)}f(\xi_t)K(ds)\right\}$$

$$= \mathbf{P}_x\left\{\int_0^t g(\xi_s)e^{-K_t(b)}e^{-K_t^s(g)}f(\xi_t)K(ds)\right\}$$

$$= \mathbf{P}_x\left\{f(\xi_t)e^{-K_t(b)}\left(1 - e^{-K_t(g)}\right)\right\}.$$

By similar calculations we have

$$\mathbf{P}_x\left\{\int_0^t g(\xi_s)e^{-K_s(b)}\mathbf{P}_{\xi_s}\left[\int_0^{t-s}e^{-K_r(\beta)}h(t - s - r, \xi_r)K(dr)\right]K(ds)\right\}$$

$$= \mathbf{P}_x\left\{\int_0^t g(\xi_s)e^{-K_s(b)}K(ds)\int_0^{t-s}e^{-K_{s+r}^s(\beta)}h(t - s - r, \xi_{s+r})K(s + dr)\right\}$$

$$= \mathbf{P}_x\left\{\int_0^t g(\xi_s)e^{-K_s(b)}K(ds)\int_s^t e^{-K_r^s(\beta)}h(t - r, \xi_r)K(dr)\right\}$$

$$= \mathbf{P}_x\left\{\int_0^t h(t - r, \xi_r)e^{-K_r(b)}K(dr)\int_0^r g(\xi_s)e^{-K_r^s(g)}K(ds)\right\}$$

$$= \mathbf{P}_x\left\{\int_0^t h(t - r, \xi_r)e^{-K_r(b)}\left(1 - e^{-K_r(g)}\right)K(dr)\right\}.$$

Then we add up both sides of the two equations and use (2.11) to get (2.12).   □

In the sequel, we assume $\beta \in b\mathscr{E}(K)$ and $f \mapsto \phi(\cdot, f)$ is an operator from $B(E)^+$ into $B(E)$ which is bounded on $B_a(E)^+$ for every $a \geq 0$. For $f \in B(E)^+$ we consider the integral evolution equation

$$v_t(x) = \mathbf{P}_x\left[e^{-K_t(\beta)}f(\xi_t)\right] - \mathbf{P}_x\left[\int_0^t e^{-K_s(\beta)}\phi(\xi_s, v_{t-s})K(ds)\right]. \quad (2.13)$$

For the convenience of statement of the results, we formulate the following conditions:

Condition 2.10 *There is a constant $L \geq 0$ so that $-\phi(x, f) \leq L\|f\|$ for $x \in E$ and $f \in B(E)^+$.*

Condition 2.11 *For every $a \geq 0$ there is a constant $L_a \geq 0$ so that*

$$\sup_{x \in E} |\phi(x, f) - \phi(x, g)| \leq L_a\|f - g\|, \qquad f, g \in B_a(E)^+.$$

Proposition 2.12 *Let $r \geq 0$ and $f \in B(E)^+$. Then $(t, x) \mapsto v_t(x)$ satisfies (2.13) for $t \geq 0$ if and only if it satisfies the equation for $0 \leq t \leq r$ and $(t, x) \mapsto v_{r+t}(x)$ satisfies*

$$v_{r+t}(x) = \mathbf{P}_x\left[e^{-K_t(\beta)}v_r(\xi_t)\right] - \mathbf{P}_x\left[\int_0^t e^{-K_s(\beta)}\phi(\xi_s, v_{r+t-s})K(ds)\right]. \quad (2.14)$$

Proof. Suppose that $(t, x) \mapsto v_t(x)$ satisfies (2.13) for $0 \leq t \leq r$ and $(t, x) \mapsto v_{r+t}(x)$ satisfies (2.14) for $t \geq 0$. Then we have

$$
\begin{aligned}
v_{r+t}(x) &= \mathbf{P}_x\left[e^{-K_t(\beta)}v_r(\xi_t)\right] - \mathbf{P}_x\left[\int_0^t e^{-K_s(\beta)}\phi(\xi_s, v_{r+t-s})K(ds)\right] \\
&= \mathbf{P}_x\left\{e^{-K_t(\beta)}\mathbf{P}_{\xi_t}\left[e^{-K_r(\beta)}f(\xi_r)\right]\right\} \\
&\quad - \mathbf{P}_x\left\{e^{-K_t(\beta)}\mathbf{P}_{\xi_t}\left[\int_0^r e^{-K_s(\beta)}\phi(\xi_s, v_{r-s})K(ds)\right]\right\} \\
&\quad - \mathbf{P}_x\left[\int_0^t e^{-K_s(\beta)}\phi(\xi_s, v_{r+t-s})K(ds)\right] \\
&= \mathbf{P}_x\left[e^{-K_{r+t}(\beta)}f(\xi_{r+t})\right] - \mathbf{P}_x\left[\int_0^r e^{-K_{t+s}(\beta)}\phi(\xi_{t+s}, v_{r-s})K(t+ds)\right] \\
&\quad - \mathbf{P}_x\left[\int_0^t e^{-K_s(\beta)}\phi(\xi_s, v_{r+t-s})K(ds)\right] \\
&= \mathbf{P}_x\left[e^{-K_{r+t}(\beta)}f(\xi_{r+t})\right] - \mathbf{P}_x\left[\int_t^{r+t} e^{-K_s(\beta)}\phi(\xi_s, v_{r+t-s})K(ds)\right] \\
&\quad - \mathbf{P}_x\left[\int_0^t e^{-K_s(\beta)}\phi(\xi_s, v_{r+t-s})K(ds)\right] \\
&= \mathbf{P}_x\left[e^{-K_{r+t}(\beta)}f(\xi_{r+t})\right] - \mathbf{P}_x\left[\int_0^{r+t} e^{-K_s(\beta)}\phi(\xi_s, v_{r+t-s})K(ds)\right].
\end{aligned}
$$

Therefore $(t, x) \mapsto v_t(x)$ satisfies (2.13) for $t \geq 0$. For the converse, suppose that (2.13) holds for $t \geq 0$. The equation certainly holds for $0 \leq t \leq r$. By calculations similar to the above we see $(t, x) \mapsto v_{r+t}(x)$ satisfies (2.14).   □

Corollary 2.13 *If for every $f \in B(E)^+$ there is a unique locally bounded positive solution $(t, x) \mapsto v_t(x, f)$ to (2.13), then the operators $V_t : f \mapsto v_t(\cdot, f)$ on $B(E)^+$ constitute a semigroup.*

Proof. Fix $r \geq 0$ and define $u_t = v_t$ for $0 \leq t \leq r$ and $u_{r+t} = v_t(\cdot, v_r)$ for $t \geq 0$. By Proposition 2.12 we see $(t, x) \mapsto u_t(x)$ solves (2.13) for $t \geq 0$. Then the uniqueness of the solution implies $u_{r+t} = v_{r+t}$ for all $t \geq 0$. That gives the semigroup property of $(V_t)_{t \geq 0}$. $\qquad\square$

Proposition 2.14 *Suppose that Condition 2.10 holds. Then there is an increasing function $t \mapsto C(t)$ on $[0, \infty)$ so that for any locally bounded positive solution $(t, x) \mapsto v_t(x, f)$ to (2.13) we have*

$$\sup_{0 \leq s \leq t} \|v_s(\cdot, f)\| \leq C(t)\|f\|, \qquad t \geq 0. \tag{2.15}$$

Proof. Let $t \mapsto l(t)$ be an increasing function so that $e^{-K_t(\beta)} \leq l(t)$ for all $t \geq 0$. By (2.13) and Condition 2.10 we have

$$\|v_t(\cdot, f)\| \leq l(t)\|f\| + Ll(t) \sup_{x \in E} \mathbf{P}_x \left[\int_0^t \|v_{t-s}(\cdot, f)\| K(\mathrm{d}s) \right].$$

It follows that

$$\sup_{0 \leq s \leq t} \|v_s(\cdot, f)\| \leq l(t)\|f\| + Lk(t)l(t) \sup_{0 \leq s \leq t} \|v_s(\cdot, f)\|.$$

Let $\delta > 0$ be sufficiently small so that $Lk(\delta)l(\delta) < 1$. For $0 \leq t \leq \delta$ the above inequality implies

$$\sup_{0 \leq s \leq t} \|v_s(\cdot, f)\| \leq l(t)[1 - Lk(t)l(t)]^{-1}\|f\|.$$

Then the desired result follows by Proposition 2.12 and a successive application of the above estimate. $\qquad\square$

Proposition 2.15 *If Condition 2.11 holds, there is at most one locally bounded positive solution $(t, x) \mapsto v_t(x, f)$ to (2.13).*

Proof. Suppose that $(t, x) \mapsto u_t(x)$ and $(t, x) \mapsto v_t(x)$ are two locally bounded positive solutions of (2.13). Let $h_t(x) = u_t(x) - v_t(x)$ and let $l(t)$ be as in the proof of Proposition 2.14. For fixed $T > 0$ we can use Proposition 2.14 to find a constant $a \geq 0$ so that $\|u_t\| \leq a$ and $\|v_t\| \leq a$ for all $0 \leq t \leq T$. By (2.13) and Condition 2.11 we have

$$\|h_t\| \leq l(t)\mathbf{P}_x \left[\int_0^t |\phi(\xi_s, u_{t-s}) - \phi(\xi_s, v_{t-s})| K(\mathrm{d}s) \right]$$
$$\leq L_a l(t) \sup_{x \in E} \mathbf{P}_x \left[\int_0^t \|h_{t-s}\| K(\mathrm{d}s) \right].$$

Then it is easy to get

$$\sup_{0 \leq s \leq t} \|h_s\| \leq L_a k(t) l(t) \sup_{0 \leq s \leq t} \|h_s\|, \quad 0 \leq t \leq T.$$

Take $0 < \delta \leq T$ so that $L_a k(\delta) l(\delta) < 1$. The above inequality implies $\|h_t\| = 0$ and hence $u_t = v_t$ for $0 \leq t \leq \delta$. Then an application of Proposition 2.12 gives the uniqueness of the solution to (2.13). $\qquad\square$

Proposition 2.16 *Let $\{\phi_n\}$ be a sequence of operators from $B(E)^+$ into $B(E)$ satisfying Conditions 2.10 and 2.11 with the constants L and L_a independent of $n \geq 1$. Suppose that $\lim_{n \to \infty} \phi_n(x, f) = \phi(x, f)$ uniformly on $E \times B_a(E)^+$ for every $a \geq 0$ and for $f_n \in B(E)^+$ there is a unique locally bounded positive solution $t \mapsto v_n(t) = v_n(t, x)$ to the equation*

$$v_n(t, x) = \mathbf{P}_x \big[e^{-K_t(\beta)} f_n(\xi_t) \big] - \mathbf{P}_x \bigg[\int_0^t e^{-K_s(\beta)} \phi_n(\xi_s, v_n(t - s)) K(ds) \bigg]. \quad (2.16)$$

If $\lim_{n \to \infty} f_n = f$ in the supremum norm, then the limit $\lim_{n \to \infty} v_n(t, x) = v_t(x)$ exists and is uniform on $[0, T] \times E$ for every $T \geq 0$. Moreover, $(t, x) \mapsto v_t(x)$ is a solution of (2.13).

Proof. Choose a sufficiently large constant $a \geq 0$ so that $\{f_n\} \subset B_a(E)^+$. By Proposition 2.14 there is an increasing $t \mapsto C(t)$ on $[0, \infty)$ so that

$$\sup_{0 \leq s \leq t} \|v_n(s)\| \leq C(t) \|f_n\| \leq a C(t), \quad t \geq 0.$$

Fix $T > 0$ and let $c = a C(T)$. For $\varepsilon > 0$ let $N = N(\varepsilon, c)$ be an integer so that $\|f_n - f\| \leq \varepsilon$ and $\|\phi_n(\cdot, h) - \phi(\cdot, h)\| \leq \varepsilon$ for $n \geq N$ and $h \in B_c(E)^+$. Let $l(t)$ be as in the proof of Proposition 2.14 and let

$$H_t(n_1, n_2) = \sup_{0 \leq s \leq t} \|v_{n_2}(s) - v_{n_1}(s)\|.$$

By (2.16) and Condition 2.11 we have

$$H_t(n_1, n_2) \leq 2l(t)[1 + k(t)]\varepsilon + L_c k(t) l(t) H_t(n_1, n_2)$$

for $0 \leq t \leq T$ and $n_1, n_2 \geq N$. Take $0 < \delta \leq T$ so that $L_c k(\delta) l(\delta) < 1$. The above inequality implies

$$H_t(n_1, n_2) \leq 2l(t)[1 + k(t)][1 - L_c k(t) l(t)]^{-1} \varepsilon$$

for $0 \leq t \leq \delta$. Then $v_n(t, x)$ converges uniformly on $[0, \delta] \times E$. By repeating the above arguments and applying Proposition 2.12 we see the limit $\lim_{n \to \infty} v_n(t, x) = v_t(x)$ exists and is uniform on $[0, T] \times E$. Then letting $n \to \infty$ in (2.16) we obtain (2.13). $\qquad\square$

2.3 Dawson–Watanabe Superprocesses

In this section we give the construction of a general class of Dawson–Watanabe superprocesses. For this purpose we need to discuss the existence of solutions of some nonlinear integral evolution equations which define cumulant semigroups. Let E be a Lusin topological space. Suppose that ξ is a Borel right process in E with transition semigroup $(P_t)_{t\geq 0}$ and $\{K(t) : t \geq 0\}$ is a continuous admissible additive functional of ξ.

Lemma 2.17 *Suppose that $b \in B(E)$ and $\gamma(x, dy)$ is a bounded kernel on E. Then for each $f \in B(E)$ there is a unique locally bounded solution $(t, x) \mapsto \pi_t f(x)$ to the linear evolution equation*

$$
\pi_t f(x) = \mathbf{P}_x f(\xi_t) + \mathbf{P}_x \left\{ \int_0^t \gamma(\xi_s, \pi_{t-s} f) K(ds) \right\}
$$
$$
- \mathbf{P}_x \left\{ \int_0^t b(\xi_s) \pi_{t-s} f(\xi_s) K(ds) \right\}, \tag{2.17}
$$

which defines a locally bounded semigroup $(\pi_t)_{t\geq 0}$ of kernels on E.

Proof. Let $b^+ = 0 \vee b$ and $b^- = 0 \vee (-b)$. By Proposition A.41 there is a unique locally bounded solution $(t, x) \mapsto \pi_t f(x)$ to the equation

$$
\pi_t f(x) = \mathbf{P}_x \left[e^{-K_t(b^+)} f(\xi_t) \right] + \mathbf{P}_x \left\{ \int_0^t e^{-K_s(b^+)} \gamma(\xi_s, \pi_{t-s} f) K(ds) \right\}
$$
$$
+ \mathbf{P}_x \left\{ \int_0^t e^{-K_s(b^+)} b^-(\xi_s) \pi_{t-s} f(\xi_s) K(ds) \right\},
$$

which defines a locally bounded semigroup $(\pi_t)_{t\geq 0}$ of kernels on E. By Proposition 2.9 the above equation is equivalent to (2.17). $\qquad\square$

Suppose that $\eta(x, dy)$ is a bounded kernel on E and $\nu(1)H(x, d\nu)$ is a bounded kernel from E to $M(E)^\circ$. We consider a function $b \in B(E)$ and an operator $f \mapsto \psi(\cdot, f)$ on $B(E)^+$ with the representation

$$
\psi(x, f) = \eta(x, f) + \int_{M(E)^\circ} \left(1 - e^{-\nu(f)} \right) H(x, d\nu). \tag{2.18}
$$

From $\eta(x, dy)$ and $H(x, d\nu)$ we can define the bounded kernel $\gamma(x, dy)$ on E by

$$
\gamma(x, dy) = \eta(x, dy) + \int_{M(E)^\circ} \nu(dy) H(x, d\nu). \tag{2.19}
$$

Let $\beta \geq 0$ be a constant so that $b(x) \leq \beta$ for all $x \in E$. For fixed $f \in B(E)^+$ set $u_0(t, x) = 0$ and define $u_n(t, x) = u_n(t, x, f)$ inductively by

$$
u_{n+1}(t, x) = \mathbf{P}_x \left[e^{-K_t(\beta)} f(\xi_t) \right] + \mathbf{P}_x \left\{ \int_0^t e^{-K_s(\beta)} \psi(\xi_s, u_n(t - s)) K(ds) \right\}
$$

$$+ \mathbf{P}_x \left\{ \int_0^t e^{-K_s(\beta)} [\beta - b(\xi_s)] u_n(t - s, \xi_s) K(ds) \right\}. \quad (2.20)$$

Proposition 2.18 *For every $f \in B(E)^+$ there is a unique locally bounded positive solution $(t, x) \mapsto u_t(x, f)$ to the evolution equation*

$$u_t(x) = \mathbf{P}_x[f(\xi_t)] + \mathbf{P}_x \left\{ \int_0^t [\psi(\xi_s, u_{t-s}) - b(\xi_s) u_{t-s}(\xi_s)] K(ds) \right\}. \quad (2.21)$$

Moreover, we have $\pi_t f(x) \geq u_t(x, f) =\uparrow \lim_{n \to \infty} u_n(t, x, f)$ for all $t \geq 0$ and $x \in E$, where $(\pi_t)_{t \geq 0}$ is the semigroup defined by (2.17).

Proof. The operator $f \mapsto \psi(\cdot, f) - bf$ clearly satisfies Condition 2.11 with $L_a = \|b\| + \|\gamma(\cdot, 1)\|$ for all $a \geq 0$. By Proposition 2.15 there is at most one locally bounded positive solution to (2.21). We next claim that

$$0 \leq u_{n-1}(t, x, f) \leq u_n(t, x, f) \leq \pi_t f(x), \qquad t \geq 0, x \in E \quad (2.22)$$

for every $n \geq 1$. By Proposition 2.9 we can also define $(t, x) \mapsto \pi_t f(x)$ by the evolution equation

$$\pi_t f(x) = \mathbf{P}_x[e^{-K_t(\beta)} f(\xi_t)] + \mathbf{P}_x \left\{ \int_0^t e^{-K_s(\beta)} \gamma(\xi_s, \pi_{t-s} f) K(ds) \right\}$$
$$+ \mathbf{P}_x \left\{ \int_0^t e^{-K_s(\beta)} [\beta - b(\xi_s)] \pi_{t-s} f(\xi_s) K(ds) \right\}. \quad (2.23)$$

Then for $n = 1$ the inequalities in (2.22) are trivial. Suppose they are true for some $n \geq 1$. By the monotonicity of the operator $f \mapsto \psi(\cdot, f) + (\beta - b)f$ we have

$$0 \leq u_n(t, x, f) \leq u_{n+1}(t, x, f) \leq v(t, x, f),$$

where

$$v(t, x, f) = \mathbf{P}_x[e^{-K_t(\beta)} f(\xi_t)] + \mathbf{P}_x \left\{ \int_0^t e^{-K_s(\beta)} \psi(\xi_s, \pi_{t-s} f) K(ds) \right\}$$
$$+ \mathbf{P}_x \left\{ \int_0^t e^{-K_s(\beta)} [\beta - b(\xi_s)] \pi_{t-s} f(\xi_s) K(ds) \right\}. \quad (2.24)$$

In view of (2.23) and (2.24) we have $v(t, x, f) \leq \pi_t f(x)$. Then (2.22) holds for all $n \geq 1$. Let $u_t(x, f) =\uparrow \lim_{n \to \infty} u_n(t, x, f)$. From (2.20) we see that $(t, x) \mapsto u_t(x, f)$ is a locally bounded positive solution of

$$u_t(x) = \mathbf{P}_x[e^{-K_t(\beta)} f(\xi_t)] + \mathbf{P}_x \left\{ \int_0^t e^{-K_s(\beta)} \psi(\xi_s, u_{t-s}) K(ds) \right\}$$
$$+ \mathbf{P}_x \left\{ \int_0^t e^{-K_s(\beta)} [\beta - b(\xi_s)] u_{t-s}(\xi_s) K(ds) \right\},$$

which is equivalent to (2.21) by Proposition 2.9. □

Proposition 2.19 *In the case where $K(\mathrm{d}s) = \mathrm{d}s$ is the Lebesgue measure, we have $u_t(x, f) = \uparrow\lim_{n\to\infty} u_n(t, x, f)$ uniformly on $[0, T] \times E \times B_a(E)^+$ for every $T \geq 0$ and $a \geq 0$.*

Proof. Let $D_n(t) = \sup_{0 \leq s \leq t} \|u_n(s) - u_{n-1}(s)\|$. From (2.20) it is easy to get

$$
\begin{aligned}
D_n(t) &\leq (\beta + \|b\| + \|\gamma(\cdot, 1)\|) \int_0^t D_{n-1}(s_1)\mathrm{d}s_1 \\
&\leq (\beta + \|b\| + \|\gamma(\cdot, 1)\|)^2 \int_0^t \mathrm{d}s_1 \int_0^{s_1} D_{n-2}(s_2)\mathrm{d}s_2 \\
&\leq \cdots \\
&\leq (\beta + \|b\| + \|\gamma(\cdot, 1)\|)^{n-1} \int_0^t \mathrm{d}s_1 \int_0^{s_1} \cdots \int_0^{s_{n-2}} \|f\|\mathrm{d}s_{n-1} \\
&\leq \frac{1}{(n-1)!}(\beta + \|b\| + \|\gamma(\cdot, 1)\|)^{n-1}t^{n-1}\|f\|,
\end{aligned}
$$

and hence

$$
D(t) := \sum_{n=1}^{\infty} D_n(t) \leq \|f\| \exp\{(\beta + \|b\| + \|\gamma(\cdot, 1)\|)t\} < \infty.
$$

Then $\lim_{n\to\infty} u_n(t, x, f) = u_t(x, f)$ uniformly on $[0, T] \times E \times B_a(E)^+$. □

Now we consider a more general operator $f \mapsto \phi(\cdot, f)$ as follows. Let $b \in B(E)$ and $c \in B(E)^+$. Let $\eta(x, \mathrm{d}y)$ be a bounded kernel on E and $H(x, \mathrm{d}\nu)$ a σ-finite kernel from E to $M(E)^{\circ}$. Suppose that

$$
\sup_{x\in E} \int_{M(E)^{\circ}} \left[\nu(1) \wedge \nu(1)^2 + \nu_x(1)\right] H(x, \mathrm{d}\nu) < \infty, \tag{2.25}
$$

where $\nu_x(\mathrm{d}y)$ denotes the restriction of $\nu(\mathrm{d}y)$ to $E \setminus \{x\}$. For $x \in E$ and $f \in B(E)^+$ write

$$
\begin{aligned}
\phi(x, f) = {}& b(x)f(x) + c(x)f(x)^2 - \int_E f(y)\eta(x, \mathrm{d}y) \\
& + \int_{M(E)^{\circ}} \left[e^{-\nu(f)} - 1 + \nu(\{x\})f(x)\right] H(x, \mathrm{d}\nu). \tag{2.26}
\end{aligned}
$$

By Taylor's expansion it is easy to see that

$$
e^{-\nu(f)} - 1 + \nu(\{x\})f(x) = -\nu_x(f) + \frac{1}{2}e^{-\theta}\nu(f)^2,
$$

where $0 \leq \theta \leq \nu(f)$. Observe also that

$$
\left|e^{-\nu(f)} - 1 + \nu(\{x\})f(x)\right| \leq \nu(f) + \nu(\{x\})f(x).
$$

Then the second integral on the right-hand side of (2.26) is bounded on $E \times B_a(E)^+$ for every $a \geq 0$. Moreover, we can rewrite (2.26) into

$$\phi(x, f) = b(x)f(x) + c(x)f(x)^2 - \int_E f(y)\gamma(x, dy)$$
$$+ \int_{M(E)^{\circ}} \left[e^{-\nu(f)} - 1 + \nu(f) \right] H(x, d\nu), \qquad (2.27)$$

where

$$\gamma(x, dy) = \eta(x, dy) + \int_{M(E)^{\circ}} \nu_x(dy) H(x, d\nu). \qquad (2.28)$$

For each integer $n \geq 1$ define

$$\phi_n(x, f) = b(x)f(x) + 2nc(x)f(x) + \int_{M(E)^{\circ}} \nu(f)h_n(\nu)H(x, d\nu)$$
$$- \int_E f(y)\gamma(x, dy) - 2n^2 c(x)(1 - e^{-f(x)/n})$$
$$- \int_{M(E)^{\circ}} \left(1 - e^{-\nu(f)} \right) h_n(\nu)H(x, d\nu), \qquad (2.29)$$

where $h_n(\nu) = 1 \wedge [n\nu(1)]$. It is easy to see $\phi_n(x, f) \to \phi(x, f)$ increasingly as $n \to \infty$. For $n \geq 1$ and $f \in B(E)^+$ we consider the equation

$$v(t, x) = \mathbf{P}_x[f(\xi_t)] - \mathbf{P}_x \left\{ \int_0^t \phi_n(\xi_s, v(t - s))K(ds) \right\}. \qquad (2.30)$$

This is clearly a special case of (2.21). By Proposition 2.18 there is a unique locally bounded positive solution $(t, x) \mapsto v_n(t, x, f)$ to (2.30).

Proposition 2.20 *Suppose that ϕ and γ are defined respectively by (2.27) and (2.28). Let $(\pi_t)_{t \geq 0}$ be defined by (2.17). Then for every $f \in B(E)^+$ there is a unique locally bounded positive solution $(t, x) \mapsto v_t(x, f)$ to*

$$v_t(x) = \mathbf{P}_x f(\xi_t) - \mathbf{P}_x \left[\int_0^t \phi(\xi_s, v_{t-s})K(ds) \right], \quad t \geq 0, x \in E. \qquad (2.31)$$

Moreover, we have $\pi_t f(x) \geq v_t(x, f) = \downarrow \lim_{n \to \infty} v_n(t, x, f)$ for $t \geq 0$ and $x \in E$.

Proof. Since $\phi_n(x, f)$ is increasing in $n \geq 1$ and $f \in B(E)^+$, by Proposition 2.18 we see $v_n(t, x, f)$ is decreasing in $n \geq 1$. Let $v_t(x, f) = \lim_{n \to \infty} v_n(t, x, f) \leq \pi_t(x, f)$. In view of (2.29) and (2.30), we conclude by dominated convergence that $(t, x) \mapsto v_t(x, f)$ is a locally bounded positive solution of (2.31). For $a \geq 0$ and $f, g \in B_a(E)^+$ we can use (2.27) to see

$$|\phi(x, f) - \phi(x, g)| \leq (\|b\| + 2a\|c\|)\|f - g\| + \gamma(x, 1)\|f - g\|$$

$$+ \int_{M(E)^\circ} |\nu(f - g) + e^{-\nu(f)} - e^{-\nu(g)}| H(x, d\nu).$$

By the mean-value theorem we have

$$\nu(f - g) + e^{-\nu(f)} - e^{-\nu(g)} = \nu(f - g)(1 - e^{-\theta}),$$

where $\nu(f \wedge g) \leq \theta \leq \nu(f \vee g) \leq a\nu(1)$. It follows that

$$|\nu(f - g) + e^{-\nu(f)} - e^{-\nu(g)}| \leq \|f - g\|(\nu(1) \wedge a\nu(1)^2).$$

Then $f \mapsto \phi(\cdot, f)$ satisfies Condition 2.11 for some constant $L_a \geq 0$ and the uniqueness of the solution of (2.31) follows by Proposition 2.15.   □

Theorem 2.21 *Let ϕ be given by (2.26) or (2.27). For every $f \in B(E)^+$ let $(t, x) \mapsto V_t f(x)$ denote the unique locally bounded positive solution of (2.31). Then the operators $(V_t)_{t \geq 0}$ constitute a cumulant semigroup.*

Proof. By (2.20) and Theorem 1.37 one checks inductively $u_n(t, x, \cdot) \in \mathscr{I}(E)$ for each $n \geq 1$. Now Corollary 1.34 and Propositions 2.18 and 2.20 imply first $u_t(x, \cdot) \in \mathscr{I}(E)$ for the solution of (2.21), and then $v_t(x, \cdot) \in \mathscr{I}(E)$ for the solution of (2.31). The semigroup property of $(V_t)_{t \geq 0}$ follows from Corollary 2.13.   □

Let ϕ be given by (2.26) or (2.27) and let $(V_t)_{t \geq 0}$ be the cumulant semigroup defined by (2.31). Then we can define a Markov transition semigroup $(Q_t)_{t \geq 0}$ on $M(E)$ by

$$\int_{M(E)} e^{-\nu(f)} Q_t(\mu, d\nu) = \exp\{-\mu(V_t f)\}, \qquad f \in B(E)^+. \qquad (2.32)$$

If X is a Markov process in $M(E)$ with transition semigroup $(Q_t)_{t \geq 0}$, we call it a *Dawson–Watanabe superprocess* with parameters (ξ, K, ϕ), or simply a (ξ, K, ϕ)-*superprocess*, where ξ is the *spatial motion*, K is the *killing functional* or *killing density*, and ϕ is the *branching mechanism*. If $K(ds) = ds$ is the Lebesgue measure, we call X a (ξ, ϕ)-*superprocess*. In this case, we can rewrite (2.31) into

$$v_t(x) = P_t f(x) - \int_0^t ds \int_E \phi(y, v_s) P_{t-s}(x, dy), \qquad x \in E, t \geq 0. \quad (2.33)$$

We say the branching mechanism is *spatially constant* if $f \mapsto \phi(\cdot, f)$ maps constant functions to constant functions. In Chapter 4 we shall give some intuitive interpretations of the superprocesses in terms of limit theorems of branching particle systems.

Theorem 2.22 *A realization $\{X_t : t \geq 0\}$ of the (ξ, K, ϕ)-superprocess is right continuous in probability.*

Proof. Let $f \in C(E)^+$. Since ξ is right continuous, the map $t \mapsto P_t f(x)$ is right continuous for every $x \in E$, so (2.31) implies $\lim_{t \to 0} V_t f(x) = f(x)$. From (2.32) we get

$$\lim_{t \to 0} \int_{M(E)} e^{-\nu(f)} Q_t(\mu, d\nu) = \exp\{-\mu(f)\}.$$

Then we have $\lim_{t \to 0} Q_t(\mu, \cdot) = \delta_\mu$ weakly. For any $\varepsilon > 0$ let $B(\mu, \varepsilon)^c = \{\nu \in M(E) : \rho(\nu, \mu) > \varepsilon\}$, where ρ is the metric on $M(E)$ defined by (1.3). Then we infer $\lim_{t \to 0} Q_t(\mu, B(\mu, \varepsilon)^c) = 0$. Using the Markov property of X and dominated convergence we get

$$\lim_{t \to r+} \mathbf{P}\{\rho(X_t, X_r) > \varepsilon\} = \lim_{t \to r+} \mathbf{P}\{Q_{t-r}(X_r, B(X_r, \varepsilon)^c)\} = 0$$

for every $r \geq 0$. Therefore $t \mapsto X_t$ is right continuous in probability. $\qquad\square$

For the (ξ, ϕ)-superprocess we can give an alternate characterization of the cumulant semigroup. Given a function $b \in B(E)$, we define a locally bounded semigroup of Borel kernels $(P_t^b)_{t \geq 0}$ on E by the following *Feynman–Kac formula*:

$$P_t^b f(x) = \mathbf{P}_x\left[e^{-\int_0^t b(\xi_s)ds} f(\xi_t)\right], \qquad x \in E, f \in B(E). \qquad (2.34)$$

Then (2.17) can be rewritten into

$$\pi_t f(x) = P_t f(x) + \int_0^t P_{t-s}(\gamma - b)\pi_s f(x)ds, \qquad t \geq 0, x \in E, \qquad (2.35)$$

which is equivalent to

$$\pi_t f(x) = P_t^b f(x) + \int_0^t P_{t-s}^b \gamma \pi_s f(x)ds, \qquad t \geq 0, x \in E. \qquad (2.36)$$

From Proposition A.41 we have

$$\pi_t f(x) = P_t^b f(x) + \sum_{n=1}^\infty \int_0^t ds_1 \cdots \int_0^{s_{n-1}} P_{t-s_1}^b \gamma P_{s_1-s_2}^b \cdots \gamma P_{s_n}^b f(x)ds_n. \qquad (2.37)$$

Let $b^+ = 0 \vee b$ and $b^- = 0 \vee (-b)$. By Proposition A.49 we have $\|\pi_t\| \leq e^{c_0 t}$ for all $t \geq 0$, where $c_0 = \|b^-\| + \|\gamma(\cdot, 1)\|$.

Theorem 2.23 *Suppose that ϕ and γ are defined respectively by (2.27) and (2.28). Let $(\pi_t)_{t \geq 0}$ be defined by (2.35). Then (2.33) is equivalent to the evolution equation*

$$v_t(x) = \pi_t f(x) - \int_0^t ds \int_E \phi_0(y, v_s)\pi_{t-s}(x, dy), \qquad (2.38)$$

where

$$\phi_0(y, f) = c(y)f(y)^2 + \int_{M(E)^\circ} [e^{-\nu(f)} - 1 + \nu(f)]H(y, d\nu). \qquad (2.39)$$

Proof. We first show (2.33) implies (2.38). By applying Proposition 2.9 to (2.33) we have

$$v_t(x) = P_t^b f(x) - \int_0^t P_{t-s}^b [\phi(v_s) - bv_s](x)\mathrm{d}s.$$

This combined with (2.36) implies

$$v_t(x) = \pi_t f(x) - \int_0^t P_{t-s}^b \phi_0(v_s)(x)\mathrm{d}s + \int_0^t P_{t-s}^b \gamma(v_s - \pi_s f)(x)\mathrm{d}s.$$

Then we use the above relation inductively to see

$$v_t(x) = \pi_t f(x) - \int_0^t P_{t-s_1}^b \phi_0(v_{s_1})(x)\mathrm{d}s_1 + w_n(t,x)$$
$$- \sum_{i=2}^n \int_0^t \mathrm{d}s_1 \cdots \int_0^{s_{i-1}} P_{t-s_1}^b \gamma P_{s_1-s_2}^b$$
$$\cdots \gamma P_{s_{i-1}-s_i}^b g_{s_i}(x)\mathrm{d}s_i, \qquad (2.40)$$

where $g_{s_i}(x) = \phi_0(x, v_{s_i})$ and

$$w_n(t,x) = \int_0^t \mathrm{d}s_1 \cdots \int_0^{s_n} P_{t-s_1}^b \gamma \cdots P_{s_n-s_{n+1}}^b \gamma(v_{s_{n+1}} - \pi_{s_{n+1}} f)(x)\mathrm{d}s_{n+1}.$$

Since $0 \le v_{s_{n+1}}(x) \le \pi_{s_{n+1}} f(x) \le \|f\| e^{c_0 s_{n+1}}$, we have

$$\|w_n(t,\cdot)\| \le \|f\|\|\gamma(\cdot,1)\|^{n+1} e^{c_0 t} \int_0^t \mathrm{d}s_1 \int_0^{s_1} \mathrm{d}s_2 \cdots \int_0^{s_n} \mathrm{d}s_{n+1}$$
$$\le \|f\|\|\gamma(\cdot,1)\|^{n+1} e^{c_0 t} \frac{t^{n+1}}{(n+1)!}.$$

Then letting $n \to \infty$ in (2.40) and using (2.37) we obtain (2.38). The uniqueness of the solution to (2.38) follows from Gronwall's inequality by standard arguments. Then the two equations are equivalent.   □

2.4 Examples of Superprocesses

The (ξ, K, ϕ)- and (ξ, ϕ)-superprocesses we have constructed are quite wide. From these one can derive the existence of various special classes of superprocesses. Some special cases of the parameters are discussed in the following examples.

Example 2.2 Let $|\cdot|$ and $(\cdot, \cdot)$ denote respectively the Euclidean norm and inner product of $\mathbb{R}^d$. For each $1 \le i \le d$ suppose that $\lambda \mapsto \phi_i(\lambda)$ is a function on $\mathbb{R}_+^d$ with the representation

$$\phi_i(\lambda) = b_i\lambda_i + c_i\lambda_i^2 + (\eta_i, \lambda) + \int_{\mathbb{R}_+^d \setminus \{0\}} \left(e^{-(\lambda,u)} - 1 + \lambda_i u_i\right) H_i(du),$$

where $c_i \geq 0$ and b_i are constants, $\eta_i \in \mathbb{R}_+^d$ is a vector, and $H_i(du)$ is a σ-finite measure on $\mathbb{R}_+^d \setminus \{0\}$ so that

$$\int_{\mathbb{R}_+^d \setminus \{0\}} \left(|u| \wedge |u|^2 + \sum_{j \neq i} u_j\right) H_i(du) < \infty.$$

By Proposition 2.20 and Theorem 2.21 for any $\lambda \in \mathbb{R}_+^d$ there is a unique locally bounded vector-valued solution $t \mapsto v(t, \lambda) \in \mathbb{R}_+^d$ to the evolution equation system

$$v_i(t, \lambda) = \lambda_i - \int_0^t \phi_i(v(s, \lambda))ds, \qquad t \geq 0, i = 1, \ldots, d, \qquad (2.41)$$

and there is a transition semigroup $(Q_t)_{t \geq 0}$ on $\mathbb{R}_+^d$ defined by

$$\int_{\mathbb{R}_+^d} e^{-(\lambda,y)} Q_t(x, dy) = e^{-(x,v(t,\lambda))}, \qquad \lambda, x \in \mathbb{R}_+^d. \qquad (2.42)$$

From (2.41) we see that $t \mapsto v_i(t, \lambda)$ is continuously differentiable. Then we can rewrite the equation into the equivalent differential form

$$\frac{dv_i}{dt}(t, \lambda) = -\phi_i(v(t, \lambda)), \quad v_i(0, \lambda) = \lambda_i, \qquad i = 1, \ldots, d.$$

A Markov process in $\mathbb{R}_+^d$ with transition semigroup $(Q_t)_{t \geq 0}$ given by (2.42) is called a *continuous-state branching process* (CB-process).

Example 2.3 By a *super-Brownian motion* we mean a superprocess with Brownian motion as underlying spatial motion. A particular super-Brownian motion is described as follows. Let ξ be a standard Brownian motion in $\mathbb{R}$. It is well-known that ξ has a continuous local time $\{2l(t, y) : t \geq 0, y \in \mathbb{R}\}$, that is,

$$\int_0^t 1_B(\xi_s)ds = \int_B 2l(t, y)dy, \qquad t \geq 0, B \in \mathscr{B}(\mathbb{R}); \qquad (2.43)$$

see, e.g., Ikeda and Watanabe (1989, p.113). Let $\rho \in M(\mathbb{R})$ and define the continuous additive functional $t \mapsto K(t)$ by

$$K(t) = \int_{\mathbb{R}} 2l(t, y)\rho(dy), \qquad t \geq 0.$$

Then we have

$$\mathbf{P}_x[K(t)] = \int_{\mathbb{R}} \rho(dy) \int_0^t g_s(y - x)ds \leq \frac{\sqrt{2t}}{\sqrt{\pi}} \rho(\mathbb{R}),$$

where

$$g_t(z) = \frac{1}{\sqrt{2\pi t}} \exp\{-z^2/2t\}, \qquad t > 0, z \in \mathbb{R}. \tag{2.44}$$

Thus $t \mapsto K(t)$ is admissible. In this case, we can rewrite (2.31) as

$$v_t(x) = P_t f(x) - \int_0^t ds \int_{\mathbb{R}} \phi(y, v_{t-s}) g_s(y - x) \rho(dy).$$

The corresponding (ξ, K, ϕ)-superprocess is called a *catalytic super-Brownian motion* with *catalyst measure* $\rho(dy)$.

Example 2.4 Let $b \in B(E)$ and $c \in B(E)^+$. Let $(u \wedge u^2)m(x, du)$ be a bounded kernel from E to $(0, \infty)$. We define a Borel function $(x, z) \mapsto \phi(x, z)$ on $E \times [0, \infty)$ by

$$\phi(x, z) = b(x)z + c(x)z^2 + \int_0^\infty (e^{-zu} - 1 + zu)m(x, du). \tag{2.45}$$

Then $(x, f) \mapsto \phi(x, f(x))$ can be represented in the form (2.26) or (2.27). In this case, we say the corresponding superprocess has a *local branching mechanism*. If there is $c \in B(E)^+$ so that $\phi(x, z) = c(x)z^2$ for all $x \in E$ and $z \geq 0$, we say the superprocess has a *binary local branching mechanism*.

Example 2.5 Let $(x, f) \mapsto \psi(x, f)$ be given by (2.18) and let $(x, z) \mapsto \phi(x, z)$ be given by (2.45). Then the operator $f \mapsto \phi(\cdot, f(\cdot)) - \psi(\cdot, f)$ can be represented in the form (2.26) or (2.27), so it defines a branching mechanism. A branching mechanism of this type is said to be *decomposable* with *local part* ϕ and *non-local part* ψ. A superprocess with such a branching mechanism is referred to as a (ξ, K, ϕ, ψ)-*superprocess*. In the special case of Lebesgue killing density, we call it a (ξ, ϕ, ψ)-*superprocess*. Of course, the expression $\phi(\cdot, f(\cdot)) - \psi(\cdot, f)$ of a decomposable branching mechanism is not unique.

Example 2.6 Let $\pi(x, dy)$ be a probability kernel on E. Suppose that $\beta \in B(E)^+$ and $un(x, du)$ is a bounded kernel from E to $(0, \infty)$. Given the function

$$\zeta(x, z) = \beta(x)z + \int_0^\infty (1 - e^{-zu})n(x, du), \qquad x \in E, z \geq 0, \tag{2.46}$$

we can define a non-local branching mechanism by

$$\psi(x, f) = \zeta(x, \pi(x, f)), \qquad x \in E, f \in B(E)^+. \tag{2.47}$$

If $\zeta(x, y, z)$ is given by (2.46) with $x \in E$ replaced by $(x, y) \in E^2$, we can define another special non-local branching mechanism by

$$\psi(x, f) = \int_E \zeta(x, y, f(y))\pi(x, dy), \qquad x \in E, f \in B(E)^+. \tag{2.48}$$

Example 2.7 Let $1 < \alpha < 2$ be a constant and let π_0 be a diffuse probability measure on E. We can define a branching mechanism on E by

$$\phi(x, f) = \int_0^1 \left[\exp\{-uf(x) - u^2 \pi_0(f)\} - 1 + uf(x) \right] \frac{du}{u^{1+\alpha}}.$$

In fact, it is easy to see

$$\phi(x, f) = \int_{M(E)^\circ} \left[e^{-\nu(f)} - 1 + \nu(\{x\}) f(x) \right] H(x, d\nu),$$

where $H(x, d\nu)$ is the image of $u^{-1-\alpha} du$ under the mapping $u \mapsto u\delta_x + u^2 \pi_0$ of $(0, 1]$ into $M(E)^\circ$. This branching mechanism cannot be decomposed into local and non-local parts.

2.5 Some Moment Formulas

In this section, we prove some moment formulas for Dawson–Watanabe superprocesses. Suppose that E is a Lusin topological space. Let ξ be a Borel right process in E with transition semigroup $(P_t)_{t \geq 0}$ and resolvent $(U^\alpha)_{\alpha > 0}$. Let $t \mapsto K(t)$ be a continuous admissible additive functional of ξ and let ϕ be a branching mechanism given by (2.26) or (2.27). Recall that $c_0 = \|b^-\| + \|\gamma(\cdot, 1)\|$.

Proposition 2.24 Let $(V_t)_{t \geq 0}$ denote the cumulant semigroup of the (ξ, K, ϕ)-superprocess represented by (2.5). Then for $t \geq 0$, $x \in E$ and $f \in B(E)$ we have

$$\pi_t f(x) = \lambda_t(x, f) + \int_{M(E)^\circ} \nu(f) L_t(x, d\nu), \tag{2.49}$$

where $(\pi_t)_{t \geq 0}$ is defined by (2.17).

Proof. For any $n \geq 1$ and $f \in B(E)^+$ we have $nv_t(x, f/n) \leq \pi_t f(x)$ by Proposition 2.20. From (2.5) we see $nv_t(x, f/n)$ is increasing in $n \geq 1$. Then we use (2.27) and (2.31) to see $\pi_t f(x) = \lim_{n \to \infty} nv_t(x, f/n)$ is the unique solution of (2.17). By (2.5) we have

$$nv_t(x, f/n) = \lambda_t(x, f) + \int_{M(E)^\circ} n\left(1 - e^{-\nu(f/n)}\right) L_t(x, d\nu).$$

Then (2.49) follows by monotone convergence. The equality for $f \in B(E)$ follows by linearity. □

Corollary 2.25 If $(V_t)_{t \geq 0}$ is the cumulant semigroup of the (ξ, K, ϕ)-superprocess represented by (2.5), then for any $f, g \in B(E)^+$ we have

$$|V_t f(x) - V_t g(x)| \leq \pi_t(x, |f - g|), \qquad t \geq 0, x \in E, \qquad (2.50)$$

where $(\pi_t)_{t \geq 0}$ is defined by (2.17).

Proof. By the canonical representation (2.5) we have

$$|V_t f(x) - V_t g(x)| \leq \lambda_t(x, |f - g|) + \int_{M(E)^\circ} |e^{-\nu(f)} - e^{-\nu(f)}| L_t(x, d\nu)$$

$$\leq \lambda_t(x, |f - g|) + \int_{M(E)^\circ} \nu(|f - g|) L_t(x, d\nu).$$

Then (2.50) follows from (2.49).   □

Corollary 2.26 *Let $(V_t)_{t \geq 0}$ be the cumulant semigroup of the (ξ, ϕ)-superprocess represented canonically by (2.5). Then (2.49) holds with $(\pi_t)_{t \geq 0}$ defined by (2.35). In particular, if ϕ is the local branching mechanism given by (2.45), the equality holds with $\pi_t = P_t^b$ for all $t \geq 0$.*

Proposition 2.27 *Let $(Q_t)_{t \geq 0}$ denote the transition semigroup of the (ξ, K, ϕ)-superprocess. Then for $t \geq 0$, $\mu \in M(E)$ and $f \in B(E)$ we have*

$$\int_{M(E)} \nu(f) Q_t(\mu, d\nu) = \mu(\pi_t f), \qquad (2.51)$$

where $(\pi_t)_{t \geq 0}$ is defined by (2.17).

Proof. This follows by differentiating both sides of (2.32) and applying Proposition 2.24.   □

Corollary 2.28 *Let $(Q_t)_{t \geq 0}$ be the transition semigroup of the (ξ, ϕ)-superprocess. Then (2.51) holds with $(\pi_t)_{t \geq 0}$ defined by (2.35). In particular, if ϕ is the local branching mechanism given by (2.45), the equality holds with $\pi_t = P_t^b$ for all $t \geq 0$.*

If $b(x) \geq \gamma(x, 1)$ for all $x \in E$, then (2.17) defines a Borel right transition semigroup $(\pi_t)_{t \geq 0}$ by Theorem A.43. In this case, we say the (ξ, K, ϕ)-superprocess is *subcritical*. In particular, if $(P_t)_{t \geq 0}$ is conservative and $b(x) = \gamma(x, 1)$ for all $x \in E$, then $(\pi_t)_{t \geq 0}$ is a conservative transition semigroup and we say the superprocess is *critical*. If $(P_t)_{t \geq 0}$ is conservative and $b(x) \leq \gamma(x, 1)$ for all $x \in E$, then $\pi_t 1(x) \geq 1$ for all $t \geq 0$ and $x \in E$ and we say the (ξ, K, ϕ)-superprocess is *supercritical*. The meanings of the notions are made clear by Proposition 2.27.

Proposition 2.29 *Let $(Q_t)_{t \geq 0}$ denote the transition semigroup of the (ξ, K, ϕ)-superprocess. Then for $t \geq 0$, $\mu \in M(E)$ and $(f, g) \in B(E)^+ \times B(E)$ we have*

$$\int_{M(E)} \nu(g) e^{-\nu(f)} Q_t(\mu, d\nu) = \exp\{-\mu(V_t f)\} \mu(V_t^g f), \qquad (2.52)$$

where $(t, x) \mapsto V_t^g f(x)$ is the unique locally bounded solution of

$$V_t^g f(x) = \mathbf{P}_x g(\xi_t) - \mathbf{P}_x \left[\int_0^t \psi(\xi_s, V_{t-s}f, V_{t-s}^g f) K(\mathrm{d}s) \right] \tag{2.53}$$

and $(f, g) \mapsto \psi(\cdot, f, g)$ is the operator from $B(E)^+ \times B(E)$ to $B(E)$ defined by

$$\psi(x, f, g) = b(x)g(x) + 2c(x)f(x)g(x) - \int_E g(y)\gamma(x, \mathrm{d}y)$$

$$+ \int_{M(E)^\circ} \nu(g)\left(1 - \mathrm{e}^{-\nu(f)}\right)H(x, \mathrm{d}\nu). \tag{2.54}$$

Proof. By Proposition 2.27 the left-hand side of (2.52) is finite. For $(f, g) \in B(E)^+ \times B(E)^+$ let $V_t^g f(x) = (\mathrm{d}/\mathrm{d}\theta)V_t(f + \theta g)(x)|_{\theta=0+}$. Then we get (2.52) and (2.53) by differentiating both sides of (2.32) and (2.31). For $(f, g) \in B(E)^+ \times B(E)$ the result follows by linearity. For any $r \geq 0$ it is not hard to show that (2.53) holds for all $t \geq 0$ if and only if it holds for $0 \leq t \leq r$ and

$$V_{r+t}^g f(x) = \mathbf{P}_x V_r^g f(\xi_t) - \mathbf{P}_x \left[\int_0^t \psi(\xi_s, V_{r+t-s}f, V_{r+t-s}^g f) K(\mathrm{d}s) \right]$$

holds for all $t \geq 0$. Based on this fact, the uniqueness of the solution to (2.53) follows by arguments similar to those in the proofs of Propositions 2.15 and 2.20. $\qquad\square$

Corollary 2.30 *Let $(f, g) \in B(E)^+ \times B(E)$ and let $(t, x) \mapsto V_t^g f(x)$ be defined by (2.53). Then we have $V_{r+t}^g f(x) = V_r^{V_t^g f} V_t f(x)$ for all $r, t \geq 0$ and $x \in E$.*

Proof. For any $(f, g) \in B(E)^+ \times B(E)$ we can use Proposition 2.29 and the semigroup property of $(Q_t)_{t \geq 0}$ to see

$$\int_{M(E)} \nu(g)\mathrm{e}^{-\nu(f)}Q_{r+t}(\mu, \mathrm{d}\nu)$$

$$= \int_{M(E)} Q_r(\mu, \mathrm{d}\eta) \int_{M(E)} \nu(g)\mathrm{e}^{-\nu(f)}Q_t(\eta, \mathrm{d}\nu).$$

By applying (2.52) to both sides for $\mu = \delta_x$ we obtain the desired equality. $\qquad\square$

Proposition 2.31 *Let $(V_t)_{t \geq 0}$ denote the cumulant semigroup of the (ξ, K, ϕ)-superprocess represented canonically by (2.5). Then for $t \geq 0$, $x \in E$ and $(f, g) \in B(E)^+ \times B(E)$ we have*

$$V_t^g f(x) = \lambda_t(x, g) + \int_{M(E)^\circ} \nu(g)\mathrm{e}^{-\nu(f)}L_t(x, \mathrm{d}\nu), \tag{2.55}$$

where the left-hand side is defined by (2.53).

Proof. Using the notation in the proof of Proposition 2.29, for $(f, g) \in B(E)^+ \times B(E)^+$ we get (2.55) by differentiating both sides of (2.5). Then the result for $g \in B(E)$ follows by linearity. $\qquad\square$

A direct proof of the existence of the solution of (2.53) can be given in the special case $K(ds) = ds$. Let us rewrite the equation into

$$V_t^g f(x) = P_t g(x) - \int_0^t ds \int_E \psi(y, V_s f, V_s^g f) P_{t-s}(x, dy). \qquad (2.56)$$

Proposition 2.32 *Let $(t, x) \mapsto V_t^g f(x)$ be defined by (2.56). Let $v_0(t, x) = 0$ and define $v_n(t, x) = v_n(t, x, f, g)$ inductively by*

$$v_{n+1}(x) = P_t g(x) - \int_0^t ds \int_E \psi(y, V_s f, v_n) P_{t-s}(x, dy). \qquad (2.57)$$

Then for every $T \geq 0$ we have $v_n(x) \to V_t^g f(x)$ uniformly on $[0, T] \times E$.

Proof. Let $D_n(t) = \sup_{0 \leq s \leq t} \|v_n(s) - v_{n-1}(s)\|$. For any $T \geq 0$, since $t \mapsto V_t f$ is locally bounded on $[0, T]$, by (2.54) and (2.57) there is a constant $L \geq 0$ so that

$$D_n(t) \leq L \int_0^t D_{n-1}(s) ds \leq \cdots \leq \frac{1}{(n-1)!} L^{n-1} t^{n-1} \|g\|, \quad 0 \leq t \leq T.$$

Then the result follows as in the proof of Proposition 2.19. $\square$

Proposition 2.33 *Let $(X_t, \mathscr{G}_t, \mathbf{P})$ be a realization of the (ξ, K, ϕ)-superprocess such that $\mathbf{P}[X_0(1)] < \infty$ and $(\pi_t)_{t \geq 0}$ is the semigroup defined by (2.17). Let $\alpha \geq 0$ and let $f \in B(E)^+$ be α-super-mean-valued for $(\pi_t)_{t \geq 0}$. Then $t \mapsto e^{-\alpha t} X_t(f)$ is a $(\mathscr{G}_t)$-supermartingale.*

Proof. Since $f \in B(E)^+$ is α-super-mean-valued for $(\pi_t)_{t \geq 0}$, by Proposition 2.27 for any $t \geq r \geq 0$ we have

$$\mathbf{P}\left[e^{-\alpha t} X_t(f) | \mathscr{G}_r\right] = e^{-\alpha t} X_r(\pi_{t-r} f) \leq e^{-\alpha r} X_r(f).$$

Therefore $t \mapsto e^{-\alpha t} X_t(f)$ is a $(\mathscr{G}_t)$-supermartingale. $\square$

Corollary 2.34 *Suppose that $(X_t, \mathscr{G}_t, \mathbf{P})$ is a realization of the (ξ, ϕ)-superprocess such that $\mathbf{P}[X_0(1)] < \infty$ and $(\pi_t)_{t \geq 0}$ is defined by (2.35). Let $\alpha \geq 0$ and let $f \in B(E)^+$ be an α-super-mean-valued function for $(P_t)_{t \geq 0}$ satisfying $\varepsilon := \inf_{x \in E} f(x) > 0$. Then for $\beta \geq \alpha + c_0 \varepsilon^{-1} \|f\|$, the process $t \mapsto e^{-2\beta t} X_t(f)$ is a $(\mathscr{G}_t)$-supermartingale.*

Proof. Since $f \in B(E)^+$ is α-super-mean-valued for $(P_t)_{t \geq 0}$, we have $P_t f(x) \leq e^{\alpha t} f(x)$. Recall that $\|\pi_t f\| \leq \|f\| e^{c_0 t}$ by Proposition A.49. Then we use (2.36) to see

$$\pi_t f(x) \leq e^{\|b^-\| t} P_t f(x) + \|\gamma(\cdot, 1)\| \|f\| e^{c_0 t} \int_0^t P_{t-s} 1(x) ds$$

$$\leq e^{(\|b^-\| + \alpha)t} f(x) + c_0 \varepsilon^{-1} \|f\| e^{c_0 t} \int_0^t P_s f(x) ds$$

$$\leq e^{\beta t} f(x) + \beta e^{\beta t} \int_0^t e^{\beta s} f(x) \mathrm{d}s \leq e^{2\beta t} f(x).$$

Then the result follows by Proposition 2.33. □

Corollary 2.35 *Let ϕ be a local branching mechanism given by (2.45). Suppose that $(X_t, \mathscr{G}_t, \mathbf{P})$ is a realization of the (ξ, ϕ)-superprocess such that $\mathbf{P}[X_0(1)] < \infty$. Let $\alpha \geq 0$ and let $f \in B(E)^+$ be an α-super-mean-valued function for $(P_t)_{t \geq 0}$. Then for any $\alpha_1 \geq \alpha + \|b^-\|$, the process $t \mapsto e^{-\alpha_1 t} X_t(f)$ is a $(\mathscr{G}_t)$-supermartingale.*

Proof. Since $f \in B(E)^+$ is α-super-mean-valued for $(P_t)_{t \geq 0}$, we have

$$P_t^b f(x) \leq e^{\|b^-\|t} P_t f(x) \leq e^{(\|b^-\|+\alpha)t} f(x) \leq e^{\alpha_1 t} f(x).$$

Then we have the result by Proposition 2.33. □

Let $\mathbf{F}$ be the set of functions $f \in B(E)$ that are finely continuous relative to ξ. Fix $\beta > 0$ and let $(A, \mathscr{D}(A))$ be the weak generator of $(P_t)_{t \geq 0}$ defined by $\mathscr{D}(A) = U^\beta \mathbf{F}$ and $Af = \beta f - g$ for $f = U^\beta g \in \mathscr{D}(A)$.

Theorem 2.36 *Suppose that $(X_t, \mathscr{G}_t, \mathbf{P})$ is a progressive realization of the (ξ, ϕ)-superprocess such that $\mathbf{P}[X_0(1)] < \infty$. Then for any $f \in \mathscr{D}(A)$, the process*

$$M_t(f) := X_t(f) - X_0(f) - \int_0^t X_s(Af + \gamma f - bf)\mathrm{d}s, \qquad t \geq 0,$$

is a $(\mathscr{G}_t)$-martingale.

Proof. Let $(\pi_t)_{t \geq 0}$ be defined by (2.35). For any $t \geq r \geq 0$ we use Corollary 2.28 and the Markov property of $\{(X_t, \mathscr{G}_t) : t \geq 0\}$ to see that

$$\mathbf{P}\big[M_t(f)\big|\mathscr{G}_r\big] = \mathbf{P}\Big[X_t(f) - X_0(f) - \int_0^t X_s(Af + \gamma f - bf)\mathrm{d}s\Big|\mathscr{G}_r\Big]$$

$$= \mathbf{P}\Big[X_t(f) - \int_0^{t-r} X_{r+s}(Af + \gamma f - bf)\mathrm{d}s\Big|\mathscr{G}_r\Big]$$

$$\qquad - X_0(f) - \int_0^r X_s(Af + \gamma f - bf)\mathrm{d}s$$

$$= X_r(\pi_{t-r}f) - \int_0^{t-r} X_r(\pi_s(A + \gamma - b)f)\mathrm{d}s$$

$$\qquad - X_0(f) - \int_0^r X_s(Af + \gamma f - bf)\mathrm{d}s$$

$$= X_r(f) - X_0(f) - \int_0^r X_s(Af + \gamma f - bf)\mathrm{d}s,$$

where we have also used Theorem A.55 for the last equality. That gives the martingale property of $\{M_t(f) : t \geq 0\}$. □

We next give some second-moment formulas. For simplicity we only consider the (ξ, ϕ)-superprocess. In this case, the semigroup $(\pi_t)_{t \geq 0}$ is defined by (2.35). We shall need the integral condition

$$\sup_{x \in E} \int_{M(E)^\circ} \nu(1)^2 H(x, \mathrm{d}\nu) < \infty. \tag{2.58}$$

Proposition 2.37 Suppose that (2.58) holds. Let $(Q_t)_{t \geq 0}$ be the transition semigroup of the (ξ, ϕ)-superprocess. Then for $t > 0$, $x \in E$ and $f \in B(E)$ we have

$$\int_{M(E)} \nu(f)^2 L_t(x, \mathrm{d}\nu) = \int_0^t \mathrm{d}s \int_E q(y, \pi_s f) \pi_{t-s}(x, \mathrm{d}y),$$

where $(\pi_t)_{t \geq 0}$ is defined by (2.35) and

$$q(y, f) = 2c(y)f(y)^2 + \int_{M(E)^\circ} \nu(f)^2 H(y, \mathrm{d}\nu). \tag{2.59}$$

Proof. We first assume $f \in B(E)^+$. By applying Proposition 1.38 to (2.5), for any $\theta > 0$ we can define the function $u'_t(x, \theta) := (\mathrm{d}/\mathrm{d}\theta)v_t(x, \theta f)$, which is given by

$$u'_t(x, \theta) = \lambda_t(x, f) + \int_{M(E)^\circ} \nu(f) \mathrm{e}^{-\theta \nu(f)} L_t(x, \mathrm{d}\nu). \tag{2.60}$$

Then we differentiate both sides of (2.38) to obtain

$$u'_t(x, \theta) = \pi_t f(x) - 2 \int_0^t \mathrm{d}s \int_E c(y) v_s(y, \theta f) u'_s(y, \theta) \pi_{t-s}(x, \mathrm{d}y)$$
$$- \int_0^t \mathrm{d}s \int_E h_s(y, \theta, f) \pi_{t-s}(x, \mathrm{d}y), \tag{2.61}$$

where

$$h_s(y, \theta, f) = \int_{M(E)^\circ} \nu(u'_s(\cdot, \theta)) \big(1 - \mathrm{e}^{-\nu(v_s(\cdot, \theta f))}\big) H(y, \mathrm{d}\nu).$$

For any $\theta > 0$ let $u''_t(x, \theta) = (\mathrm{d}^2/\mathrm{d}\theta^2)v_t(x, \theta f)$. By Proposition 1.38,

$$u''_t(x, \theta) = - \int_{M(E)^\circ} \nu(f)^2 \mathrm{e}^{-\theta \nu(f)} L_t(x, \mathrm{d}\nu). \tag{2.62}$$

On the other hand, from (2.61) we have

$$u''_t(x, \theta) = -2 \int_0^t \mathrm{d}s \int_E c(y) \big[u'_s(y, \theta)^2 + v_s(y, \theta f) u''_s(y, \theta)\big] \pi_{t-s}(x, \mathrm{d}y)$$
$$- \int_0^t \mathrm{d}s \int_E h'_s(y, \theta, f) \pi_{t-s}(x, \mathrm{d}y)$$

where

$$h'_s(y, \theta, f) = \int_{M(E)^\circ} \nu(u'_s(\cdot, \theta))^2 e^{-\nu(v_s(\cdot, \theta f))} H(y, d\nu)$$

$$+ \int_{M(E)^\circ} \nu(u''_s(\cdot, \theta))\big(1 - e^{-\nu(v_s(\cdot, \theta f))}\big) H(y, d\nu).$$

By dominated convergence we have

$$\lim_{\theta \to 0} u''_t(x, \theta) = -\int_0^t ds \int_E q(y, \pi_s f) \pi_{t-s}(x, dy).$$

From this and (2.62) we get the desired equality for $f \in B(E)^+$. The extension to $f \in B(E)$ is elementary. □

Proposition 2.38 *Suppose that (2.58) holds. Let $(Q_t)_{t\geq 0}$ be the transition semigroup of the (ξ, ϕ)-superprocess. Then for $t \geq 0$, $\mu \in M(E)$ and $f \in B(E)$ we have*

$$\int_{M(E)} \nu(f)^2 Q_t(\mu, d\nu) = \mu(\pi_t f)^2 + \int_0^t ds \int_E q(y, \pi_s f) \mu \pi_{t-s}(dy), \quad (2.63)$$

where $(\pi_t)_{t\geq 0}$ is defined by (2.35) and $q(y, f)$ is defined by (2.59).

Proof. Let $u'_t(x, \theta)$ and $u''_t(x, \theta)$ be defined as in the proof of Proposition 2.37. In view of (2.32), we have

$$\int_{M(E)} \nu(f)^2 e^{-\theta \nu(f)} Q_t(\mu, d\nu)$$
$$= \big[\mu(u'_t(\cdot, \theta))^2 - \mu(u''_t(\cdot, \theta))\big] \exp\{-\mu(v_t(\cdot, \theta))\}.$$

By letting $\theta \to 0$ in the above equation we obtain (2.63), first for $f \in B(E)^+$ and then for $f \in B(E)$. □

Corollary 2.39 *Let $(Q_t)_{t\geq 0}$ be the transition semigroup of the (ξ, ϕ)-superprocess with local branching mechanism given by (2.45) and assume*

$$x \mapsto \phi''(x, 0) := 2c(x) + \int_0^\infty u^2 m(x, du) \quad (2.64)$$

is bounded on E. Then for $t \geq 0$, $\mu \in M(E)$ and $f \in B(E)$ we have

$$\int_{M(E)} \nu(f)^2 Q_t(\mu, d\nu) = \mu(P_t^b f)^2 + \int_0^t ds \int_E \phi''(x, 0) P_s^b f(x)^2 \mu P_{t-s}^b(dx).$$

Example 2.8 Suppose that $X = (W, \mathscr{G}, \mathscr{G}_t, X_t, \mathbf{Q}_\mu)$ is a (ξ, ϕ)-superprocess with binary local branching mechanism $\phi(x, z) = c(x) z^2 / 2$. Let $(V_t)_{t\geq 0}$ denote the cumulant semigroup of X. Fix $f \in B(E)^+$ and define

$$v_t^{(n)}(x) = (-1)^{n-1} \frac{\partial^n}{\partial \theta^n} V_t(\theta f)(x)\Big|_{\theta=0+}.$$

Then we have $v_t^{(1)}(x) = P_t f(x)$ and

$$v_t^{(n)}(x) = \sum_{k=1}^{n-1} \binom{n-1}{k} \int_0^t P_{t-s}(c v_s^{(k)} v_s^{(n-k)})(x) ds$$

for $n = 2, 3, \ldots$. The moments of X are determined by $\mathbf{Q}_\mu[X_t(f)] = \mu(P_t f)$ and

$$\mathbf{Q}_\mu[X_t(f)^n] = \sum_{k=0}^{n-1} \binom{n-1}{k} \mu(v_t^{(n-k)}) \mathbf{Q}_\mu[X_t(f)^k].$$

2.6 Notes and Comments

The one-to-one correspondence stated in Theorem 2.4 was established in Watanabe (1968) under some stronger assumptions. Theorem 2.5 was also proved in Watanabe (1968). Jiřina (1964) studied the extinction problem of discrete-time branching processes taking values of finite measures on the positive half line. A class of superprocesses over compact metric spaces were constructed in Watanabe (1968), where it was shown those processes arise as high-density limits of branching particle systems. Silverstein (1969) constructed more general superprocesses with decomposable branching mechanisms; see also Dawson et al. (2002c) and Dynkin (1993a). Some inhomogeneous superprocesses with general branching mechanisms were constructed in Dynkin (1994), who assumed the existence of a càdlàg realization of the underlying spatial motion and a technical condition on the tail behavior of the kernel in the expression of the branching mechanism. The superprocesses constructed in Dynkin (1994) are not necessarily conservative. Dawson et al. (1998) proved that a general class of local branching (ξ, ϕ, K)-superprocesses with a fixed underlying spatial motion ξ depend on the parameters (ϕ, K) continuously and the superprocesses with Lebesgue killing density constitute a dense subset of the class. Leduc (2000) constructed some Hunt superprocesses under a second-moment condition.

Our assumptions on the branching mechanism guarantee that the corresponding superprocesses have finite first-moments in the sense of (2.51). Let $X = (W, \mathscr{G}, \mathscr{G}_t, X_t, \mathbf{Q}_\mu)$ be a realization of the (ξ, K, ϕ)-superprocess. For $t \geq 0$ and $\mu \in M(E)$ we can define the mean measure $I_{\mu,t}$ on E by

$$I_{\mu,t}(B) = \mathbf{Q}_\mu[X_t(B)], \qquad B \in \mathscr{B}(E).$$

The *Campbell measure* of the random measure X_t is the unique finite measure $R_{\mu,t}$ on $E \times M(E)$ such that

$$R_{\mu,t}(B \times A) = \mathbf{Q}_\mu[X_t(B)1_A(X_t)], \quad B \in \mathscr{B}(E), A \in \mathscr{B}(M(E)).$$

In view of (2.52) we have

$$\int_E \int_{M(E)} g(x) \mathrm{e}^{-\nu(f)} R_{\mu,t}(\mathrm{d}x, \mathrm{d}\nu) = \exp\{-\mu(V_t f)\}\mu(V_t^g f),$$

where $f \in B(E)^+$ and $g \in B(E)$. By the existence of regular conditional probabilities, there is a probability kernel $J_{\mu,t}(x, \mathrm{d}\nu)$ from E to $M(E)$ so that

$$R_{\mu,t}(\mathrm{d}x, \mathrm{d}\nu) = I_{\mu,t}(\mathrm{d}x) J_{\mu,t}(x, \mathrm{d}\nu), \quad x \in E, \nu \in M(E).$$

The probability measures $\{J_{\mu,t}(x, \cdot) : x \in E\}$ are called *Palm distributions* of X_t. If (η, Y) is a random variable on $E \times M(E)$ distributed according to the Campbell measure $R_{\mu,t}$, then η is chosen according to the random measure Y and $J_{\mu,t}(x, \cdot)$ is the conditional distribution of Y given $\eta = x$. See Dawson (1993) and Dawson and Perkins (1991) for some applications of the Campbell measure and the Palm distributions in the study of the superprocess.

Example 2.1 was given by Dynkin et al. (1994). Rhyzhov and Skorokhod (1970) and Watanabe (1969) constructed CB-processes under conditions on the branching mechanism weaker than those of Example 2.2. Moment formulas for superprocesses as in Example 2.8 were established in Dynkin (1989a) and Konno and Shiga (1988). A construction for super-Brownian motions was given in Ren (2001) under a weaker admissibility assumption on the killing additive functional. A super-stable process with infinite mean was constructed in Fleischmann and Sturm (2004) by a passage to the limit.

The catalyst measure $\rho(\mathrm{d}y)$ in Example 2.3 can be time dependent. In fact, it can be replaced by a measure-valued process $\{\rho_t : t \geq 0\}$. The study of superprocesses with measure-valued catalysts was initiated by Dawson and Fleischmann (1991, 1992). A binary local branching super-Brownian motion with super-Brownian catalyst was constructed in Dawson and Fleischmann (1997a). The property of persistence (no loss of expected mass in the long-time behavior) of the process with underlying dimensions $d \leq 3$ was proved in Dawson and Fleischmann (1997a, 1997b) and Etheridge and Fleischmann (1998). This phenomenon is in contrast to the super-Brownian motion with Lebesgue catalyst, where persistence only holds in high dimensions. A construction of catalytic super-Brownian motion via collision local times was given in Mörters and Vogt (2005). The long-time behavior of a branching random walk in a random catalytic medium was investigated in Greven et al. (1999). Engländer (2007) gave a survey of some recent topics in spatial branching processes in deterministic and random media.

There is another important class of measure-valued Markov processes, the so-called *Fleming–Viot superprocesses*. A Fleming–Viot superprocess takes values of probability measures and describes the evolution of a genetic system involving mutation, selection and recombination. The Saint-Flour lecture notes of Dawson (1993) provide a complete survey of the literature before 1992 on both Dawson–Watanabe and Fleming–Viot superprocesses. For a survey of the latter see also Ethier and

Kurtz (1993). It was conjectured in Ethier and Kurtz (1993) that a Fleming–Viot superprocess is reversible if and only if its mutation operator is of the uniform jump type. This was proved in Li et al. (1999); see also Handa (2002) and Schmuland and Sun (2002). A nice introduction of the theory of superprocesses was given by Etheridge (2000), where Brownian spatial motion was mainly considered. The connections between Dawson–Watanabe and Fleming–Viot superprocesses were investigated in Etheridge and March (1991), Perkins (1992) and Shiga (1990). The two classes of superprocesses model large population systems in which branching or splitting occurs. The dual phenomenon is coalescent or coagulation. Bertoin (2006) gave a comprehensive account of stochastic models involving fragmentation and coagulation. A kind of generalized Fleming–Viot superprocesses arising from coalescent processes were studied in Bertoin and Le Gall (2003, 2005, 2006). Feng (2010) provided an up-to-date account of Fleming–Viot superprocesses and Poisson–Dirichlet type distributions. Durrett (2008) and Ewens (2004) gave comprehensive coverage of mathematical population genetics.

Chapter 3
One-Dimensional Branching Processes

A one-dimensional CB-process is a Markov process with branching property taking values from the positive half line. A more general model is the CB-process with immigration which deals with the situation where immigrants may come from outer sources. In this chapter, we first give some characterizations of the extinction probabilities and evolution rates of CB-processes. Then we prove some conditional limit theorems for the processes that extinguish with strictly positive probability. In particular, we shall see that a class of CB-processes with immigration can be obtained from those without immigration by conditioning on non-extinction. The proofs of those theorems are based on the asymptotic analysis of the cumulant semigroup and are easier than their discrete-state counterparts given as in Athreya and Ney (1972). The greater tractability of the CB-processes arises because both their time and state spaces are smooth, and the distributions which appear are infinitely divisible. In this sense, the continuous-state models provide a more economical way to establish the nicest conditional limit theorems for branching processes. We also show that the CB-process with immigration arises naturally as the scaling limit of a sequence of discrete Galton–Watson branching processes with immigration. The contents of this chapter are helpful for the reader in developing the intuitions of Dawson–Watanabe superprocesses.

3.1 Continuous-State Branching Processes

In this section we prove some basic properties of the one-dimensional CB-process, which is a special case of the process considered in Example 2.2. Suppose that ϕ is a branching mechanism defined by

$$\phi(z) = bz + cz^2 + \int_0^\infty \left(e^{-zu} - 1 + zu \right) m(du), \quad z \geq 0, \qquad (3.1)$$

Z. Li, *Measure-Valued Branching Markov Processes*,
Probability and Its Applications, DOI 10.1007/978-3-642-15004-3_3,
© Springer-Verlag Berlin Heidelberg 2011

where $c \geq 0$ and b are constants and $(u \wedge u^2)m(du)$ is a finite measure on $(0, \infty)$. A one-dimensional CB-process has the transition semigroup $(Q_t)_{t \geq 0}$ defined by

$$\int_0^\infty e^{-\lambda y} Q_t(x, dy) = e^{-xv_t(\lambda)}, \qquad \lambda \geq 0, x \geq 0, \qquad (3.2)$$

where $t \mapsto v_t(\lambda)$ is the unique positive solution of

$$v_t(\lambda) = \lambda - \int_0^t \phi(v_s(\lambda))ds, \qquad t \geq 0. \qquad (3.3)$$

As in the case of Dawson–Watanabe superprocesses, the infinite divisibility of the transition semigroup $(Q_t)_{t \geq 0}$ implies that the *cumulant semigroup* $(v_t)_{t \geq 0}$ can be expressed canonically as

$$v_t(\lambda) = h_t \lambda + \int_0^\infty (1 - e^{-\lambda u})l_t(du), \qquad t \geq 0, \lambda \geq 0, \qquad (3.4)$$

where $h_t \geq 0$ and $ul_t(du)$ is a finite measure on $(0, \infty)$. From (3.3) we see that $t \mapsto v_t(\lambda)$ is first continuous and then continuously differentiable. Moreover, we have the backward differential equation:

$$\frac{\partial}{\partial t} v_t(\lambda) = -\phi(v_t(\lambda)), \qquad v_0(\lambda) = \lambda. \qquad (3.5)$$

By (3.5) and the semigroup property $v_r \circ v_t = v_{r+t}$ for $r, t \geq 0$ we also have the forward differential equation

$$\frac{\partial}{\partial t} v_t(\lambda) = -\phi(\lambda) \frac{\partial}{\partial \lambda} v_t(\lambda), \qquad v_0(\lambda) = \lambda. \qquad (3.6)$$

From a moment formula for general superprocesses we have

$$\int_0^\infty y Q_t(x, dy) = xe^{-bt}, \qquad t \geq 0, x \geq 0. \qquad (3.7)$$

We say the CB-process is *critical, subcritical* or *supercritical* according as $b = 0$, ≥ 0 or ≤ 0.

Proposition 3.1 *For every $t \geq 0$ the function $\lambda \mapsto v_t(\lambda)$ is strictly increasing on* $[0, \infty)$.

Proof. By the continuity of $t \mapsto v_t(\lambda)$, for any $\lambda_0 > 0$ there is $t_0 > 0$ so that $v_t(\lambda_0) > 0$ for $0 \leq t \leq t_0$. Then (3.2) implies $Q_t(x, \{0\}) < 1$ for $x > 0$ and $0 \leq t \leq t_0$, and so $\lambda \mapsto v_t(\lambda)$ is strictly increasing for $0 \leq t \leq t_0$. By the semigroup property of $(v_t)_{t \geq 0}$ we infer $\lambda \mapsto v_t(\lambda)$ is strictly increasing for all $t \geq 0$. $\qquad \square$

Corollary 3.2 *The transition semigroup $(Q_t)_{t \geq 0}$ defined by (3.2) is a Feller semigroup.*

Proof. By Proposition 3.1 for $t \geq 0$ and $\lambda > 0$ we have $v_t(\lambda) > 0$. From (3.2) we see the operator Q_t maps $\{x \mapsto e^{-\lambda x} : \lambda > 0\}$ to itself. By the Stone–Weierstrass theorem, the linear span of $\{x \mapsto e^{-\lambda x} : \lambda > 0\}$ is dense in $C_0(\mathbb{R}_+)$ in the supremum norm. Then Q_t maps $C_0(\mathbb{R}_+)$ to itself. The Feller property of $(Q_t)_{t\geq 0}$ follows by the continuity of $t \mapsto v_t(\lambda)$. □

Proposition 3.3 *Suppose that $\lambda > 0$ and $\phi(\lambda) \neq 0$. Then the equation $\phi(z) = 0$ has no root between λ and $v_t(\lambda)$. Moreover, we have*

$$\int_{v_t(\lambda)}^{\lambda} \phi(z)^{-1}\mathrm{d}z = t, \qquad t \geq 0. \tag{3.8}$$

Proof. By (3.1) we see $\phi(0) = 0$ and $z \mapsto \phi(z)$ is a convex function. Since $\phi(\lambda) \neq 0$ for some $\lambda > 0$ according to the assumption, the equation $\phi(z) = 0$ has at most one root in $(0, \infty)$. Suppose that $\lambda_0 \geq 0$ is a root of $\phi(z) = 0$. Then (3.6) implies $v_t(\lambda_0) = \lambda_0$ for all $t \geq 0$. By Proposition 3.1 we have $v_t(\lambda) > \lambda_0$ for $\lambda > \lambda_0$ and $0 < v_t(\lambda) < \lambda_0$ for $0 < \lambda < \lambda_0$. Then $\lambda > 0$ and $\phi(z) \neq 0$ imply there is no root of $\phi(z) = 0$ between λ and $v_t(\lambda)$. From (3.5) we get (3.8). □

Proposition 3.4 *For any $t \geq 0$ and $\lambda \geq 0$ let $v_t'(\lambda) = (\partial/\partial\lambda)v_t(\lambda)$. Then we have*

$$v_t'(\lambda) = \exp\left\{-\int_0^t \phi'(v_s(\lambda))\mathrm{d}s\right\}, \tag{3.9}$$

where

$$\phi'(z) = b + 2cz + \int_0^{\infty} u(1 - e^{-zu})m(\mathrm{d}u). \tag{3.10}$$

Proof. Based on (3.3) and (3.5) it is elementary to see that

$$\frac{\partial}{\partial t}v_t'(\lambda) = -\phi'(v_t(\lambda))v_t'(\lambda) = \frac{\partial}{\partial\lambda}\frac{\partial}{\partial t}v_t(\lambda).$$

It follows that

$$\frac{\partial}{\partial t}\big[\log v_t'(\lambda)\big] = v_t'(\lambda)^{-1}\frac{\partial}{\partial t}v_t'(\lambda) = -\phi'(v_t(\lambda)).$$

Then we have (3.9) since $v_0'(\lambda) = 1$. □

Since $(Q_t)_{t\geq 0}$ is a Feller semigroup by Corollary 3.2, the CB-process has a Hunt realization $X = (\Omega, \mathscr{F}, \mathscr{F}_t, x(t), \mathbf{Q}_x)$. Let $\tau_0 := \inf\{s \geq 0 : x(s) = 0\}$ denote the *extinction time* of the CB-process.

Theorem 3.5 *For every $t \geq 0$ the limit $\bar{v}_t = \uparrow\lim_{\lambda\to\infty} v_t(\lambda)$ exists in $(0, \infty]$. Moreover, the mapping $t \mapsto \bar{v}_t$ is decreasing and for any $t \geq 0$ and $x > 0$ we have*

$$\mathbf{Q}_x\{\tau_0 \leq t\} = \mathbf{Q}_x\{x(t) = 0\} = \exp\{-x\bar{v}_t\}. \tag{3.11}$$

Proof. By Proposition 3.1 the limit $\bar{v}_t = \uparrow\lim_{\lambda\to\infty} v_t(\lambda)$ exists in $(0, \infty]$ for every $t \geq 0$. For $t \geq r \geq 0$ we have

$$\bar{v}_t = \uparrow \lim_{\lambda\to\infty} v_r(v_{t-r}(\lambda)) = v_r(\bar{v}_{t-r}) \leq \bar{v}_r. \qquad (3.12)$$

Since zero is a trap for the CB-process, we get (3.11) by letting $\lambda \to \infty$ in (3.2). $\quad\square$

For the convenience of statement of the results in the sequel, we formulate the following condition on the branching mechanism:

Condition 3.6 *There is some constant $\theta > 0$ so that*

$$\phi(z) > 0 \text{ for } z \geq \theta \text{ and } \int_\theta^\infty \phi(z)^{-1} dz < \infty.$$

Theorem 3.7 *We have $\bar{v}_t < \infty$ for some and hence all $t > 0$ if and only if Condition 3.6 holds.*

Proof. By (3.12) it is simple to see that $\bar{v}_t = \uparrow\lim_{\lambda\to\infty} v_t(\lambda) < \infty$ for all $t > 0$ if and only if this holds for some $t > 0$. If Condition 3.6 holds, we can let $\lambda \to \infty$ in (3.8) to obtain

$$\int_{\bar{v}_t}^\infty \phi(z)^{-1} dz = t \qquad (3.13)$$

and hence $\bar{v}_t < \infty$ for $t > 0$. For the converse, suppose that $\bar{v}_t < \infty$ for some $t > 0$. By (3.5) there exists some $\theta > 0$ so that $\phi(\theta) > 0$, for otherwise we would have $v_t(\lambda) \geq \lambda$, yielding a contradiction. Then $\phi(z) > 0$ for all $z \geq \theta$ by the convexity of the branching mechanism. As in the above we see that (3.13) still holds, so Condition 3.6 is satisfied. $\quad\square$

Theorem 3.8 *Let $\bar{v} = \downarrow\lim_{t\to\infty} \bar{v}_t \in [0, \infty]$. Then for any $x > 0$ we have*

$$\mathbf{Q}_x\{\tau_0 < \infty\} = \exp\{-x\bar{v}\}. \qquad (3.14)$$

Moreover, we have $\bar{v} < \infty$ if and only if Condition 3.6 holds, and in this case $\bar{v}$ is the largest root of $\phi(z) = 0$.

Proof. The first assertion follows immediately from Theorem 3.5. By Theorem 3.7 we have $\bar{v}_t < \infty$ for some and hence all $t > 0$ if and only if Condition 3.6 holds. This is clearly equivalent to $\bar{v} < \infty$. From (3.13) it is easy to see that $\bar{v}$ is the largest root of $\phi(z) = 0$. $\quad\square$

Corollary 3.9 *Suppose that Condition 3.6 holds. Then for any $x > 0$ we have $\mathbf{Q}_x\{\tau_0 < \infty\} = 1$ if and only if $b \geq 0$.*

Let $(Q_t^\circ)_{t\geq 0}$ be the restriction to $(0, \infty)$ of the semigroup $(Q_t)_{t\geq 0}$. The special case of the canonical representation (3.4) with $h_t = 0$ for all $t > 0$ is particularly interesting. In this case, we have

$$v_t(\lambda) = \int_0^\infty (1 - e^{-\lambda u}) l_t(du), \qquad t > 0, \lambda \geq 0. \tag{3.15}$$

Theorem 3.10 *The cumulant semigroup admits the representation (3.15) if and only if*

$$\phi'(\infty) := 2c \cdot \infty + \int_0^\infty u\, m(du) = \infty \tag{3.16}$$

with $0 \cdot \infty = 0$ by convention. If condition (3.16) is satisfied, then $(l_t)_{t>0}$ is an entrance law for the restricted semigroup $(Q_t^\circ)_{t \geq 0}$.

Proof. From (3.10) it is clear that the limit $\phi'(\infty) = \lim_{z \to \infty} \phi'(z)$ always exists in $(-\infty, \infty]$. By (3.4) we have

$$v_t'(\lambda) = h_t + \int_0^\infty u e^{-\lambda u} l_t(du), \qquad t \geq 0, \lambda \geq 0. \tag{3.17}$$

From (3.9) and (3.17) it follows that

$$h_t = v_t'(\infty) = \exp\left\{ -\int_0^t \phi'(\bar{v}_s) ds \right\}. \tag{3.18}$$

Then $h_t = 0$ for any $t > 0$ implies $\phi'(\infty) = \infty$. For the converse, assume that $\phi'(\infty) = \infty$. If Condition 3.6 holds, by Theorem 3.7 for every $t > 0$ we have $\bar{v}_t < \infty$, so $h_t = 0$ by (3.4). If Condition 3.6 does not hold, then $\bar{v}_t = \infty$ for $t > 0$ by Theorem 3.7. Then (3.18) implies $h_t = 0$ for $t > 0$. That proves the first assertion of the theorem. If $(v_t)_{t>0}$ admits the representation (3.15), we can use the semigroup property of $(v_t)_{t \geq 0}$ to see

$$\int_0^\infty (1 - e^{-\lambda u}) l_{r+t}(du) = \int_0^\infty (1 - e^{-u v_t(\lambda)}) l_r(du)$$

$$= \int_0^\infty l_r(dx) \int_0^\infty (1 - e^{-\lambda u}) Q_t^\circ(x, du)$$

for $r, t > 0$ and $\lambda \geq 0$. Then $(l_t)_{t>0}$ is an entrance law for $(Q_t^\circ)_{t \geq 0}$. $\qquad \square$

Corollary 3.11 *If Condition 3.6 holds, the cumulant semigroup admits the representation (3.15) and $t \mapsto \bar{v}_t = l_t(0, \infty)$ is the minimal solution of the differential equation*

$$\frac{d}{dt} \bar{v}_t = -\phi(\bar{v}_t), \qquad t > 0 \tag{3.19}$$

with singular initial condition $\bar{v}_{0+} = \infty$.

Proof. Under Condition 3.6, for every $t > 0$ we have $\bar{v}_t < \infty$ by Theorem 3.7. Moreover, the condition and the convexity of $z \mapsto \phi(z)$ imply $\phi'(\infty) = \infty$. Then we have the representation (3.15) by Theorem 3.10. The semigroup property of

$(v_t)_{t\geq 0}$ implies $\bar{v}_{t+s} = v_t(\bar{v}_s)$ for $t > 0$ and $s > 0$. Then $t \mapsto \bar{v}_t$ satisfies (3.19). From (3.13) it is easy to see $\bar{v}_{0+} = \infty$. It is simple to see that $t \mapsto \bar{v}_t$ is the minimal solution of the singular initial value problem.   □

Corollary 3.12 *Suppose that Condition 3.6 holds. Then for any $t > 0$ the function $\lambda \mapsto v_t(\lambda)$ is strictly increasing and concave on $[0, \infty)$, and $\bar{v}$ is the largest solution of the equation $v_t(\lambda) = \lambda$. Moreover, we have $\bar{v} = \uparrow\lim_{t\to\infty} v_t(\lambda)$ for $0 < \lambda < \bar{v}$ and $\bar{v} = \downarrow\lim_{t\to\infty} v_t(\lambda)$ for $\lambda > \bar{v}$.*

Proof. By Corollary 3.11 we have the canonical representation (3.15) for every $t > 0$. Since $\lambda \mapsto v_t(\lambda)$ is strictly increasing by Proposition 3.1, the measure $l_t(du)$ is non-trivial, so $\lambda \mapsto v_t(\lambda)$ is strictly concave. The equality $\bar{v} = v_t(\bar{v})$ follows by letting $s \to \infty$ in $\bar{v}_{t+s} = v_t(\bar{v}_s)$, where $\bar{v}_{t+s} \leq \bar{v}_s$. Then $\bar{v}$ is clearly the largest solution to $v_t(\lambda) = \lambda$. When $b \geq 0$, we have $\bar{v} = 0$ by Theorem 3.8 and Corollary 3.9. Furthermore, since $\phi(z) \geq 0$, from (3.5) we see $t \mapsto v_t(\lambda)$ is decreasing, and hence $\downarrow\lim_{t\to\infty} v_t(\lambda) = \downarrow\lim_{t\to\infty} \bar{v}_t = 0$. If $b < 0$ and $0 < \lambda < \bar{v}$, we have $\lambda \leq v_t(\lambda) < v_t(\bar{v}) = \bar{v}$ for all $t \geq 0$. Then the limit $v_\infty(\lambda) = \uparrow\lim_{t\uparrow\infty} v_t(\lambda)$ exists. From the relation $v_t(v_s(\lambda)) = v_{t+s}(\lambda)$ we have $v_t(v_\infty(\lambda)) = v_\infty(\lambda)$, and hence $v_\infty(\lambda) = \bar{v}$ since $\bar{v}$ is the unique solution to $v_t(\lambda) = \lambda$ in $(0, \infty)$. The assertion for $b < 0$ and $\lambda > \bar{v}$ can be proved similarly.   □

We remark that in Theorem 3.10 one usually cannot extend $(l_t)_{t>0}$ to a σ-finite entrance law for the semigroup $(Q_t)_{t\geq 0}$ on $\mathbb{R}_+$. For example, let us assume Condition 3.6 holds and $(\bar{l}_t)_{t>0}$ is such an extension. For any $0 < r < \varepsilon < t$ we have

$$\bar{l}_t(\{0\}) \geq \int_0^\infty Q_{t-r}(x, \{0\})l_r(dx) \geq \int_0^\infty e^{-x\bar{v}_{t-\varepsilon}}l_r(dx)$$
$$= \bar{v}_r - \int_0^\infty (1 - e^{-u\bar{v}_{t-\varepsilon}})l_r(du) = \bar{v}_r - v_r(\bar{v}_{t-\varepsilon}).$$

The right-hand side tends to infinity as $r \to 0$. Then $\bar{l}_t(dx)$ cannot be a σ-finite measure on $\mathbb{R}_+$.

Example 3.1 Suppose that there are constants $c > 0$, $0 < \alpha \leq 1$ and b so that $\phi(z) = cz^{1+\alpha} + bz$. Then Condition 3.6 is satisfied. Let $q_\alpha^0(t) = \alpha t$ and

$$q_\alpha^b(t) = b^{-1}(1 - e^{-\alpha bt}), \qquad b \neq 0.$$

By solving the equation

$$\frac{\partial}{\partial t}v_t(\lambda) = -cv_t(\lambda)^{1+\alpha} - bv_t(\lambda), \qquad v_0(\lambda) = \lambda$$

we get

$$v_t(\lambda) = \frac{e^{-bt}\lambda}{\left[1 + cq_\alpha^b(t)\lambda^\alpha\right]^{1/\alpha}}, \qquad t \geq 0, \lambda \geq 0. \tag{3.20}$$

Thus $\bar{v}_t = c^{-1/\alpha}e^{-bt}q_\alpha^b(t)^{-1/\alpha}$ for $t > 0$. In particular, if $\alpha = 1$, then (3.15) holds with

$$l_t(du) = \frac{e^{-bt}}{c^2 q_1^b(t)^2} \exp\left\{ -\frac{u}{cq_1^b(t)} \right\} du, \qquad t > 0, u > 0.$$

3.2 Long-Time Evolution Rates

In this section we study the long-time asymptotic behavior of the CB-process. This makes sense only in the event of non-extinction of course. Let $(Q_t)_{t \geq 0}$ denote the transition semigroup defined by (3.2) and (3.3). Let $X = (\Omega, \mathscr{F}, \mathscr{F}_t, x(t), \mathbf{Q}_x)$ be a Hunt realization of the CB-process. To avoid triviality, we assume $z \mapsto \phi(z)$ is strictly convex throughout this section.

Recall that $\bar{v}_t = \uparrow \lim_{\lambda \to \infty} v_t(\lambda) \in (0, \infty]$ and $\bar{v} = \downarrow \lim_{t \to \infty} \bar{v}_t \in [0, \infty]$. By Proposition 3.1 the function $\lambda \mapsto v_t(\lambda)$ is strictly increasing on $[0, \infty)$ for each $t \geq 0$. Let $v \mapsto \eta_t(v)$ denote its inverse, which is a strictly increasing function on $[0, \bar{v}_t)$. It is easy to show that $\eta_s(\eta_t(z)) = \eta_{s+t}(z)$ for any $s, t \geq 0$ and $0 \leq z < \bar{v}_{s+t}$. For $0 < \lambda < \bar{v}_t$ such that $\phi(\eta_t(\lambda)) \neq 0$ we get from (3.8) that

$$\int_\lambda^{\eta_t(\lambda)} \frac{dz}{\phi(z)} = -\int_{\eta_t(\lambda)}^\lambda \frac{dz}{\phi(z)} = t. \tag{3.21}$$

Let $\theta_0 = \inf\{z > 0 : \phi(z) \geq 0\} \in [0, \infty]$. Then the continuity of $z \mapsto \phi(z)$ implies $\phi(\theta_0) = 0$ if $\theta_0 < \infty$. By Theorem 3.8 we have $\theta_0 = \bar{v}$ when $\bar{v} < \infty$.

Theorem 3.13 Let $0 < \lambda < \bar{v}$ and define $W_t = \eta_t(\lambda)x(t)$ for $t \geq 0$. Then $t \mapsto e^{-W_t}$ is an $(\mathscr{F}_t)$-martingale and

$$\mathbf{Q}_x[e^{-\theta W_t}] = e^{-x f_t(\theta)}, \qquad t \geq 0, x \geq 0, \theta \geq 0, \tag{3.22}$$

where $f_t(\theta) = v_t(\theta \eta_t(\lambda))$. If $\eta_t(\lambda)$ and $\theta \eta_t(\lambda)$ belong to $(0, \theta_0)$, we have

$$\int_\lambda^{f_t(\theta)} \frac{dz}{\phi(z)} = \int_{\eta_t(\lambda)}^{\theta \eta_t(\lambda)} \frac{dz}{\phi(z)}. \tag{3.23}$$

Proof. For any $s, t \geq 0$ we can use the Markov property of $\{x(t) : t \geq 0\}$ to get

$$\mathbf{Q}_x[e^{-W_{s+t}} | \mathscr{F}_s] = \mathbf{Q}_x[e^{-\eta_{s+t}(\lambda)x(s+t)} | \mathscr{F}_s] = e^{-v_t(\eta_{s+t}(\lambda))x(s)} = e^{-\eta_s(\lambda)x(s)}.$$

Then $t \mapsto e^{-W_t}$ is an $(\mathscr{F}_t)$-martingale. By similar calculations we get (3.22). If $\eta_t(\lambda)$ and $\theta \eta_t(\lambda)$ belong to $(0, \theta_0)$, then $\phi(z) = 0$ has no root between $\eta_t(\lambda)$ and $\theta \eta_t(\lambda)$. Observe that

$$\int_\lambda^{f_t(\theta)} \frac{dz}{\phi(z)} = \int_\lambda^{\eta_t(\lambda)} \frac{dz}{\phi(z)} + \int_{\eta_t(\lambda)}^{\theta \eta_t(\lambda)} \frac{dz}{\phi(z)} + \int_{\theta \eta_t(\lambda)}^{v_t(\theta \eta_t(\lambda))} \frac{dz}{\phi(z)}.$$

Then (3.23) follows by (3.8) and (3.21). $\qquad\square$

Proposition 3.14 *Suppose that $b < 0$ and $0 < \lambda < \theta_0$. Then $\eta_t(\lambda) = Ke^{bt} + o(e^{bt})$ as $t \to \infty$ for some constant $K = K(\lambda) > 0$ if and only if $\int_1^\infty u \log u \, m(du) < \infty$.*

Proof. We first note that $\phi(z) = bz + o(z)$ as $z \to 0$. Observe also that $\phi(\lambda) < 0$ and (3.21) implies $\eta_t(\lambda) \to 0$ decreasingly as $t \to \infty$. Moreover, we get

$$\int_{\eta_t(\lambda)}^{\lambda} \left(\frac{b}{\phi(z)} - \frac{1}{z} \right) dz = -bt + \log \frac{\eta_t(\lambda)}{\lambda} = \log \frac{\eta_t(\lambda)}{\lambda e^{bt}},$$

where the integrand is positive because of the convexity of $z \mapsto \phi(z)$. Then $t \mapsto e^{-bt} \eta_t(\lambda)$ increases, and it remains bounded if and only if

$$\int_0^\lambda \left(\frac{b}{\phi(z)} - \frac{1}{z} \right) dz < \infty,$$

which is equivalent to

$$\int_0^\lambda \frac{\phi(z) - bz}{z^2} dz < \infty.$$

By (3.1) the value on the left-hand side is equal to

$$c\lambda + \int_0^\lambda dz \int_0^\infty \left(e^{-zu} - 1 + zu \right) \frac{1}{z^2} m(du) = c\lambda + \int_0^\infty u h_\lambda(u) m(du),$$

where

$$h_\lambda(u) = \frac{1}{u} \int_0^\lambda \left(e^{-zu} - 1 + zu \right) \frac{dz}{z^2} = \int_0^{\lambda u} \left(e^{-y} - 1 + y \right) \frac{dy}{y^2}$$

is equivalent to $\lambda u / 2$ as $u \to 0$ and equivalent to $\log u$ as $u \to \infty$. Then we have the desired result. $\square$

Theorem 3.15 *Suppose that $b < 0$ and $0 < \lambda < \theta_0$. Let $W_t = \eta_t(\lambda) x(t)$ for $t \geq 0$. Then the limit $W := \lim_{t \to \infty} W_t$ exists a.s. and $\mathbf{Q}_x\{W = 0\} = e^{-x\theta_0}$ for any $x > 0$.*

Proof. By Theorem 3.13 and martingale theory, the limit $Y := \lim_{t \to \infty} e^{-W_t}$ a.s. exists; see, e.g., Dellacherie and Meyer (1982, p.72). As observed in the proof of Proposition 3.14 we have $\phi(\lambda) < 0$ and $\eta_t(\lambda) \to 0$ decreasingly as $t \to \infty$. Then for any $\theta > 0$ we get from (3.23) that

$$\lim_{t \to \infty} \int_\lambda^{f_t(\theta)} \frac{dz}{\phi(z)} = \lim_{t \to \infty} \int_{\eta_t(\lambda)}^{\theta \eta_t(\lambda)} \frac{dz}{\phi(z)} = \lim_{t \to \infty} \int_{\eta_t(\lambda)}^{\theta \eta_t(\lambda)} \frac{dz}{bz} = \frac{1}{b} \log \theta,$$

so the limit $f(\theta) := \lim_{t \to \infty} f_t(\theta)$ exists and

$$\int_\lambda^{f(\theta)} \frac{dz}{\phi(z)} = \frac{1}{b} \log \theta.$$

This equality implies $\lim_{\theta \to 0} f(\theta) = 0$ and $\lim_{\theta \to \infty} f(\theta) = \theta_0$. By (3.22) and dominated convergence, for any $x > 0$ we have

$$\mathbf{Q}_x[Y^\theta] = \lim_{t \to \infty} \mathbf{Q}_x[e^{-\theta W_t}] = \lim_{t \to \infty} e^{-x f_t(\theta)} = e^{-x f(\theta)}.$$

It follows that

$$\mathbf{Q}_x\{Y = 0\} = \lim_{\theta \to 0} \mathbf{Q}_x[1 - Y^\theta] = \lim_{\theta \to 0}[1 - e^{-x f(\theta)}] = 0$$

and

$$\mathbf{Q}_x\{Y = 1\} = \lim_{\theta \to \infty} \mathbf{Q}_x[Y^\theta] = \lim_{\theta \to \infty} e^{-x f(\theta)} = e^{-x \theta_0}.$$

Then the desired result follows with $W = -\log Y$. $\qquad\qquad\qquad\qquad\square$

Corollary 3.16 *Suppose that $b < 0$ and $\int_1^\infty u \log u\, m(du) < \infty$. Then the limit $Z := \lim_{t \to \infty} e^{bt} x(t)$ exists a.s. and $\mathbf{Q}_x\{Z = 0\} = e^{-x \theta_0}$ for any $x > 0$.*

Theorem 3.15 and Corollary 3.16 characterize the long-time evolution rate of the supercritical branching CB-process. In particular, Corollary 3.16 gives a necessary and sufficient condition for the exponential evolution rate. To get similar results in the critical and subcritical case, we consider a special form of the branching mechanism. If $c = 0$ and if $u m(du)$ is a finite measure on $(0, \infty)$, we can rewrite (3.1) into

$$\phi(z) = b_1 z - \int_0^\infty (1 - e^{-zu}) m(du), \qquad\qquad (3.24)$$

where

$$b_1 = b + \int_0^\infty u m(du). \qquad\qquad (3.25)$$

In this case, we have $\phi(z) = b_1 z + o(z)$ as $z \to \infty$, so Theorem 3.8 implies $\bar{v} = \infty$. The following results can be proved by arguments similar to those for the supercritical case.

Proposition 3.17 *Suppose that ϕ is given by (3.24) and (3.25) with $b \geq 0$. For any fixed $\lambda > 0$ we have $\eta_t(\lambda) = K e^{b_1 t} + o(e^{b_1 t})$ as $t \to \infty$ for some constant $K > 0$ if and only if $\int_0^1 u \log(1/u) m(du) < \infty$.*

Theorem 3.18 *Suppose that ϕ is given by (3.24) and (3.25) with $b \geq 0$. Fix $\lambda > 0$ and let $W_t = \eta_t(\lambda) x(t)$ for $t \geq 0$. Then the limit $W := \lim_{t \to \infty} W_t$ exists a.s. and $\mathbf{Q}_x\{W = 0\} = 0$ for any $x > 0$.*

Corollary 3.19 *Suppose that ϕ is given by (3.24) and (3.25) with $b \geq 0$ and $\int_0^1 u \log(1/u) m(du) < \infty$. Then the limit $Z := \lim_{t \to \infty} e^{b_1 t} x(t)$ a.s. exists and $\mathbf{Q}_x\{Z = 0\} = 0$ for any $x > 0$.*

3.3 Immigration and Conditioned Processes

We first consider a generalization of the CB-process. Let $(Q_t)_{t\geq 0}$ be the transition semigroup defined by (3.2) and (3.3). Suppose that $\psi \in \mathscr{I}$ is a function with the representation

$$\psi(z) = \beta z + \int_0^\infty (1 - e^{-zu})n(du), \qquad z \geq 0, \tag{3.26}$$

where $\beta \geq 0$ is a constant and $(1 \wedge u)n(du)$ is a finite measure on $(0, \infty)$. By Theorems 1.35 and 1.37 one may see that

$$\int_0^\infty e^{-\lambda y}\gamma_t(dy) = \exp\left\{ -\int_0^t \psi(v_s(\lambda))ds \right\}, \qquad \lambda \geq 0 \tag{3.27}$$

defines a family of infinitely divisible probability measures $(\gamma_t)_{t\geq 0}$ on $[0, \infty)$. Then we can define the probability measures

$$Q_t^\gamma(x, \cdot) := Q_t(x, \cdot) * \gamma_t(\cdot), \qquad t, x \geq 0. \tag{3.28}$$

It is easily seen that

$$\int_0^\infty e^{-\lambda y}Q_t^\gamma(x, dy) = \exp\left\{ -xv_t(\lambda) - \int_0^t \psi(v_s(\lambda))ds \right\}. \tag{3.29}$$

Moreover, the kernels $(Q_t^\gamma)_{t\geq 0}$ form a Feller transition semigroup on $\mathbb{R}_+$. A Markov process is called a *continuous-state branching process with immigration* (CBI-process) with *branching mechanism* ϕ and *immigration mechanism* ψ if it has transition semigroup $(Q_t^\gamma)_{t\geq 0}$.

Theorem 3.20 *Suppose that $b \geq 0$ and $\phi(z) \neq 0$ for $z > 0$. Then $Q_t^\gamma(x, \cdot)$ converges to a probability measure η on $[0, \infty)$ as $t \to \infty$ if and only if*

$$\int_0^\lambda \frac{\psi(z)}{\phi(z)}dz < \infty \text{ for some } \lambda > 0. \tag{3.30}$$

If (3.30) holds, the Laplace transform of η is given by

$$L_\eta(\lambda) = \exp\left\{ -\int_0^\infty \psi(v_s(\lambda))ds \right\}, \qquad \lambda \geq 0. \tag{3.31}$$

Proof. Since $\phi(z) \geq 0$ for all $z \geq 0$, from (3.5) we see $t \mapsto v_t(\lambda)$ is decreasing. Then (3.8) implies $\lim_{t\to\infty} v_t(\lambda) = 0$. By (3.29) we have

$$\lim_{t\to\infty} \int_0^\infty e^{-\lambda y}Q_t^\gamma(x, dy) = \exp\left\{ -\int_0^\infty \psi(v_s(\lambda))ds \right\} \tag{3.32}$$

for every $\lambda \geq 0$. A further application of (3.5) gives

$$\int_0^t \psi(v_s(\lambda))\mathrm{d}s = \int_{v_t(\lambda)}^\lambda \frac{\psi(z)}{\phi(z)}\mathrm{d}z.$$

It follows that

$$\int_0^\infty \psi(v_s(\lambda))\mathrm{d}s = \int_0^\lambda \frac{\psi(z)}{\phi(z)}\mathrm{d}z,$$

which is a continuous function of $\lambda \geq 0$ if and only if (3.30) holds. Then the result follows by (3.32) and Theorem 1.20. $\qquad\square$

Corollary 3.21 *Suppose that $b > 0$. Then $Q_t^\gamma(x, \cdot)$ converges to a probability measure η on $[0, \infty)$ as $t \to \infty$ if and only if $\int_1^\infty \log u n(\mathrm{d}u) < \infty$. In this case, the Laplace transform of η is given by (3.31).*

Proof. We have $\phi(z) = bz + o(z)$ as $z \to 0$. Thus (3.30) holds if and only if

$$\int_0^\lambda \frac{\psi(z)}{z}\mathrm{d}z < \infty \quad \text{for some } \lambda > 0,$$

which is equivalent to

$$\int_0^\lambda \frac{\mathrm{d}z}{z} \int_0^\infty (1 - \mathrm{e}^{-zu})n(\mathrm{d}u) = \int_0^\infty n(\mathrm{d}u) \int_0^{\lambda u} \frac{1 - \mathrm{e}^{-y}}{y}\mathrm{d}y < \infty$$

for some $\lambda > 0$. The latter holds if and only if $\int_1^\infty \log u n(\mathrm{d}u) < \infty$. Then we have the result by Theorem 3.20. $\qquad\square$

In the situation of Theorem 3.20, it is easy to show that η is a stationary distribution for $(Q_t^\gamma)_{t\geq 0}$. The fact that the CBI-process may have a non-trivial stationary distribution makes it a more interesting model in many respects than the CB-process without immigration.

Example 3.2 Suppose that $c > 0$, $0 < \alpha \leq 1$ and b are constants and let $\phi(z) = cz^{1+\alpha} + bz$ for $z \geq 0$. In this case the cumulant semigroup $(v_t)_{t\geq 0}$ is given by (3.20). Let $\beta \geq 0$ and let $\psi(z) = \beta z^\alpha$ for $z \geq 0$. We can use (3.29) to define the transition semigroup $(Q_t^\gamma)_{t\geq 0}$. It is easy to show that

$$\int_0^\infty \mathrm{e}^{-\lambda y} Q_t^\gamma(x, \mathrm{d}y) = \frac{1}{\left[1 + cq_\alpha^b(t)\lambda^\alpha\right]^{\beta/c\alpha}} \mathrm{e}^{-xv_t(\lambda)}, \qquad \lambda \geq 0.$$

In the special case of $\alpha = 1$, the corresponding CBI-process $\{y(t) : t \geq 0\}$ is a diffusion process defined by the stochastic differential equation

$$\mathrm{d}y(t) = \sqrt{2cy(t)}\mathrm{d}B(t) + (\beta - by(t))\mathrm{d}t, \qquad t \geq 0, \qquad (3.33)$$

where $\{B(t) : t \geq 0\}$ is a standard Brownian motion; see Ikeda and Watanabe (1989, pp.235–236). Let $C^2(\mathbb{R}_+)$ denote the set of bounded continuous real functions on $\mathbb{R}_+$ with bounded continuous derivatives up to the second order. Then this diffusion process has generator A given by

$$Af(x) = c\frac{\mathrm{d}^2}{\mathrm{d}x^2}f(x) + (\beta - bx)\frac{\mathrm{d}}{\mathrm{d}x}f(x), \qquad f \in C^2(\mathbb{R}_+). \tag{3.34}$$

In particular, for $\beta = 0$ the solution of (3.33) is called *Feller's branching diffusion*.

Recall that $(Q_t^\circ)_{t\geq 0}$ is the restriction to $(0, \infty)$ of the semigroup $(Q_t)_{t\geq 0}$. It is easy to check that $Q_t^b(x, \mathrm{d}y) := e^{bt}x^{-1}yQ_t^\circ(x, \mathrm{d}y)$ defines a Markov semigroup on $(0, \infty)$. Let $q_t(\lambda) = e^{bt}v_t(\lambda)$ and let $q_t'(\lambda) = (\partial/\partial\lambda)q_t(\lambda)$. Recall that $z \mapsto \phi'(z)$ is defined by (3.10). From (3.9) we have

$$q_t'(\lambda) = \exp\left\{ -\int_0^t \phi_0'(v_s(\lambda))\mathrm{d}s \right\}, \tag{3.35}$$

where $\phi_0'(z) = \phi'(z) - b$. By differentiating both sides of (3.2) we see

$$\int_0^\infty e^{-\lambda y}Q_t^b(x, \mathrm{d}y) = \exp\{-xv_t(\lambda)\}q_t'(\lambda), \quad \lambda \geq 0. \tag{3.36}$$

It follows that

$$\int_0^\infty e^{-\lambda y}Q_t^b(x, \mathrm{d}y) = \exp\left\{ -xv_t(\lambda) - \int_0^t \phi_0'(v_s(\lambda))\mathrm{d}s \right\}. \tag{3.37}$$

Using (3.37) it is easy to extend $(Q_t^b)_{t\geq 0}$ to a Feller semigroup on $[0, \infty)$, which is a special case of the semigroup defined by (3.29).

Theorem 3.22 *Suppose that $b > 0$ and $\phi'(z) \to \infty$ as $z \to \infty$. Let $(l_t)_{t>0}$ be defined by (3.15). Then for any $t > 0$ the probability measure $Q_t^b(0, \cdot)$ is supported by $(0, \infty)$ and $Q_t^b(0, \mathrm{d}u) = ue^{bt}l_t(\mathrm{d}u)$ for $u > 0$.*

Proof. We first note that $(v_t)_{t>0}$ really has the representation (3.15) by Theorem 3.10. Then

$$q_t(\lambda) = \int_0^\infty \left(1 - e^{-\lambda u}\right)e^{bt}l_t(\mathrm{d}u), \tag{3.38}$$

and

$$q_t'(\lambda) = \int_0^\infty ue^{-\lambda u}e^{bt}l_t(\mathrm{d}u). \tag{3.39}$$

From (3.36) and (3.39) we see $Q_t^b(0, \mathrm{d}u) = ue^{bt}l_t(\mathrm{d}u)$ for $u > 0$. $\qquad\square$

Theorem 3.23 *Suppose that $b > 0$. Then for every $\lambda \geq 0$ the limit $q'(\lambda) := \downarrow \lim_{t\to\infty} q_t'(\lambda)$ exists and is given by*

$$q'(\lambda) = \exp\left\{ -\int_0^\infty \phi_0'(v_s(\lambda))\mathrm{d}s \right\}, \qquad \lambda \geq 0. \tag{3.40}$$

Moreover, we have $q'(0+) = q'(0) = 1$ if and only if $\int_1^\infty u \log um(du) < \infty$. The last condition is also equivalent to $q'(\lambda) > 0$ for some and hence all $\lambda > 0$.

Proof. The first assertion is immediate in view of (3.35). By Corollary 3.21, we have $\int_1^\infty u \log um(du) < \infty$ if and only if $\lambda \mapsto q'(\lambda)$ is the Laplace transform of a probability η on $[0, \infty)$. Then the other two assertions hold obviously. □

Theorem 3.24 *Suppose that $b > 0$ and $\phi'(z) \to \infty$ as $z \to \infty$. Then $Q_t^b(x, \cdot)$ converges as $t \to \infty$ to a probability η on $(0, \infty)$ if and only if $\int_1^\infty u \log um(du) < \infty$. If the condition holds, then η has Laplace transform $L_\eta = q'$ given by (3.40).*

Proof. By Corollary 3.21 and Theorem 3.23 we have the results with η being a probability measure on $[0, \infty)$. By Theorem 3.22 the measure $Q_t^b(0, \cdot)$ is supported by $(0, \infty)$, hence $Q_t^b(x, \cdot)$ is supported by $(0, \infty)$ for every $x \geq 0$. From (3.39) we have $L_\eta(\infty) \leq q_t'(\infty) = 0$ for $t > 0$. That implies $\eta(\{0\}) = 0$. □

Now let $X = (\Omega, \mathscr{F}, \mathscr{F}_t, x(t), \mathbf{Q}_x)$ be a Hunt realization of the CB-process with transition semigroup $(Q_t)_{t \geq 0}$. Let $\tau_0 := \inf\{s \geq 0 : x(s) = 0\}$ denote the extinction time of X.

Theorem 3.25 *Suppose that $b \geq 0$ and Condition 3.6 holds. Then for any $t \geq 0$ and $x > 0$, the distribution of $x(t)$ under $\mathbf{Q}_x\{\cdot | r + t < \tau_0\}$ converges as $r \to \infty$ to $Q_t^b(x, \cdot)$.*

Proof. Since zero is a trap for the CB-process, for any $r > 0$ we can use the Markov property of $\{x(t) : t \geq 0\}$ to see

$$\mathbf{Q}_x\left[e^{-\lambda x(t)} | r + t < \tau_0\right] = \frac{\mathbf{Q}_x\left[e^{-\lambda x(t)} 1_{\{r+t<\tau_0\}}\right]}{\mathbf{Q}_x[1_{\{r+t<\tau_0\}}]}$$
$$= \lim_{\theta \to \infty} \frac{\mathbf{Q}_x\left[e^{-\lambda x(t)}(1 - e^{-\theta x(r+t)})\right]}{\mathbf{Q}_x\left[(1 - e^{-\theta x(r+t)})\right]}$$
$$= \frac{\mathbf{Q}_x\left[e^{-\lambda x(t)}(1 - e^{-x(t)\bar{v}_r})\right]}{1 - e^{-x\bar{v}_{r+t}}}. \tag{3.41}$$

Recall that $\bar{v}_{r+t} = v_t(\bar{v}_r)$ and $v_t'(0) = e^{-bt}$. By Theorem 3.8 and Corollary 3.9 we have $\lim_{r \to \infty} \bar{v}_r = 0$. Then

$$\lim_{r \to \infty} \mathbf{Q}_x\left[e^{-\lambda x(t)} | r + t < \tau_0\right] = \lim_{r \to \infty} \frac{\mathbf{Q}_x\left[e^{-\lambda x(t)} \bar{v}_r^{-1}(1 - e^{-x(t)\bar{v}_r})\right]}{\bar{v}_r^{-1}(1 - e^{-xv_t(\bar{v}_r)})}$$
$$= \frac{1}{x} e^{bt} \mathbf{Q}_x[x(t) e^{-\lambda x(t)}].$$

That gives the desired convergence result. □

The above theorem shows that in the critical and subcritical cases $Q_t^b(x, \cdot)$ is intuitively the law of $x(t)$ conditioned on large extinction times. Some more conditional limit theorems of the CB-process will be given in the next section.

3.4 More Conditional Limit Theorems

Throughout this section, we assume Condition 3.6 is satisfied. Let us consider a Hunt realization $X = (\Omega, \mathscr{F}, \mathscr{F}_t, x(t), \mathbf{Q}_x)$ of the transition semigroup $(Q_t)_{t\geq 0}$ defined by (3.2) and (3.3). Then $\bar{v}_t := \uparrow \lim_{\lambda \to \infty} v_t(\lambda) \in (0, \infty)$ by Theorem 3.7 and $\bar{v} := \downarrow \lim_{t\to\infty} \bar{v}_t \in [0, \infty)$ by Theorem 3.8. Let $\tau_0 := \inf\{s \geq 0 : x(s) = 0\}$ denote the extinction time. Recall that $(Q_t^\circ)_{t\geq 0}$ denotes the restriction to $(0, \infty)$ of the semigroup $(Q_t)_{t\geq 0}$.

Theorem 3.26 *Suppose that $b > 0$. Then the limit $g(\lambda) := \uparrow \lim_{t\to\infty} \bar{v}_t^{-1} v_t(\lambda)$ exists for every $\lambda \geq 0$ and $0 = g(0) = g(0+) \leq g(\lambda) \leq g(\infty) = 1$. Consequently, $\bar{v}_t^{-1} l_t$ converges as $t \to \infty$ to a probability measure π_0 on $(0, \infty)$ with Laplace transform $L_{\pi_0}(\lambda) = 1 - g(\lambda)$.*

Proof. Let $g_t(\lambda) = \bar{v}_t^{-1} v_t(\lambda)$ and $h_t(\lambda) = \lambda^{-1} v_t(\lambda)$ for $\lambda \geq 0$. Then $0 \leq g_t(\lambda) \leq 1$ and

$$g_{t+s}(\lambda) = v_s(\bar{v}_t)^{-1} v_s(v_t(\lambda)) = h_s(\bar{v}_t)^{-1} h_s(v_t(\lambda)) g_t(\lambda) \qquad (3.42)$$

for $s, t > 0$. Since $v_t(0) = 0$ and $\lambda \mapsto v_t(\lambda)$ is a concave function, we have $h_t'(\lambda) = \lambda^{-2}[v_t'(\lambda)\lambda - v_t(\lambda)] \leq 0$, so $\lambda \mapsto h_t(\lambda)$ is decreasing. Thus $t \mapsto g_t(\lambda)$ is increasing by (3.42). Consequently, the limit $g(\lambda) = \uparrow \lim_{t\to\infty} g_t(\lambda)$ exists and $0 = g(0) \leq g(\lambda) \leq g(\infty) = 1$. Observe also that

$$g_t(v_s(\lambda)) = \bar{v}_t^{-1} v_{t+s}(\lambda) = \bar{v}_{t+s}^{-1} v_{t+s}(\lambda) \bar{v}_t^{-1} v_s(\bar{v}_t) = g_{t+s}(\lambda) h_s(\bar{v}_t). \quad (3.43)$$

By Theorem 3.8 we have $\lim_{t\to\infty} \bar{v}_t = 0$, so $\lim_{t\to\infty} h_s(\bar{v}_t) = v_s'(0) = e^{-bs}$. Taking $t \to \infty$ in (3.43) gives

$$g(v_s(\lambda)) = e^{-bs} g(\lambda), \qquad \lambda \geq 0, s \geq 0. \qquad (3.44)$$

Then we must have $g(0+) = g(0) = 0$. From the relation

$$\lim_{t\to\infty} \int_0^\infty e^{-\lambda u} \bar{v}_t^{-1} l_t(du) = 1 - \lim_{t\to\infty} \bar{v}_t^{-1} v_t(\lambda) = 1 - g(\lambda)$$

we see that $\bar{v}_t^{-1} l_t$ converges as $t \to \infty$ to a probability measure π_0 on $[0, \infty)$ with Laplace transform $1 - g(\lambda)$. Since $g(\infty) = 1$, we have $\pi_0(\{0\}) = 0$. $\qquad \square$

Theorem 3.27 *Suppose that $b > 0$. Then for any $0 \leq \lambda \leq \infty$, the limit $q(\lambda) := \downarrow \lim_{t\to\infty} q_t(\lambda)$ exists (with $q_t(\infty) = e^{bt} \bar{v}_t$ by convention). Moreover, we have $q(\lambda) > 0$ for some and hence all $0 < \lambda \leq \infty$ if and only if $\int_1^\infty u \log u\, m(du) < \infty$.*

Proof. By Theorem 3.23 and dominated convergence it is easy to see $q(\lambda) = \downarrow \lim_{t\to\infty} q_t(\lambda)$ for all $0 \leq \lambda < \infty$, where

$$q(\lambda) := \int_0^\lambda q'(u) du.$$

This convergence can be extended to $\lambda = \infty$ because Condition 3.6 holds. Since $q(\infty) \geq q(\lambda)$ for all $0 < \lambda < \infty$, the second assertion is immediate. $\qquad\square$

Corollary 3.28 *Suppose that $b > 0$. Then $e^{bt} l_t$ converges as $t \to \infty$ to $q(\infty)\pi_0$, which is non-trivial if and only if $\int_1^\infty u \log u m(du) < \infty$.*

Theorem 3.29 *Suppose that $b > 0$. Then for $r \geq 0$ and $x > 0$ the distribution of $x(t)$ under $\mathbf{Q}_x\{\cdot | r + t < \tau_0\}$ converges as $t \to \infty$ to a probability measure π_r on $(0, \infty)$ independent of x. Moreover, π_0 is also the limit distribution of $\bar{v}_t^{-1} l_t$ given by Theorem 3.26 and $\pi_0 Q_t^\circ = e^{-bt}\pi_0$ for all $t \geq 0$.*

Proof. For $r > 0$ we can use the calculations in (3.41) to see

$$\mathbf{Q}_x[e^{-\lambda x(t)} | r + t < \tau_0] = \frac{e^{-x v_t(\lambda)} - e^{-x v_t(\lambda + \bar{v}_r)}}{1 - e^{-x \bar{v}_{r+t}}}. \tag{3.45}$$

By letting $r \to 0$ we obtain

$$\mathbf{Q}_x[e^{-\lambda x(t)} | t < \tau_0] = \frac{e^{-x v_t(\lambda)} - e^{-x \bar{v}_t}}{1 - e^{-x \bar{v}_t}} = 1 - \frac{1 - e^{-x v_t(\lambda)}}{1 - e^{-x \bar{v}_t}}.$$

Let π_0 be the probability measure on $(0, \infty)$ given by Theorem 3.26. It follows that

$$\lim_{t \to \infty} \mathbf{Q}_x[e^{-\lambda x(t)} | t < \tau_0] = 1 - \lim_{t \to \infty} \bar{v}_t^{-1} v_t(\lambda) = L_{\pi_0}(\lambda), \tag{3.46}$$

so the distribution of $x(t)$ under $\mathbf{Q}_x\{\cdot | t < \tau_0\}$ converges to π_0. In view of (3.44) we have

$$\int_0^\infty \left(1 - e^{-v_t(\lambda)u}\right) \pi_0(du) = \int_0^\infty \left(1 - e^{-\lambda u}\right) e^{-bt} \pi_0(du),$$

and hence $\pi_0 Q_t^\circ = e^{-bt}\pi_0$. On the other hand, as $t \to \infty$ the right-hand side of (3.45) is equivalent to

$$\bar{v}_{r+t}^{-1}(v_t(\lambda + \bar{v}_r) - v_t(\lambda)) = v_t(\bar{v}_r)^{-1}(v_t(\lambda + \bar{v}_r) - v_t(\lambda)).$$

Using the canonical representation (3.15) we may write this into

$$\left(\int_0^\infty \left(1 - e^{-\bar{v}_r u}\right) l_t(du)\right)^{-1} \int_0^\infty e^{-\lambda u}\left(1 - e^{-\bar{v}_r u}\right) l_t(du),$$

which converges as $t \to \infty$ to

$$\left(\int_0^\infty \left(1 - e^{-\bar{v}_r u}\right) \pi_0(du)\right)^{-1} \int_0^\infty e^{-\lambda u}\left(1 - e^{-\bar{v}_r u}\right) \pi_0(du), \tag{3.47}$$

giving the Laplace transform of a probability π_r on $(0, \infty)$. $\qquad\square$

Corollary 3.30 *Suppose that $b > 0$. Let π_0 be given by Theorem 3.26. Then we have $\int_0^\infty u\pi_0(\mathrm{d}u) = q(\infty)^{-1}$. Consequently, π_0 has finite mean if and only if $\int_1^\infty u \log u m(\mathrm{d}u) < \infty$.*

Proof. By Theorem 3.26 the measure π_0 has Laplace transform $1 - g(\lambda)$. Letting $\lambda \to \infty$ in (3.44) we have $g(\bar{v}_s) = \mathrm{e}^{-bs}$. By Theorem 3.8 and Corollary 3.9 we have $\lim_{s\to\infty} \bar{v}_s = 0$. It follows that

$$g'(0) = \lim_{s\to\infty} \bar{v}_s^{-1} g(\bar{v}_s) = \lim_{s\to\infty} \bar{v}_s^{-1} \mathrm{e}^{-bs} = \lim_{s\to\infty} q_s(\infty)^{-1} = q(\infty)^{-1},$$

which together with Theorem 3.27 gives the desired conclusion. □

There are counterparts of the above conditional limit theorems in the supercritical case. In fact, there is a symmetry in the limit theorems between the subcritical and supercritical processes if we use suitable conditioning. In the strictly supercritical case, we have $b < 0$ and

$$\mathbf{Q}_x[\mathrm{e}^{-\lambda x(t)}|\tau_0 < \infty] = \mathrm{e}^{-xw_t(\lambda)}, \quad \lambda \geq 0, \tag{3.48}$$

where $w_t(\lambda) = v_t(\lambda + \bar{v}) - \bar{v}$. Setting $\psi(\lambda) = \phi(\lambda + \bar{v})$ one may see that $w_t(\lambda)$ satisfies

$$\frac{\partial}{\partial t} w_t(\lambda) = -\psi(w_t(\lambda)), \quad w_0(\lambda) = \lambda. \tag{3.49}$$

Recall that $\bar{v} > 0$ is the largest root of $\phi(z) = 0$. Then it is simple to check that $\psi(\lambda) = \phi(\lambda + \bar{v}) - \phi(\bar{v})$ has the representation (3.1) with parameters $b_1 := \phi'(\bar{v}) > 0$, $c_1 := c$ and $m_1(\mathrm{d}u) := \mathrm{e}^{-\bar{v}u} m(\mathrm{d}u)$. Thus (3.48) implies that $\{x(t) : t \geq 0\}$ conditioned on $\tau_0 < \infty$ is a strictly subcritical CB-process with cumulant semigroup $(w_t)_{t\geq 0}$. By Corollary 3.12 we have $v_t(\bar{v}) = \bar{v}$, so $w_t(\lambda) = v_t(\lambda + \bar{v}) - v_t(\bar{v})$ has the representation (3.15) with canonical measure $\mathrm{e}^{-\bar{v}u} l_t(\mathrm{d}u)$ for $t > 0$. From Theorems 3.24, 3.25 and 3.29 we derive the following:

Theorem 3.31 *Suppose that $b < 0$. Then for any $t \geq 0$ and $x > 0$ the distribution of $x(t)$ under $\mathbf{Q}_x\{\cdot | r+t < \tau_0 < \infty\}$ converges as $r \to \infty$ to a probability measure $\bar{Q}_t^b(x, \cdot)$ on $(0, \infty)$ given by*

$$\int_0^\infty \mathrm{e}^{-\lambda y} \bar{Q}_t^b(x, \mathrm{d}y) = \exp\left\{ - xw_t(\lambda) - \int_0^t \psi_0'(w_s(\lambda))\mathrm{d}s \right\},$$

where $\psi_0'(z) = \phi'(z + \bar{v}) - b_1$. Moreover, $\bar{Q}_t^b(x, \cdot)$ converges as $t \to \infty$ to a probability measure η on $(0, \infty)$.

Theorem 3.32 *Suppose that $b < 0$. Then for $r \geq 0$ and $x > 0$ the distribution of $x(t)$ under $\mathbf{Q}_x\{\cdot | r+t < \tau_0 < \infty\}$ converges as $t \to \infty$ to a probability measure π_r on $(0, \infty)$ which is independent of x. Moreover, $\pi_0(\mathrm{d}u)$ is also the limit distribution of $(\bar{v}_t - \bar{v})^{-1} \mathrm{e}^{-\bar{v}u} l_t(\mathrm{d}u)$.*

We now consider the critical CB-process. In this case, we shall see that suitable conditioning of the process may lead to some universal limit laws independent of the explicit form of the branching mechanism.

Theorem 3.33 *Suppose that $b = 0$ and $\sigma^2 := \phi''(0) < \infty$. Then as $t \to \infty$ we have*

$$\frac{1}{t}\left(\frac{1}{v_t(\lambda)} - \frac{1}{\lambda}\right) \to \frac{1}{2}\sigma^2$$

uniformly in $0 < \lambda \leq \infty$ with the convention $1/\infty = 0$. In particular, we have $tv_t(\lambda) \to 2/\sigma^2$ for $0 < \lambda \leq \infty$ as $t \to \infty$.

Proof. Since Condition 3.6 holds, we have $\sigma^2 > 0$. For $0 < \lambda \leq \infty$ and $t > 0$, we may use the backward equation (3.5) to see that

$$\frac{1}{t}\left(\frac{1}{v_t(\lambda)} - \frac{1}{\lambda}\right) = -\frac{1}{t}\int_0^t \frac{1}{v_s(\lambda)^2}\frac{\partial}{\partial s}v_s(\lambda)ds = \frac{1}{t}\int_0^t \frac{\phi(v_s(\lambda))}{v_s(\lambda)^2}ds. \quad (3.50)$$

By l'Hôpital's rule,

$$\lim_{z\to 0+}\phi(z)/z^2 = \lim_{z\to 0+}\phi''(z)/2 = \sigma^2/2. \quad (3.51)$$

But by Theorem 3.8 and Corollary 3.9, we have $\lim_{t\to\infty}\bar{v}_t = 0$, and hence $\lim_{t\to\infty}v_t(\lambda) = 0$ uniformly on $0 < \lambda \leq \infty$. Then the assertion follows from (3.50) and (3.51). $\qquad\square$

Corollary 3.34 *Suppose that $b = 0$ and $\sigma^2 := \phi''(0) < \infty$. Then for any $\lambda \geq 0$ we have*

$$\lim_{t\to\infty}v_t'(\lambda/t) = (1 + \sigma^2\lambda/2)^{-2}.$$

Proof. For $\lambda = 0$ the above limit relation holds trivially. For $\lambda > 0$ we can use (3.5) and (3.6) to get $v_t'(\lambda/t) = \phi(\lambda/t)^{-1}\phi(v_t(\lambda/t))$. Then the result follows by Theorem 3.33. $\qquad\square$

Theorem 3.35 *Suppose that $b = 0$ and $\sigma^2 := \phi''(0) < \infty$. Let $\{y(t) : t \geq 0\}$ be a Markov process with transition semigroup $(Q_t^b)_{t\geq 0}$ given by (3.37). Then the distribution of $y(t)/t$ converges as $t \to \infty$ to the one on $(0,\infty)$ with density $4\sigma^{-4}xe^{-2x/\sigma^2}$.*

Proof. In this critical case, we have $q_t'(\lambda) = v_t'(\lambda)$. Since $\lim_{t\to\infty}v_t(\lambda/t) = 0$, by (3.36) and Corollary 3.34 we see that

$$\lim_{t\to\infty}\int_0^\infty e^{-\lambda y/t}Q_t^b(x,dy) = \lim_{t\to\infty}v_t'(\lambda/t) = \frac{1}{(1 + \sigma^2\lambda/2)^2},$$

which is the Laplace transform of the desired limit distribution. $\qquad\square$

Theorem 3.36 *Suppose that* $b = 0$ *and* $\sigma^2 := \phi''(0) < \infty$. *Then for any fixed* $r \geq 0$ *and* $x > 0$ *we have*

$$\lim_{t \to \infty} \mathbf{Q}_x\{x(t)/t > z | r + t < \tau_0\} = e^{-2z/\sigma^2}, \quad z \geq 0. \tag{3.52}$$

Proof. For any $t > 0$ we get from (3.45) that

$$\mathbf{Q}_x\left[e^{-\lambda x(t)/t} | r + t < \tau_0\right] = \frac{e^{-x v_t(\lambda/t)} - e^{-x v_t(\lambda/t + \bar{v}_r)}}{1 - e^{-x \bar{v}_{r+t}}},$$

which is still correct for $r = 0$ if we understand $\bar{v}_0 = \infty$. The right-hand side is equivalent to

$$\bar{v}_{r+t}^{-1}(v_t(\lambda/t + \bar{v}_r) - v_t(\lambda/t))$$

as $t \to \infty$. By Theorem 3.33 we have

$$\lim_{t \to \infty} t\bar{v}_{r+t} = 2/\sigma^2 \quad \text{and} \quad \lim_{t \to \infty} t v_t(\lambda/t) = (1/\lambda + \sigma^2/2)^{-1}. \tag{3.53}$$

From the uniform convergence we get

$$\lim_{t \to \infty} \frac{1}{t v_t(\lambda/t + \bar{v}_r)} = \lim_{t \to \infty} \frac{1}{t}\left(\frac{1}{v_t(\lambda/t + \bar{v}_r)} - \frac{1}{\lambda/t + \bar{v}_r}\right) = \frac{\sigma^2}{2}.$$

Then it follows immediately that

$$\lim_{t \to \infty} \mathbf{Q}_x\left[e^{-\lambda x(t)/t} | r + t < \tau_0\right] = \frac{\sigma^2}{2}\left(\frac{2}{\sigma^2} - \frac{1}{1/\lambda + \sigma^2/2}\right) = \frac{1}{1 + \sigma^2\lambda/2}.$$

The right-hand side gives the Laplace transform of the desired limit distribution. □

Theorem 3.37 *Suppose that* $b = 0$ *and* $\sigma^2 := \phi''(0) < \infty$. *Then for any* $x > 0$ *and* $a \geq 0$ *the distribution of* $x(t)/t$ *under* $\mathbf{Q}_x\{\cdot | (1+a)t < \tau_0\}$ *converges as* $t \to \infty$ *to the one on* $(0, \infty)$ *with density*

$$2\sigma^{-2}(1 + a)e^{-2x/\sigma^2}[1 - e^{-2x/a\sigma^2}] \tag{3.54}$$

with $e^{-\infty} = 0$ *by convention.*

Proof. For any $t > 0$ we use (3.45) to get

$$\mathbf{Q}_x\left[e^{-\lambda x(t)/t} | (1+a)t < \tau_0\right] = \frac{e^{-x v_t(\lambda/t)} - e^{-x v_t(\lambda/t + \bar{v}_{at})}}{1 - e^{-x \bar{v}_{(1+a)t}}}$$

under the convention $\bar{v}_0 = \infty$. The right-hand side is equivalent to

$$\bar{v}_{(1+a)t}^{-1}(v_t(\lambda/t + \bar{v}_{at}) - v_t(\lambda/t))$$

as $t \to \infty$. By (3.53) and the uniform convergence stated in Theorem 3.33 we have

$$\frac{1}{2}\sigma^2 = \lim_{t\to\infty} \frac{1}{t}\left(\frac{1}{v_t(\lambda/t + \bar{v}_{at})} - \frac{1}{\lambda/t + \bar{v}_{at}}\right)$$

$$= \lim_{t\to\infty}\left(\frac{1}{tv_t(\lambda/t + \bar{v}_{at})} - \frac{1}{\lambda + t\bar{v}_{at}}\right)$$

$$= \lim_{t\to\infty} \frac{1}{tv_t(\lambda/t + \bar{v}_{at})} - \frac{1}{\lambda + 2/a\sigma^2}.$$

It follows that

$$\lim_{t\to\infty} tv_t(\lambda/t + \bar{v}_{at}) = \left(\frac{\sigma^2}{2} + \frac{1}{\lambda + 2/a\sigma^2}\right)^{-1} = \frac{2 + a\sigma^2\lambda}{\sigma^2(1 + a + a\sigma^2\lambda/2)}.$$

Then one shows easily

$$\lim_{t\to\infty} \mathbf{Q}_x\left[e^{-\lambda x(t)/t}|(1+a)t < \tau_0\right] = \frac{1+a}{(1+\sigma^2\lambda/2)(1+a+a\sigma^2\lambda/2)},$$

which is the Laplace transform of the distribution with density (3.54). □

3.5 Scaling Limits of Discrete Processes

Let g and h be two probability generating functions. Suppose that $\{\xi_{n,i} : n = 0, 1, 2, \ldots; i = 1, 2, \ldots\}$ and $\{\eta_n : n = 0, 1, 2, \ldots\}$ are independent families of positive integer-valued i.i.d. random variables with distributions given by g and h, respectively. Given another positive integer-valued random variable $y(0)$ independent of $\{\xi_{n,i}\}$ and $\{\eta_n\}$, we define inductively

$$y(n+1) = \sum_{i=1}^{y(n)} \xi_{n,i} + \eta_n, \qquad n = 0, 1, 2, \ldots. \tag{3.55}$$

It is easy to show that $\{y(n) : n = 0, 1, 2, \ldots\}$ is a discrete-time positive integer-valued Markov chain with transition matrix $Q(i, j)$ determined by

$$\sum_{j=0}^{\infty} Q(i,j)z^j = g(z)^i h(z), \qquad |z| \leq 1. \tag{3.56}$$

The random variable $y(n)$ can be thought of as the number of individuals in generation $n \geq 0$ of an evolving particle system. After one unit time, each of the $y(n)$ particles splits independently of others into a random number of offspring according to the distribution given by g and a random number of immigrants are added to the system according to the probability law given by h. The n-step transition matrix $Q^n(i, j)$ of $\{y(n) : n = 0, 1, 2, \ldots\}$ is given by

$$\sum_{j=0}^{\infty} Q^n(i,j)z^j = g^n(z)^i \prod_{j=1}^{n} h(g^{j-1}(z)), \qquad |z| \le 1, \tag{3.57}$$

where $g^n(z)$ is defined by $g^n(z) = g(g^{n-1}(z))$ successively with $g^0(z) = z$. We call any positive integer-valued Markov chain with transition probabilities given by (3.56) or (3.57) a *Galton–Watson branching process with immigration* (GWI-process) with parameters (g, h). If $g'(1-) < \infty$ and $h'(1-) < \infty$, then the discrete probability distribution $\{Q^n(i,j) : j = 0, 1, 2, \ldots\}$ has the first-moment given by

$$\sum_{j=1}^{\infty} jQ^n(i,j) = ig'(1-)^n + \sum_{j=1}^{n} h'(1-)g'(1-)^{j-1}, \tag{3.58}$$

which can be obtained by differentiating both sides of (3.57). In the special case where $h(z) \equiv 1$, we simply call $\{y(n) : n = 0, 1, 2, \ldots\}$ a *Galton–Watson branching process* (GW-process).

Suppose that for each integer $k \ge 1$ we have a GWI-process $\{y_k(n) : n \ge 0\}$ with parameters (g_k, h_k). Let $z_k(n) = y_k(n)/k$. Then $\{z_k(n) : n \ge 0\}$ is a Markov chain with state space $E_k := \{0, 1/k, 2/k, \ldots\}$ and n-step transition probability $Q_k^n(x, dy)$ determined by

$$\int_{E_k} e^{-\lambda y} Q_k^n(x, dy) = g_k^n(e^{-\lambda/k})^{kx} \prod_{j=1}^{n} h(g_k^{j-1}(e^{-\lambda/k})), \qquad \lambda \ge 0. \tag{3.59}$$

Suppose that $\{\gamma_k\}$ is a positive real sequence so that $\gamma_k \to \infty$ increasingly as $k \to \infty$. Let $[\gamma_k t]$ denote the integer part of $\gamma_k t \ge 0$. We are interested in the asymptotic behavior of the continuous-time process $\{z_k([\gamma_k t]) : t \ge 0\}$ as $k \to \infty$. For any $z \ge 0$ define

$$H_k(z) = \gamma_k[1 - h_k(e^{-z/k})] \tag{3.60}$$

and

$$G_k(z) = k\gamma_k[g_k(e^{-z/k}) - e^{-z/k}]. \tag{3.61}$$

For the convenience of statement of the results, we formulate the following conditions:

Condition 3.38 *There is a function ψ on $[0, \infty)$ such that $H_k(z) \to \psi(z)$ uniformly on $[0, a]$ for every $a \ge 0$ as $k \to \infty$.*

Condition 3.39 *The sequence $\{G_k\}$ is uniformly Lipschitz on $[0, a]$ for every $a \ge 0$ and there is a function ϕ on $[0, \infty)$ such that $G_k(z) \to \phi(z)$ uniformly on $[0, a]$ for every $a \ge 0$ as $k \to \infty$.*

Proposition 3.40 *If Condition 3.38 holds, the limit function ψ has the representation (3.26). If Condition 3.39 holds, then ϕ has the representation (3.1).*

Proof. The representation (3.26) for ψ follows by a modification of the proof of Theorem 1.43. Since ϕ is locally Lipschitz, by Proposition 1.45 it suffices to show the function has the representation (1.39). This could be done by modifying the proofs of Theorems 1.43 and 1.44. We here give a derivation of the representation by considering the sequence

$$\phi_k(z) = k\gamma_k[g_k(1 - z/k) - (1 - z/k)], \qquad 0 \le z \le k.$$

Fix the constant $a \ge 0$. By the mean-value theorem, for $k \ge a$ and $0 \le z \le a$ we have

$$G_k(z) - \phi_k(z) = k\gamma_k[g'_k(\eta_k) - 1](e^{-z/k} - 1 + z/k), \qquad (3.62)$$

where

$$1 - a/k \le 1 - z/k \le \eta_k \le e^{-z/k} \le 1.$$

Choose $k_0 \ge a$ so that $e^{-2a/k_0} \le 1 - a/k_0$. Then for $k \ge k_0$ we have $e^{-2a/k} \le 1 - a/k$ and hence

$$\gamma_k|g'_k(\eta_k) - 1| \le \sup_{0 \le \lambda \le 2a} \gamma_k|g'_k(e^{-\lambda/k}) - 1|.$$

Since $\{G_k\}$ is uniformly Lipschitz on $[0, 2a]$, the sequence

$$G'_k(z) = \gamma_k e^{-z/k}[1 - g'_k(e^{-z/k})]$$

is uniformly bounded on $[0, 2a]$. Thus $\{\gamma_k|g'_k(\eta_k) - 1| : k \ge k_0\}$ is a bounded sequence and (3.62) implies

$$\phi(z) = \lim_{k \to \infty} G_k(z) = \lim_{k \to \infty} \phi_k(z).$$

Then we can use Theorem 1.44 to see ϕ has the representation (1.39). $\qquad\square$

We shall work with the Laplace transform of the process $\{z_k([\gamma_k t]) : t \ge 0\}$. In view of (3.59), given $z_k(0) = x$ the conditional distribution $Q_k^{[\gamma_k t]}(x, \cdot)$ of $z_k([\gamma_k t])$ on E_k is determined by

$$\int_{E_k} e^{-\lambda y} Q_k^{[\gamma_k t]}(x, \mathrm{d}y)$$

$$= \exp\left\{ -xv_k(t, \lambda) - \int_0^{\frac{[\gamma_k t]}{\gamma_k}} \bar{H}_k(v_k(s, \lambda))\mathrm{d}s \right\}, \qquad (3.63)$$

where

$$v_k(t, \lambda) = -k \log g_k^{[\gamma_k t]}(e^{-\lambda/k}) \qquad (3.64)$$

and

$$\bar{H}_k(\lambda) = -\gamma_k \log h_k(e^{-\lambda/k}), \qquad \lambda \geq 0.$$

Lemma 3.41 *Suppose that the sequence $\{G_k\}$ defined by (3.61) is uniformly Lipschitz on $[0, 1]$. Then there are constants $B \geq 0$ and $N \geq 1$ such that $v_k(t, \lambda) \leq \lambda e^{Bt}$ for every $t, \lambda \geq 0$ and $k \geq N$.*

Proof. Let $b_k := G_k'(0+)$ for $k \geq 1$. Since $\{G_k\}$ is uniformly Lipschitz on $[0, 1]$, the sequence $\{b_k\}$ is bounded. Let $B \geq 0$ be a constant such that $2|b_k| \leq B$ for all $k \geq 1$. In view of (3.63), there is a probability kernel $P_k^{[\gamma_k t]}(x, dy)$ on E_k such that

$$\int_{E_k} e^{-\lambda y} P_k^{[\gamma_k t]}(x, dy) = \exp\{-x v_k(t, \lambda)\}, \qquad \lambda \geq 0. \tag{3.65}$$

From (3.61) we have $b_k = \gamma_k[1 - g_k'(1-)]$. It is not hard to obtain

$$\int_{E_k} y P_k^{[\gamma_k t]}(x, dy) = x g_k'(1-)^{[\gamma_k t]} = x \left(1 - \frac{b_k}{\gamma_k}\right)^{[\gamma_k t]}.$$

Since $\gamma_k \to \infty$ as $k \to \infty$, there is $N \geq 1$ so that

$$0 \leq \left(1 - \frac{b_k}{\gamma_k}\right)^{\frac{\gamma_k}{B}} \leq \left(1 + \frac{B}{2\gamma_k}\right)^{\frac{\gamma_k}{B}} \leq e, \qquad k \geq N.$$

It follows that, for $t \geq 0$ and $k \geq N$,

$$\int_{E_k} y P_k^{[\gamma_k t]}(x, dy) \leq x \exp\left\{\frac{B}{\gamma_k}[\gamma_k t]\right\} \leq x e^{Bt}. \tag{3.66}$$

Then the desired estimate follows from (3.65), (3.66) and Jensen's inequality. ☐

Theorem 3.42 *Suppose that Condition 3.39 is satisfied. Let $(t, \lambda) \mapsto v_t(\lambda)$ be the unique locally bounded positive solution of (3.3). Then for every $a \geq 0$ we have $v_k(t, \lambda) \to v_t(\lambda)$ uniformly on $[0, a]^2$ as $k \to \infty$.*

Proof. It suffices to show $v_k(t, \lambda)$ converges uniformly on $[0, a]^2$ for every $a \geq 0$ and the limit solves (3.3). Let

$$\bar{G}_k(z) = k\gamma_k \log\left[g_k(e^{-z/k})e^{z/k}\right], \qquad z \geq 0.$$

For any integer $n \geq 0$ we may write

$$\begin{aligned}
\log g_k^{n+1}(e^{-\lambda/k}) &= \log\left[g_k(g_k^n(e^{-\lambda/k}))g_k^n(e^{-\lambda/k})^{-1}\right] + \log g_k^n(e^{-\lambda/k}) \\
&= (k\gamma_k)^{-1}\bar{G}_k\left(-k\log g_k^n(e^{-\lambda/k})\right) + \log g_k^n(e^{-\lambda/k}).
\end{aligned}$$

From this and (3.64) it follows that

$$v_k(t + \gamma_k^{-1}, \lambda) = v_k(t, \lambda) - \gamma_k^{-1}\bar{G}_k(v_k(t, \lambda)), \qquad t \geq 0.$$

By applying the above equation to $t = 0, 1/\gamma_k, \ldots, ([\gamma_k t] - 1)/\gamma_k$ and adding the resulting equations we obtain

$$v_k(t, \lambda) = \lambda - \sum_{i=1}^{[\gamma_k t]} \gamma_k^{-1} \bar{G}_k \big(v_k(\gamma_k^{-1}(i-1), \lambda) \big).$$

Then we have

$$v_k(t, \lambda) = \lambda + \varepsilon_k(t, \lambda) - \int_0^t \bar{G}_k(v_k(s, \lambda)) ds, \qquad (3.67)$$

where

$$\varepsilon_k(t, \lambda) = \big(t - \gamma_k^{-1}[\gamma_k t] \big) \bar{G}_k \big(v_k(\gamma_k^{-1}[\gamma_k t], \lambda) \big).$$

It is elementary to see

$$\bar{G}_k(z) = k\gamma_k \log \big[1 + (k\gamma_k)^{-1} G_k(z) e^{z/k} \big].$$

Let $B \geq 0$ and $N \geq 1$ be chosen as in Lemma 3.41. Under Condition 3.39 one can show $\bar{G}_k(z) \to \phi(z)$ uniformly on every bounded interval. Then for any $0 < \varepsilon \leq 1$ we can enlarge the constant $N \geq 1$ so that

$$|\bar{G}_k(z) - \phi(z)| \leq \varepsilon, \qquad 0 \leq z \leq a e^{Ba}, \ k \geq N. \qquad (3.68)$$

It follows that

$$|\varepsilon_k(t, \lambda)| \leq \gamma_k^{-1} M, \qquad 0 \leq t, \lambda \leq a, \qquad (3.69)$$

where

$$M = 1 + \sup_{0 \leq z \leq a e^{Ba}} |\phi(z)|.$$

For $n \geq k \geq N$ let

$$K_{k,n}(t, \lambda) = \sup_{0 \leq s \leq t} |v_n(s, \lambda) - v_k(s, \lambda)|.$$

By (3.67), (3.68) and (3.69) we obtain

$$K_{k,n}(t, \lambda) \leq 2(\gamma_k^{-1} M + \varepsilon a) + L \int_0^t K_{k,n}(s, \lambda) ds, \qquad 0 \leq t, \lambda \leq a,$$

where $L = \sup_{0 \leq z \leq a e^{Ba}} |\phi'(z)|$. By Gronwall's inequality,

$$K_{k,n}(t, \lambda) \leq 2(\gamma_k^{-1} M + \varepsilon a) e^{Lt}, \qquad 0 \leq t, \lambda \leq a.$$

Then $v_k(t, \lambda) \to$ some $v_t(\lambda)$ uniformly on $[0, a]^2$ as $k \to \infty$. In view of (3.69) and (3.67) we have (3.3). $\qquad \square$

Let $D([0, \infty), \mathbb{R}_+)$ denote the space of càdlàg paths from $[0, \infty)$ to $\mathbb{R}_+$ furnished with the Skorokhod topology. The main limit theorem of this section is the following:

Theorem 3.43 *Suppose that Conditions 3.38 and 3.39 are satisfied. Let $\{y(t) : t \geq 0\}$ be a càdlàg CBI-process with transition semigroup $(Q_t^\gamma)_{t\geq 0}$ defined by (3.29). If $z_k(0)$ converges to $y(0)$ in distribution, then $\{z_k([\gamma_k t]) : t \geq 0\}$ converges to $\{y(t) : t \geq 0\}$ in distribution on $D([0, \infty), \mathbb{R}_+)$.*

Proof. For $\lambda > 0$ and $x \geq 0$ set $e_\lambda(x) = e^{-\lambda x}$. We denote by D_1 the linear span of $\{e_\lambda : \lambda > 0\}$. It is easy to see that D_1 is an algebra strongly separating the points of $\mathbb{R}_+$ in the sense of Ethier and Kurtz (1986, pp.112–113). Let $C_0(\mathbb{R}_+)$ be the space of continuous functions on $\mathbb{R}_+$ vanishing at infinity. Then D_1 is uniformly dense in $C_0(\mathbb{R}_+)$ by the Stone–Weierstrass theorem; see, e.g., Hewitt and Stromberg (1965, pp.98–99). By Proposition 3.1 it is easy to see that the function $t \mapsto v_t(\lambda)$ is locally bounded away from zero. Under Condition 3.38 we have $\bar{H}_k(z) \to \psi(z)$ uniformly on every bounded interval. Then one can use (3.29), (3.63) and Theorem 3.42 to show

$$\lim_{k\to\infty} \sup_{x\in E_k} \left| Q_k^{[\gamma_k t]} e_\lambda(x) - Q_t^\gamma e_\lambda(x) \right| = 0$$

for every $t \geq 0$. It follows that

$$\lim_{k\to\infty} \sup_{x\in E_k} \left| Q_k^{[\gamma_k t]} f(x) - Q_t^\gamma f(x) \right| = 0$$

for every $t \geq 0$ and $f \in C_0(\mathbb{R}_+)$. By Ethier and Kurtz (1986, p.226 and pp.233–234) we conclude that $\{z_k([\gamma_k t]) : t \geq 0\}$ converges to the CBI-process $\{y(t) : t \geq 0\}$ in distribution on $D([0, \infty), \mathbb{R}_+)$. $\square$

The theorem above gives an interpretation of the CBI-process as the limit of a sequence of rescaled GWI-processes. The following examples describe some typical situations where Conditions 3.38 and 3.39 are satisfied.

Example 3.3 Suppose that h is a probability generating function so that $\beta := h'(1-) < \infty$. Let $\gamma_k = k$ and $h_k(z) = h(z)$. Then the sequence $H_k(z)$ defined by (3.60) converges to βz as $k \to \infty$.

Example 3.4 For any $0 < \alpha \leq 1$ let $\gamma_k = k^\alpha$ and $h_k(z) = 1 - (1 - z)^\alpha$. Then the sequence $H_k(z)$ defined by (3.60) converges to z^α as $k \to \infty$.

Example 3.5 Suppose that g is a probability generating function so that $g'(1-) = 1$ and $c := g''(1-)/2 < \infty$. Let $\gamma_k = k$ and $g_k(z) = g(z)$. By Taylor's expansion one sees that the sequence $G_k(z)$ defined by (3.61) converges to cz^2 as $k \to \infty$.

Example 3.6 For any $1 \leq \alpha \leq 2$ let $\gamma_k = \alpha k^{\alpha-1}$ and $g_k(z) = z + \alpha^{-1}(1 - z)^\alpha$. Then the sequence $G_k(z)$ defined by (3.61) converges to z^α as $k \to \infty$.

We now give some applications of Theorem 3.43 to the characterizations of local times. Let $X^0 = (\Omega, \mathscr{F}, \mathscr{F}_t, X_t^0, \mathbf{P}_x)$ be a standard one-dimensional Brownian motion. Given any constant $a \in \mathbb{R}$ we define $X_t^a = X_t^0 - at$ for $t \geq 0$. Then $X^a = (\Omega, \mathscr{F}, \mathscr{F}_t, X_t^a, \mathbf{P}_x)$ is a *Brownian motion with drift* $-a$. It is well-known that there is a positive continuous two-parameter process $\{l^a(t, y) : t \geq 0, y \in \mathbb{R}\}$ such that

$$\int_0^t 1_A(X_s^a)ds = \int_A 2l^a(t, y)dy, \qquad A \in \mathscr{B}(\mathbb{R}).$$

The process $\{2l^a(t, y) : t \geq 0, y \in \mathbb{R}\}$ is the local time of $\{X_t^a : t \geq 0\}$. Let $\tau_x^a = \inf\{t > 0 : X_t^a = x\}$ denote the hitting time of $x \in \mathbb{R}$ by the Brownian motion with drift. By a δ-*downcrossing* of $\{X_t^a : t \geq 0\}$ at $y \in \mathbb{R}$ before time $T > 0$ we mean an interval $[u, v] \subset [0, T)$ such that $X_u^a = y + \delta$, $X_v^a = y$ and $y < X_t^a < y + \delta$ for all $u < t < v$.

We first consider the special case $a = 0$. It is well-known that $\mathbf{P}_x\{l^0(t, y) \to \infty$ as $t \to \infty\} = 1$ for all $x, y \in \mathbb{R}$. Then for every $u \geq 0$ we have $\mathbf{P}_x$-a.s.

$$\sigma^0(u) := \inf\{t \geq 0 : l^0(t, 0) \geq u\} < \infty.$$

The following theorem is the well-known *Ray–Knight theorem* on Brownian local times.

Theorem 3.44 *For any $x \geq 0$, under the probability law $\mathbf{P}_x$ we have:*

(1) $\{l^0(\sigma^0(u), -t) : t \geq 0\}$ *and* $\{l^0(\sigma^0(u), x + t) : t \geq 0\}$ *are CB-processes with branching mechanism* $\phi(z) = z^2$;
(2) $\{l^0(\sigma^0(u), t) : 0 \leq t \leq x\}$ *is a CBI-process with branching mechanism* $\phi(z) = z^2$ *and immigration mechanism* $\psi(z) = z$.

Proof. (1) Let ξ_k denote the number of $(1/k)$-downcrossings at x before time $\tau_{x-1/k}^0$. By the property of independent increments of $\{X_t^0 : t \geq 0\}$ we have

$$\mathbf{P}_0[z^{\xi_k}] = \sum_{i=0}^{\infty} \frac{z^i}{2^{i+1}} = \frac{1}{2 - z}, \qquad |z| \leq 1.$$

For $k \geq 1$ and $i \geq 0$ let $Z_k(i)$ denote the number of $(1/k)$-downcrossings of $\{X_t^0 : t \geq 0\}$ at $x_i = x + i/k$ before time $\sigma^0(u)$. It is easy to see that $Z_k(i + 1)$ is the sum of $Z_k(i)$ independent copies of ξ_k. Thus $\{Z_k(i) : i = 0, 1, \ldots\}$ is a GW-process determined by the probability generating function

$$g(z) = \frac{1}{2 - z}, \qquad |z| \leq 1. \tag{3.70}$$

By the approximation of the local time by downcrossing numbers, for every $t \geq 0$ we have $Z_k([kt])/k \to l^0(\sigma^0(u), x + t)$ in probability as $k \to \infty$; see, e.g, Revuz and Yor (1999, p.227). On the other hand, it is easy to show

$$\phi(z) := \lim_{k \to \infty} k^2 [g(e^{-z/k}) - e^{-z/k}] = z^2.$$

By Theorem 3.43 one can see $\{l^0(\sigma^0(u), x + t) : t \geq 0\}$ is a CB-process with branching mechanism $\phi(z) = z^2$. The result for $\{l^0(\sigma^0(u), -t) : t \geq 0\}$ follows in a similar way.

(2) This is similar to the first part of the proof, so we only give the sketch. For $k \geq 1$ and $0 \leq i \leq [kx]$ let $Y_k(i)$ denote the number of $(1/k)$-downcrossings of $\{X_t^0 : t \geq 0\}$ at $z_i = i/k$ before time $\sigma^0(u)$. One can see that $Y_k(i + 1) - 1$ is the sum of $Y_k(i)$ independent copies of ξ_k. Then $\{Y_k(i) : i = 0, 1, \ldots, [kx]\}$ is a GWI-process determined by the pair of generating functions (g, h), where $g(z)$ is given by (3.70) and $h(z) = z$. For any $0 \leq t \leq x$ we have $Y_k([kt])/k \to l^0(\sigma^0(u), t)$ in probability as $k \to \infty$. Then the result follows by Theorem 3.43.   □

Let $\{2l(t, y) : t \geq 0, y \geq 0\}$ denote the local time of the reflecting Brownian motion $\{|X_t^0| : t \geq 0\}$. Then $\{l(t, y) : t \geq 0, y \geq 0\}$ is a positive continuous two-parameter process such that

$$\int_0^t 1_A(|X_s^0|)ds = 2 \int_A l(t, y)dy, \qquad A \in \mathscr{B}(\mathbb{R}_+).$$

For any $x \geq 0$ and $u \geq 0$ we have $\mathbf{P}_x$-a.s.

$$\sigma(u) := \inf\{t \geq 0 : l(t, 0) \geq u\} < \infty.$$

By modifying the arguments in the proof of Theorem 3.44 one can show the following:

Theorem 3.45 *For any $x \geq 0$, under the probability law $\mathbf{P}_x$ we have:*

(1) $\{l(\sigma(u), x+t) : t \geq 0\}$ *is a CB-process with branching mechanism $\phi(z) = z^2$;*
(2) $\{l(\sigma(u), t) : 0 \leq t \leq x\}$ *is a CBI-process with branching mechanism $\phi(z) = z^2$ and immigration mechanism $\psi(z) = z$.*

We next consider the case $a > 0$. In this case we have $\mathbf{P}_x\{\tau_0^a < \infty\} = 1$ for every $x > 0$. For $\delta > 0$ and $|x| \leq \delta$ let $u_\delta(x) = \mathbf{P}_x\{\tau_{-\delta}^a < \tau_\delta^a\}$. Then $x \mapsto u_\delta(x)$ solves the differential equation

$$\frac{1}{2}u''(x) - au'(x) = 0, \qquad |x| \leq \delta$$

with boundary conditions $u(\delta) = 0$ and $u(-\delta) = 1$. By solving the above boundary value problem we find

$$u_\delta(x) = \frac{e^{2a\delta} - e^{2ax}}{e^{2a\delta} - e^{-2a\delta}}, \qquad |x| \leq \delta. \tag{3.71}$$

The following theorem slightly generalizes the Ray–Knight theorem.

Theorem 3.46 *Suppose that $a > 0$ and $x > 0$. Then under $\mathbf{P}_x$ we have:*

(1) $\{l^a(\tau_0^a, x + t) : t \geq 0\}$ is a CB-process with branching mechanism $\phi(z) = z^2 + 2az$;

(2) $\{l^a(\tau_0^a, t) : 0 \leq t \leq x\}$ is a CBI-process with branching mechanism $\phi(z) = z^2 + 2az$ and immigration mechanism $\psi(z) = z$.

Proof. The arguments are modifications of those in the proof of Theorem 3.44, so we only describe the difference. For $k \geq 1$ and $i \geq 0$ let $Z_k(i)$ denote the number of $(1/k)$-downcrossings of $\{X_t^a : t \geq 0\}$ at $x_i = x + i/k$ before time τ_0^a. Then $\{Z_k(i) : i = 0, 1, \ldots\}$ is a GW-process corresponding to the generating function

$$g_k(z) = \sum_{i=0}^{\infty} p_k(q_k z)^i = \frac{p_k}{1 - q_k z}, \qquad |z| \leq 1, \tag{3.72}$$

where $p_k = u_{1/k}(0)$ and $q_k = 1 - p_k$. From (3.71) we get

$$p_k = \frac{e^{2a/k} - 1}{e^{2a/k} - e^{-2a/k}} = \frac{1}{2} + \frac{a}{2k} + o\left(\frac{1}{k}\right)$$

and

$$q_k = \frac{1 - e^{-2a/k}}{e^{2a/k} - e^{-2a/k}} = \frac{1}{2} - \frac{a}{2k} + o\left(\frac{1}{k}\right)$$

as $k \to \infty$. Then we use (3.72) to see

$$\begin{aligned}
\phi(z) &:= \lim_{k \to \infty} k^2 [g_k(e^{-z/k}) - e^{-z/k}] \\
&= \lim_{k \to \infty} \frac{k^2 [1 - e^{-z/k} - q_k(1 - e^{-2z/k})]}{1 - q_k e^{-z/k}} \\
&= z^2 + 2az.
\end{aligned}$$

For $t \geq 0$ we have $Z_k([kt])/k \to l^a(\tau_0^a, x + t)$ in probability as $k \to \infty$. Thus the assertion (1) follows by Theorem 3.43. Similarly one obtains (2). $\qquad\square$

3.6 Notes and Comments

The convergence of rescaled Galton–Watson branching processes to diffusion processes was first studied by Feller (1951). Jiřina (1958) introduced CB-processes in both discrete and continuous times. Lamperti (1967a) showed that the continuous-time processes are weak limits of rescaled Galton–Watson branching processes. Lamperti (1967b) characterized the CB-processes by random time changes of Lévy processes. New proofs of the result were given recently by Caballero et al. (2009).

For the sake of simplicity, we have assumed the branching mechanism is given by (3.1). Kawazu and Watanabe (1971) constructed more general CBI-processes that are not necessarily conservative. The "if" part of Theorem 3.10 was proved in

Silverstein (1967/8). It seems the "only if" part is a new result. Most other results on extinction probabilities and growth rates in Sections 3.1 and 3.2 can be found in Grey (1974). It is simple to check that if $\{X_t : t \geq 0\}$ is a Dawson–Watanabe super-process with conservative underlying spatial motion and spatially constant branching mechanism, then the total mass $\{X_t(1) : t \geq 0\}$ is a CB-process. The properties of local extinction and growth rate for superprocesses were studied in Engländer and Kyprianou (2004), Liu et al. (2009) and Pinsky (1995, 1996). A zero-one law on the local extinction for a super-Brownian motion was given in Zhou (2008). See also Engländer (2007) and the references therein.

Theorem 3.20 and Corollary 3.21 were given in Pinsky (1972). Other results in Sections 3.3 and 3.4 can be found in Li (2000). The result of Theorem 3.29 was already expected by Pakes (1988, p.86); see also Pakes and Trajstman (1985). A number of conditional limit theorems for Galton–Watson processes were proved in Pakes (1999) by introducing some general conditioning events. Theorems 3.36 and 3.37 treat the two simplest special cases of the conditional events of Pakes (1999). Some of the results in Section 3.4 were proved in Lambert (2007) by different methods; see also Kyprianou and Pardo (2008).

Conditions 3.38 and 3.39 were given in Aliev (1985) and Aliev and Shchurenkov (1982). Slightly different forms of the two conditions can be found in Li (2006). The proof of Theorem 3.42 follows Aliev and Shchurenkov (1982). The convergence in distribution on the path space $D([0, \infty), \mathbb{R}_+)$ of Theorem 3.43 was established in Li (2006) by proving the convergence of the generators of the rescaled GWI-processes; see also Ma (2009). Theorem 3.44 was originally proved by Knight (1963) and Ray (1963). There are many generalizations of the Ray–Knight theorem; see, e.g., Borodin and Salminen (1996).

If $(v_t)_{t>0}$ admits the representation (3.15), then each $l_t(\mathrm{d}u)$ is a diffuse measure on $(0, \infty)$; see Bertoin and Le Gall (2000). CBI-processes were used by Bertoin and Le Gall (2000, 2006) in studying the coalescent processes with multiple collisions of Pitman (1999) and Sagitov (1999). See also Limic and Sturm (2006) and Schweinsberg (2000, 2003) for some related results. Using the results for self-similar CBI-processes, Patie (2009) gave a characterization of the density of the law of an exponential functional associated to some one-sided Lévy processes.

The genealogical structures of Galton–Watson branching processes can be represented by Galton–Watson trees. Those trees can be coded by two kinds of discrete paths called height functions and contour functions. The basic idea of the Ray–Knight theorem is to code the genealogical structures of Feller's branching diffusion by the Brownian paths. Le Gall and Le Jan (1998a, 1998b) proposed an approach of coding the genealogy of a general subcritical branching CB-process using a spectrally positive Lévy process, which corresponds to the reflecting Brownian motion in the case of Feller's branching diffusion. A key contribution of Le Gall and Le Jan (1998a, 1998b) is an explicit expression for the height process as a functional of the Lévy process whose Laplace exponent is precisely the branching mechanism. This suggests that many problems concerning the genealogies of CB-processes can be restated and solved in terms of spectrally positive Lévy processes, for which there is a rich literature; see, e.g., Bertoin (1996) and Sato (1999).

In view of Theorem 3.43, one may want to look for limit theorems of branching models involving genealogical structures. Some limit theorems of that type were established in Duquesne and Le Gall (2002) in terms of height processes and contour processes. Pitman (2006) studied various combinatorial models of random partitions and trees, and the asymptotics of these models related to stochastic processes. See Aldous (1991, 1993) for the early work on continuum random trees. The method of Gromov–Hausdorff distance was developed in Evans et al. (2006) and Evans and Winter (2006) to study the asymptotic behavior of random trees when the number of vertices goes to infinity. Evans (2008) and Winter (2007) gave surveys of the relevant backgrounds and applications; see also Le Gall (2005). The genealogical structures of catalytic branching models were studied in Greven et al. (2009).

A natural generalization of the CBI-process is described as follows. Let $m \geq 0$ and $n \geq 0$ be integers and define $D = \mathbb{R}_+^m \times \mathbb{R}^n$ and $U = \mathbb{C}_-^m \times (i\mathbb{R})^n$, where $\mathbb{C}_- = \{a + ib : a \leq 0, b \in \mathbb{R}\}$ and $i\mathbb{R} = \{ib : b \in \mathbb{R}\}$. Let $(\cdot, \cdot)$ denote the duality between D and U. A transition semigroup $(P_t)_{t \geq 0}$ on D is called an *affine semigroup* if its characteristic function has the representation

$$\int_D e^{(u,\xi)} P_t(x, d\xi) = \exp\{(x, \psi(t, u)) + \phi(t, u)\}, \qquad u \in U, \qquad (3.73)$$

where $u \mapsto \psi(t, u)$ is a continuous mapping of U into itself and $u \mapsto \phi(t, u)$ is a continuous function on U satisfying $\phi(t, 0) = 0$. A Markov process in D is called an *affine process* if it has affine transition semigroup. The process reduces to an m-dimensional CBI-process when $n = 0$. Affine processes have been used widely in mathematical finance; see Duffie et al. (2003) and the references therein.

The affine semigroup defined by (3.73) is called *regular* if it is stochastically continuous and the right derivatives $\psi_t'(0, u)$ and $\phi_t'(0, u)$ exist for all $u \in U$ and are continuous at $u = 0$. A number of characterizations of regular affine processes were given in Duffie et al. (2003). By a result of Kawazu and Watanabe (1971), a stochastically continuous CBI-process is automatically regular. The regularity of affine processes was studied in Dawson and Li (2006) under a moment assumption. The regularity problem was settled in Keller-Ressel et al. (2010), where it was proved that any stochastically continuous affine process is regular.

Chapter 4
Branching Particle Systems

Branching particle systems arise from applications in a number of subjects. Typical examples of those systems are biological populations in isolated regions, families of neutrons in nuclear reactions, cosmic-ray showers and so on. In this chapter, we show that suitable scaling limits of those particle systems lead to the Dawson–Watanabe superprocesses in finite-dimensional distributions, giving intuitive interpretations for the superprocesses. To show the ideas in a simple and clear way, we shall first develop the results in detail for local branching particle systems. After that we show how the argument can be modified to general non-local branching models.

4.1 Particle Systems with Local Branching

In this section, we introduce a special class of branching particle systems, which can be regarded as the discrete-state counterpart of the local branching Dawson–Watanabe superprocesses. Let E be a Lusin topological space and let $N(E)$ denote the space of integer-valued finite measures on E. Let $\xi = (\Omega, \mathscr{F}, \mathscr{F}_t, \xi_t, \mathbf{P}_x)$ be a Borel right process in E with conservative transition semigroup $(P_t)_{t\geq 0}$. We assume the sample path $\{\xi_t : t \geq 0\}$ is right continuous in both the original and Ray topologies and has left limits $\{\xi_{t-} : t > 0\}$ in the Ray–Knight completion $\bar{E}$ of E. Let $\gamma \geq 0$ be a constant. Suppose that $g \in B(E \times [-1, 1])$ and $g(x, \cdot)$ is a probability generating function for each $x \in E$, that is,

$$g(x, z) = \sum_{k=0}^{\infty} p_k(x)z^k, \qquad |z| \leq 1,$$

where $p_k(x) \geq 0$ and $\sum_{k=0}^{\infty} p_k(x) = 1$. Moreover, we assume

$$\sup_{x \in E} g_z'(x, 1-) = \sup_{x \in E} \sum_{k=1}^{\infty} kp_k(x) < \infty. \tag{4.1}$$

Z. Li, *Measure-Valued Branching Markov Processes*,
Probability and Its Applications, DOI 10.1007/978-3-642-15004-3_4,
© Springer-Verlag Berlin Heidelberg 2011

Let $g(x, z) = 1$ for $x \in \bar{E} \setminus E$ and $|z| \le 1$. We consider a particle system on E characterized by the following properties:

(1) The particles in E move independently according to the law given by the transition probabilities of ξ.
(2) For a particle which is alive at time $r \ge 0$ and follows the path $\{\xi_s : s \ge r\}$, the conditional probability of survival in the time interval $[r, t)$ is $\exp\{-(t - r)\gamma\}$.
(3) When a particle following the path $\{\xi_s : s \ge r\}$ dies at time $t > r$, it gives birth to a random number of offspring at $\xi_t \in E$ according to the probability distribution given by the generating function $g(\xi_{t-}, \cdot)$. The offspring then start to move from their common birth site.

In addition, we assume that the lifetimes and the branchings of different particles are independent. By a *branching particle system* with parameters (ξ, γ, g) we mean the measure-valued process $\{X_t : t \ge 0\}$, where $X_t(B)$ denotes the number of particles in $B \in \mathscr{B}(E)$ that are alive at time $t \ge 0$. A construction of the branching particle system with initial value $\sigma \in N(E)$ is given as follows.

Let $\mathscr{A}$ be the set of all finite strings of the form $\alpha = n_0 n_1 \cdots n_{l(\alpha)}$ for integers $l(\alpha) \ge 0$ and $n_i \ge 1$. We provide $\mathscr{A}$ with the arboreal ordering. Then $m_0 m_1 \cdots m_p \prec n_0 n_1 \cdots n_q$ if and only if $p \le q$ and $m_0 = n_0, m_1 = n_1, \ldots, m_p = n_p$. The particles will be labeled by the strings in $\alpha \in \mathscr{A}$. The integer $l(\alpha)$ is interpreted as the generation number of the particle with label $\alpha \in \mathscr{A}$. Then this particle has exactly $l(\alpha)$ predecessors, which we denote respectively by $\alpha \setminus 1, \alpha \setminus 2, \ldots, \alpha \setminus l(\alpha)$. For example, if $\alpha = 12436$, then $\alpha \setminus 1 = 1243$ and $\alpha \setminus 3 = 12$. Suppose we are given a probability space $(W, \mathscr{G}, \mathbf{P})$ on which the following family of independent random elements are defined:

$$\{\xi_\alpha(x), S_\alpha, \eta_\alpha(x) : \alpha \in \mathscr{A}, x \in E\}, \tag{4.2}$$

where each $\xi_\alpha(x) = \{\xi_\alpha(x, t) : t \ge 0\}$ is a Markov process with transition semigroup $(P_t)_{t \ge 0}$ and $\xi_\alpha(x, 0) = x$, each S_α is an exponential random variable with parameter γ, and each $\eta_\alpha(x)$ is an integer-valued random variable with distribution defined by the probability generating function $g(x, \cdot)$.

Given a finite set $\{x_1, \ldots, x_n\} \subset E$, the branching particle system $\{X_t^\sigma : t \ge 0\}$ with initial state $\sigma = \sum_{i=1}^n \delta_{x_i}$ is constructed as follows. Let ∂ be a point that is not in E. For all labels $\alpha \in \mathscr{A}$ we shall define the birth times β_α, birth places b_α, death times ζ_α and trajectories $\xi_\alpha = \{\xi_\alpha(t) : t \ge 0\}$ of the corresponding particles in an inductive way. If $\alpha = n_0 \in \mathscr{A}$ has generation number $l(\alpha) = 0$, we define the birth time and place by

$$\beta_{n_0} = \begin{cases} 0 & \text{if } n_0 \le n \\ \infty & \text{if } n_0 > n \end{cases} \quad \text{and} \quad b_{n_0} = \begin{cases} x_{n_0} & \text{if } n_0 \le n \\ \partial & \text{if } n_0 > n. \end{cases}$$

The death time is defined by $\zeta_{n_0} = \beta_{n_0} + S_{n_0}$ and the trajectory by

$$\xi_{n_0}(t) = \begin{cases} \xi_{n_0}(b_{n_0}, t - \beta_{n_0}) & \text{if } \beta_{n_0} \le t < \infty \\ \partial & \text{if } 0 \le t < \beta_{n_0}. \end{cases}$$

Suppose now the birth times, birth places, death times and trajectories are already defined for the particles in the $(k-1)$-th generation. For $\alpha = n_0 n_1 \cdots n_{l(\alpha)} \in \mathscr{A}$ with generation number $l(\alpha) = k \geq 1$, we define the birth time and place of the particle by

$$\beta_\alpha = \begin{cases} \zeta_{(\alpha\backslash 1)} & \text{if } n_k \leq \eta_{(\alpha\backslash 1)}(\xi_{(\alpha\backslash 1)}(\zeta_{(\alpha\backslash 1)}-)) \\ \infty & \text{if } n_k > \eta_{(\alpha\backslash 1)}(\xi_{(\alpha\backslash 1)}(\zeta_{(\alpha\backslash 1)}-)) \end{cases}$$

and

$$b_\alpha = \begin{cases} \xi_{(\alpha\backslash 1)}(\zeta_{(\alpha\backslash 1)}) & \text{if } n_k \leq \eta_{(\alpha\backslash 1)}(\xi_{(\alpha\backslash 1)}(\zeta_{(\alpha\backslash 1)}-)) \\ \partial & \text{if } n_k > \eta_{(\alpha\backslash 1)}(\xi_{(\alpha\backslash 1)}(\zeta_{(\alpha\backslash 1)}-)). \end{cases}$$

The death time of the particle is defined by $\zeta_\alpha = \beta_\alpha + S_\alpha$ and the trajectory by

$$\xi_\alpha(t) = \begin{cases} \xi_\alpha(b_\alpha, t - \beta_\alpha) & \text{if } \beta_\alpha \leq t < \infty \\ \partial & \text{if } 0 \leq t < \beta_\alpha. \end{cases}$$

Now the branching particle system generated by the initial mass σ is constructed as

$$X_t^\sigma = \sum_{\alpha \in \mathscr{A}} 1_{[\beta_\alpha, \zeta_\alpha)}(t)\delta_{\xi_\alpha(t)}, \qquad t \geq 0. \tag{4.3}$$

At this moment, it is not clear if $X_t^\sigma(E)$ is a.s. finite, so we think of $\{X_t^\sigma : t \geq 0\}$ as a process taking values from $N(E) \cup \{\Delta\}$, where Δ denotes the infinity. The independence of the family (4.2) implies

$$\mathbf{P} \exp\{-X_t^\sigma(f)\} = \exp\{-\sigma(u_t)\}, \qquad f \in B(E)^{++}, \tag{4.4}$$

where

$$u_t(x) \equiv u_t(x, f) = -\log \mathbf{P} \exp\{-X_t^x(f)\}$$

and $\{X_t^x : t \geq 0\}$ is a system with $X_0^x = \delta_x$. Here and in the sequel, we make the convention that $\Delta(f) = \infty$ for $f \in B(E)^{++}$. Moreover, we have the following renewal equation:

$$e^{-u_t(x)} = \mathbf{P}_x[e^{-\gamma t - f(\xi_t)}] + \int_0^t \gamma e^{-\gamma s} \mathbf{P}_x[g(\xi_s, e^{-u_{t-s}(\xi_s)})]ds. \tag{4.5}$$

This follows as we think about the Laplace functional of the random measure X_t^x produced by a single particle that starts moving from the point $x \in E$. Suppose the particle is labeled by $\alpha \in \mathscr{A}$ with $l(\alpha) = 0$. Then it has birth time $\beta_\alpha = 0$, birth place $b_\alpha = x$, death time $\zeta_\alpha = S_\alpha$ and trajectory $\xi_\alpha = \{\xi_\alpha(x, t) : t \geq 0\}$. In view of (4.4), the Laplace functional of X_t^x is given by the left-hand side of (4.5). By the independence of ζ_α and ξ_α we have

$$\mathbf{P}[1_{\{\zeta_\alpha > t\}}e^{-X_t^x(f)}] = \mathbf{P}\{1_{\{\zeta_\alpha > t\}}\mathbf{P}[e^{-X_t^x(f)}|\zeta_\alpha, \xi_\alpha]\}$$

$$= \mathbf{P}\left\{1_{\{\zeta_\alpha > t\}} e^{-f(\xi_\alpha(x,t))}\right\} = \mathbf{P}_x[e^{-\gamma t - f(\xi_t)}],$$

where the expectations related to $\{X_t^x : t \geq 0\}$ and $\{\xi_\alpha(x,t) : t \geq 0\}$ are taken on the probability space $(W, \mathscr{G}, \mathbf{P})$ and the one related to $\{\xi_t : t \geq 0\}$ is taken on $(\Omega, \mathscr{F}, \mathbf{P}_x)$. That gives the first term on the right-hand side of (4.5). By property (3), if it happens that $0 < \zeta_\alpha \leq t$, then the particle dies at $\xi_\alpha(x, \zeta_\alpha-)$ and gives birth to a random number of offspring at $\xi_\alpha(x, \zeta_\alpha)$ according to the probability law given by the generating function $g(\xi_\alpha(x, \zeta_\alpha-), \cdot)$. With those considerations we compute

$$\mathbf{P}\left[1_{\{0 < \zeta_\alpha \leq t\}} e^{-X_t^x(f)}\right]$$
$$= \mathbf{P}\left\{1_{\{0 < \zeta_\alpha \leq t\}} \mathbf{P}\left[e^{-X_t^x(f)}\big|\zeta_\alpha, \xi_\alpha\right]\right\}$$
$$= \mathbf{P}\left[1_{\{0 < \zeta_\alpha \leq t\}} \sum_{k=1}^\infty p_k(\xi_\alpha(x, \zeta_\alpha-))e^{-ku_{t-\zeta_\alpha}(\xi_\alpha(x,\zeta_\alpha))}\right]$$
$$= \mathbf{P}\left[1_{\{0 < \zeta_\alpha \leq t\}} g\big(\xi_\alpha(x, \zeta_\alpha-), e^{-u_{t-\zeta_\alpha}(\xi_\alpha(x,\zeta_\alpha))}\big)\right]$$
$$= \mathbf{P}_x\left[\int_0^t \gamma e^{-\gamma s} g\big(\xi_{s-}, e^{-u_{t-s}(\xi_s)}\big) ds\right],$$

where we used again the independence of ζ_α and $\{\xi_\alpha(x,t) : t \geq 0\}$. That leads to the second term on the right-hand side of (4.5) because $\xi_{s-} \neq \xi_s$ for at most countably many $s > 0$. By Proposition 2.9 the solution $(t, x) \mapsto e^{-u_t(x)}$ of (4.5) also solves

$$e^{-u_t(x)} = \mathbf{P}_x e^{-f(\xi_t)} + \int_0^t \gamma \mathbf{P}_x\left[g(\xi_s, e^{-u_{t-s}(\xi_s)}) - e^{-u_{t-s}(\xi_s)}\right] ds. \qquad (4.6)$$

Then it is easy to see that

$$v(t, x) = 1 - \exp\{-u_t(x)\} \qquad (4.7)$$

is a solution of

$$v(t, x) = \mathbf{P}_x\left[1 - e^{-f(\xi_t)}\right] + \int_0^t \gamma \mathbf{P}_x[1 - v(t - s, \xi_s)] ds$$
$$- \int_0^t \gamma \mathbf{P}_x[g(\xi_s, 1 - v(t - s, \xi_s))] ds. \qquad (4.8)$$

The arguments above actually proved the existence of the solution $(t, x) \mapsto v(t, x)$ to (4.8). The uniqueness of the solution follows by a standard application of Gronwall's inequality.

Proposition 4.1 *We have* $\mathbf{P}\{X_t^\sigma \in N(E)\} = 1$ *and*

$$\mathbf{P}[X_t^\sigma(f)] = \sigma(P_t^b f), \qquad t \geq 0, f \in B(E), \qquad (4.9)$$

where $P_t^b f$ *is defined by* (2.34) *with* $b(x) = \gamma[1 - g_z'(x, 1-)]$ *for* $x \in E$.

Proof. We first assume $f \in B(E)^{++}$. For any $k \geq 1$ we have

$$w_k(t, x) := \mathbf{P}[1 - e^{-X_t^x(f/k)}] = 1 - e^{-u_t(x, f/k)}. \tag{4.10}$$

From (4.8) we get

$$w_k(t, x) = \mathbf{P}_x\left[1 - e^{-f(\xi_t)/k}\right] - \int_0^t \mathbf{P}_x[\phi(\xi_s, w_k(t - s, \xi_s))]ds, \tag{4.11}$$

where

$$\phi(x, z) = \gamma[g(x, 1 - z) - (1 - z)].$$

It is easy to see

$$\phi(x, z) \geq \phi'(x, 0+)z = b(x)z, \qquad x \in E, 0 \leq z \leq 1.$$

Then (4.11) implies

$$\|w_k(t, \cdot)\| \leq k^{-1}\|f\| + \|b\| \int_0^t \|w_k(s, \cdot)\|ds,$$

so an application of Gronwall's inequality shows $\|w_k(t, \cdot)\| \leq k^{-1}\|f\|e^{\|b\|t}$. From (4.10) one sees that $kw_k(t, x)$ is increasing in $k \geq 1$. Thus the limit $\pi_t(x) := \lim_{k \to \infty} kw_k(t, x)$ exists and solves

$$\pi_t(x) = P_t f(x) - \int_0^t ds \int_E b(y)\pi_s(y)P_{t-s}(x, dy),$$

so $\pi_t(x) = P_t^b f(x)$. From (4.10) it follows that

$$\mathbf{P}[X_t^x(f)] = \lim_{k \to \infty} \mathbf{P}[k(1 - e^{-X_t^x(f/k)})] = P_t^b f(x).$$

Similarly, from (4.4) we get (4.9) for $f \in B(E)^{++}$. In particular,

$$\mathbf{P}[X_t^\sigma(1)] = \sigma(P_t^b 1) < \infty,$$

which implies $\mathbf{P}\{X_t^\sigma \in N(E)\} = 1$. The extension of (4.9) to $f \in B(E)$ is immediate. □

From the assumptions on the family (4.2) it follows that $\{X_t^\sigma : t \geq 0\}$ is a Markov process in $N(E)$. By (4.4) the process has transition semigroup $(Q_t)_{t \geq 0}$ given by

$$\int_{N(E)} e^{-\nu(f)}Q_t(\sigma, d\nu) = \exp\{-\sigma(u_t)\}, \qquad f \in B(E)^+, \tag{4.12}$$

where $(t, x) \mapsto u_t(x)$ is determined by (4.7) and (4.8).

4.2 Scaling Limits of Local Branching Systems

In this section, we prove a scaling limit theorem for a sequence of branching particle systems, which leads to a superprocess with local branching. The limit theorem gives interpretations for the parameters of the superprocess. For each integer $k \geq 1$, let $\{Y_k(t) : t \geq 0\}$ be a branching particle system with parameters (ξ, γ_k, g_k) and let $X_k(t) = k^{-1}Y_k(t)$. It is easy to see that $\{X_k(t) : t \geq 0\}$ is a Markov process in

$$N_k(E) := \{\nu \in M(E) : k\nu \in N(E)\}.$$

Let $Q_\nu^{(k)}$ denote the conditional law given $X_k(0) = \nu \in N_k(E)$. Let $(t, x, f) \mapsto u_t(k, x, f)$ be defined by (4.7) and (4.8) with (γ, g) replaced by (γ_k, g_k). From (4.12) we have

$$Q_\nu^{(k)} \exp\left\{ - \langle X_k(t), f \rangle \right\} = \exp\left\{ - \langle \nu, u_k(t) \rangle \right\}, \tag{4.13}$$

where $u_k(t, x) = k u_t(k, x, f/k)$. By the discussions in the first section, we can also define $u_k(t, x)$ by

$$v_k(t, x) = k[1 - \exp\{-u_k(t, x)/k\}] \tag{4.14}$$

and the evolution equation

$$v_k(t, x) = k\mathbf{P}_x\left[1 - e^{-f(\xi_t)/k}\right] - \int_0^t \mathbf{P}_x\left[\phi_k(\xi_s, v_k(t - s, \xi_s))\right]ds, \tag{4.15}$$

where

$$\phi_k(x, z) = k\gamma_k[g_k(x, 1 - z/k) - (1 - z/k)], \qquad 0 \leq z \leq k. \tag{4.16}$$

Condition 4.2 For each $a \geq 0$ the sequence $\{\phi_k(x, z)\}$ is Lipschitz with respect to z uniformly on $E \times [0, a]$ and $\phi_k(x, z)$ converges to some $\phi(x, z)$ uniformly on $E \times [0, a]$ as $k \to \infty$.

Proposition 4.3 If Condition 4.2 is satisfied, the limit function ϕ has the representation (2.45).

Proof. By applying Corollary 1.46 for fixed $x \in E$ we get the representation (2.45), where $c(x) \geq 0$ and $b(x)$ are constants, and $(u \wedge u^2)m(x, du)$ is a finite measure on $(0, \infty)$. By dominated convergence we can differentiate both sides of (2.45) with respect to $z \geq 0$ to obtain

$$\phi'(x, z) = b(x) + 2c(x)z + \int_0^\infty u\left(1 - e^{-zu}\right)m(x, du), \tag{4.17}$$

which is bounded on the set $E \times [0, a]$ for every $a \geq 0$. Then $\phi(x, z)$ is Lipschitz in z uniformly on $E \times [0, a]$ for each $a \geq 0$. In particular, $x \mapsto b(x) = \phi'(x, 0+)$ is bounded on E. By taking $z = 1$ in (4.17) we see that

$$x \mapsto c(x) + \int_0^\infty (u \wedge u^2) m(x, du)$$

is bounded on E. By applications of dominated convergence again, for any $a > 0$ and $\lambda \geq 0$ we have

$$\phi''(x, a + \lambda) = 2c(x) + \int_0^\infty e^{-\lambda u} u^2 e^{-au} m(x, du) \qquad (4.18)$$

and

$$\phi^{(3)}(x, a + \lambda) = -\int_0^\infty e^{-\lambda u} u^3 e^{-au} m(x, du).$$

Then the finite measure $u^3 e^{-au} m(x, du)$ has Laplace transform $\lambda \mapsto -\phi^{(3)}(x, a + \lambda)$. By Theorem 1.22 we see $u^3 e^{-au} m(x, du)$ is a finite kernel from E to $(0, \infty)$, so $m(x, du)$ is a σ-finite kernel from E to $(0, \infty)$. Now (4.18) implies that $x \mapsto c(x)$ is measurable and hence (4.17) implies $x \mapsto b(x)$ is measurable. $\qquad \square$

Proposition 4.4 *For any function ϕ with the representation (2.45) there is a sequence $\{\phi_k\}$ in the form of (4.16) satisfying Condition 4.2.*

Proof. To simplify the formulations we decompose the function ϕ into two parts. Let $\phi_0(x, z) = \phi(x, z) - b(x)z$. We first define

$$\gamma_{0,k} = 1 + 2k\|c\| + \sup_{x \in E} \int_0^\infty u(1 - e^{-ku}) m(x, du)$$

and

$$g_{0,k}(x, z) = z + k^{-1} \gamma_{0,k}^{-1} \phi_0(x, k(1 - z)), \quad x \in E, |z| \leq 1.$$

It is easy to see that $z \mapsto g_{0,k}(x, z)$ is an analytic function in $(-1, 1)$ satisfying $g_{0,k}(x, 1) = 1$ and

$$\frac{d^n}{dz^n} g_{0,k}(x, 0) \geq 0, \qquad x \in E, n \geq 0.$$

Therefore $g_{0,k}(x, \cdot)$ is a probability generating function. Let $\phi_{0,k}$ be defined by (4.16) with (γ_k, g_k) replaced by $(\gamma_{0,k}, g_{0,k})$. Then $\phi_{0,k}(x, z) = \phi_0(x, z)$ for $0 \leq z \leq k$. That completes the proof if $\|b\| = 0$. In the case $\|b\| > 0$, we set

$$g_{1,k}(x, z) = \frac{1}{2}\left(1 + \frac{b(x)}{\|b\|}\right) + \frac{1}{2}\left(1 - \frac{b(x)}{\|b\|}\right) z^2.$$

Let $\gamma_{1,k} = \|b\|$ and let $\phi_{1,k}(x, z)$ be defined by (4.16) with (γ_k, g_k) replaced by $(\gamma_{1,k}, g_{1,k})$. Then we have

$$\phi_{1,k}(x, z) = b(x)z + \frac{1}{2k}[\|b\| - b(x)]z^2.$$

Finally, let $\gamma_k = \gamma_{0,k} + \gamma_{1,k}$ and $g_k = \gamma_k^{-1}(\gamma_{0,k}g_{0,k} + \gamma_{1,k}g_{1,k})$. Then the sequence $\phi_k(x, z)$ defined by (4.16) is equal to $\phi_{0,k}(x, z) + \phi_{1,k}(x, z)$ which satisfies the required condition. $\qquad\square$

Proposition 4.5 *If Condition 4.2 holds, for each $T \geq 0$ both $v_k(t, x)$ and $u_k(t, x)$ converge uniformly on $[0, T] \times E$ to the unique locally bounded positive solution $(t, x) \mapsto V_t f(x)$ of the evolution equation*

$$V_t f(x) = P_t f(x) - \int_0^t \mathrm{d}s \int_E \phi(y, V_{t-s}f(y))P_s(x, \mathrm{d}y). \qquad (4.19)$$

Proof. By Proposition 2.20, there is a unique locally bounded positive solution $t \mapsto V_t f$ to (4.19). Let

$$b_k(x) := \frac{\mathrm{d}}{\mathrm{d}z}\phi_k(x, 0+) = \gamma_k\left[1 - \frac{\mathrm{d}}{\mathrm{d}z}g_k(x, 1-)\right], \qquad x \in E.$$

The uniformly local Lipschitz assumption on $\{\phi_k\}$ implies

$$B := \sup_{k \geq 1} \|b_k\| < \infty.$$

We may then extend the definition of ϕ_k by setting

$$\phi_k(x, z) = \phi_k(x, k) + \frac{\mathrm{d}}{\mathrm{d}z}\phi_k(x, k-)(z - k), \qquad x \in E, z > k.$$

Since $z \mapsto \phi_k(x, z)$ is a convex function, we have $-\phi_k(x, z) \leq Bz$ for all $z \geq 0$. Then the convergence of $v_k(t, x, f)$ is true by Proposition 2.16. The convergence of $u_k(t, x, f)$ follows by the relation (4.14). $\qquad\square$

Let $\mu \in M(E)$ and let $\mathbf{Q}_{(\mu)}^{(k)}$ denote the conditional law given that $Y_k(0) = kX_k(0)$ is a Poisson random measure on E with intensity $k\mu$. By (4.13) and Theorem 1.25 it is not hard to show that

$$\mathbf{Q}_{(\mu)}^{(k)} \exp\left\{-\langle X_k(t), f\rangle\right\} = \exp\left\{-\langle \mu, v_k(t)\rangle\right\}. \qquad (4.20)$$

By Theorem 2.21 the solution of (4.19) defines a cumulant semigroup $(V_t)_{t \geq 0}$.

Theorem 4.6 *If Condition 4.2 is satisfied, the finite-dimensional distributions of $\{X_k(t) : t \geq 0\}$ under $\mathbf{Q}_{(\mu)}^{(k)}$ converge to those of the (ξ, ϕ)-superprocess $\{X_t : t \geq 0\}$ with initial value $X_0 = \mu$.*

Proof. Let $\mathbf{Q}_\mu$ denote the conditional law of the (ξ, ϕ)-superprocess $\{X_t : t \geq 0\}$ given $X_0 = \mu$. To get the desired convergence of the finite-dimensional distributions of $\{X_k(t) : t \geq 0\}$ it suffices to prove

$$\lim_{k \to \infty} \mathbf{Q}_{(\mu)}^{(k)} \exp\left\{-\sum_{i=1}^n \langle X_k(t_i), f_i\rangle\right\} = \mathbf{Q}_\mu \exp\left\{-\sum_{i=1}^n \langle X_{t_i}, f_i\rangle\right\} \qquad (4.21)$$

for all $\{t_1 < \cdots < t_n\} \subset [0, \infty)$ and $\{f_1, \ldots, f_n\} \subset B(E)^+$ and use Theorem 1.24. For $n = 1$ this follows by Proposition 4.5. Now suppose (4.21) holds when n is replaced by $n-1$. For any $n \geq 2$ the Markov property of $\{X_k(t) : t \geq 0\}$ implies that

$$Q_{(\mu)}^{(k)} \exp \left\{ -\sum_{i=1}^{n} \langle X_k(t_i), f_i \rangle \right\}$$

$$= Q_{(\mu)}^{(k)} \exp \left\{ -\sum_{i=1}^{n-1} \langle X_k(t_i), f_i \rangle - \langle X_k(t_{n-1}), u_k(\Delta t_n) \rangle \right\}, \quad (4.22)$$

where $\Delta t_n = t_n - t_{n-1}$. Let $B \geq 0$ and $b_k \in B(E)$ be defined as in the proof of Proposition 4.5. For $f \in B(E)^+$ one uses Proposition 4.1 and a property of the Poisson random measure to see

$$Q_{(\mu)}^{(k)}[\langle X_k(t), f \rangle] = Q_{(\mu)}^{(k)}[\langle Y_k(0), P_t^{b_k}(f/k) \rangle] = \langle \mu, P_t^{b_k} f \rangle \leq e^{Bt} \langle \mu, P_t f \rangle.$$

It then follows that

$$Q_{(\mu)}^{(k)} \left[\exp \left\{ -\sum_{i=1}^{n-1} \langle X_k(t_i), f_i \rangle \right\} \middle| \exp \left\{ -\langle X_k(t_{n-1}), u_k(\Delta t_n) \rangle \right\} \right.$$

$$\left. - \exp \left\{ -\langle X_k(t_{n-1}), V_{\Delta t_n} f_n \rangle \right\} \middle| \right]$$

$$\leq Q_{(\mu)}^{(k)} \left[\left\langle X_k(t_{n-1}), |u_k(\Delta t_n) - V_{\Delta t_n} f_n| \right\rangle \right]$$

$$\leq e^{Bt_{n-1}} \left\langle \mu, P_{t_{n-1}} |u_k(\Delta t_n) - V_{\Delta t_n} f_n| \right\rangle.$$

By Proposition 4.5, the right-hand side goes to zero as $k \to \infty$. Since (4.21) holds when n is replaced by $n-1$, we have

$$\lim_{k \to \infty} Q_{(\mu)}^{(k)} \exp \left\{ -\sum_{i=1}^{n} \langle X_k(t_i), f_i \rangle \right\}$$

$$= \lim_{k \to \infty} Q_{(\mu)}^{(k)} \exp \left\{ -\sum_{i=1}^{n-1} \langle X_k(t_i), f_i \rangle - \langle X_k(t_{n-1}), V_{\Delta t_n} f_n \rangle \right\}$$

$$= Q_\mu \exp \left\{ -\sum_{i=1}^{n-1} \langle X_{t_i}, f_i \rangle - \langle X_{t_{n-1}}, V_{\Delta t_n} f_n \rangle \right\}$$

$$= Q_\mu \exp \left\{ -\sum_{i=1}^{n} \langle X_{t_i}, f_i \rangle \right\},$$

so (4.21) follows by induction in $n \geq 1$. $\qquad\square$

The above theorem gives the heuristical interpretations for the parameters of the superprocess. That is, the process ξ gives the law of migration of the "particles" and ϕ arises from the branching rate and the generating function determining the distribution of the offspring production.

In the remainder of this section, we consider the special case where (E, d) is a complete separable metric space. Let $D_E := D([0, \infty), E)$ be the space of càdlàg paths from $[0, \infty)$ to E. We fix a metric q on D_E for the Skorokhod topology. By definition, a *stopped path* is a pair (w, z), where $z \geq 0$ and $w \in D_E$ satisfies $w(t) = w(t \wedge z)$ for all $t \geq 0$. Let S be the set of all stopped paths and let ρ be the metric on S defined by

$$\rho((w_1, z_1), (w_2, z_2)) = |z_1 - z_2| + q(w_1, w_2).$$

Then (S, ρ) is a complete separable metric space. Let $\xi = (\Omega, \mathscr{F}, \mathscr{F}_t, \xi_t, \mathbf{P}_x)$ be a càdlàg Borel right process in E satisfying:

Condition 4.7 *For every $\varepsilon > 0$ we have*

$$\lim_{t \to 0} \sup_{x \in E} \mathbf{P}_x \left\{ \sup_{0 \leq s \leq t} d(x, \xi_s) \geq \varepsilon \right\} = 0.$$

Let $(u, y) \in S$ and let $b \geq a \geq 0$ be constants satisfying $a \leq y$. With those parameters we define a probability measure $R_{a,b}((u, y), \mathrm{d}(w, z))$ on S by the following prescriptions:

(1) $R_{a,b}((u, y), \mathrm{d}(w, z))$-a.s. $z = b$ and $w(t) = u(t)$ for $0 \leq t \leq a$;
(2) the law of $\{w(a + t) : 0 \leq t \leq b - a\}$ under $R_{a,b}((u, y), \mathrm{d}(w, z))$ coincides with that of $\{\xi_t : 0 \leq t \leq b - a\}$ under $\mathbf{P}_{u(a)}$.

For $s > 0$ and $y \geq 0$ let $\beta = \{\beta_t : t \geq 0\}$ be a reflecting Brownian motion starting at $\beta_0 = y$ and let $\gamma_s^y(\mathrm{d}a, \mathrm{d}b)$ be the distribution of $(\inf_{0 \leq r \leq s} \beta_r, \beta_s)$ on $\mathbb{R}_+^2$. The reflection principle gives that

$$\gamma_s^y(\mathrm{d}a, \mathrm{d}b) = \frac{2(y + b - 2a)}{\sqrt{2\pi s^3}} \exp\left\{ -\frac{(y + b - 2a)^2}{2s} \right\} 1_{\{0 < a < b \wedge y\}} \mathrm{d}a \, \mathrm{d}b$$
$$+ \frac{2}{\sqrt{2\pi s}} \exp\left\{ -\frac{(y + b)^2}{2s} \right\} 1_{\{0 < b\}} \delta_0(\mathrm{d}a) \mathrm{d}b.$$

A Markov process with state space S is called a ξ-*Brownian snake* if it has transition semigroup $(Q_s)_{s \geq 0}$ defined by

$$Q_s((u, y), \mathrm{d}(w, z)) = \int_{\mathbb{R}_+^2} \gamma_s^y(\mathrm{d}a, \mathrm{d}b) R_{a,b}((u, y), \mathrm{d}(w, z)). \qquad (4.23)$$

The semigroup property of $(Q_s)_{s \geq 0}$ follows from the Markov properties of the processes β and ξ. Suppose that $\{(\eta_s, \zeta_s) : s \geq 0\}$ is a realization of the ξ-Brownian snake. Then, heuristically, the process $\{\eta_s(t) : t \geq 0\}$ is a realization of $\{\xi_{t \wedge \zeta_s} : t \geq 0\}$ ending at time $\zeta_s \geq 0$. The ending time $\{\zeta_s : s \geq 0\}$ evolves as a reflecting Brownian. When ζ_s decreases, the path $\{\eta_s(t) : 0 \leq t \leq \zeta_s\}$ is erased from its final point, and when ζ_s increases this path is extended according to the law of ξ.

Example 4.1 Let $x \in E$ and $\{(\eta_s, \zeta_s) : s \geq 0\}$ be a ξ-Brownian snake started with $(w_0, 0)$, where $w_0(t) = x$ for all $t \geq 0$. Then $\eta_s(0) = x$ for all $s \geq 0$ and $\{\zeta_s : s \geq 0\}$ is a reflecting Brownian motion with $\zeta_0 = 0$. Using Condition 4.7 it is easy to show $s \mapsto (\eta_s, \zeta_s)$ is right continuous in probability at $s = 0$. Then the process $\{(\eta_s, \zeta_s) : s \geq 0\}$ is continuous in probability by Lemma 1 of Le Gall (1999, p.56). Let $\{2l_s(y) : s \geq 0, y \geq 0\}$ be the local time of $\{\zeta_s : s \geq 0\}$. For any $u \geq 0$ let

$$\sigma(u) = \inf\{s \geq 0 : l_s(0) \geq u\}.$$

For $k \geq 1$ let $[a_1(t), b_1(t)], \ldots, [a_{n_t}(t), b_{n_t}(t)]$ be the excursion intervals of the stopped path $\{\zeta_s : 0 \leq s \leq \sigma(u)\}$ above $t \geq 0$ with height $> 1/k$. This means that $[a_i(t), b_i(t)] \subset [0, \sigma(u)]$, $\zeta_{a_i(t)} = \zeta_{b_i(t)} = t$, $\zeta_s > t$ for $a_i(t) < s < b_i(t)$ and

$$\sup\{\zeta_s : a_i(t) < s < b_i(t)\} > t + 1/k.$$

By an observation of Le Gall (1993), the measure-valued process

$$Y_k(t) = \sum_{i=1}^{n_t} \delta_{\eta_{a_i(t)}(t)}, \qquad t \geq 0$$

is a branching particle system with parameters $(\xi, g, 2k)$ in the sense of Section 4.1, where $g(z) = 1/2 + z^2/2$. Note that n_0 is a Poisson random variable with mean ku by Itô's excursion theory. Then the continuity in probability of the process $s \mapsto \eta_s$ and the approximation of the Brownian local time by upcrossing numbers imply $k^{-1}Y_k(t) \to X_t$ as $k \to \infty$ for a random measure X_t on E defined by

$$X_t(f) = \int_0^{\sigma(u)} f(\eta_s(t)) \mathrm{d}l_s(t), \qquad f \in B(E), \tag{4.24}$$

where $\mathrm{d}l_s(t)$ denotes the integration with respect to the increasing function $s \mapsto l_s(t)$. By Theorem 4.6 we conclude $\{X_t : t \geq 0\}$ is a Dawson–Watanabe superprocess with spatial motion ξ and binary local branching mechanism $\phi(z) = z^2$. This gives a representation of the superprocess in terms of the ξ-Brownian snake. When E is a singleton, the representation reduces to the first result of Theorem 3.45.

4.3 General Branching Particle Systems

In this section, we consider a model of branching particle systems that generalizes the system introduced in the first section. The high-density limits of these systems will lead to Dawson–Watanabe superprocesses with decomposable branching mechanisms. Let E be a Lusin topological space and let $N(E)$ denote the space of integer-valued finite measures on E. Let $\xi = (\Omega, \mathscr{F}, \mathscr{F}_t, \xi_t, \mathbf{P}_x)$ be a Borel right process with state space E and conservative transition semigroup $(P_t)_{t \geq 0}$. We assume the sample path $\{\xi_t : t \geq 0\}$ is right continuous in both the original and the Ray topologies and has left limits $\{\xi_{t-} : t > 0\}$ in the Ray–Knight completion $\bar{E}$. Let $\{K(t) : t \geq 0\}$ be a continuous admissible additive functional of ξ. Let $\alpha \in B(E)^+$ and let $F(x, d\nu)$ be a Markov kernel from E to $N(E)$ such that

$$\sup_{x \in E} \int_{N(E)} \nu(1) F(x, d\nu) < \infty. \tag{4.25}$$

For $x \in \bar{E} \setminus E$ let $F(x, d\nu)$ be the unit mass at δ_x. A general branching particle system is characterized by the following properties:

(1) The particles in E move independently according to the law given by the transition probabilities of ξ.
(2) For a particle which is alive at time $r \geq 0$ and follows the path $\{\xi_s : s \geq r\}$, the conditional probability of survival in the time interval $[r, t)$ is $\exp\{-\int_r^t \alpha(\xi_s) K(ds)\}$.
(3) When a particle following the path $\{\xi_s : s \geq r\}$ dies at time $t > r$, it gives birth to a random number of offspring in E according to the probability kernel $F(\xi_{t-}, d\nu)$. The offspring then start to move from their birth places.

We also assume that the lifetimes and the branchings of different particles are independent. Let $X_t(B)$ denote the number of particles in $B \in \mathcal{B}(E)$ that are alive at time $t \geq 0$. If we assume $X_0(E) < \infty$, then $\{X_t : t \geq 0\}$ is a Markov process with state space $N(E)$, which will be referred to as the *general branching particle system* with parameters (ξ, K, α, F). We are not going to give the rigorous construction of the general branching system, which involves the same ideas as the construction in the first section but is considerably more complicated.

Let $\sigma \in N(E)$ and let $\{X_t^\sigma : t \geq 0\}$ be a general branching particle system with parameters (ξ, K, α, F) and initial state $X_0 = \sigma$. Suppose that the process is defined on the probability space $(W, \mathcal{G}, \mathbf{P})$. The above properties imply

$$\mathbf{P} \exp\{-X_t^\sigma(f)\} = \exp\{-\sigma(u_t)\}, \qquad f \in B(E)^+, \tag{4.26}$$

where $u_t(x) \equiv u_t(x, f)$ is determined by the renewal equation

$$e^{-u_t(x)} = \mathbf{P}_x \exp\left\{ - f(\xi_t) - \int_0^t \alpha(\xi_s) K(ds) \right\}$$
$$+ \mathbf{P}_x \left[\int_0^t e^{-\int_0^s \alpha(\xi_r) dr} \alpha(\xi_s) K(ds) \int_{N(E)} e^{-\nu(u_{t-s})} F(\xi_s, d\nu) \right].$$

This equation is derived by arguments similar to those used for (4.5). By Proposition 2.9 the above equation implies

$$e^{-u_t(x)} = \mathbf{P}_x e^{-f(\xi_t)} - \mathbf{P}_x \left[\int_0^t \alpha(\xi_s) e^{-u_{t-s}(\xi_s)} K(ds) \right]$$
$$+ \mathbf{P}_x \left[\int_0^t \alpha(\xi_s) K(ds) \int_{N(E)} e^{-\nu(u_{t-s})} F(\xi_s, d\nu) \right]. \tag{4.27}$$

For the general branching particle system, it is natural to treat separately from others the offspring that start their migration from the death sites of their parents. To this end, we need to introduce some additional parameters as follows. Let $g \in B(E \times [-1, 1])$ be such that for each $x \in E$,

$$g(x, z) = \sum_{i=0}^{\infty} p_i(x) z^i, \qquad |z| \leq 1,$$

is a probability generating function with $\sup_x g_z'(x, 1-) < \infty$. Recall that $P(E)$ denotes the space of probability measures on E. Let $G(x, d\pi)$ be a probability kernel from E to $P(E)$ and let $h \in B(E \times P(E) \times [-1, 1])$ be such that for each $(x, \pi) \in E \times P(E)$,

$$h(x, \pi, z) = \sum_{i=0}^{\infty} q_i(x, \pi) z^i, \qquad |z| \leq 1,$$

is a probability generating function with $\sup_{x, \pi} h_z'(x, \pi, 1-) < \infty$. Now we can define the probability kernels $F_0(x, d\nu)$ and $F_1(x, d\nu)$ from E to $N(E)$ by

$$\int_{N(E)} e^{-\nu(f)} F_0(x, d\nu) = \sum_{i=0}^{\infty} p_i(x) e^{-if(x)} = g(x, e^{-f(x)})$$

and

$$\int_{N(E)} e^{-\nu(f)} F_1(x, d\nu) = \int_{P(E)} h(x, \pi, \pi(e^{-f})) G(x, d\pi).$$

Suppose we have the decomposition $\alpha(x) = \gamma(x) + \rho(x)$ for $\gamma, \rho \in B(E)^+$. Let

$$F(x, d\nu) = \frac{1}{\alpha(x)} \left[\gamma(x) F_0(x, d\nu) + \rho(x) F_1(x, d\nu) \right] \qquad (4.28)$$

if $\alpha(x) > 0$ and let $F(x, d\nu) = F_0(x, d\nu)$ if $\alpha(x) = 0$. For the kernel $F(x, d\nu)$ given by (4.28), the general branching particle system is determined by the parameters $(\xi, K, \gamma, g, \rho, h, G)$. Intuitively, as a particle dies at $x \in E$, the branching is of local type with probability $\gamma(x)/\alpha(x)$ and is of non-local type with probability $\rho(x)/\alpha(x)$. If the local branching type is chosen, the particle gives birth to a number of offspring at its death site x according to the distribution $\{p_i(x)\}$. If non-local branching occurs, an offspring-location-distribution $\pi \in P(E)$ is first selected according to the probability kernel $G(x, d\pi)$, the particle then gives birth to a random number of offspring according to the distribution $\{q_i(x, \pi)\}$, and those offspring choose their locations in E independently of each other according to the distribution $\pi(dy)$. Therefore the locations of non-locally displaced offspring involve two sources of randomness. From (4.27) and (4.28) we obtain

$$e^{-u_t(x)} = \mathbf{P}_x e^{-f(\xi_t)} - \mathbf{P}_x\left[\int_0^t \alpha(\xi_s)e^{-u_{t-s}(\xi_s)}K(ds)\right]$$

$$+ \mathbf{P}_x\left[\int_0^t \rho(\xi_s)K(ds)\int_{P(E)} h(\xi_s, \pi, \pi(e^{-u_{t-s}}))G(\xi_s, d\pi)\right]$$

$$+ \mathbf{P}_x\left[\int_0^t \gamma(\xi_s)g(\xi_s, e^{-u_{t-s}(\xi_s)})K(ds)\right].$$

Setting $v(t, x) = 1 - \exp\{-u_t(x)\}$ we have

$$v(t, x) = \mathbf{P}_x\left[1 - e^{-f(\xi_t)}\right] - \mathbf{P}_x\left[\int_0^t \alpha(\xi_s)v(t - s, \xi_s)K(ds)\right]$$

$$+ \mathbf{P}_x\left\{\int_0^t \rho(\xi_s)\int_{P(E)} \left[1 - h(\xi_s, \pi, \pi(e^{-u_{t-s}}))\right]G(\xi_s, d\pi)K(ds)\right\}$$

$$- \mathbf{P}_x\left\{\int_0^t \gamma(\xi_s)\left[g(\xi_s, e^{-u_{t-s}(\xi_s)}) - 1\right]K(ds)\right\}.$$

Then $(t, x) \mapsto v(t, x)$ solves the equation

$$v(t, x) = \mathbf{P}_x\left[1 - e^{-f(\xi_t)}\right] - \mathbf{P}_x\left[\int_0^t \rho(\xi_s)v(t - s, \xi_s)K(ds)\right]$$

$$- \mathbf{P}_x\left[\int_0^t \phi(\xi_s, v(t - s, \xi_s))K(ds)\right]$$

$$+ \mathbf{P}_x\left[\int_0^t \rho(\xi_s)\psi(\xi_s, v(t - s))K(ds)\right], \tag{4.29}$$

where

$$\phi(x, z) = \gamma(x)[g(x, 1 - z) - (1 - z)]$$

and

$$\psi(x, f) = \int_{P(E)} [1 - h(x, \pi, 1 - \pi(f))]G(x, d\pi).$$

The uniqueness of the solution of (4.29) follows by a standard application of Gronwall's inequality.

4.4 Scaling Limits of General Branching Systems

In this section, we prove that a Dawson–Watanabe superprocess with local and non-local branching mechanisms arises as the small particle limit of a sequence of the general branching particle systems introduced in Section 4.3. The arguments are similar to those used in the local branching case. For each integer $k \geq 1$, let $\{Y_k(t) :$

$t \geq 0\}$ be a sequence of general branching particle systems determined by the parameters $(\xi, K, \gamma_k, g_k, \rho_k, h_k, G)$. Recall that $N_k(E) = \{k^{-1}\nu : \nu \in N(E)\}$. Let $X_k(t) = k^{-1}Y_k(t)$ for $t \geq 0$. Then $\{X_k(t) : t \geq 0\}$ is a Markov process in $N_k(E)$. For $0 \leq z \leq k$ let

$$\phi_k(x, z) = k\gamma_k(x)[g_k(x, 1 - z/k) - (1 - z/k)] \tag{4.30}$$

and

$$\zeta_k(x, \pi, z) = k[1 - h_k(x, \pi, 1 - z/k)]. \tag{4.31}$$

For $f \in B(E)^+$ let

$$\psi_k(x, f) = \int_{P(E)} \zeta_k(x, \pi, \pi(f))G(x, d\pi), \tag{4.32}$$

which makes sense when $k \geq 1$ is sufficiently large. Let $\mathbf{Q}_\nu^{(k)}$ denote the conditional law given $X_k(0) = \nu \in N_k(E)$. By (4.26) for any $f \in B(E)^+$ we have

$$\mathbf{Q}_\nu^{(k)} \exp\left\{ - \langle X_k(t), f \rangle \right\} = \exp\left\{ - \langle \nu, u_k(t) \rangle \right\}, \tag{4.33}$$

where $u_k(t, x)$ is determined by

$$v_k(t, x) = k[1 - \exp\{-u_k(t, x)/k\}]$$

and

$$\begin{aligned}
v_k(t, x) = \mathbf{P}_x k\left[1 - e^{-f(\xi_t)/k}\right] - \mathbf{P}_x\left[\int_0^t \rho_k(\xi_s)v_k(t - s, \xi_s)K(ds)\right] \\
- \mathbf{P}_x\left[\int_0^t \phi_k(\xi_s, v_k(t - s, \xi_s))K(ds)\right] \\
+ \mathbf{P}_x\left[\int_0^t \rho_k(\xi_s)\psi_k(\xi_s, v_k(t - s))K(ds)\right].
\end{aligned} \tag{4.34}$$

For the convenience of statement of the results in this section, we formulate the following conditions:

Condition 4.8 For each $a \geq 0$ the sequence $\{\phi_k(x, z)\}$ is Lipschitz with respect to z uniformly on $E \times [0, a]$ and $\phi_k(x, z)$ converges to some $\phi(x, z)$ uniformly on $E \times [0, a]$ as $k \to \infty$.

Condition 4.9 For each $k \geq 1$ we have $(d/dz)h_k(x, \pi, 1-) \leq 1$ uniformly on $E \times P(E)$.

Condition 4.10 For each $a \geq 0$ the sequence $\zeta_k(x, \pi, z)$ converges to some $\zeta(x, \pi, z)$ uniformly on $E \times P(E) \times [0, a]$ as $k \to \infty$.

Under Condition 4.8, one can see as in the proof of Proposition 4.3 that $\phi(x, z)$ has the representation (2.45).

Proposition 4.11 *If Conditions 4.9 and 4.10 hold, then $\zeta(x, \pi, z)$ has the representation*

$$\zeta(x, \pi, z) = \beta(x, \pi)z + \int_0^\infty (1 - e^{-zu})n(x, \pi, du), \qquad (4.35)$$

where $\beta \in B(E \times P(E))^+$ and $un(x, \pi, du)$ is a bounded kernel from $E \times P(E)$ to $(0, \infty)$ satisfying

$$\beta(x, \pi) + \int_0^\infty un(x, \pi, du) \le 1, \qquad x \in E, \pi \in P(E). \qquad (4.36)$$

Proof. We first apply Theorem 1.43 for fixed $(x, \pi) \in E \times P(E)$ to get the representation (4.35), where $\beta(x, \pi) \ge 0$ is a constant and $(1 \wedge u)n(x, \pi, du)$ is a finite measure on $(0, \infty)$. From Condition 4.9 we have

$$\frac{d}{dz}\zeta_k(x, \pi, z) = \frac{d}{dz}h_k(x, \pi, 1 - z/k) \le 1,$$

and hence $\zeta_k(x, \pi, z) \le z$ for $0 \le z \le k$. It then follows that $\zeta(x, \pi, z) \le z$ for all $z \ge 0$. Therefore

$$\beta(x, \pi) + \int_0^\infty un(x, \pi, du) = \frac{d}{dz}\zeta(x, \pi, 0+) \le 1$$

uniformly on $E \times P(E)$. As in the proof of Proposition 4.3 one sees $\beta \in B(E \times P(E))^+$ and $n(x, \pi, du)$ is a kernel from $E \times P(E)$ to $(0, \infty)$. $\qquad \square$

Proposition 4.12 *To each function $\zeta(x, \pi, z)$ given by (4.35) and (4.36) there corresponds a sequence $\{\zeta_k(x, \pi, z)\}$ in form (4.31) so that Conditions 4.9 and 4.10 are satisfied.*

Proof. We can give a direct construction of the sequence as in the proof of Proposition 4.4. For $(x, \pi, z) \in E \times P(E) \times [-1, 1]$ and $k \ge 1$ set

$$h_k(x, \pi, z) = 1 + \beta(x, \pi)(z - 1) + \frac{1}{k}\int_0^\infty (e^{ku(z-1)} - 1)n(x, \pi, du).$$

Clearly, the function $z \mapsto h_k(x, \pi, z)$ is analytic in $(-1, 1)$ and $h_k(x, \pi, 1) = 1$. Observe also that

$$h_k(x, \pi, 0) \ge 1 - \beta(x, \pi) - \int_0^\infty un(x, \pi, du) \ge 0$$

and

$$\frac{d^i}{dz^i}h_k(x, \pi, 0) \ge 0, \qquad i = 1, 2, \dots.$$

Therefore $h_k(x, \pi, \cdot)$ is a probability generating function. Let $\zeta_k(x, \pi, z)$ be defined by (4.31). Then $\zeta_k(x, \pi, z) = \zeta(x, \pi, z)$ for $0 \le z \le k$. □

Proposition 4.13 *Suppose* $\zeta(x, \pi, z)$ *is given by* (4.35) *and* (4.36) *and* $G(x, d\pi)$ *is a probability kernel from* E *to* $P(E)$. *Let*

$$\psi(x, f) = \int_{P(E)} \zeta(x, \pi, \pi(f)) G(x, d\pi), \qquad x \in E, f \in B(E)^+. \quad (4.37)$$

Then $\psi(x, f)$ *admits the representation* (2.18) *with*

$$\eta(x, 1) + \int_{M(E)^\circ} \nu(1) H(x, d\nu) \le 1, \qquad x \in E. \quad (4.38)$$

Proof. We get (2.18) from (4.35) and (4.37) by changing the variables of integration. The boundedness (4.38) follows from (4.36). □

Let $\mu \in M(E)$ and let $\mathbf{Q}_{(\mu)}^{(k)}$ denote the conditional law given that $Y_k(0) = kX_k(0)$ is a Poisson random measure on E with intensity $k\mu$. From (4.33) and Theorem 1.25 we get

$$\mathbf{Q}_{(\mu)}^{(k)} \exp\{-\langle X_k(t), f\rangle\} = \exp\{-\langle \mu, v_k(t)\rangle\}. \quad (4.39)$$

By modifications of the arguments in Section 4.2 we have:

Proposition 4.14 *Suppose that* $\rho_k \to \rho \in B(E)^+$ *in the supremum norm and Conditions 4.8, 4.9 and 4.10 are satisfied. Then for each* $a \ge 0$ *both* $v_k(t, x, f)$ *and* $u_k(t, x, f)$ *converge uniformly on the set* $[0, a] \times E \times B_a(E)^+$ *of* (t, x, f) *to the unique locally bounded positive solution* $(t, x) \mapsto V_t f(x)$ *of the evolution equation*

$$\begin{aligned}
V_t f(x) = P_t f(x) &- \mathbf{P}_x\left[\int_0^t \rho(\xi_s) V_{t-s} f(\xi_s) K(ds)\right] \\
&- \mathbf{P}_x\left[\int_0^t \phi(\xi_s, V_{t-s} f(\xi_s)) K(ds)\right] \\
&+ \mathbf{P}_x\left[\int_0^t \rho(\xi_s) \psi(\xi_s, V_{t-s} f) K(ds)\right]. \quad (4.40)
\end{aligned}$$

Moreover, the operators $(V_t)_{t \ge 0}$ *constitute a cumulant semigroup.*

Theorem 4.15 *Suppose that* $\rho_k \to \rho \in B(E)^+$ *in the supremum norm and Conditions 4.8, 4.9 and 4.10 are satisfied. Then the finite-dimensional distributions of* $\{X_k(t) : t \ge 0\}$ *under* $\mathbf{Q}_{(\mu)}^{(k)}$ *converge to those of a Dawson–Watanabe superprocess* $\{X_t : t \ge 0\}$ *with initial state* $X_0 = \mu$ *and with cumulant semigroup given by* (4.40).

The Dawson–Watanabe superprocess with cumulant semigroup given by (4.40) has local branching mechanism $(x, z) \mapsto \rho(x)z + \phi(x, z)$ and non-local branching mechanism $(x, f) \mapsto \rho(x)\psi(x, f)$. Note that conditions (4.36) and (4.38) actually put no restriction on the non-local branching mechanism because of the multiplying factor $\rho(x)$. Heuristically, the local branching mechanism describes the death and birth of particles at $x \in E$ and $\zeta(x, \pi, \cdot)$ describes the birth of particles at $x \in E$ that are displaced into E following the distribution $\pi \in P(E)$, which is selected according to the kernel $G(x, d\pi)$. More specifically, we can explain the non-local branching mechanisms given by (2.47) and (2.48) as follows. In the first case, the non-locally displaced offspring born at $x \in E$ are produced according to $\zeta(x, \cdot)$ and they choose their locations independently according to the distribution $\pi(x, dy)$. In the second case, once a particle dies at $x \in E$, a point $y \in E$ is first chosen following the distribution $\pi(x, dy)$ and then the non-locally displaced offspring are produced at this point according to the law given by $\zeta(x, y, \cdot)$.

Example 4.2 If there is $c \in B(E)^+$ so that $\gamma_k(x) = kc(x)$ and $g_k(x, z) = (1 + z^2)/2$, from (4.30) we have $\phi_k(x, z) = c(x)z^2/2$, which gives a binary local branching mechanism for the corresponding superprocess.

The (ξ, K, ϕ)-superprocess with general branching mechanism given by (2.26) or (2.27) also arises as the high-density limits of branching particle systems in finite-dimensional distributions. We leave the consideration to the reader.

4.5 Notes and Comments

Silverstein (1968) proved an existence theorem for branching particle systems as measure-valued processes. A different but equivalent formulation of branching systems was given in Ikeda et al. (1968a, 1968b, 1969). The construction given in Section 4.1 follows that of Walsh (1986).

Watanabe (1968) established a rescaling limit theorem of discrete-time branching particle systems, which gave a super-Brownian motion as the limit process. See also Dawson (1975) and Ethier and Kurtz (1986, pp.400–407). The main references of this chapter are Dawson et al. (2002c) and Dynkin (1993a), where non-local branching superprocesses were obtained. For a Hunt spatial motion process, one can prove the weak convergence of the rescaled branching systems in the space of càdlàg paths; see, e.g., Schied (1999). The construction of binary branching superprocesses using Brownian snakes was originally given by Le Gall (1991). The proof of (4.24) was given in Le Gall (1993) for a continuous spatial motion. A Brownian snake representation for catalytic branching superprocesses was provided in Klenke (2003).

The super-Brownian motion also arises in limit theorems of other rescaled interacting particle systems. For example, such limit theorems were proved for contact processes in Durrett and Perkins (1999), for high-dimensional percolation in Hara and Slade (2000) and van der Hofstad and Slade (2003), for voter models in Cox et al. (2000), for interacting diffusions in Cox and Klenke (2003), for Lotka–Volterra models in Cox and Perkins (2005, 2008) and for epidemic models in Lalley (2009).

See Slade (2002) for a nice introduction of the explorations in the subject. The Donsker invariance principle is deeply involved in those results. Generally speaking, if the relevant spatial transition kernel has finite variance, a super-Brownian motion would arise as the limit process. He (2010) considered the case where the transition kernel is in the contraction domain of a stable law and obtained a super-stable process in the limit. One motivation of the study is to use the limit theorems to investigate the asymptotic properties of the approximating systems; see, e.g., Cox and Perkins (2004, 2007). For the general backgrounds of interacting particle systems see Chen (2004), Durrett (1995) and Liggett (1985, 1999).

Chapter 5
Basic Regularities of Superprocesses

In this chapter we prove some basic regularities of Dawson–Watanabe superprocesses. We shall develop the theory in the Borel right setting, which is particularly suitable for the applications of various transformations. We shall see that if the underlying spatial motion ξ is a Borel right process, the (ξ, ϕ)-superprocess is a Borel right process with quasi-left continuous natural filtration; and if ξ is a Hunt process, so is the superprocess. We also give a characterization of the so-called occupation times of the superprocess.

5.1 Right Continuous Realizations

Suppose that E is a Lusin topological space and d is a metric for its topology so that the d-completion of E is compact. Let $\xi = (\Omega, \mathscr{F}, \mathscr{F}_t, \xi_t, \mathbf{P}_x)$ be a conservative Borel right process in E with transition semigroup $(P_t)_{t\geq 0}$ and resolvent $(U^\alpha)_{\alpha>0}$. Let ϕ be a branching mechanism on E given by (2.26) or (2.27) and let $(Q_t)_{t\geq 0}$ denote the transition semigroup on $M(E)$ of the (ξ, ϕ)-superprocess defined by (2.32) and (2.33).

Let $\mathscr{D}$ be a countable and uniformly dense subset of $C_u(E)^{++}$ and assume $1 \in \mathscr{D}$. Let $\mathscr{R}$ be the countable Ray cone for $(P_t)_{t\geq 0}$ constructed from $\mathscr{D}$. We clearly have $\mathscr{R} \setminus \{0\} \subset B(E)^{++}$. Note also that for each $f \in \mathscr{R}$ there is a constant $\alpha = \alpha(f) > 0$ so that f is an α-excessive function relative to $(P_t)_{t\geq 0}$. Let $\bar{E}$ be the Ray–Knight completion of E defined from $\mathscr{R}$. Then $\bar{E}$ is a compact metric space and $E \in \mathscr{B}(\bar{E})$ by Proposition A.30. Let $(\bar{U}^\alpha)_{\alpha>0}$ be the Ray extension of $(U^\alpha)_{\alpha>0}$. We denote the Ray extension of $(P_t)_{t\geq 0}$ by $(\bar{P}_t)_{t\geq 0}$, which is a Borel semigroup on $\bar{E}$.

Let E_ρ denote the set E furnished with the Ray topology inherited from $\bar{E}$. Then each $f \in \mathscr{R}$ is uniformly continuous on E_ρ and admits a unique continuous extension $\bar{f}$ to $\bar{E}$. Let $\bar{\mathscr{R}}$ denote the class of those extensions. By Proposition A.27 the collection $\bar{\mathscr{R}} - \bar{\mathscr{R}} := \{\bar{f} - \bar{g} : \bar{f}, \bar{g} \in \bar{\mathscr{R}}\}$ is uniformly dense in $C(\bar{E})$. By Proposition A.30 we have $\mathscr{B}(E_\rho) = \mathscr{B}(E)$. Let $D = \{x \in \bar{E} : \bar{P}_0(x, \cdot) = \delta_x(\cdot)\}$

and $B = \{x \in \bar{E} : \bar{P}_0(x, \cdot) \neq \delta_x(\cdot)\}$ denote respectively the sets of non-branch points and branch points for $(\bar{P}_t)_{t \geq 0}$. Clearly, we have $D \in \mathscr{B}(\bar{E})$ and $D \supset E$. By Theorem A.24 the restriction of $(\bar{P}_t)_{t \geq 0}$ to D is a Borel right semigroup. We shall use the same notation for the restriction. Since $\bar{E}$ is a compact metric space, the space $M(\bar{E})$ is locally compact, separable and metrizable. We fix a metric on $M(\bar{E})$ compatible with its topology and regard $M(E_\rho)$ and $M(D)$ as topological subspaces of $M(\bar{E})$ comprising measures supported by E_ρ and D, respectively.

We extend $f \mapsto \phi(\cdot, f)$ to an operator $\bar{f} \mapsto \bar{\phi}(\cdot, \bar{f})$ from $B(\bar{E})^+$ to $B(\bar{E})$ by setting $\bar{\phi}(x, \bar{f}) = \phi(x, f)$ for $x \in E$ and $\bar{\phi}(x, \bar{f}) = 0$ for $x \in \bar{E} \setminus E$, where $f = \bar{f}|_E$ is the restriction to E of $\bar{f} \in B(\bar{E})^+$. Then $\bar{\phi}$ has the canonical representation (2.26) with the parameters $(\bar{b}, \bar{c}, \bar{\eta}, \bar{H})$ defined in obvious ways. The construction of the Dawson–Watanabe superprocess given by (2.32) and (2.33) certainly applies to the restrictions of $(\bar{P}_t)_{t \geq 0}$ and $\bar{\phi}$ to D. We can even extend the construction to the space $\bar{E}$. More precisely, for every $\bar{f} \in B(\bar{E})^+$ there is a unique locally bounded positive solution $t \mapsto \bar{V}_t \bar{f}$ to the equation

$$\bar{V}_t \bar{f}(x) = \bar{P}_t \bar{f}(x) - \int_0^t \mathrm{d}s \int_{\bar{E}} \bar{\phi}(y, \bar{V}_s \bar{f}) \bar{P}_{t-s}(x, \mathrm{d}y), \quad t \geq 0, x \in \bar{E}, \quad (5.1)$$

which defines a cumulant semigroup $(\bar{V}_t)_{t \geq 0}$ with underlying space $\bar{E}$. In fact, by Proposition A.23 for every $t \geq 0$ and every $x \in \bar{E}$ the probability measure $\bar{P}_t(x, \cdot)$ is supported by D, so we can first solve the equation on D and then define $\bar{V}_t \bar{f}(x) = \bar{P}_0 \bar{V}_t \bar{f}(x)$ for $x \in \bar{E}$. Thus the function $\bar{V}_t \bar{f}$ is actually independent of the values of $\bar{f}$ on $\bar{E} \setminus D$. Let $(\bar{Q}_t)_{t \geq 0}$ be the transition semigroup on $M(\bar{E})$ defined by (2.3) from $(\bar{V}_t)_{t \geq 0}$. Then for every $t \geq 0$ and every $\mu \in M(\bar{E})$ the measure $\bar{Q}_t(\mu, \cdot) = \bar{Q}_t(\mu \bar{P}_0, \cdot)$ is carried by $M(D)$. Since the restriction of $(\bar{P}_t)_{t \geq 0}$ to D is a Borel right semigroup, the restriction of $(\bar{Q}_t)_{t \geq 0}$ to $M(D)$ is a normal Borel transition semigroup. We denote this restriction by the same notation $(\bar{Q}_t)_{t \geq 0}$. Note also that the restriction of $(\bar{Q}_t)_{t \geq 0}$ to $M(E)$ coincides with the transition semigroup $(Q_t)_{t \geq 0}$ of the (ξ, ϕ)-superprocess.

Let $\bar{W}$ denote the space of all right continuous paths from $[0, \infty)$ into $M(D)$ with left limits in $M(\bar{E})$. Let $\{X_t : t \geq 0\}$ denote the coordinate process of $\bar{W}$ and let $(\mathscr{G}^0, \mathscr{G}_t^0)$ denote its natural σ-algebras. The following theorem gives a right continuous realization of the semigroup $(\bar{Q}_t)_{t \geq 0}$ on $M(D)$.

Theorem 5.1 *For each $\mu \in M(D)$ there is a unique probability measure $\bar{\mathbf{Q}}_\mu$ on $(\bar{W}, \mathscr{G}^0)$ such that $\bar{\mathbf{Q}}_\mu\{X_0 = \mu\} = 1$ and $\{X_t : t \geq 0\}$ is a Markov process in $M(D)$ relative to $(\mathscr{G}^0, \mathscr{G}_t^0, \bar{\mathbf{Q}}_\mu)$ with transition semigroup $(\bar{Q}_t)_{t \geq 0}$.*

Proof. Let $(\Omega, \mathscr{A}^0, \mathscr{A}_t^0, Z_t, \bar{\mathbf{P}}_\mu)$ be a Markov process with state space $M(D)$ and transition semigroup $(\bar{Q}_t)_{t \geq 0}$. Recall that each $\bar{f} \in \bar{\mathscr{R}}$ is α-excessive relative to $(\bar{P}_t)_{t \geq 0}$ for some constant $\alpha = \alpha(\bar{f}) > 0$. By Corollary 2.34 there exists $\alpha_1 > 0$ so that $t \mapsto e^{-\alpha_1 t} Z_t(\bar{f})$ is an $(\mathscr{A}_t^0)$-supermartingale. By Dellacherie and Meyer (1982, pp.66–67), there is $\Omega_1 \in \mathscr{A}^0$ with $\bar{\mathbf{P}}_\mu(\Omega_1) = 1$ such that $\{Z_t(\omega, \bar{f}) : t \geq 0\}$ possesses finite right and left limits along rationals for $\omega \in \Omega_1$ and $\bar{f} \in \bar{\mathscr{R}}$. For $t \geq 0$ and $\bar{f} \in \bar{\mathscr{R}}$ let

$$Z_{t+}(\omega, \bar{f}) = \begin{cases} \lim_{\mathrm{rat.}r \to t+} Z_r(\omega, \bar{f}) & \text{if } \omega \in \Omega_1, \\ 0 & \text{if } \omega \in \Omega \setminus \Omega_1. \end{cases}$$

Since $(\bar{P}_t)_{t \geq 0}$ is a Borel right semigroup on D, by Theorem 2.22 the process $\{Z_t : t \geq 0\}$ is right continuous in probability under $\bar{\mathbf{P}}_\mu$. Then we have

$$\bar{\mathbf{P}}_\mu\{Z_{t+}(\bar{f}) = Z_t(\bar{f}) \text{ for all } \bar{f} \in \bar{\mathscr{R}}\} = 1, \qquad t \geq 0. \tag{5.2}$$

Recall that $\bar{\mathscr{R}} - \bar{\mathscr{R}}$ is dense in $C(\bar{E})$ with the supremum norm. Thus for every $t \geq 0$ and $\omega \in \Omega$ there is a unique measure $Y_t(\omega, \cdot) \in M(\bar{E})$ such that $Y_t(\omega, \bar{f}) = Z_{t+}(\omega, \bar{f})$ for all $\bar{f} \in \bar{\mathscr{R}}$. It is easy to see that $t \mapsto Y_t(\omega)$ is càdlàg in $M(\bar{E})$ for every $\omega \in \Omega$. In view of (5.2), we have

$$\bar{\mathbf{P}}_\mu\{Y_t(\bar{f}) = Z_t(\bar{f}) \text{ for all } \bar{f} \in \bar{\mathscr{R}}\} = 1, \qquad t \geq 0. \tag{5.3}$$

It follows that the random measure Y_t is $\bar{\mathbf{P}}_\mu$-a.s. supported by D and $\{Y_t|_D : t \geq 0\}$ is a Markov process in $M(D)$ with transition semigroup $(\bar{Q}_t)_{t \geq 0}$ relative to $(\mathscr{A}^0, \mathscr{A}^0_t, \bar{\mathbf{P}}_\mu)$. Let $(\mathscr{A}^\mu, \mathscr{A}^\mu_t)$ be the augmentation of $(\mathscr{A}^0, \mathscr{A}^0_t)$ by $\bar{\mathbf{P}}_\mu$. Then $\{Y_t|_D : t \geq 0\}$ is also a Markov process with semigroup $(\bar{Q}_t)_{t \geq 0}$ relative to $(\mathscr{A}^\mu, \mathscr{A}^\mu_t, \bar{\mathbf{P}}_\mu)$. Recall that $D = \{x \in E : \bar{P}_0(x, \cdot) = \delta_x\}$ and $B = \bar{E} \setminus D$. Then (5.3) implies that

$$\bar{\mathbf{P}}_\mu\{Y_t(\bar{f}) = Y_t(\bar{P}_0 \bar{f}) \text{ for all } \bar{f} \in \bar{\mathscr{R}}\} = 1, \qquad t \geq 0. \tag{5.4}$$

For $\bar{f} \in \bar{\mathscr{R}}$ which is α-excessive for $(\bar{P}_t)_{t \geq 0}$ let $\bar{f}_k = k(\bar{f} - k\bar{U}^{k+\alpha}\bar{f})$. Then $\bar{f}_k \in C(\bar{E})^+$ and $\bar{U}^\alpha \bar{f}_k = k\bar{U}^{k+\alpha}\bar{f} \to \bar{P}_0 \bar{f}$ increasingly as $k \to \infty$. Since $\bar{U}^\alpha \bar{f}_k \in C(\bar{E})^+$ is α-excessive for $(\bar{P}_t)_{t \geq 0}$, by Corollary 2.34 there exists $\alpha_1 > 0$ so that $t \mapsto e^{-\alpha_1 t} Y_t(\bar{U}^\alpha \bar{f}_k)$ is a right continuous $(\mathscr{A}^\mu_t)$-supermartingale, and by Dellacherie and Meyer (1982, p.69) it is also an $(\mathscr{A}^\mu_{t+})$-supermartingale. Since $(\mathscr{A}^\mu, \mathscr{A}^\mu_{t+}, \bar{\mathbf{P}}_\mu)$ satisfies the usual hypotheses, $t \mapsto e^{-\alpha_1 t} Y_t(\bar{P}_0 \bar{f})$ and hence $t \mapsto Y_t(\bar{P}_0 \bar{f})$ is $\bar{\mathbf{P}}_\mu$-a.s. right continuous; see Dellacherie and Meyer (1982, p.79) or Sharpe (1988, p.390). Then (5.4) and the right continuity of $\{Y_t : t \geq 0\}$ imply that

$$\bar{\mathbf{P}}_\mu\{Y_t(\bar{f}) = Y_t(\bar{P}_0 \bar{f}) \text{ for all } t \geq 0 \text{ and } \bar{f} \in \bar{\mathscr{R}}\} = 1.$$

Therefore we must have $\bar{\mathbf{P}}_\mu\{Y_t(B) = 0 \text{ for all } t \geq 0\} = 1$. Now we can simply let $\bar{Q}_\mu$ be the image of $\bar{\mathbf{P}}_\mu$ under the mapping $\omega \mapsto \{Y_t(\omega)|_D : t \geq 0\}$, and the theorem is proved. $\qquad \square$

5.2 The Strong Markov Property

Let $\bar{X} = (\bar{W}, \bar{\mathscr{G}}^0, \bar{\mathscr{G}}^0_t, X_t, \bar{Q}_\mu)$ be the Markov process in $M(D)$ given by Theorem 5.1. We write X^ρ_{t-} for the left limit of the process in $M(\bar{E})$ at $t > 0$. It is not hard to show that for any $G \in b\bar{\mathscr{G}}^0$, the map $\mu \mapsto \bar{Q}_\mu(G)$ is $\mathscr{B}(M(D))$-measurable.

Given a probability measure K on $M(D)$ we can define the probability measure $\bar{Q}_K$ on $(\bar{W}, \bar{\mathscr{G}}^0)$ by

$$\bar{Q}_K(G) = \int_{M(D)} \bar{Q}_\mu(G)K(\mathrm{d}\mu), \qquad G \in \mathrm{b}\bar{\mathscr{G}}^0. \tag{5.5}$$

Let $(\bar{\mathscr{G}}^K, \bar{\mathscr{G}}_t^K)$ be the $\bar{Q}_K$-augmentation of $(\bar{\mathscr{G}}^0, \bar{\mathscr{G}}_t^0)$, and simply write $(\bar{\mathscr{G}}^\mu, \bar{\mathscr{G}}_t^\mu)$ in the special case of $K = \delta_\mu$ for $\mu \in M(D)$. Let $\bar{\mathscr{G}} = \cap_K \bar{\mathscr{G}}^K$ and $\bar{\mathscr{G}}_t = \cap_K \bar{\mathscr{G}}_t^K$, where the intersections are taken over all probability measures K on $M(D)$.

Let $\gamma(x, \mathrm{d}y)$ be the kernel on E defined by (2.28) and let $c_0 = \|b^-\| + \|\gamma(\cdot, 1)\|$. Let $\bar{\gamma}(x, \mathrm{d}y)$ be the extension of $\gamma(x, \mathrm{d}y)$ to $\bar{E}$ so that $\bar{\gamma}(x, \bar{E} \setminus E) = 0$ for $x \in E$ and $\bar{\gamma}(x, \bar{E}) = 0$ for $x \in \bar{E} \setminus E$. Let $(\bar{\pi}_t)_{t\geq 0}$ be the semigroup of kernels on D defined by (2.35) from $(\bar{P}_t)_{t\geq 0}$ and $\bar{\gamma}(x, \mathrm{d}y)$. Let $(\bar{R}^\alpha)_{\alpha > c_0}$ be defined by (A.42) from $(\bar{\pi}_t)_{t\geq 0}$. By the formula $\bar{\pi}_t = \bar{P}_0\bar{\pi}_t$ we can extend $(\bar{\pi}_t)_{t\geq 0}$ to a semigroup of kernels on $\bar{E}$. Let $(\bar{R}^\alpha)_{\alpha > c_0}$ be extended accordingly.

Proposition 5.2 *Let* $\mu \in M(D)$ *and* $\bar{f} \in B(D)$. *For any bounded* $(\bar{\mathscr{G}}_{t+}^\mu)$-*stopping time* T, *we have*

$$\bar{Q}_\mu[X_{T+t}(\bar{f})|\bar{\mathscr{G}}_{T+}^\mu] = X_T(\bar{\pi}_t\bar{f}), \quad t \geq 0. \tag{5.6}$$

If T *is predictable in addition, then*

$$\bar{Q}_\mu[X_{T+t}(\bar{f})|\bar{\mathscr{G}}_{T-}^\mu] = X_{T-}^\rho(\bar{\pi}_t\bar{f}), \quad t \geq 0. \tag{5.7}$$

Proof. Step 1. Suppose that $\bar{g} \in \bar{\mathscr{R}}$ is α-excessive for $(\bar{P}_t)_{t\geq 0}$. By Corollary 2.34 and the Markov property of $(X_t, \bar{\mathscr{G}}_t^\mu)$, there exists $\alpha_1 > 0$ so that $t \mapsto \mathrm{e}^{-\alpha_1 t}X_t(\bar{g})$ is a right continuous $(\bar{\mathscr{G}}_t^\mu)$-supermartingale. Then it is a strong $(\bar{\mathscr{G}}_{t+}^\mu)$-supermartingale by Dellacherie and Meyer (1982, p.69 and p.74). In particular, for any $(\bar{\mathscr{G}}_{t+}^\mu)$-stopping time T with upper bound $U > 0$ we have

$$\bar{Q}_\mu[X_{T+t}(\bar{g})] \leq \mathrm{e}^{\alpha_1(U+t)}\mu(\bar{g}). \tag{5.8}$$

Step 2. In view of (5.8), for any fixed $G \in \mathrm{pb}\bar{\mathscr{G}}_{T+}^\mu$ we can define the finite measure $\mu_t = \bar{Q}_\mu[GX_{T+t}] \in M(D)$. For $\beta > c_0$ and $\bar{g} \in \bar{\mathscr{R}}$ let $\bar{f} = \bar{U}^\beta\bar{g} \in C(\bar{E})^+$. Theorem 2.36 implies that

$$M_t(\bar{f}) := X_t(\bar{f}) - X_0(\bar{f}) - \int_0^t X_s(\beta\bar{f} - \bar{g} + \bar{\gamma}\bar{f} - \bar{b}\bar{f})\mathrm{d}s \tag{5.9}$$

is a right continuous $(\bar{\mathscr{G}}_t^\mu)$-martingale. By Dellacherie and Meyer (1982, p.69 and p.74) we conclude that (5.9) is a strong $(\bar{\mathscr{G}}_{t+}^\mu)$-martingale. It follows that

$$\mu_t(\bar{f}) = \mu_0(\bar{f}) + \int_0^t \mu_s(\beta\bar{f} - \bar{g} + \bar{\gamma}\bar{f} - \bar{b}\bar{f})\mathrm{d}s.$$

Then $t \mapsto \mu_t(\bar{f})$ is continuous. Let $\hat{\mu}^\beta(\bar{f}) = \int_0^\infty e^{-\beta t} \mu_t(\bar{f}) dt$. The above equation implies that

$$\hat{\mu}^\beta(\bar{f}) = \beta^{-1}\mu_0(\bar{f}) + \beta^{-1}\hat{\mu}^\beta(\beta\bar{f} - \bar{g} + \bar{\gamma}\bar{f} - \bar{b}\bar{f}),$$

and so

$$\hat{\mu}^\beta(\bar{g} - \bar{\gamma}\bar{U}^\beta\bar{g} + \bar{b}\bar{U}^\beta\bar{g}) = \mu_0(\bar{U}^\beta\bar{g}). \tag{5.10}$$

Since $\bar{\mathscr{R}} - \bar{\mathscr{R}}$ is uniformly dense in $C(\bar{E})$, we also have (5.10) for all $\bar{g} \in C(\bar{E})$ and hence for all $\bar{g} \in B(D)$.

Step 3. For $\beta > c_0$ and $\bar{g} \in \bar{\mathscr{R}}$ let $\bar{f} = \bar{U}^\beta\bar{g}$ and $\bar{h} = \bar{f} + (\bar{\gamma} - \bar{b})\bar{R}^\beta\bar{f}$. By Proposition A.50 we have

$$\bar{U}^\beta\bar{h} = \bar{U}^\beta\bar{f} + \bar{U}^\beta(\bar{\gamma} - \bar{b})\bar{R}^\beta\bar{f} = \bar{R}^\beta\bar{f}.$$

Then we can apply (5.10) to the function $\bar{h} \in B(D)$ to see

$$\hat{\mu}^\beta(\bar{f}) = \hat{\mu}^\beta(\bar{h} - (\bar{\gamma} - \bar{b})\bar{R}^\beta\bar{f}) = \mu_0(\bar{R}^\beta\bar{f}).$$

By Proposition A.42 and the uniqueness of Laplace transforms we get $\mu_t(\bar{f}) = \mu_0(\bar{\pi}_t\bar{f})$, and hence $\mu_t(\beta\bar{U}^\beta\bar{g}) = \mu_0(\beta\bar{\pi}_t\bar{U}^\beta\bar{g})$. Then we also have $\mu_t(\bar{g}) = \mu_0(\bar{\pi}_t\bar{g})$ because $\beta\bar{U}^\beta\bar{g} \to \bar{g}$ as $\beta \to \infty$. Since $\bar{g} \in \bar{\mathscr{R}}$ was arbitrary, we get $\mu_t(\bar{f}) = \mu_0(\bar{\pi}_t\bar{f})$ for all $f \in B(D)$. That proves (5.6).

Step 4. For $\beta > c_0$ and $\bar{g} \in \bar{\mathscr{R}}$ we have $\bar{f} := \bar{U}^\beta\bar{g} \in C(\bar{E})^+$. Then the right continuous martingale $\{M_t(\bar{f}) : t \geq 0\}$ defined by (5.9) has predictable projection $\{M_{t-}(\bar{f}) : t \geq 0\}$; see Dellacherie and Meyer (1982, pp.106–107). It follows that $\{X_t(\bar{f}) : t \geq 0\}$ has predictable projection $\{X_{t-}^\rho(\bar{f}) : t \geq 0\}$. From Dellacherie and Meyer (1982, p.103), we have $\bar{\mathbf{Q}}_\mu[X_T(\bar{f})|\mathscr{G}_{T-}^\mu] = X_{T-}^\rho(\bar{f})$, and hence

$$\bar{\mathbf{Q}}_\mu[X_T(\bar{U}^\beta\bar{g})G] = \bar{\mathbf{Q}}_\mu[X_{T-}^\rho(\bar{U}^\beta\bar{g})G]$$

for every $G \in \mathrm{pb}\mathscr{G}_{T-}^\mu$. Because $\beta\bar{U}^\beta\bar{g}(x) \to \bar{g}(x)$ for every $x \in D$ and $\beta\bar{U}^\beta\bar{g}(x) \to \bar{P}_0\bar{g}(x) = \bar{\pi}_0\bar{g}(x)$ for every $x \in \bar{E}$ as $\beta \to \infty$, the above equation implies

$$\bar{\mathbf{Q}}_\mu[X_T(\bar{g})G] = \bar{\mathbf{Q}}_\mu[X_{T-}^\rho(\bar{\pi}_0\bar{g})G].$$

Since $\bar{g} \in \bar{\mathscr{R}}$ was arbitrary, we obtain

$$\bar{\mathbf{Q}}_\mu[X_T(\bar{f})G] = \bar{\mathbf{Q}}_\mu[X_{T-}^\rho(\bar{\pi}_0\bar{f})G]$$

for every $\bar{f} \in B(D)$. Then (5.7) holds for $t = 0$. For $t \geq 0$ we may appeal to (5.6) to get

$$\bar{\mathbf{Q}}_\mu[X_{T+t}(\bar{f})|\mathscr{G}_{T-}^\mu] = \bar{\mathbf{Q}}_\mu\{\bar{\mathbf{Q}}_\mu[X_{T+t}(\bar{f})|\mathscr{G}_{T+}^\mu]|\mathscr{G}_{T-}^\mu\}$$
$$= \bar{\mathbf{Q}}_\mu[X_T(\bar{\pi}_t\bar{f})|\mathscr{G}_{T-}^\mu] = X_{T-}^\rho(\bar{\pi}_t\bar{f}).$$

That completes the proof of the result.   □

Corollary 5.3 *For any* $\bar{f} \in B(D)$, *the process* $t \mapsto X_t(\bar{f})$ *has* $(\bar{\mathscr{G}}_{t+}^\mu)$*-predictable projection* $t \mapsto X_{t-}^\rho(\bar{\pi}_0 \bar{f}) = X_{t-}^\rho(\bar{P}_0 \bar{f})$.

Proposition 5.4 *Let* $\mu \in M(D)$ *and* $\bar{f} \in B(D)$. *For any bounded* $(\bar{\mathscr{G}}_{t+}^\mu)$*-stopping time* T *satisfying* $0 \leq T \leq u$, *we have*

$$\bar{\mathbf{Q}}_\mu[X_u(\bar{f})|\bar{\mathscr{G}}_{T+}^\mu] = X_T(\bar{\pi}_{u-T}\bar{f}). \tag{5.11}$$

If T *is predictable in addition, then*

$$\bar{\mathbf{Q}}_\mu[X_u(\bar{f})|\bar{\mathscr{G}}_{T-}^\mu] = X_{T-}^\rho(\bar{\pi}_{u-T}\bar{f}). \tag{5.12}$$

Proof. Let $\{S_k\}$ be the sequence of random times defined by $S_k(w) = i/2^k$ for $(i-1)/2^k \leq u - T(w) < i/2^k$. For any $t \geq 0$ we have

$$\{T + S_k \leq t\} = \bigcup_{i=1}^{\infty} \{T + i/2^k \leq t\} \cap \{(i-1)/2^k \leq u - T < i/2^k\}$$

$$= \bigcup_{i=1}^{\infty} \{T \leq t - i/2^k\} \cap \{u - i/2^k < T \leq u - (i-1)/2^k\},$$

which belongs to $\bar{\mathscr{G}}_{t+}^\mu$. Then $T + S_k$ is a $(\bar{\mathscr{G}}_{t+}^\mu)$-stopping time. Clearly, $S_k \to u - T$ decreasingly as $k \to \infty$. By the proof of Proposition 5.2 for any $\bar{g} \in \bar{\mathscr{R}}$ there exists $\alpha_1 > 0$ so that $t \mapsto e^{-\alpha_1 t} X_t(\bar{g})$ is a right continuous $(\bar{\mathscr{G}}_{t+}^\mu)$-supermartingale. Then the family

$$e^{-\alpha_1(T+S_k)} X_{T+S_k}(\bar{g}), \qquad k = 1, 2, \ldots,$$

is $\bar{\mathbf{Q}}_\mu$-uniformly integrable; see Dellacherie and Meyer (1982, p.24) or Sharpe (1988, p.390). It follows that for each $\bar{f} \in C(D)$ the sequence $\{X_{T+S_k}(\bar{f})\}$ is $\bar{\mathbf{Q}}_\mu$-uniformly integrable. Since $\{S_k = i/2^k\} \in \bar{\mathscr{G}}_{T+}^\mu$, for any $G \in b\bar{\mathscr{G}}_{T+}^\mu$ we see by Proposition 5.2 that

$$\bar{\mathbf{Q}}_\mu[X_u(\bar{f})G] = \lim_{k \to \infty} \bar{\mathbf{Q}}_\mu[X_{T+S_k}(\bar{f})G]$$

$$= \lim_{k \to \infty} \sum_{i=1}^{\infty} \bar{\mathbf{Q}}_\mu[X_{T+i/2^k}(\bar{f})G 1_{\{S_k=i/2^k\}}]$$

$$= \lim_{k \to \infty} \sum_{i=1}^{\infty} \bar{\mathbf{Q}}_\mu\{\bar{\mathbf{Q}}_\mu[X_{T+i/2^k}(\bar{f})|\bar{\mathscr{G}}_{T+}^\mu]G 1_{\{S_k=i/2^k\}}\}$$

$$= \lim_{k \to \infty} \sum_{i=1}^{\infty} \bar{\mathbf{Q}}_\mu[X_T(\bar{\pi}_{i/2^k}\bar{f})G 1_{\{S_k=i/2^k\}}]$$

$$= \lim_{k \to \infty} \bar{\mathbf{Q}}_\mu[X_T(\bar{\pi}_{S_k}\bar{f})G]$$

$$= \bar{\mathbf{Q}}_\mu[X_T(\bar{\pi}_{u-T}\bar{f})G],$$

where we have used (5.8), the pointwise right continuity of $t \mapsto \bar{\pi}_t \bar{f}$ and the dominated convergence theorem for the last equality. That gives (5.11) for $\bar{f} \in C(D)$, and the extension to $\bar{f} \in B(D)$ is trivial. The proof of (5.12) is similar. □

For any $\mu \in M(D)$ the space $(\bar{W}, \bar{\mathscr{G}}^\mu, \bar{\mathscr{G}}^\mu_{t+}, \bar{\mathbf{Q}}_\mu)$ clearly satisfies the usual hypotheses. If $(s, x) \mapsto h_s(x)$ is a bounded and uniformly continuous function on $[0, \infty) \times D$, then $s \mapsto X_s(h_s)$ is right continuous and hence $(\bar{\mathscr{G}}^\mu_{t+})$-optional. Now Proposition A.1 implies that $s \mapsto X_s(h_s)$ is also $(\bar{\mathscr{G}}^\mu_{t+})$-optional for any bounded Borel function $(s, x) \mapsto h_s(x)$ on $[0, \infty) \times D$. By Proposition 5.4 the process $\{X_s(\bar{\pi}_{t-s} \bar{f}) : 0 \leq s \leq t\}$ is a strong $(\bar{\mathscr{G}}^\mu_{t+})$-martingale for every $\bar{f} \in B(D)^+$. Then $\{X_s(\bar{\pi}_{t-s} \bar{f}) : 0 \leq s \leq t\}$ is $\bar{\mathbf{Q}}_\mu$-a.s. right continuous; see Dellacherie and Meyer (1982, p.109) or Sharpe (1988, pp.389–390). Let $\bar{b}(x) = 1_E(x) b(x)$ and $\bar{\phi}_0(x, \bar{f}) = \bar{\phi}(x, \bar{f}) - \bar{b}(x) \bar{f}(x)$. By Theorem 2.23 we can rewrite (5.1) into

$$\bar{V}_t \bar{f}(x) = \bar{\pi}_t \bar{f}(x) - \int_0^t \bar{\pi}_{t-s} \bar{\phi}_0(\cdot, \bar{V}_s \bar{f})(x) ds, \qquad t \geq 0, x \in D.$$

Using the equation above it is easy to see that $\{X_s(\bar{V}_{t-s} \bar{f}) : 0 \leq s \leq t\}$ is $\bar{\mathbf{Q}}_\mu$-a.s. right continuous.

Theorem 5.5 *For every $t \geq 0$, every initial law K and every $F \in B(M(D))$, the process $\{\bar{Q}_{t-s} F(X_s) : 0 \leq s \leq t\}$ is a $\bar{\mathbf{Q}}_K$-a.s. right continuous $(\bar{\mathscr{G}}^K_{t+})$-martingale.*

Proof. Let us fix $t \geq 0$ and the initial law K on $M(D)$. By (5.5) and the above analysis, the process $\{\bar{Q}_{t-s} F(X_s) : 0 \leq s \leq t\}$ is $\bar{\mathbf{Q}}_K$-a.s. right continuous if $F(\nu) = e^{-\nu(\bar{f})}$ for some $\bar{f} \in \bar{\mathscr{R}}$. The Markov property of $\{X_t : t \geq 0\}$ implies that $\{\bar{Q}_{t-s} F(X_s) : 0 \leq s \leq t\}$ is a $(\bar{\mathscr{G}}^K_t)$-martingale. Then $\{\bar{Q}_{t-s} F(X_s) : 0 \leq s \leq t\}$ is also a $(\bar{\mathscr{G}}^K_{t+})$-martingale; see Dellacherie and Meyer (1982, p.69). We choose a compatible metric on $M(\bar{E})$ so that its completion coincides with its one-point compactification. By the Stone–Weierstrass theorem, the linear span of $\{\nu \mapsto e^{-\nu(\bar{f})} : \bar{f} \in \bar{\mathscr{R}}\}$ is uniformly dense in $C_u(M(\bar{E}))$. Since each $F \in C_u(M(D))$ has an extension in $C_u(M(\bar{E}))$, we infer that $\{\bar{Q}_{t-s} F(X_s) : 0 \leq s \leq t\}$ is a $\bar{\mathbf{Q}}_K$-a.s. right continuous $(\bar{\mathscr{G}}^K_{t+})$-martingale for every $F \in C_u(M(D))$. By Proposition A.1 we conclude that $\{\bar{Q}_{t-s} F(X_s) : 0 \leq s \leq t\}$ is a $\bar{\mathbf{Q}}_K$-a.s. right continuous $(\bar{\mathscr{G}}^K_{t+})$-martingale for every $F \in B(M(D))$; see Dellacherie and Meyer (1982, p.79) or Sharpe (1988, p.390). □

Theorem 5.6 *The filtrations $(\bar{\mathscr{G}}_t)$ and $(\bar{\mathscr{G}}^K_t)$ are right continuous and the process $\bar{X} = (\bar{W}, \bar{\mathscr{G}}, \bar{\mathscr{G}}_t, X_t, \bar{\mathbf{Q}}_K)$ satisfies the strong Markov property, that is, for every $t \geq 0$, every $(\bar{\mathscr{G}}_t)$-stopping time T, every initial law K and every function $F \in B(M(D))$, we have*

$$\bar{\mathbf{Q}}_K \big[F(X_{T+t}) 1_{\{T < \infty\}} \big| \bar{\mathscr{G}}_T \big] = \bar{Q}_t F(X_T) 1_{\{T < \infty\}}. \tag{5.13}$$

Proof. Since $(\bar{Q}_t)_{t \geq 0}$ is a Borel semigroup on $M(D)$, by Theorem A.16 the property stated in Theorem 5.5 is equivalent to the strong Markov property of (X_t)

relative to $(\bar{\mathscr{G}}_{t+})$. Then the natural filtrations $(\bar{\mathscr{G}}_t)$ and $(\bar{\mathscr{G}}_t^K)$ are right continuous by Corollary A.17. $\qquad\square$

5.3 Borel Right Superprocesses

By Theorems 5.5 and 5.6, the system $\bar{X} = (\bar{W}, \bar{\mathscr{G}}, \bar{\mathscr{G}}_t, X_t, \bar{\mathbf{Q}}_\mu)$ is a Borel right process in $M(D)$. In particular, for every $\mu \in M(D)$ the space $(\bar{W}, \bar{\mathscr{G}}^\mu, \bar{\mathscr{G}}_t^\mu, \bar{\mathbf{Q}}_\mu)$ satisfies the usual hypotheses. Let $\xi = (\Omega, \mathscr{F}, \mathscr{F}_t, \xi_t, \mathbf{P}_x)$ be a Borel right realization of the spatial motion process.

Theorem 5.7 *For every $\mu \in M(E)$ we have $\bar{\mathbf{Q}}_\mu\{X_t(D\backslash E) = 0 \, for \, all \, t \geq 0\} = 1$.*
Consequently, $\{X_t : t \geq 0\}$ is $\bar{\mathbf{Q}}_\mu$-a.s. right continuous in $M(E_\rho)$.

Proof. For $\alpha \geq c_0$ and $t \geq 0$ define $\bar{\pi}_t^\alpha = \mathrm{e}^{-\alpha t}\bar{\pi}_t$. Since $\alpha + \bar{b}(x) \geq \bar{\gamma}(x, 1)$ for all $x \in \bar{E}$, by Theorem A.43 and Proposition A.49 we see that $(\bar{\pi}_t^\alpha)_{t \geq 0}$ is a Borel right semigroup on D. Let T be a bounded $(\bar{\mathscr{G}}_t^\mu)$-stopping time and define $\nu \in M(D)$ by letting $\nu(h) = \bar{\mathbf{Q}}_\mu[\mathrm{e}^{-\alpha T} X_T(h)]$ for $h \in B(D)^+$. By Proposition 5.2 we have

$$
\begin{aligned}
\nu(\bar{R}^\alpha h) &= \int_0^\infty \bar{\mathbf{Q}}_\mu\big[\mathrm{e}^{-\alpha(t+T)} X_T(\bar{\pi}_t h)\big]\mathrm{d}t \\
&= \int_0^\infty \bar{\mathbf{Q}}_\mu\big\{\mathrm{e}^{-\alpha(t+T)} \bar{\mathbf{Q}}_\mu\big[X_{t+T}(h)\big|\bar{\mathscr{G}}_T^\mu\big]\big\}\mathrm{d}t \\
&= \int_0^\infty \bar{\mathbf{Q}}_\mu\big[\mathrm{e}^{-\alpha(t+T)} X_{t+T}(h)\big]\mathrm{d}t \\
&= \bar{\mathbf{Q}}_\mu\bigg[\int_T^\infty \mathrm{e}^{-\alpha t} X_t(h)\mathrm{d}t\bigg],
\end{aligned}
$$

and hence

$$
\nu(\bar{R}^\alpha h) \leq \bar{\mathbf{Q}}_\mu\bigg[\int_0^\infty \mathrm{e}^{-\alpha t} X_t(h)\mathrm{d}t\bigg] = \mu(\bar{R}^\alpha h).
$$

Since $\mu \in M(E)$, the above inequality means that $\nu\bar{R}^\alpha$ is dominated by a potential for the Borel right semigroup $(\pi_t^\alpha)_{t \geq 0}$. Consequently, $\nu\bar{R}^\alpha$ is supported by E and it is an excessive measure for $(\pi_t^\alpha)_{t \geq 0}$. By Getoor (1990, p.42), there is a unique measure $\lambda \in M(E)$ such that $\nu\bar{R}^\alpha = \lambda R^\alpha$. By applying Getoor (1990, p.12) or Sharpe (1988, p.195) to the Borel right semigroup $(\bar{\pi}_t^\alpha)_{t \geq 0}$ on D, we conclude $\nu = \lambda$ and hence $\bar{\mathbf{Q}}_\mu[X_T(D\backslash E)] = 0$. Since T was arbitrary, we have $\bar{\mathbf{Q}}_\mu\{X_t(D\backslash E) = 0$ for all $t \geq 0\} = 1$ by an application of the optional section theorem; see Dellacherie and Meyer (1978, p.138) or Sharpe (1988, p.388). Then $\{X_t : t \geq 0\}$ is $\bar{\mathbf{Q}}_\mu$-a.s. right continuous in $M(E_\rho)$. $\qquad\square$

Let $\bar{W}_1$ denote the space of sample paths $w \in \bar{W}$ satisfying $X_t(w) = w_t \in M(E)$ for all $t \geq 0$. By (5.5) and Theorem 5.7 the set $\bar{W}_1$ has $\bar{\mathbf{Q}}_K$-outer measure one for every initial distribution K on $M(E_\rho)$. Let $(\mathscr{G}^0, \mathscr{G}_t^0, \mathbf{Q}_K)$ be the traces of

$(\bar{\mathscr{G}}^0, \bar{\mathscr{G}}_t^0, \bar{\mathbf{Q}}_K)$ on $\bar{W}_1$. It is not hard to show $(\mathscr{G}^0, \mathscr{G}_t^0)$ coincide with the natural σ-algebras generated by the coordinate process of $\bar{W}_1$. Let $(\mathscr{G}^K, \mathscr{G}_t^K)$ be the augmentation of $(\mathscr{G}^0, \mathscr{G}_t^0)$ by $\mathbf{Q}_K$, and write $(\mathscr{G}^\mu, \mathscr{G}_t^\mu)$ simply if K is the unit mass at $\mu \in M(E_\rho)$. Let $\mathscr{G} = \cap_K \mathscr{G}^K$ and $\mathscr{G}_t = \cap_K \mathscr{G}_t^K$, where the intersections are taken over all probability laws K on $M(E_\rho)$.

Theorem 5.8 *The system* $X = (\bar{W}_1, \mathscr{G}, \mathscr{G}_t, X_t, \mathbf{Q}_\mu)$ *is a Borel right process in* $M(E_\rho)$ *with transition semigroup* $(Q_t)_{t\geq 0}$. *Moreover, for every initial distribution* K *on* $M(E_\rho)$ *the filtration* $(\mathscr{G}_t^K)$ *is quasi-left continuous.*

Proof. The property stated in Theorem 5.5 remains true for the system X. Then X is a Borel right process in $M(E_\rho)$ by Theorem A.16 and Corollary A.17. Fix an initial law K and a $(\mathscr{G}_t^K)$-predictable time T. The σ-algebra $\mathscr{G}_T^K$ can be generated by $\mathscr{G}_{T-}^K$ and X_T; see Sharpe (1988, p.118). Moreover, a left-handed version of the arguments in proofs of Theorems 5.5 and 5.6 shows that

$$\mathbf{Q}_K\big[F(X_{T+t})1_{\{T<\infty\}}\big|\mathscr{G}_{T-}^K\big] = \bar{Q}_t F(X_{T-}^\rho)1_{\{T<\infty\}} \tag{5.14}$$

for every $t \geq 0$ and every function $F \in B(M(\bar{E}))$. In particular, for every $\bar{f} \in C(\bar{E})^+$ we have

$$\mathbf{Q}_K\big[\mathrm{e}^{-X_T(\bar{f})}1_{\{T<\infty\}}\big|\mathscr{G}_{T-}^K\big] = \mathrm{e}^{-X_{T-}^\rho(\bar{P}_0\bar{f})}1_{\{T<\infty\}},$$

where we have used the equality $\bar{V}_0\bar{f}(x) = \bar{P}_0\bar{f}(x)$ for $x \in \bar{E}$. It is then easy to show

$$\mathbf{Q}_K\big\{\big[\mathrm{e}^{-X_T(\bar{f})} - \mathrm{e}^{-X_{T-}^\rho(\bar{P}_0\bar{f})}\big]^2 1_{\{T<\infty\}}\big\} = 0.$$

Consequently, we have $\mathbf{Q}_K$-a.s. $X_T(\bar{f}) = X_{T-}^\rho(\bar{P}_0\bar{f})$ on $\{T < \infty\}$. It follows that $\mathscr{G}_{T-}^K = \mathscr{G}_T^K$ and hence $(\mathscr{G}_t^K)$ is quasi-left continuous for every K; see Sharpe (1988, p.220). That proves the second assertion. $\qquad\square$

Proposition 5.9 *Let* $\mu \in M(E)$ *and* $f \in B(E)$. *If* f *is finely continuous relative to* ξ, *then* $\mathbf{Q}_\mu$-a.s. $t \mapsto X_t(f)$ *is right continuous. If* $t \mapsto f(\xi_t)$ *a.s. has left limits on* $(0, \infty)$, *then* $\mathbf{Q}_\mu$-a.s. $t \mapsto X_t(f)$ *has left limits on* $(0, \infty)$.

Proof. Let $\{T_n\}$ be a decreasing sequence of bounded $(\mathscr{G}_t^\mu)$-stopping times with limit T. For $\alpha > c_0$ define $\nu \in M(E)$ by $\nu(f) = \mathbf{Q}_\mu\{\mathrm{e}^{-\alpha T}X_T(f)\}$ and define $\nu_n \in M(E)$ analogously with T_n replacing T. For $f \in B(E)^+$ the calculations in the proof of Theorem 5.7 imply

$$\nu_n(R^\alpha f) = \mathbf{Q}_\mu\bigg[\int_0^\infty \mathrm{e}^{-\alpha(t+T_n)}X_{t+T_n}(f)\mathrm{d}t\bigg], \tag{5.15}$$

which converges increasingly as $n \to \infty$ to

$$\nu(R^\alpha f) = \mathbf{Q}_\mu\bigg[\int_0^\infty \mathrm{e}^{-\alpha(t+T)}X_{t+T}(f)\mathrm{d}t\bigg]. \tag{5.16}$$

As observed in Section A.6, a right process realization $\tilde{\xi}$ of $(\pi_t^\alpha)_{t\geq0}$ can be obtained by concatenating a countable number of copies of the subprocess $\hat{\xi}$ constructed from ξ and the multiplicative functional

$$t \mapsto m_t := \exp\left\{ -\alpha t - \int_0^t b(\xi_s)\mathrm{d}s \right\}. \tag{5.17}$$

Since $t \mapsto m_t$ is continuous, a function $f \in B(E)$ finely continuous relative to $(P_t)_{t\geq0}$ is also finely continuous relative to $(\pi_t^\alpha)_{t\geq0}$. By a result of Fitzsimmons (1988) we have $\nu_n(f) \to \nu(f)$; see Theorem A.22. A monotone class argument shows that $\{\mathrm{e}^{-\alpha t}X_t(f) : t \geq 0\}$ is optional. Then $t \mapsto \mathrm{e}^{-\alpha t}X_t(f)$ is $\mathbf{Q}_\mu$-a.s. right continuous by Dellacherie and Meyer (1982, p.109) or Sharpe (1988, p.389). That proves the first assertion. From the construction of the process $\tilde{\xi}$ it is also clear that if $t \mapsto f(\xi_t)$ a.s. has left limits in $(0,\infty)$, so does $t \mapsto f(\tilde{\xi}_t)$. Then the second assertion follows by arguments similar to those in the above.   □

Proposition 5.10 *Let $f \in B(E)$. If $t \mapsto f(\xi_t)$ is quasi-left continuous, so is $t \mapsto X_t(f)$.*

Proof. Step 1. Let $\mu \in M(E)$ and let $\{T_n\}$ be a uniformly bounded increasing sequence of $(\mathscr{G}_t^\mu)$-stopping times with limit T. Let ν and ν_n be defined as in the proof of Proposition 5.9. By (5.15) and (5.16) we have $\nu_n R^\alpha \to \nu R^\alpha$ decreasingly as $n \to \infty$. Since the multiplicative functional $\{m_t : t \geq 0\}$ defined by (5.17) is continuous and strictly positive, the lifetime of $\hat{\xi}$ is totally inaccessible. Then the function $f \in B(E)$ quasi-left continuous relative to ξ is also quasi-left continuous relative to $\hat{\xi}$. By a result of Fitzsimmons (1988) we have $\nu_n(f) \to \nu(f)$ as $n \to \infty$; see Theorem A.22. Consequently, the left limits $\lim_{s\to t-} \mathrm{e}^{-\alpha s}X_s(f)$ exist $\mathbf{Q}_\mu$-a.s. and the predictable projection of $t \mapsto \mathrm{e}^{-\alpha t}X_t(f)$ is indistinguishable from $t \mapsto \lim_{s\to t-} \mathrm{e}^{-\alpha s}X_s(f)$; see Dellacherie and Meyer (1982, pp.113–114). For any $(\mathscr{G}_t^\mu)$-predictable time S we have $\mathscr{G}_{S-}^\mu = \mathscr{G}_S^\mu$ by Theorem 5.8. Then $\mathbf{Q}_\mu$-a.s.

$$\lim_{s\to S-} \mathrm{e}^{-\alpha s}X_s(f) = \mathbf{Q}_\mu[\mathrm{e}^{-\alpha S}X_S(f)|\mathscr{G}_{S-}^\mu]$$
$$= \mathbf{Q}_\mu[\mathrm{e}^{-\alpha S}X_S(f)|\mathscr{G}_S^\mu] = \mathrm{e}^{-\alpha S}X_S(f)$$

on $\{S < \infty\}$. It follows that $\mathbf{Q}_\mu$-a.s. $\lim_{s\to S-} X_s(f) = X_S(f)$ on $\{S < \infty\}$.

Step 2. Let $\{T_n\}$ be a general increasing sequence of $(\mathscr{G}_t^\mu)$-stopping times with limit T. Then $X_{T_n}(f) \to X_T(f)$ certainly holds on $\{T < \infty\} \cap \{T_n = T$ for some $n\}$. Let

$$S(w) = \begin{cases} T(w) & \text{if } T_n(w) < T(w) \text{ for all } n, \\ \infty & \text{if } T_n(w) = T(w) \text{ for some } n. \end{cases}$$

It is simple to check that S is a predictable time with announcing sequence $\{S_n\}$ defined by

$$S_n(w) = \begin{cases} T_n(w) & \text{if } T_n(w) < T(w), \\ \infty & \text{if } T_n(w) = T(w). \end{cases}$$

By the first step we have $\mathbf{Q}_\mu$-a.s. $\lim_{n\to\infty} X_{S_n}(f) = X_S(f)$ on $\{S < \infty\}$, which implies $\mathbf{Q}_\mu$-a.s. $\lim_{n\to\infty} X_{T_n}(f) = X_T(f)$ on $\{T < \infty\}$. Since $\mu \in M(E)$ was arbitrary, the quasi-left continuity of $t \mapsto X_t(f)$ follows in view of (5.5). □

Since $C_u(E)$ is separable in the supremum norm, by Proposition 5.9 the process $t \mapsto X_t$ is a.s. right continuous in the original topology of $M(E)$. By deleting from $\bar{W}_1$ the paths not right continuous in $M(E)$ we obtain the subspace W of paths that are right continuous in both $M(E)$ and $M(E_\rho)$ and have left limits in $M(\bar{E})$. We equip W with the σ-algebras and probability measures inherited from those on $\bar{W}_1$ without changing the notation.

Theorem 5.11 *The system $X = (W, \mathscr{G}, \mathscr{G}_t, X_t, \mathbf{Q}_\mu)$ is a Borel right process in $M(E)$ with transition semigroup $(Q_t)_{t\geq 0}$. If, in addition, ξ is a Hunt process, then X is a Hunt process in $M(E)$.*

Proof. It is simple to see that X is a Borel right process in $M(E)$ with transition semigroup $(Q_t)_{t\geq 0}$. If ξ is a Hunt process, then each $f \in C_u(E)$ is quasi-left continuous relative to ξ. By (5.5) and Proposition 5.10 one sees $t \mapsto X_t(f)$ is quasi-left continuous. Since $C_u(E)$ is separable, we infer that $t \mapsto X_t$ is quasi-left continuous. □

Theorem 5.12 *For a general Borel right spatial motion, the (ξ, ϕ)-superprocess has a right realization in $M(E)$. If ξ is a Hunt process, the (ξ, ϕ)-superprocess has a Hunt realization in $M(E)$.*

Proof. Let $(\tilde{P}_t)_{t\geq 0}$ be the conservative Borel right extension of $(P_t)_{t\geq 0}$ on the Lusin topological space $\tilde{E} := E \cup \{\partial\}$ with ∂ being an isolated cemetery. For $\tilde{f} \in B(\tilde{E})^+$ let $\tilde{\phi}(\partial, \tilde{f}) = 0$ and let $\tilde{\phi}(x, \tilde{f}) = \phi(x, \tilde{f}|_E)$ if $x \in E$. Let $(\tilde{V}_t)_{t\geq 0}$ be defined by (2.33) from those extended ingredients and let $(\tilde{Q}_t)_{t\geq 0}$ be the corresponding transition semigroup on $M(\tilde{E})$. For any $f \in B(E)^+$ we extend its definition to $\tilde{E}$ by setting $f(\partial) = 0$. Since ∂ is a cemetery of $(\tilde{P}_t)_{t\geq 0}$, we have $\tilde{V}_t f(\partial) = \tilde{P}_t f(\partial) = 0$ and

$$\tilde{V}_t f(x) = P_t f(x) - \int_0^t ds \int_E \phi(y, \tilde{V}_s f) P_{t-s}(x, dy), \quad t \geq 0, x \in E.$$

The uniqueness of the solution of (2.33) now implies $\tilde{V}_t f(x) = V_t f(x)$ for $t \geq 0$ and $x \in E$. Let $\psi(\mu)$ denote the restriction of the measure $\mu \in M(\tilde{E})$ to E. For $\mu \in M(\tilde{E})$ and $f \in B(E)^+$ it is easy to show

$$\int_{M(\tilde{E})} e^{-\psi(\nu)(f)} \tilde{Q}_t(\mu, d\nu) = \int_{M(E)} e^{-\nu(f)} Q_t(\psi(\mu), d\nu), \tag{5.18}$$

where $(Q_t)_{t\geq 0}$ is the transition semigroup of the (ξ, ϕ)-superprocess. Let $\mathscr{L}$ be the set of functions $F \in b\mathscr{B}^u(M(E))$ such that

$$\tilde{Q}_t(F \circ \psi)(\mu) = (Q_t F) \circ \psi(\mu), \quad t \geq 0, \mu \in M(\tilde{E}).$$

By (5.18) and a monotone class argument one shows $\mathscr{L} \supset b\mathscr{B}(M(E))$. It is then easily seen that $\mathscr{L} = b\mathscr{B}^u(M(E))$. Let $\tilde{X}$ be the Borel right realization of $(\tilde{Q}_t)_{t\geq 0}$ provided by Theorem 5.11. Since ∂ is isolated from E, the path $t \mapsto \psi(\tilde{X}_t)$ is right continuous in $M(E)$. By Theorem A.21 one sees $X := \psi(\tilde{X})$ with the augmented natural σ-algebras is a right realization of the (ξ, ϕ)-superprocess in $M(E)$. The second assertion follows by a further application of Theorem 5.11. $\square$

Corollary 5.13 *If E is a locally compact separable metric space and ξ has Feller transition semigroup, then the (ξ, ϕ)-superprocess has a Hunt realization in $M(E)$.*

We have proved that the (ξ, ϕ)-superprocess with transition semigroup $(Q_t)_{t\geq 0}$ defined by (2.32) and (2.33) has a Borel right process realization. Then every right continuous realization of the superprocess with the augmented natural σ-algebras is a right process; see Theorem A.33. In particular, the (ξ, ϕ)-superprocess can be realized canonically on the space of right continuous paths from $[0, \infty)$ to $M(E)$.

Given a general continuous admissible additive functional $\{K(t) : t \geq 0\}$, we can define the semigroup of kernels $(\pi_t)_{t\geq 0}$ by (2.17). Let $\pi_t^\beta = e^{-\beta t}\pi_t$ for $\beta \geq 0$ and $t \geq 0$. If there exists $\beta \geq 0$ so that $(\pi_t^\beta)_{t\geq 0}$ is a Borel right semigroup, one can use this as a replacement of $(P_t)_{t\geq 0}$ to show that the (ξ, K, ϕ)-superprocess has a right process realization with state space $M(E)$. By Theorem A.43, this is true if $b(x) \geq \gamma(x, 1)$ for every $x \in E$. Similarly, if there exists $\beta \geq 0$ so that $(\pi_t^\beta)_{t\geq 0}$ has a Hunt realization, so does the (ξ, K, ϕ)-superprocess.

5.4 Weighted Occupation Times

In this section, we give some characterizations of a class of linear functionals of the (ξ, ϕ)-superprocess. It will be more convenient to start the underlying spatial motion and the superprocess from the arbitrary initial time $r \geq 0$. Let $\xi = (\Omega, \mathscr{F}, \mathscr{F}_{r,t}, \xi_t, \mathbf{P}_{r,x})$ and $X = (W, \mathscr{G}, \mathscr{G}_{r,t}, X_t, \mathbf{Q}_{r,\mu})$ be right continuous realizations of those processes. In view of (2.32) and (2.33), for any $t \geq r \geq 0$ and $f \in B(E)^+$ we have

$$\mathbf{Q}_{r,\mu} \exp\{-X_t(f)\} = \exp\{-\mu(u_r)\}, \tag{5.19}$$

where $u_r(x) := V_{t-r}f(x)$ satisfies

$$u_r(x) + \int_r^t \mathbf{P}_{r,x}[\phi(\xi_s, u_s)]ds = \mathbf{P}_{r,x}[f(\xi_t)], \quad 0 \leq r \leq t, x \in E. \tag{5.20}$$

Proposition 5.14 *Suppose that $\{s_1 < \cdots < s_n\} \subset [0, \infty)$ and $\{f_1, \ldots, f_n\} \subset B(E)^+$. Then we have*

$$\mathbf{Q}_{r,\mu} \exp\left\{ -\sum_{j=1}^n X_{s_j}(f_j)1_{\{r\leq s_j\}} \right\} = \exp\{-\mu(u_r)\}, \quad 0 \leq r \leq s_n, \tag{5.21}$$

where $(r, x) \mapsto u_r(x)$ is a bounded positive solution on $[0, s_n] \times E$ of

$$u_r(x) + \int_r^{s_n} \mathbf{P}_{r,x}[\phi(\xi_s, u_s)]ds = \sum_{j=1}^n \mathbf{P}_{r,x}[f_j(\xi_{s_j})]1_{\{r \le s_j\}}. \quad (5.22)$$

Proof. We shall give the proof by induction in $n \ge 1$. For $n = 1$ the result follows from (5.19) and (5.20). Now supposing (5.21) and (5.22) are satisfied when n is replaced by $n-1$, we prove they are also true for n. It is clearly sufficient to consider the case with $0 \le r \le s_1 < \cdots < s_n$. By the Markov property of X,

$$\mathbf{Q}_{r,\mu} \exp\left\{ -\sum_{j=1}^n X_{s_j}(f_j) \right\} = \mathbf{Q}_{r,\mu} \exp\left\{ -X_{s_1}(f_1) - X_{s_1}(v_{s_1}) \right\},$$

where $(r, x) \mapsto v_r(x)$ is a bounded positive Borel function on $[0, s_n] \times E$ satisfying

$$v_r(x) + \int_r^{s_n} \mathbf{P}_{r,x}[\phi(\xi_s, v_s)]ds = \sum_{j=2}^n \mathbf{P}_{r,x}[f_j(\xi_{s_j})]1_{\{r \le s_j\}}. \quad (5.23)$$

Then the result for $n = 1$ implies that

$$\mathbf{Q}_{r,\mu} \exp\left\{ -\sum_{j=1}^n X_{s_j}(f_j) \right\} = \exp\{-\mu(u_r)\}$$

with $(r, x) \mapsto u_r(x)$ being a bounded positive Borel function on $[0, s_1] \times E$ satisfying

$$u_r(x) + \int_r^{s_1} \mathbf{P}_{r,x}[\phi(\xi_s, u_s)]ds = \mathbf{P}_{r,x}[f_1(\xi_{s_1})] + \mathbf{P}_{r,x}[v_{s_1}(\xi_{s_1})]. \quad (5.24)$$

Setting $u_r = v_r$ for $s_1 < r \le s_n$, from (5.23) and (5.24) one checks that $(r, x) \mapsto u_r(x)$ is a bounded positive solution on $[0, s_n] \times E$ of (5.22). $\square$

Theorem 5.15 *Suppose that $t \ge 0$ and $\lambda \in M([0, t])$. Let $(s, x) \mapsto f_s(x)$ be a bounded positive Borel function on $[0, t] \times E$. Then we have*

$$\mathbf{Q}_{r,\mu} \exp\left\{ -\int_{[r,t]} X_s(f_s)\lambda(ds) \right\} = \exp\{-\mu(u_r)\}, \quad 0 \le r \le t, \quad (5.25)$$

where $(r, x) \mapsto u_r(x)$ is the unique bounded positive solution on $[0, t] \times E$ of

$$u_r(x) + \int_r^t \mathbf{P}_{r,x}[\phi(\xi_s, u_s)]ds = \int_{[r,t]} \mathbf{P}_{r,x}[f_s(\xi_s)]\lambda(ds). \quad (5.26)$$

Proof. Step 1. We first assume $(s, x) \mapsto f_s(x)$ is uniformly continuous on $[0, t] \times E$. For any integer $n \ge 1$ let

$$\lambda_n(ds) = \sum_{k=0}^{\infty} \lambda((t - \gamma_n(k+1), t - \gamma_n(k)] \cap [0, t]) \delta_{t-\gamma_n(k)}(ds),$$

where $\gamma_n(k) = k/2^n$. From Proposition 5.14 we see that

$$\mathbf{Q}_{r,\mu} \exp\left\{ -\int_{[r,t]} X_s(f_s)\lambda_n(ds) \right\} = \exp\{-\mu(u_n(r))\}, \qquad (5.27)$$

where $(r, x) \mapsto u_n(r, x)$ is a bounded positive solution on $[0, t] \times E$ to

$$u_n(r, x) + \int_r^t \mathbf{P}_{r,x}\left[\phi(\xi_s, u_n(s))\right]ds = \mathbf{P}_{r,x}\left[\int_{[r,t]} f_s(\xi_s)\lambda_n(ds) \right]. \quad (5.28)$$

For any $0 \le s \le t$ let $p_n(s) = t - \gamma_n([(t-s)2^n]+1)$ and $q_n(s) = t - \gamma_n([(t-s)2^n])$, where $[(t - s)2^n]$ denotes the integer part of $(t - s)2^n$. Then we have $s - 2^{-n} \le p_n(s) < s \le q_n(s) < s + 2^{-n}$. It is easy to see that

$$\int_{[r,t]} f_s(\xi_s)\lambda_n(ds) = \int_{[r,t]} f_{q_n(s)}(\xi_{q_n(s)})\lambda(ds)$$
$$+ f_{q_n(r)}(\xi_{q_n(r)})\lambda((p_n(r), r) \cap [0, t])$$

and the second term on the right-hand side tends to zero as $n \to \infty$. By the right continuity of $s \mapsto \xi_s$ and the uniform continuity of $(s, x) \mapsto f_s(x)$ we have

$$\lim_{n \to \infty} \int_{[r,t]} f_s(\xi_s)\lambda_n(ds) = \int_{[r,t]} f_s(\xi_s)\lambda(ds).$$

A similar argument shows that

$$\lim_{n \to \infty} \int_{[r,t]} X_s(f_s)\lambda_n(ds) = \int_{[r,t]} X_s(f_s)\lambda(ds).$$

From (5.27) we see the limit $u_r(x) = \lim_{n \to \infty} u_n(r, x)$ exists and (5.25) holds. It is not hard to show that $\{u_n\}$ is uniformly bounded on $[0, t] \times E$. Then we get (5.26) by letting $n \to \infty$ in (5.28).

Step 2. Let $B_1 \subset B([0, t] \times E)^+$ be the set of functions $(s, x) \mapsto f_s(x)$ for which there exist bounded positive solutions $(r, x) \mapsto u_r(x)$ of (5.26) such that (5.25) holds. It is easy to show that B_1 is closed under bounded pointwise convergence. The result of the first step shows that B_1 contains all uniformly continuous functions in $B([0, t] \times E)^+$, so we have $B_1 = B([0, t] \times E)^+$ by Proposition 1.3.

Step 3. To show the uniqueness of the solution of (5.26), suppose that $(r, x) \mapsto v_r(x)$ is another bounded positive Borel function on $[0, t] \times E$ satisfying this equation. It is easy to find a constant $K \ge 0$ such that

$$\|u_r - v_r\| \le \int_r^t \|\phi(\cdot, u_s) - \phi(\cdot, v_s)\|ds \le K \int_r^t \|u_s - v_s\|ds.$$

We may rewrite the above inequality into

$$\|u_{t-r} - v_{t-r}\| \le K \int_0^r \|u_{t-s} - v_{t-s}\| ds, \qquad 0 \le r \le t,$$

so Gronwall's lemma implies $\|u_{t-r} - v_{t-r}\| = 0$ for every $0 \le r \le t$. □

Suppose that $\lambda(ds)$ is a locally bounded Borel measure on $[0, \infty)$ and $(s, x) \mapsto f_s(x)$ is a locally bounded positive Borel function on $[0, \infty) \times E$. For any $t \ge r \ge 0$ we can define the positive random variable

$$A[r, t] := \int_{[r,t]} X_s(f_s) \lambda(ds),$$

which is called a *weighted occupation time* of the superprocess on $[r, t]$. By replacing f_s with θf_s in Theorem 5.15 for $\theta \ge 0$ we get a characterization of the Laplace transform of $A[r, t]$.

Theorem 5.16 *Let $t \ge 0$ be given. Let $f \in B(E)^+$ and let $(s, x) \mapsto g_s(x)$ be a bounded positive Borel function on $[0, t] \times E$. Then for $0 \le r \le t$ we have*

$$\mathbf{Q}_{r,\mu} \exp\left\{ - X_t(f) - \int_r^t X_s(g_s) ds \right\} = \exp\{-\mu(u_r)\}, \qquad (5.29)$$

where $(r, x) \mapsto u_r(x)$ is the unique bounded positive solution on $[0, t] \times E$ of

$$u_r(x) + \int_r^t \mathbf{P}_{r,x}[\phi(\xi_s, u_s)] ds = \mathbf{P}_{r,x}[f(\xi_t)] + \int_r^t \mathbf{P}_{r,x}[g_s(\xi_s)] ds. \quad (5.30)$$

Proof. This follows by an application of Theorem 5.15 to the measure $\lambda(ds) = ds + \delta_t(ds)$ and the function $f_s(x) = 1_{\{s<t\}} g_s(x) + 1_{\{s=t\}} f(x)$. □

Corollary 5.17 *Let $X = (W, \mathscr{G}, \mathscr{G}_t, X_t, \mathbf{Q}_\mu)$ be a right realization of the (ξ, ϕ)-superprocess started from time zero. Then for $t \ge 0$ and $f, g \in B(E)^+$ we have*

$$\mathbf{Q}_\mu \exp\left\{ - X_t(f) - \int_0^t X_s(g) ds \right\} = \exp\{-\mu(v_t)\}, \qquad (5.31)$$

where $(t, x) \mapsto v_t(x)$ is the unique locally bounded positive solution of

$$v_t(x) + \int_0^t ds \int_E \phi(y, v_s) P_{t-s}(x, dy) = P_t f(x) + \int_0^t P_s g(x) ds. \quad (5.32)$$

Corollary 5.18 *Suppose that ϕ_1 and ϕ_2 are two branching mechanisms given by (2.26) or (2.27) satisfying $\phi_1(x, f) \ge \phi_2(x, f)$ for all $x \in E$ and $f \in B(E)^+$. Let $(t, x) \mapsto v_i(t, x)$ be the solution of (2.33) with ϕ replaced by ϕ_i. Then $v_1(t, x) \le v_2(t, x)$ for all $t \ge 0$ and $x \in E$.*

Proof. Fix $t \geq 0$ and let $u_i(r, x) = v_i(t - r, x)$ for $0 \leq r \leq t$ and $x \in E$. Then $(r, x) \mapsto u_i(r, x)$ is the unique bounded positive solution of

$$u(r, x) + \int_r^t \mathbf{P}_{r,x}[\phi_i(\xi_s, u(s))]\mathrm{d}s = \mathbf{P}_{r,x}[f(\xi_t)], \qquad i = 1, 2.$$

One can see that $(r, x) \mapsto u_2(r, x)$ is also the unique bounded positive solution of

$$u(r, x) + \int_r^t \mathbf{P}_{r,x}[\phi_1(\xi_s, u(s))]\mathrm{d}s = \mathbf{P}_{r,x}[f(\xi_t)] + \int_r^t \mathbf{P}_{r,x}[g_s(\xi_s)]\mathrm{d}s,$$

where

$$g_s(x) = \phi_1(x, u_2(s)) - \phi_2(x, u_2(s))$$

is a bounded positive Borel function on $[0, t] \times E$. By Theorem 5.16 one can see $u_1(t, x) \leq u_2(t, x)$ for all $0 \leq r \leq t$ and $x \in E$. $\square$

5.5 A Counterexample

In this section we provide a counterexample showing that the (ξ, ϕ)-superprocess usually does not have a Hunt realization if the underlying spatial motion ξ is not a Hunt process. Let $E_1 := (0, 1)$ and let $\mu(\mathrm{d}x)$ be a probability measure on E_1. We define a Borel transition semigroup $(P_t)_{t \geq 0}$ on E_1 by

$$P_t f(x) = f(x - t)1_{\{0 \leq t < x\}} + \int_{E_1} f(x + x_1 - t)1_{\{x \leq t < x + x_1\}}\mu(\mathrm{d}x_1)$$
$$+ \sum_{n=2}^{\infty} \int_{E_1} \cdots \int_{E_1} f(s_n - t)1_{\{s_{n-1} \leq t < s_n\}}\mu(\mathrm{d}x_1)\cdots\mu(\mathrm{d}x_n), \quad (5.33)$$

where $s_n = x + \sum_{i=1}^n x_i$ and $f \in B(E_1)$. The corresponding Markov process ξ is intuitively described as follows. Starting from $x \in E_1$ the process moves to the left at the unit speed until it reaches zero; at that moment it takes a new position in E_1 according to the distribution $\mu(\mathrm{d}y)$; then it starts moving to the left again and so on. Clearly, the process ξ has a right realization, but none of its realizations is càdlàg. Thus ξ has no Hunt process realization. From (5.33) we have the equation

$$P_t f(x) = f(x - t)1_{\{0 \leq t < x\}} + \mu(P_{t-x}f)1_{\{t \geq x\}}, \quad t \geq 0, x \in E_1. \quad (5.34)$$

Let $C_u(E_1)$ denote the set of uniformly continuous functions on E_1. For $f \in C_u(E_1)$ it is easy to see that

$$P_t f(t) = \mu(f) \text{ and } P_{t-}f(t) = f(0+), \qquad t \in E_1. \quad (5.35)$$

Proposition 5.19 *Let $(U^\alpha)_{\alpha>0}$ be the resolvent of $(P_t)_{t\geq0}$. Then $U^\alpha C_u(E_1) \subset C_u(E_1)$ for every $\alpha > 0$ and the Ray topology of ξ is coarser than the original topology of E_1.*

Proof. In view of (5.34) for any $f \in C_u(E_1)$ we have

$$U^\alpha f(x) = \int_0^x e^{-\alpha t} f(x-t) dt + e^{-\alpha x} \int_0^\infty e^{-\alpha t} \mu(P_t f) dt. \qquad (5.36)$$

Then $U^\alpha f \in C_u(E_1)$. By Proposition A.28 the Ray topology of ξ is coarser than the original topology. $\qquad\square$

Suppose that ϕ is a spatially constant branching mechanism defined by (3.1) and $X = (W, \mathscr{G}, \mathscr{G}_t, X_t, \mathbf{Q}_\mu)$ is a right realization of the (ξ, ϕ)-superprocess. Let $x(t) = X_t(1)$ for $t \geq 0$. It is not hard to show that $\{x(t) : t \geq 0\}$ is a CB-process with cumulant semigroup defined by (3.3).

Proposition 5.20 *For any $z \in E_1$ and $a > 0$ we have $\mathbf{Q}_{a\delta_z}\{X_t = x(t)\delta_{z-t}$ for $0 \leq t < z$ and $X_z = x(z)\mu\} = 1$.*

Proof. Let $g_s(x) = 1_{\{x \neq z-s\}}$ for $s \geq 0$ and $x \in E_1$. By Theorem 5.16 we have

$$\mathbf{Q}_\mu \exp\left\{ -\int_0^z X_s(g_s) ds \right\} = \exp\{-\mu(u_0)\}, \qquad (5.37)$$

where $(r, x) \mapsto u_r(x)$ is the unique bounded positive solution of

$$u_r(x) + \int_r^z \mathbf{P}_{r,x}[\phi(u_s(\xi_s))] ds = \int_r^z \mathbf{P}_{r,x}[g_s(\xi_s)] ds, \quad 0 \leq r \leq z, x \in E_1.$$

From the above equation it follows that

$$u_r(z-r) + \int_r^z \phi(u_s(z-s)) ds = 0, \qquad 0 \leq r < z.$$

Then Gronwall's inequality implies $u_r(z-r) = 0$ for $0 \leq r < z$. In view of (5.37) we get

$$\mathbf{Q}_{a\delta_z} \exp\left\{ -\int_0^z X_s(g_s) ds \right\} = 1. \qquad (5.38)$$

Let $g_s^{(k)}(x) = 1 \wedge (k|z-s-x|)$ for $s \geq 0$ and $x \in E_1$. Then $g_s^{(k)}(x) \to g_s(x)$ increasingly as $k \to \infty$. From (5.38) it follows that

$$\mathbf{Q}_{a\delta_z} \exp\left\{ -\int_0^z X_s(g_s^{(k)}) ds \right\} = 1.$$

Then the right continuity of $s \mapsto X_s$ yields $\mathbf{Q}_{a\delta_z}\{X_s(g_s^{(k)}) = 0$ for all $0 \leq s < z$ and all integer $k \geq 1\} = 1$. That implies $\mathbf{Q}_{a\delta_z}\{X_t = x(t)\delta_{z-t}$ for $0 \leq t < z\} = 1$.

Let $(v_t)_{t \geq 0}$ be the cumulant semigroup of the CB-process defined by (3.3). Fix $f \in C_u(E_1)$ and let $h_r = V_r f(r)$ for $0 < r < 1$. From (2.33) and (5.35) we have

$$h_r = \mu(f) - \int_0^r \phi(h_s) \mathrm{d}s, \qquad 0 < r < 1,$$

and so $h_r = v_r(\mu(f))$ by the uniqueness of the solution of (3.3). Then (2.32) implies

$$\mathbf{Q}_{a\delta_z} \exp\{-X_z(f)\} = \exp\{-a v_z(\mu(f))\}. \tag{5.39}$$

By (3.2) and (5.39) one sees that $X_z(f)$ has the distribution of $x(z)\mu(f)$ under $\mathbf{Q}_{a\delta_z}$. Then we must have $\mathbf{Q}_{a\delta_z}\{X_z = x(z)\mu\} = 1$. $\qquad\square$

By Proposition 5.20 we have $\lim_{t \to z-} X_t = x(z)\delta_0$ by the weak convergence in $M([0,1))$. However, Theorem 3.5 implies $x(z) > 0$ with strictly positive probability. Then any realization of the (ξ, ϕ)-superprocess cannot be càdlàg in $M(E_1)$, so the superprocess has no Hunt process realization in $M(E_1)$. This superprocess does not even have a Hunt realization in $M(E_\rho)$, where E_ρ denotes the set E_1 equipped with the Ray topology of ξ. To show this, let $\mathcal{D} = \{1, h_1, h_2, \ldots\}$ be a countable and uniformly dense subset of $C_u(E_1)^+$, where $h_1 \in C_u(E_1)^+$ is a non-constant function satisfying $U^1 h_1(0+) = 1$. From $\mathcal{D}$ we can construct the countable Ray cone $\mathcal{R}$ for $(P_t)_{t \geq 0}$. Then $g_1 := U^1 h_1 \in \mathcal{R}$ and $g_0 := 1 \wedge g_1 \in \mathcal{R}$ are both continuous in the topology of E_ρ. By Proposition 5.19 they are also continuous in the original topology of E_1.

Corollary 5.21 *Suppose that μ has support* $\mathrm{supp}(\mu) = E_1$. *Then* $t \mapsto X_t(g_0)$ *is not quasi-left continuous.*

Proof. By (5.36) it is easy to show that $U^\alpha f(0+) = \mu(U^\alpha f)$ for every $\alpha > 0$ and $f \in C_u(E_1)$. In particular, we have $1 = g_1(0+) = \mu(g_1)$. Since g_1 is not a constant, we have

$$\mu(g_0) < \min\{\mu(1), \mu(g_1)\} = 1. \tag{5.40}$$

For each $n \geq 1$ let $T_n = \inf\{s \geq 0 : X_s((0, 1/n]) > 0\}$. Let $T = z$ if $x(z) > 0$ and let $T = \infty$ otherwise. Then $\{T_n\}$ is an increasing sequence of stopping times and Proposition 5.20 implies $\mathbf{Q}_{a\delta_z}$-a.s. $T_n \to T$. Moreover, we have $\mathbf{Q}_{a\delta_z}$-a.s.

$$\lim_{t \to z-} X_t(g_0) = x(z)g_0(0+) = x(z) \quad \text{and} \quad X_z(g_0) = x(z)\mu(g_0).$$

Since $x(z) > 0$ with strictly positive probability, by (5.40) we see $t \mapsto X_t(g_0)$ cannot be quasi-left continuous at the stopping time T. $\qquad\square$

By Corollary 5.21, any realization of the (ξ, ϕ)-superprocess cannot be quasi-left continuous in $M(E_\rho)$, so the superprocess has no Hunt realization in $M(E_\rho)$. Then it seems the last assertion in Theorem 2.20 of Fitzsimmons (1988, p.347) requires some additional condition. On the other hand, Theorem 5.8 implies that X has a Hunt realization in its own Ray topology; see Sharpe (1988, p.220). Thus the Ray topology of the (ξ, ϕ)-superprocess on $M(E_1)$ is different from the topology of $M(E_\rho)$.

5.6 Notes and Comments

The proofs in the first three sections follow those of Fitzsimmons (1988, 1992), where local branching mechanisms were considered. Some different potential theoretical methods for the regularity of superprocesses were given in Beznea (2010). For càdlàg spatial motions, the existence of right realizations of superprocesses was studied in Dynkin (1993b), Kuznetsov (1994), Leduc (2000) and Schied (1999). In particular, Leduc (2000) constructed a class of Hunt superprocesses under a second-moment condition on the kernel $H(x, d\nu)$ in the expression of the branching mechanism and proved that any Hunt MB-process satisfying certain assumptions has a version in his class. The weighted occupation times were first introduced by Iscoe (1986) for super-stable processes. They were then used in Iscoe (1988) to study supporting properties of super-Brownian motions. The results and their proofs in Section 5.4 are reorganizations of those of Dynkin (1993a).

We may think of (5.31) as a Feynman–Kac formula for the (ξ, ϕ)-superprocess X. The formula gives a characterization of the *subprocess* of X obtained from the decreasing multiplicative functional

$$t \mapsto \exp\left\{ -\int_0^t X_s(g)\mathrm{d}s \right\}.$$

In view of (2.33) and (5.32), this subprocess can also be regarded as a superprocess with branching mechanism $f \mapsto \phi(\cdot, f) - g$. This type of branching mechanism was considered in Dynkin (1994) under the technical condition

$$\lim_{\varepsilon \to 0} \sup_{x \in E} \int_{\{\nu(1) \leq \varepsilon\}} \left[\nu_x(1) + \nu(\{x\})^2\right] H(x, \mathrm{d}\nu) = 0,$$

where $\nu_x(\mathrm{d}y)$ denotes the restriction of $\nu(\mathrm{d}y)$ to $E \setminus \{x\}$.

Chapter 6
Constructions by Transformations

In this chapter, we give the construction of several classes of superprocesses by transformations. In particular, we extend the state space of the superprocess to some σ-finite measures. Other classes we shall construct include multitype superprocesses, age-structured superprocesses, conditioned superprocesses and time-inhomogeneous superprocesses. The constructions give not only the existence but also the regularity of those superprocesses. The setting of Borel right processes we have chosen is particularly suitable for the applications of those transformations.

6.1 Spaces of Tempered Measures

In this section, we construct some Dawson–Watanabe superprocesses in a space of infinite measures. Suppose that E is a Lusin topological space. We fix a strictly positive function $h \in p\mathscr{B}(E)$. Recall that $M_h(E)$ is the space of tempered measures μ on E satisfying $\mu(h) < \infty$. Let $M_h(E)^\circ = M_h(E) \setminus \{0\}$. The topology on $M_h(E)$ is defined by the convention:

$$\mu_n \to \mu \text{ if and only if } \mu_n(hf) \to \mu(hf) \text{ for all } f \in C(E).$$

Suppose that $\xi = (\Omega, \mathscr{F}, \mathscr{F}_t, \xi_t, \mathbf{P}_x)$ is a Borel right process in E with transition semigroup $(P_t)_{t\geq0}$. We here assume $(\mathscr{F}, \mathscr{F}_t)$ are the augmentations of the natural σ-algebras $(\mathscr{F}^0, \mathscr{F}_t^0)$ generated by the sample path $\{\xi_t : t \geq 0\}$. Let $(\mathscr{F}^u, \mathscr{F}_t^u)$ be the natural σ-algebras on Ω generated by $\{\xi_t : t \geq 0\}$ as random variables in E furnished with the universal σ-algebra $\mathscr{B}^u(E)$. Let $t \mapsto K(t)$ be a continuous additive functional of ξ and assume each $\omega \mapsto K_t(\omega)$ is measurable with respect to $\mathscr{F}^0$. Let $\rho \in p\mathscr{B}(E)$ be a strictly positive function so that $\rho \leq h$ and define the continuous additive functional

$$J(t) = \int_0^t \rho(\xi_s) h(\xi_s)^{-1} K(\mathrm{d}s), \qquad t \geq 0.$$

Z. Li, *Measure-Valued Branching Markov Processes*,
Probability and Its Applications, DOI 10.1007/978-3-642-15004-3_6,
© Springer-Verlag Berlin Heidelberg 2011

Suppose that there is a constant $\alpha \geq 0$ so that, as $t \to 0+$,

$$\mathbf{P}_x\left[e^{-\alpha J(t)}h(\xi_t)\right] \to h(x), \qquad x \in E \tag{6.1}$$

increasingly and

$$h(x)^{-1}\mathbf{P}_x\left[\int_0^t e^{-\alpha J(s)}\rho(\xi_s)K(\mathrm{d}s)\right] \to 0 \tag{6.2}$$

uniformly on E. Let $b \in \mathscr{B}(E)$, $c \in p\mathscr{B}(E)$ and let $\eta(x, \mathrm{d}y)$ be a σ-finite kernel on E and $H(x, \mathrm{d}\nu)$ a σ-finite kernel from E to $M_h(E)^\circ$ such that

$$\sup_{x \in E} \rho(x)^{-1}\Big\{|b(x)|h(x) + c(x)h(x)^2 + \eta(x, h)$$
$$+ \int_{M_h(E)^\circ} \left[\nu(h) \wedge \nu(h)^2 + \nu_x(h)\right]H(x, \mathrm{d}\nu)\Big\} < \infty, \tag{6.3}$$

where $\nu_x(\mathrm{d}y)$ denotes the restriction of $\nu(\mathrm{d}y)$ to $E \setminus \{x\}$. Recall that $B_h(E)$ is the set of functions $f \in \mathscr{B}(E)$ satisfying $|f| \leq \text{const} \cdot h$. Let $B_\rho(E)$ be defined similarly with ρ replacing h. We consider the operator $f \mapsto \phi(\cdot, f)$ from $B_h(E)^+$ to $B_\rho(E)$ with the representation

$$\phi(x, f) = b(x)f(x) + c(x)f(x)^2 - \int_E f(y)\eta(x, \mathrm{d}y)$$
$$+ \int_{M_h(E)^\circ} \left[e^{-\nu(f)} - 1 + \nu(\{x\})f(x)\right]H(x, \mathrm{d}\nu). \tag{6.4}$$

Theorem 6.1 *For each $f \in B_h(E)^+$ there is a unique positive solution $(t, x) \mapsto v_t(x, f) = V_t f(x)$ to the evolution equation*

$$v_t(x) = \mathbf{P}_x[f(\xi_t)] - \mathbf{P}_x\left[\int_0^t \phi(\xi_s, v_{t-s})K(\mathrm{d}s)\right], \quad t \geq 0, x \in E, \tag{6.5}$$

so that $t \mapsto \|h^{-1}v_t(\cdot, f)\|$ is bounded on each bounded interval $[0, T]$. Moreover,

$$\int_{M_h(E)} e^{-\nu(f)}Q_t(\mu, \mathrm{d}\nu) = \exp\{-\mu(V_t f)\}, \qquad f \in B_h(E)^+, \tag{6.6}$$

defines a transition semigroup $(Q_t)_{t \geq 0}$ on $M_h(E)$.

A realization of the transition semigroup $(Q_t)_{t \geq 0}$ defined by (6.6) is naturally called a (ξ, K, ϕ)-*superprocess* with state space $M_h(E)$. The proof of the above theorem is based on a number of transformations. Since $(P_t)_{t \geq 0}$ is Borel and each $\omega \mapsto J_t(\omega)$ is measurable with respect to the natural σ-algebra $\mathscr{F}^0$, we can define a Borel right semigroup $(P_t^\alpha)_{t \geq 0}$ on E by

$$P_t^\alpha f(x) = \mathbf{P}_x\left[e^{-\alpha J(t)}f(\xi_t)\right], \quad x \in E, f \in B(E).$$

Let ζ denote the lifetime of ξ. By the discussions in Sharpe (1988, pp.286–287), for every initial law μ on E there exists a probability measure $\mathbf{P}_\mu^\alpha$ on $(\Omega, \mathscr{F}^u)$ so that

$$\mathbf{P}_\mu^\alpha(H1_{\{t<\zeta\}}) = \mathbf{P}_\mu\big[e^{-\alpha J(t)}H1_{\{t<\zeta\}}\big], \qquad H \in b\mathscr{F}_t^u \tag{6.7}$$

and $\xi^\alpha = (\Omega, \mathscr{F}^\alpha, \mathscr{F}_t^\alpha, \xi_t, \mathbf{P}_x^\alpha)$ is a right process with transition semigroup $(P_t^\alpha)_{t\geq 0}$, where $(\mathscr{F}^\alpha, \mathscr{F}_t^\alpha)$ is the augmentation of $(\mathscr{F}^u, \mathscr{F}_t^u)$ by the system $\{\mathbf{P}_\mu^\alpha : \mu$ is a probability on $E\}$. From (6.1) we see that h is an excessive function for $(P_t^\alpha)_{t\geq 0}$. Then we can define another Borel semigroup $(\tilde{P}_t)_{t\geq 0}$ by

$$\tilde{P}_t f(x) = h(x)^{-1} P_t^\alpha(x, hf), \qquad x \in E, f \in B(E).$$

By the discussions in Sharpe (1988, pp.296–299), there is a unique probability kernel $\tilde{\mathbf{P}}_x(dw)$ from $(E, \mathscr{B}^u(E))$ to $(\Omega, \mathscr{F}^u)$ rendering $\{\xi_t : t \geq 0\}$ Markov with transition semigroup $(\tilde{P}_t)_{t\geq 0}$ and $\tilde{\mathbf{P}}_x\{\xi_0 = x\} = 1$. In addition, we have

$$\tilde{\mathbf{P}}_x(H1_{\{t<\zeta\}}) = h(x)^{-1}\mathbf{P}_x\big[e^{-\alpha J(t)}h(\xi_t)H\big], \qquad t \geq 0, H \in b\mathscr{F}_t^u. \tag{6.8}$$

For each initial law μ on E define $\tilde{\mathbf{P}}_\mu$ as usual and let $(\tilde{\mathscr{F}}, \tilde{\mathscr{F}}_t)$ be the augmentation of $(\mathscr{F}^u, \mathscr{F}_t^u)$ by $\{\tilde{\mathbf{P}}_\mu : \mu$ is an initial law on $E\}$. Then $\tilde{\xi} = (\Omega, \tilde{\mathscr{F}}, \tilde{\mathscr{F}}_t, \xi_t, \tilde{\mathbf{P}}_x)$ is a right process.

Lemma 6.2 *For any $t \geq 0$ and $f \in B(E)$ we have*

$$\mathbf{P}_x\left[\int_0^t f(\xi_s)J(ds)e^{-\alpha J(t)}h(\xi_t)\right]$$
$$= \mathbf{P}_x\left[\int_0^t f(\xi_s)e^{-\alpha J(s)}\rho(\xi_s)K(ds)\right]. \tag{6.9}$$

Proof. We first assume $f \in B_h(E)^+$. Let $0 = t_0 < t_1 < \cdots < t_n = t$ be a partition of $[0, t]$ and write

$$\text{l.h.s. of (6.9)} = \sum_{i=1}^n \mathbf{P}_x\left[\int_{t_{i-1}}^{t_i} f(\xi_s)J(ds)e^{-\alpha J(t)}h(\xi_t)\right].$$

One can use (6.8) to see

$$\mathbf{P}_x\big[Ge^{-\alpha J(t)}h(\xi_t)\big] = \mathbf{P}_x\big[Ge^{-\alpha J(t_i)}h(\xi_{t_i})\big], \qquad G \in p\mathscr{F}_{t_i}.$$

In particular, we get

$$\text{l.h.s. of (6.9)} = \sum_{i=1}^n \mathbf{P}_x\left[\int_{t_{i-1}}^{t_i} f(\xi_s)\rho(\xi_s)h(\xi_s)^{-1}K(ds)e^{-\alpha J(t_i)}h(\xi_{t_i})\right]$$
$$= \mathbf{P}_x\left[\int_0^t f(\xi_s)\rho(\xi_s)e^{-\alpha J(\tau_n(s))}h(\xi_{\tau_n(s)})h(\xi_s)^{-1}K(ds)\right],$$

where $\tau_n(s) = t_i$ for $t_{i-1} < s \leq t_i$. Since h is an excessive function for $(P_t^\alpha)_{t \geq 0}$, it is finely continuous relative to this semigroup. From (6.7) we see that h is also finely continuous relative to $(P_t)_{t \geq 0}$, so $t \mapsto h(\xi_t)$ is $\mathbf{P}_x$-a.s. right continuous. Then we get (6.9) by taking limits in the right-hand side of the above equation. By monotone convergence, we see (6.9) remains true for $f \in B(E)^+$. The equality for $f \in B(E)$ follows by linearity. $\square$

Proof (of Theorem 6.1). If we write $\psi(x, f) = \rho(x)^{-1}\phi(x, hf)$, then the operator $f \mapsto \psi(\cdot, f) - \alpha f$ satisfies the assumptions on the branching mechanism in Theorem 2.21. From (6.2) we see that $t \mapsto J(t)$ is an admissible additive functional of $\tilde{\xi}$. By Theorem 2.21 for each $f \in B(E)^+$ there is a unique locally bounded positive solution $(t, x) \mapsto u_t(x, f)$ to

$$u_t(x) = \tilde{\mathbf{P}}_x[f(\xi_t)] - \tilde{\mathbf{P}}_x\left\{ \int_0^t [\psi(\xi_s, u_{t-s}) - \alpha u_{t-s}(\xi_s)] J(\mathrm{d}s) \right\}$$

and the operators $U_t : f \mapsto u_t(\cdot, f)$ constitute a cumulant semigroup. Let $(Q_t^h)_{t \geq 0}$ be the transition semigroup of the Dawson–Watanabe superprocess in $M(E)$ corresponding to $(U_t)_{t \geq 0}$. By Lemma 6.2 we can rewrite the above equation into

$$h(x)u_t(x) = \mathbf{P}_x\left[e^{-\alpha J(t)}h(\xi_t)f(\xi_t)\right] - \mathbf{P}_x\left[\int_0^t \psi(\xi_s, u_{t-s})e^{-\alpha J(s)}\rho(\xi_s)K(\mathrm{d}s)\right]$$
$$+ \mathbf{P}_x\left[\int_0^t \alpha u_{t-s}(\xi_s)e^{-\alpha J(s)}\rho(\xi_s)K(\mathrm{d}s)\right].$$

A careful application of Proposition 2.9 shows the above equation is equivalent to

$$h(x)u_t(x) = \mathbf{P}_x\left[h(\xi_t)f(\xi_t)\right] - \mathbf{P}_x\left[\int_0^t \psi(\xi_s, u_{t-s})\rho(\xi_s)K(\mathrm{d}s)\right].$$

Let $v_t(x, f) = h(x)u_t(x, h^{-1}f)$ for $f \in B_h(E)^+$. Then $(t, x) \mapsto v_t(x, f)$ is the unique positive solution of (6.5) so that $t \mapsto \|h^{-1}v_t(\cdot, f)\|$ is bounded on each bounded interval $[0, T]$. Now $(Q_t^h)_{t \geq 0}$ induces a transition semigroup $(Q_t)_{t \geq 0}$ on $M_h(E)$ by the homeomorphism $\mu(\mathrm{d}x) \mapsto h(x)^{-1}\mu(\mathrm{d}x)$ from $M(E)$ to $M_h(E)$. It is easy to see that $(Q_t)_{t \geq 0}$ is characterized by (6.6). $\square$

Now let us consider the special case where $h \in p\mathscr{B}(E)$ is a strictly positive α-excessive function for $(P_t)_{t \geq 0}$ for some $\alpha \geq 0$. Then we can define a Borel right process $(\tilde{P}_t)_{t \geq 0}$ on E by

$$\tilde{P}_t f(x) = h(x)^{-1}e^{-\alpha t}P_t(x, hf), \quad x \in E, f \in B(E). \qquad (6.10)$$

Theorem 6.3 *Suppose that (6.3) is satisfied for $\rho = h$. Then for each $f \in B_h(E)^+$ there is a unique positive solution $(t, x) \mapsto v_t(x, f) = V_t f(x)$ to the evolution equation*

$$v_t(x) = P_t f(x) - \int_0^t \mathrm{d}s \int_E \phi(y, v_s)P_{t-s}(x, \mathrm{d}y), \quad t \geq 0, x \in E, \qquad (6.11)$$

so that $t \mapsto \|h^{-1}v_t(\cdot, f)\|$ *is bounded on each bounded interval* $[0, T]$. *Moreover, a Borel right semigroup* $(Q_t)_{t \geq 0}$ *on* $M_h(E)$ *is defined by* (6.6). *If, in addition, the semigroup* $(\tilde{P}_t)_{t \geq 0}$ *given by* (6.10) *has a Hunt realization, then* $(Q_t)_{t \geq 0}$ *has a Hunt realization.*

Proof. The first assertion is a special case of Theorem 6.1. By Theorem 5.12 the semigroup $(Q_t^h)_{t \geq 0}$ constructed in the proof of Theorem 6.1 has a right realization. Then $(Q_t)_{t \geq 0}$ has a right realization by Theorem A.21. If $(\tilde{P}_t)_{t \geq 0}$ has a Hunt realization, then $(Q_t^h)_{t \geq 0}$ has a Hunt realization by Theorem 5.12. From the proof of Theorem 6.1 one can see $(Q_t)_{t \geq 0}$ has a Hunt realization.   □

A typical situation where the above theorems apply is described as follows. Let **F** be the set of functions $f \in B(E)$ that are finely continuous relative to ξ. Fix $\beta > 0$ and let $(A, \mathscr{D}(A))$ be the weak generator of $(P_t)_{t \geq 0}$ defined by $\mathscr{D}(A) = U^\beta \mathbf{F}$ and $Af = \beta f - g$ for $f = U^\beta g \in \mathscr{D}(A)$. Take a constant $\alpha > 0$ and a strictly positive function $h \in \mathscr{D}(A)$ satisfying $Ah(x) \leq \alpha h(x)$ for all $x \in E$. By Theorem A.46 and integration by parts we have

$$e^{-\alpha t}P_t h(x) = h(x) + \int_0^t e^{-\alpha s}[P_s Ah(x) - \alpha P_s h(x)]ds \leq h(x). \quad (6.12)$$

Then h is an α-excessive function for $(P_t)_{t \geq 0}$.

Example 6.1 Consider the d-dimensional Euclidean space $\mathbb{R}^d$. Let $C^2(\mathbb{R}^d)$ denote the set of bounded continuous real functions on $\mathbb{R}^d$ with bounded continuous derivatives up to the second order. Suppose that ξ is a diffusion process in $\mathbb{R}^d$ with generator A defined by

$$Af(x) = \sum_{i,j=1}^d a_{ij}(x)\frac{\partial^2 f}{\partial x_i \partial x_j}(x) + \sum_{j=1}^d b_j(x)\frac{\partial f}{\partial x_j}(x), \quad f \in C^2(\mathbb{R}^d),$$

where $x \mapsto a_{ij}(x)$ and $x \mapsto b_j(x)$ are bounded Hölder continuous functions on $\mathbb{R}^d$ satisfying the *uniform elliptic condition*. That is, there is a constant $\theta_0 > 0$ so that

$$\sum_{i,j=1}^d a_{ij}(x)u_i u_j \geq \theta_0 \sum_{i=1}^d u_i^2, \quad x \in \mathbb{R}^d, u_i \in \mathbb{R}, i = 1, \dots, d.$$

Fix $p > 0$ and let $h(x) = (1+|x|^2)^{-p/2}$ for $x \in \mathbb{R}^d$, where $|\cdot|$ denotes the Euclidean norm. It is easy to find a constant $\alpha > 0$ so that $|Ah(x)| \leq \alpha h(x)$ for all $x \in \mathbb{R}^d$.

Example 6.2 Let ξ be the standard one-dimensional Brownian motion killed at the origin. Then ξ has state space $\mathbb{R}^\circ := \mathbb{R} \setminus \{0\}$. Let $(P_t)_{t \geq 0}$ denote the transition semigroup of ξ. For any $t > 0$ the sub-Markov kernel $P_t(x, dy)$ has density

$$p_t(x, y) = \begin{cases} g_t(x-y) - g_t(x+y) & \text{if } xy > 0, \\ 0 & \text{otherwise,} \end{cases}$$

where $g_t(z)$ is given by (2.44). It is easy to show that $h(x) \equiv |x|$ is an invariant function for $(P_t)_{t \geq 0}$. Let $\phi(x, f) = |x|^{-\sigma} f(x)^{1+\beta}$ for constants $0 < \beta \leq 1$ and $\beta \leq \sigma \leq 1 + \beta$. Then (6.3) is satisfied with $\rho(x) = |x|^{1+\beta-\sigma}$. Moreover, since $0 \leq 1 + \beta - \sigma \leq 1$, we have

$$
\begin{aligned}
\int_0^t \mathbf{P}_x[\rho(\xi_s)] ds &\leq \int_0^t \mathbf{P}_x[1 + |\xi_s|] ds \\
&\leq \int_0^t \frac{ds}{\sqrt{2\pi s}} \int_{-|x|}^{|x|} e^{-z^2/2s} dz + t|x| \\
&\leq \left(\frac{2\sqrt{2t}}{\sqrt{\pi}} + t \right) h(x).
\end{aligned}
$$

By Theorem 6.1 we can define a cumulant semigroup $(V_t)_{t \geq 0}$ on $B_h(\mathbb{R}^\circ)^+$ by

$$
V_t f(x) = \mathbf{P}_x[f(\xi_t)] - \int_0^t \mathbf{P}_x\{|\xi_s|^{-\sigma} V_{t-s} f(\xi_s)^{1+\beta}\} ds, \quad t \geq 0, x \in \mathbb{R}^\circ.
$$

That gives a (ξ, ϕ)-superprocess X in $M_h(\mathbb{R}^\circ)$. By Proposition 2.27 and the construction in the proof of Theorem 6.1 we have the moment formula

$$
\mathbf{Q}_\mu[X_t(f)] = \mu(P_t f), \quad t \geq 0, \mu \in M_h(\mathbb{R}^\circ), f \in B_h(\mathbb{R}^\circ).
$$

Then we can also take $M(\mathbb{R}^\circ)$ as the state space of X and the above formula remains true for $\mu \in M(\mathbb{R}^\circ)$ and $f \in B(\mathbb{R}^\circ)$. It is simple to see that $\{X_t(1) : t \geq 0\}$ is a supermartingale but not a martingale unless $X_0 = 0$.

6.2 Multitype Superprocesses

In this section, we derive the existence of some multitype superprocesses from the non-local branching superprocess. For simplicity we only consider Lebesgue killing densities. Suppose that E and I are Lusin topological spaces. Let $\xi = \{\Omega, \mathscr{F}, \mathscr{F}_t, (\xi_t, \alpha_t), \mathbf{P}_{(x,a)}\}$ be a Borel right process with state space $E \times I$. Let $\phi = \phi(x, a, z)$ and $\zeta = \zeta(x, a, z)$ be given respectively by (2.45) and (2.46) with $x \in E$ replaced by $(x, a) \in E \times I$. Let $\pi(x, a, db)$ be a probability kernel from $E \times I$ to I. By Theorem 5.12, there is a Borel right superprocess $X = (W, X_t, \mathscr{G}, \mathscr{G}_t, \mathbf{Q}_\mu)$ with state space $M(E \times I)$ and with transition probabilities determined by

$$
\mathbf{Q}_\mu \exp\{-X_t(f)\} = \exp\{-\mu(V_t f)\}, \quad t \geq 0, f \in B(E \times I)^+, \quad (6.13)
$$

where $t \mapsto V_t f$ is the unique locally bounded positive solution of

$$
\begin{aligned}
V_t f(x, a) = \mathbf{P}_{(x,a)}[f(\xi_t, \alpha_t)] &- \int_0^t \mathbf{P}_{(x,a)}[\phi(\xi_s, \alpha_s, V_{t-s} f(\xi_s, \alpha_s))] ds \\
&+ \int_0^t \mathbf{P}_{(x,a)}[\zeta(\xi_s, \alpha_s, \pi(\xi_s, \alpha_s, V_{t-s} f(\xi_s, \cdot)))] ds. \quad (6.14)
\end{aligned}
$$

We may call X a *multitype superprocess* with type space I. Heuristically, $\{\xi_t : t \geq 0\}$ gives the law of migration of the particles, $\{\alpha_t : t \geq 0\}$ represents the mutation of their types, $\phi(x, a, \cdot)$ describes the death and birth of particles of type $a \in I$ at $x \in E$, and $\zeta(x, a, \cdot)$ describes the amount of the offspring born by a parent of type $a \in I$ at $x \in E$ that change into new types randomly according to the kernel $\pi(x, a, db)$. In this model, the offspring may change their types, but they all start migrating from the death sites of their parents. Note that the migration process $\{\xi_t : t \geq 0\}$ and the mutation process $\{\alpha_t : t \geq 0\}$ are not necessarily independent.

From the non-local branching superprocess, we can derive another multitype superprocess. Suppose that $\zeta = \zeta(x, a, b, z)$ is given by (2.46) with $x \in E$ replaced by $(x, a, b) \in E \times I^2$. Instead of (6.14), we may also define $t \mapsto V_t f$ by

$$V_t f(x, a) = \mathbf{P}_{(x,a)} [f(\xi_t, \alpha_t)] - \int_0^t \mathbf{P}_{(x,a)} [\phi(\xi_s, \alpha_s, V_{t-s} f(\xi_s, \alpha_s))] ds$$

$$+ \int_0^t \mathbf{P}_{(x,a)} \left[\int_I \zeta(\xi_s, \alpha_s, b, V_{t-s} f(\xi_s, b)) \pi(\xi_s, \alpha_s, db) \right] ds. \quad (6.15)$$

The resulting multitype superprocess X can be interpreted in the same way as the above except the mutations of the particles. In this case, when a particle of type $a \in I$ dies at point $x \in E$, a new type index $b \in I$ is first chosen following the distribution $\pi(x, a, \cdot)$, then offspring of type b are produced according to the law given by $\zeta(x, a, b, \cdot)$.

Example 6.3 Let us consider the case where $I = \mathbb{R}_+$ and $\alpha_t = \alpha_0 + t$ for all $t \geq 0$. Suppose that $\xi = (\Omega, \mathscr{F}, \mathscr{F}_t, \xi_t, \mathbf{P}_x)$ is a Borel right process with state space E. Let $\rho \in B(E \times \mathbb{R}_+)$ and let $\zeta = \zeta(x, a, z)$ be given by (2.46) with $x \in E$ replaced by $(x, a) \in E \times \mathbb{R}_+$. In addition, we assume $\sup_{x,a} \zeta_z'(x, a, 0+) \leq 1$. A special form of (6.14) is the equation

$$V_t f(x, a) = \mathbf{P}_x [f(\xi_t, a + t)] - \int_0^t \mathbf{P}_x [\rho(\xi_s, a + s) V_{t-s} f(\xi_s, a + s)] ds$$

$$+ \int_0^t \mathbf{P}_x [\rho(\xi_s, a + s) \zeta(\xi_s, a + s, V_{t-s} f(\xi_s, 0))] ds. \quad (6.16)$$

The corresponding multitype superprocess in $M(E \times \mathbb{R}_+)$ is called an *age-structured superprocess*. Clearly, we can also get (6.16) as a special case of (4.40) with $\phi = 0$. Heuristically, ξ_t represents the location of a "particle" and α_t represents its age. At its branching time a particle gives birth to a random number of offspring whose spatial motions start from the branching site and whose ages start from zero. See also the explanations following Theorem 4.15.

In many cases, we only consider multitype superprocesses with finite or countable type spaces. For simplicity, let $I = \{1, 2, \ldots, k\}$. Suppose for each $i \in I$ we have:

- a Borel right process ξ_i in E with transition semigroup $(P_i(t))_{t \geq 0}$;
- a function $\phi_i = \phi_i(x, z)$ belonging to the class given by (2.45);
- a discrete probability distribution $p_i(x) = \{p_{ij}(x) : j \in I\}$ on I with $p_{ij} \in B(E)^+$;
- a function $\zeta_i = \zeta_i(x, z)$ belonging to the class given by (2.46).

The next two theorems deal with processes with state space $M(E)^I$. We shall write $\mu = (\mu_i : i \in I)$ and $Y_t = (Y_i(t) : i \in I)$.

Theorem 6.4 *There is a Borel right process* $Y = (W, Y_t, \mathscr{G}, \mathscr{G}_t, \mathbf{Q}_\mu)$ *with state space* $M(E)^I$ *and with transition probabilities defined by*

$$\mathbf{Q}_\mu \exp\left\{ -\sum_{i \in I} \langle Y_i(t), f_i \rangle \right\} = \exp\left\{ -\sum_{i \in I} \langle \mu_i, v_i(t) \rangle \right\}, \qquad (6.17)$$

where $f_i \in B(E)^+$ *and* $v_i(t) = v_i(t, x)$ *is determined by the evolution equation*

$$v_i(t, x) = P_i(t) f_i(x) - \int_0^t ds \int_E \phi_i(y, v_i(t - s, y)) P_i(s, x, dy)$$

$$+ \int_0^t ds \int_E \zeta_i\left(y, \sum_{j \in I} p_{ij}(y) v_j(t - s, y)\right) P_i(s, x, dy). \quad (6.18)$$

Proof. Let ξ be the Borel right process in the product space $E \times I$ with transition semigroup $(P_t)_{t \geq 0}$ defined by

$$P_t f(x, i) = \int_E f(y, i) P_i(t, x, dy), \qquad (x, i) \in E \times I.$$

Let $\phi(x, i, z) = \phi_i(x, z)$ and let $\pi(x, i, \cdot)$ be the Markov kernel from $E \times I$ to I defined by

$$\pi(x, i, \cdot) = \sum_{j \in I} p_{ij}(x) \delta_j(\cdot),$$

where $\delta_j(\cdot)$ stands for the unit mass at $j \in I$. Then we have a multitype superprocess $\{X_t : t \geq 0\}$ in $M(E \times I)$ defined by (6.13) and (6.14). For $i \in I$ and $\mu \in M(E \times I)$ we define $U_i\mu \in M(E)$ by $U_i\mu(B) = \mu(B \times \{i\})$ for $B \in \mathscr{B}(E)$. Then $\mu \mapsto (U_i\mu : i \in I)$ is a homeomorphism between $M(E \times I)$ and $M(E)^I$. Let $Y_i(t) = U_i X_t$. It is clear that $\{(Y_i(t) : i \in I) : t \geq 0\}$ is a Markov process in $M(E)^I$ with transition probabilities defined by (6.17) and (6.18). By Theorem A.21 this process has a right process realization. $\qquad\square$

The heuristical meaning of the process $\{(Y_i(t) : i \in I) : t \geq 0\}$ constructed in Theorem 6.4 is described as follows. The process ξ_i gives the law of migration of the particles of type $i \in I$, $\phi_i(x, \cdot)$ describes the death and birth of the particles of type $i \in I$ at $x \in E$, and $\zeta_i(x, \cdot)$ describes the amount of the offspring born by a parent of type $i \in I$ at $x \in E$ that change into new types randomly according to the discrete distribution $p_i(x) = \{p_{ij}(x) : j \in I\}$.

For $i \in I$ let the parameters (ξ_i, ϕ_i, p_i) be given as in the paragraph before Theorem 6.4. Let $\zeta_{ij} = \zeta_{ij}(x, z)$ be given by (2.46) depending on the parameters $i, j \in I$. Then we have the following:

Theorem 6.5 *There is a Borel right process* $Y = (W, Y_t, \mathscr{G}, \mathscr{G}_t, \mathbf{Q}_\mu)$ *with state space* $M(E)^I$ *and with transition probabilities defined by* (6.17) *with* $v_i(t) = v_i(t, x)$ *determined by the evolution equation*

$$v_i(t, x) = P_i(t)f_i(x) - \int_0^t ds \int_E \phi_i(y, v_i(t - s, y))P_i(s, x, dy)$$

$$+ \int_0^t ds \int_E \sum_{j \in I} p_{ij}(y)\zeta_{ij}(y, v_j(t - s, y))P_i(s, x, dy). \quad (6.19)$$

The proof of the above theorem is based on (6.13) and (6.15) and is similar to that of Theorem 6.4. The process constructed in this way can be interpreted as the one given in Theorem 6.4. The difference is that when a particle of type $i \in I$ dies at $x \in E$, a new label $j \in I$ is first chosen according to the distribution $p_i(x) = \{p_{ij}(x) : j \in I\}$, then offspring of this type are produced at $x \in E$ according to the law given by $\zeta_{ij}(x, \cdot)$.

6.3 A Two-Type Superprocess

A two-type superprocess can be constructed by a conditioning argument. For $i = 1$ and 2, let ξ_i be a Borel right process in E with transition semigroup $(P_i(t))_{t \geq 0}$ and let $\phi_i = \phi_i(x, z)$ be a function belonging to the class given by (2.45). Given $f_i \in B(E)^+$ let $v_i(t) = v_i(t, x)$ be defined by

$$v_i(t, x) = P_i(t)f_i(x) - \int_0^t ds \int_E \phi_i(y, v_i(t - s, y))P_i(s, x, dy). \quad (6.20)$$

Let $X = \{W, (X_1(t), X_2(t)), \mathscr{G}, \mathscr{G}_t, \mathbf{Q}_{(\mu_1, \mu_2)}\}$ be a Markov process in $M(E) \times M(E)$ so that $(X_1(t), \mathscr{G}_t)$ under $\mathbf{Q}_{(\mu_1, \mu_2)}$ is a (ξ_1, ϕ_1)-superprocess and $(X_2(t), \mathscr{G}_t)$ under the conditional probability $\mathbf{Q}_{(\mu_1, \mu_2)}\{\cdot | X_1(t) : t \geq 0\}$ is an inhomogeneous Markov process in $M(E)$ with transition semigroup $(Q_{r,t}^{X_1})_{t \geq r}$ determined by

$$\int_{M(E)} e^{-\langle \nu, f \rangle} Q_{r,t}^{X_1}(\mu_2, d\nu)$$

$$= \exp\left\{ - \langle \mu_2, v_2(t - r) \rangle - \int_r^t \langle X_1(s), v_2(t - s) \rangle ds \right\}. \quad (6.21)$$

Using (6.21) and Theorem 5.16 we obtain

$$\mathbf{Q}_{(\mu_1, \mu_2)} \exp\left\{ - \langle X_1(t), f_1 \rangle - \langle X_2(t), f_2 \rangle \right\}$$

$$= \mathbf{Q}_{(\mu_1, \mu_2)} \exp\left\{ - \langle X_1(t), f_1 \rangle - \langle \mu_2, v_2(t) \rangle - \int_0^t \langle X_1(s), v_2(t - s) \rangle ds \right\}$$

$$= \exp\left\{ - \langle \mu_1, u_1(0, t) \rangle - \langle \mu_2, v_2(t) \rangle \right\}$$

where $u_1(r, t) = u_1(r, t, x)$ is the solution of

$$u_1(r, t, x) + \int_r^t ds \int_E \phi_1(y, u_1(s, t, y)) P_1(s - r, x, dy)$$
$$= P_1(t - r) f_1(x) + \int_r^t ds \int_E v_2(t - s, y) P_1(s - r, x, dy). \quad (6.22)$$

Let $w_1(t) = w_1(t, x)$ be defined by

$$w_1(t, x) + \int_0^t ds \int_E \phi_1(y, w_1(t - s, y)) P_1(s, x, dy)$$
$$= P_1(t) f_1(x) + \int_0^t ds \int_E v_2(t - s, y) P_1(s, x, dy). \quad (6.23)$$

It is not hard to show that $u_1(r, t, x) = w_1(t - r, x)$ for all $t \geq r \geq 0$. Obviously, the system (6.20) and (6.23) can be regarded as a special case of (6.18) or (6.19). Then X is in fact a special two-type superprocess. In this model, particles of type one can produce particles of type two, but particles of type two cannot produce those of type one.

6.4 Change of the Probability Measure

Let E be a Lusin topological space and let ξ be a conservative Borel right process in E with transition semigroup $(P_t)_{t \geq 0}$. For simplicity we consider a local branching mechanism $(x, z) \mapsto \phi(x, z)$ given by (2.45) with constant function $b(x) \equiv b \geq 0$. Let $(Q_t)_{t \geq 0}$ denote the transition semigroup of the (ξ, ϕ)-superprocess defined by (2.32) and (2.33). By Corollary 2.28,

$$\int_{M(E)} \nu(1) Q_t(\mu, d\nu) = e^{-bt} \mu(1)$$

for $t \geq 0$ and $\mu \in M(E)$. Then we can define a Borel transition semigroup $(\tilde{Q}_t)_{t \geq 0}$ on $M(E)^\circ$ by

$$\tilde{Q}_t(\mu, d\nu) = e^{bt} \mu(1)^{-1} \nu(1) Q_t(\mu, d\nu). \quad (6.24)$$

This formula is a simple variation of the h-transform of Doob; see, e.g., Sharpe (1988, p.298). A realization of $(\tilde{Q}_t)_{t \geq 0}$ can be obtained by a change of the probability measure in the (ξ, ϕ)-superprocess.

Let W be the space of paths $w : [0, \infty) \to M(E)$ that are right continuous in both $M(E)$ and $M(E_\rho)$ and have left limits in $M(\bar{E})$, where $\bar{E}$ is a Ray–Knight completion of E with respect to ξ and E_ρ denotes the set E with the Ray topology inherited from $\bar{E}$. Let W_0 be the set of paths $w \in W$ that have zero as a trap. Let $X = (W_0, \mathcal{G}, \mathcal{G}_t, X_t, \mathbf{Q}_\mu)$ be the canonical Borel right realization of the (ξ, ϕ)-

superprocess. Let $\tau_0 = \inf\{t \geq 0 : X_t(1) = 0\}$ denote the *extinction time* of X. It is easy to show that

$$m_t := e^{bt} X_0(1)^{-1} X_t(1), \qquad t \geq 0, \tag{6.25}$$

defines a positive martingale multiplicative functional of the restriction of X on $M(E)^\circ$. Let $(\mathscr{G}^u, \mathscr{G}_t^u)$ be the natural σ-algebras on W_0 generated by $\{X_t : t \geq 0\}$ as random variables on $M(E)$ furnished with the universal σ-algebra $\mathscr{B}^u(M(E))$. By the results in Sharpe (1988, p.296), for each $\mu \in M(E)^\circ$ there is a unique probability measure $\tilde{\mathbf{Q}}_\mu$ on $(W_0, \mathscr{G}^u)$ so that $\{X_t : t \geq 0\}$ is a Markov process with transition semigroup $(\tilde{Q}_t)_{t \geq 0}$ and $\tilde{\mathbf{Q}}_\mu\{X_0 = \mu\} = 1$. In addition, we have

$$\tilde{\mathbf{Q}}_\mu(H1_{\{t < \tau_0\}}) = \mu(1)^{-1} \mathbf{Q}_\mu[e^{bt} X_t(1) H], \quad t \geq 0, H \in b\mathscr{G}_t^u.$$

For each initial law K on $M(E)^\circ$ define $\tilde{\mathbf{Q}}_K$ in the usual way. Then the system $\tilde{X} = (W_0, \mathscr{G}, \mathscr{G}_t, X_t, \tilde{\mathbf{Q}}_\mu)$ is a Borel right process, where $(\mathscr{G}, \mathscr{G}_t)$ is the augmentation of $(\mathscr{G}^u, \mathscr{G}_t^u)$ by $\{\tilde{\mathbf{Q}}_K : K \text{ is an initial law on } M(E)^\circ\}$. Consequently, we have the following:

Theorem 6.6 *The semigroup $(\tilde{Q}_t)_{t \geq 0}$ on $M(E)^\circ$ has a right realization.*

The process $\tilde{X}$ defined above is called the *subprocess* of the (ξ, ϕ)-superprocess X generated by the martingale multiplicative functional $\{m_t : t \geq 0\}$. Let

$$\phi_0'(x, z) = 2c(x)z + \int_0^\infty u(1 - e^{-zu}) m(x, du), \quad x \in E, z \geq 0. \tag{6.26}$$

The next theorem gives a characterization of the transition semigroup $(\tilde{Q}_t)_{t \geq 0}$.

Theorem 6.7 *For every $t \geq 0$, $\mu \in M(E)^\circ$ and $f \in B(E)^+$ we have*

$$\int_{M(E)^\circ} e^{-\nu(f)} \tilde{Q}_t(\mu, d\nu) = \exp\{-\mu(V_t f)\} \hat{\mu}(U_t f), \tag{6.27}$$

where $\hat{\mu} = \mu(1)^{-1}\mu$ and $(t, x) \mapsto U_t f(x)$ is the unique locally bounded positive solution to

$$U_t f(x) = 1 - \int_0^t ds \int_E \phi_0'(y, V_s f(y)) U_s f(y) P_{t-s}(x, dy). \tag{6.28}$$

Proof. We first use Proposition 2.29 to see

$$\int_{M(E)} \nu(1) e^{-\nu(f)} Q_t(\mu, d\nu) = \exp\{-\mu(V_t f)\} \mu(V_t^1 f),$$

where $(t, x) \mapsto V_t^1 f(x)$ is the unique locally bounded positive solution of

$$V_t^1 f(x) = 1 - \int_0^t ds \int_E [b + \phi_0'(y, V_s f(y))] V_s^1 f(y) P_{t-s}(x, dy).$$

By Proposition 2.9 the above equation is equivalent to

$$V_t^1 f(x) = e^{-bt} - \int_0^t ds \int_E \phi_0'(y, V_s f(y)) V_s^1 f(y) P_{t-s}^b(x, dy).$$

Then we have (6.27) and (6.28) with $U_t f(x) = e^{bt} V_t^1 f(x)$.   □

By a modification of the proof of Proposition 2.9 it is not hard to show that the solution of (6.28) can be expressed in terms of the spatial motion process as

$$U_t f(x) = \mathbf{P}_x \left[\exp \left\{ - \int_0^t \phi_0'(\xi_s, V_{t-s} f(\xi_s)) ds \right\} \right].$$

Using Theorems 1.35 and 1.37 we see that the quantity under expectation gives the Laplace functional of an infinitely divisible probability measure on $M(E)$. Then for each $\mu \in M(E)^\circ$ a probability measure $N_t(\mu, d\nu)$ on $M(E)$ is defined by

$$\int_{M(E)} e^{-\nu(f)} N_t(\mu, d\nu) = \hat{\mu}(U_t f), \qquad f \in B(E)^+.$$

Now (6.27) implies

$$\tilde{Q}_t(\mu, \cdot) = Q_t(\mu, \cdot) * N_t(\mu, \cdot). \tag{6.29}$$

This decomposition describes an interesting structure of the semigroup $(\tilde{Q}_t)_{t \geq 0}$. Recall that $\tau_0 = \inf\{t \geq 0 : X_t(1) = 0\}$. The next theorem shows that in the subcritical case we can understand $\tilde{X}$ as the *conditioned superprocess* given the null event $\{\tau_0 = \infty\}$. The proof is very similar to that of Theorem 3.25 and is left to the reader.

Theorem 6.8 *Suppose that the branching mechanism ϕ is spatially constant and satisfies Condition 3.6. Then for any $t \geq 0$ and $\mu \in M(E)^\circ$, the distribution of X_t under $\mathbf{Q}_\mu\{\cdot | r + t < \tau_0\}$ converges to $\tilde{Q}_t(\mu, \cdot)$ as $r \to \infty$.*

6.5 Time-Inhomogeneous Superprocesses

Suppose that $I \subset \mathbb{R}_+$ is an interval and F is a Lusin topological space. Let $\tilde{E}$ be a Borel subset of $I \times F$. For any $t \in I$ let $I_t = [0, t] \cap I$ and $E_t = \{x \in F : (t, x) \in \tilde{E}\}$. We fix an abstract point $\partial \notin I \times F$ and assume all functions on $\tilde{E} \subset I \times F$ have been extended trivially to $\tilde{E}^c \cup \{\partial\}$. Let us consider an inhomogeneous Borel right transition semigroup $(P_{r,t} : t \geq r \in I)$ with global state space $\tilde{E}$. Let $(\tilde{P}_t)_{t \geq 0}$ be the corresponding time–space semigroup on $\tilde{E}$ defined by (A.44). Suppose that $\tilde{\xi} =$

$(\Omega, \tilde{\mathscr{F}}, \tilde{\mathscr{F}}_t, (\alpha_t, y_t), \mathbf{P}_{r,x})$ is a right process realizing $(\tilde{P}_t)_{t \geq 0}$, where $\alpha_t = \alpha_0 + t$ for all $t \geq 0$. For $\omega \in \Omega$ define

$$\xi_t(\omega) = \begin{cases} y_{t-\alpha_0(\omega)}(\omega) & \text{if } t \in I \cap [\alpha_0(\omega), \infty), \\ \partial & \text{if } t \in I \cap [0, \alpha_0(\omega)). \end{cases}$$

Let $\mathscr{F} = \sigma(\{\xi_t : t \in I\})$ and let $\mathscr{F}_{r,t} = \sigma(\{\xi_s : r \leq s \leq t\})$ for $t \geq r \in I$. By Theorem A.59, the system $\xi = (\Omega, \mathscr{F}, \mathscr{F}_{r,t}, \xi_t, \mathbf{P}_{r,x})$ is a right continuous inhomogeneous Markov process realizing $(P_{r,t} : t \geq r \in I)$.

Lemma 6.9 *The set $\tilde{M} := \{(t, \mu) : t \in I, \mu \in M(E_t)\}$ with the topology inherited from $I \times M(F)$ is a Lusin topological space.*

Proof. We here understand $\tilde{M} = \{(t, \mu) \in I \times M(F) : \mu(F \setminus E_t) = 0\}$. For $(t, \mu) \in I \times M(F)$ define $\gamma_t \in M(I \times F)$ by $\gamma_t(B) = \mu(\{x \in F : (t, x) \in B\})$, where $B \in \mathscr{B}(I \times F)$. Let

$$M_0 = \{\gamma \in M(I \times F) : \gamma((I \setminus \{t\}) \times F) = 0 \text{ for some } t \in I\}.$$

Let $Q = \{0, r_1, r_2, \dots\}$ be the set of positive rationals. For $r \in Q$ and $n \geq 1$ let

$$M_{n,r} = \{\gamma \in M(I \times F) : \gamma((I \setminus [r, r + 1/n]) \times F) = 0\}.$$

Then $M_0 = \cap_{n \geq 1} \cup_{r \in Q} M_{n,r}$, so M_0 is a Borel subset of $M(I \times F)$. It is easy to see that the mapping $(t, \mu) \mapsto \gamma_t$ induces a homeomorphism between $\tilde{M}$ and $M_0 \cap M(\tilde{E})$. Therefore $\tilde{M}$ is a Lusin topological space. $\quad\square$

Let $b \in B(\tilde{E})$ and $c \in B(\tilde{E})^+$. Let $\eta(s, x, \mathrm{d}y)$ be a bounded kernel on $\tilde{E}$ and let $H(s, x, \mathrm{d}\nu)$ be a σ-finite kernel from $\tilde{E}$ to $M(\tilde{E})^\circ$. For every $(s, x) \in \tilde{E}$ we assume $\eta(s, x, \mathrm{d}y)$ is supported by $\{s\} \times E_s$ and $H(s, x, \mathrm{d}\nu)$ is supposed by $M(\{s\} \times E_s)^\circ$. Then we can regard $\eta(s, x, \mathrm{d}y)$ as a measure on E_s and regard $H(s, x, \mathrm{d}\nu)$ as a measure on $M(E_s)^\circ$. In addition, we assume

$$\sup_{(s,x) \in \tilde{E}} \left[|b(s, x)| + c(s, x) + \eta(s, x, E_s) \right. \\ \left. + \int_{M(E_s)^\circ} [\nu(1) \wedge \nu(1)^2 + \nu_x(1)] H(s, x, \mathrm{d}\nu) \right] < \infty, \quad (6.30)$$

where $\nu_x(\mathrm{d}y)$ denotes the restriction of $\nu(\mathrm{d}y)$ to $E_s \setminus \{x\}$. For $(s, x) \in \tilde{E}$ and $f \in B(E_s)^+$ define

$$\phi(s, x, f) = b(s, x)f(x) + c(s, x)f(x)^2 - \int_{E_s} f(y)\eta(s, x, \mathrm{d}y) \\ + \int_{M(E_s)^\circ} \left[e^{-\nu(f)} - 1 + \nu(\{x\})f(x) \right] H(s, x, \mathrm{d}\nu). \quad (6.31)$$

The following theorem establishes the existence and regularity of a class of *time-inhomogeneous superprocesses*.

Theorem 6.10 *For every* $t \in I$ *and* $f \in B(E_t)^+$ *there is a unique bounded positive solution* $(r, x) \mapsto v_{r,t}(x) = V_{r,t}f(x)$ *to the integral equation*

$$v_{r,t}(x) = \mathbf{P}_{r,x}[f(\xi_t)] - \int_r^t \mathbf{P}_{r,x}[\phi(s, \xi_s, v_{s,t})]\mathrm{d}s, \quad r \in I_t, x \in E_r. \quad (6.32)$$

Moreover, we can define an inhomogeneous Borel right transition semigroup $(Q_{r,t} : t \geq r \in I)$ *with global state space* $\tilde{M}$ *by*

$$\int_{M(E_t)} \mathrm{e}^{-\nu(f)} Q_{r,t}(\mu, \mathrm{d}\nu) = \exp\{-\mu(V_{r,t}f)\}, \quad f \in B(E_t)^+. \quad (6.33)$$

Proof. Given $\tilde{f} \in B(\tilde{E})^+$ we can apply Theorem 2.21 to the time–space process $\tilde{\xi}$ to see there is a unique locally bounded positive solution $(t, r, x) \mapsto \tilde{v}_t(r, x) = \tilde{V}_t\tilde{f}(r, x)$ to the evolution equation

$$\tilde{v}_t(r, x) = \mathbf{P}_{r,x}[\tilde{f}(r + t, y_t)] - \int_0^t \mathbf{P}_{r,x}[\phi(r + s, y_s, \tilde{v}_{t-s})]\mathrm{d}s, \quad (6.34)$$

where $t \geq 0$ and $(r, x) \in \tilde{E}$. Moreover, the family of operators $(\tilde{V}_t)_{t \geq 0}$ constitute a cumulant semigroup. By Theorem 5.12, the corresponding superprocess in $M(\tilde{E})$ has a right realization $\tilde{X} = (\tilde{W}, \tilde{\mathscr{G}}, \tilde{\mathscr{G}}_t, \tilde{X}_t, \tilde{\mathbf{Q}}_\mu)$. We define a bounded kernel on $\tilde{E}$ by

$$\gamma(s, x, \mathrm{d}y) = \eta(s, x, \mathrm{d}y) + \int_{M(\{s\} \times E_s)^\circ} \nu_{(s,x)}(\mathrm{d}y)H(s, x, \mathrm{d}\nu),$$

where $\nu_{(s,x)}(\mathrm{d}y)$ denotes the restriction of $\nu(\mathrm{d}y)$ to $\tilde{E} \setminus \{(s, x)\}$. For any $(s, x) \in \tilde{E}$ we can also regard $\gamma(s, x, \mathrm{d}y)$ as a measure on E_s. Let $(\tilde{\pi}_t)_{t \geq 0}$ be the semigroup of linear operators on $B(\tilde{E})$ defined by

$$\tilde{\pi}_t\tilde{f}(r, x) = \mathbf{P}_{r,x}[\tilde{f}(r + t, y_t)] - \int_0^t \mathbf{P}_{r,x}[b(r + s, y_s)\tilde{\pi}_{t-s}\tilde{f}(r + s, y_s)]\mathrm{d}s$$
$$+ \int_0^t \mathbf{P}_{r,x}[\gamma(r + s, y_s, \tilde{\pi}_{t-s}\tilde{f}(r + s, \cdot))]\mathrm{d}s.$$

By the construction given in Proposition A.41 it is not hard to see that for any $t \geq 0$ and $(r, x) \in \tilde{E}$ the finite measure $\tilde{\pi}_t(r, x, \cdot)$ is supported by $\{r + t\} \times E_{r+t} \subset \tilde{E}$. From the moment formula in Proposition 2.27 one can see for any $\mu \in M(\tilde{E})$ carried by $\{r\} \times E_r$ the random measure $\tilde{X}_t \in M(\tilde{E})$ is $\tilde{\mathbf{Q}}_\mu$-a.s. carried by $\{r + t\} \times E_{r+t}$. In particular, for any $f \in B(E_{r+t})^+$ we have $\tilde{\pi}_t(s, x, 1_{\{r+t\}}f) = 0$ if $s \in I \setminus \{r\}$. Then we can use the result of Proposition 2.20 to see

$$\tilde{V}_t(1_{\{r+t\}}f)(s, x) = 1_{\{r\}}(s)\tilde{V}_t(1_{\{r+t\}}f)(s, x), \quad (s, x) \in \tilde{E}. \quad (6.35)$$

Let $\bar{X} = (\bar{W}, \bar{\mathscr{G}}, \bar{\mathscr{G}}_t, (\bar{\alpha}_t, \bar{X}_t), \bar{\mathbf{Q}}_{r,\mu})$ be a Borel right time–space process in $I \times M(\tilde{E})$ associated with $\tilde{X}$. The existence of $\bar{X}$ follows from Theorem A.58. For $(s, \mu) \in I \times M(\tilde{E})$ let $\psi(s, \mu) = (s, \pi_s\mu)$, where $\pi_s\mu \in M(E_s)$ is defined by $\pi_s\mu(B) = \mu(\{s\} \times B)$ for $B \in \mathscr{B}(E_s)$. Then ψ is a surjective Borel map from $I \times M(\tilde{E})$ to $\tilde{M}$. From $\bar{X}$ and ψ we can use Theorem A.21 to obtain a Borel right process $\hat{X} = (\hat{W}, \hat{\mathscr{G}}, \hat{\mathscr{G}}_t, (\hat{\alpha}_t, \hat{X}_t), \hat{\mathbf{Q}}_{r,\mu})$ with state space $\tilde{M}$. For $(r, x) \in \tilde{E}$, $r \le t \in I$ and $f \in B(E_t)^+$ let $V_{r,t}f(x) = \tilde{V}_{t-r}(1_{\{t\}}f)(r, x)$. From (6.34) one may see that $(r, x) \mapsto V_{r,t}f(x)$ solves the equation (6.32). On the other hand, starting from any solution to (6.32) one can also construct a solution to (6.34). Then the uniqueness of the solution to (6.32) follows from that of (6.34). From (6.35) and the semigroup property of $(\tilde{V}_t)_{t \ge 0}$ it follows that $V_{r,s}V_{s,t} = V_{r,t}$ for $r \le t \in I$. For $r \le r + t \in I$, $\mu \in M(E_r)$ and $f \in B(E_{r+t})^+$ it is simple to see

$$\hat{\mathbf{Q}}_{r,\mu}e^{-\langle \hat{X}_t, f\rangle} = \bar{\mathbf{Q}}_{r,\delta_r \times \mu}e^{-\langle \bar{X}_t, 1_{\{r+t\}}f\rangle}$$
$$= e^{-\langle \mu, \tilde{V}_t(1_{\{r+t\}}f)(r, \cdot)\rangle} = e^{-\langle \mu, V_{r,r+t}f\rangle}.$$

Then (6.33) defines an inhomogeneous Borel transition semigroup $(Q_{r,t} : r \le t \in I)$ with global space $\tilde{M}$ and $\hat{X}$ is a right realization of the corresponding homogeneous time–space semigroup. That gives the desired result. $\qquad\square$

Example 6.4 Let E be a complete separable metric space. Suppose that $(P_t)_{t \ge 0}$ is a Borel right semigroup on E with a càdlàg realization. Let $D_E := D([0, \infty), E)$ be the space of càdlàg paths from $[0, \infty)$ to E furnished with the Skorokhod metric. Let $\xi = (D_E, \mathscr{F}^0, \mathscr{F}^0_t, \xi_t, \mathbf{P}_x)$ be the canonical realization of $(P_t)_{t \ge 0}$ and let $\bar{\xi} = (D_E, \mathscr{F}^0, \mathscr{F}^0_{r,t}, \bar{\xi}_t, \bar{\mathbf{P}}_{r,y})$ be the path process of ξ defined in Example A.2. Then $\bar{\xi}$ is an inhomogeneous càdlàg Markov process with global state space $S := \{(t, y) \in [0, \infty) \times D_E : y = y^t\}$. For $t \ge 0$ let $S_t = \{(s, y) \in S : s \le t\}$ and let D_E^t be defined as in Example A.2. We regard $M(D_E^t)$ as a subspace of $M(D_E)$ and endow

$$\tilde{M} := \{(t, \mu) \in [0, \infty) \times M(D_E) : \mu(D_E \setminus D_E^t) = 0\}$$

with the topology inherited from $[0, \infty) \times M(D_E)$. Suppose that ϕ is a local branching mechanism on E given by (2.45) and let

$$\bar{\phi}(s, y, z) = \phi(y(s), z), \quad (s, y) \in S, z \ge 0.$$

The $(\bar{\xi}, \bar{\phi})$-superprocess is an inhomogeneous Markov process with global state space $\tilde{M}$ and Borel right transition semigroup $(\bar{Q}_{r,t} : t \ge r \ge 0)$ given by

$$\int_{M(D_E^t)} e^{-\nu(f)}\bar{Q}_{r,t}(\mu, d\nu) = \exp\{-\mu(\bar{V}_{r,t}f)\}, \quad f \in B(D_E^t)^+, \quad (6.36)$$

where $(r, y) \mapsto \bar{v}_{r,t}(y) = \bar{V}_{r,t}f(y)$ is the unique bounded positive solution on S_t of the equation

$$\bar{v}_{r,t}(y) = \bar{\mathbf{P}}_{r,y}[f(\bar{\xi}_t)] - \int_r^t \bar{\mathbf{P}}_{r,y}\big[\bar{\phi}(s, \bar{\xi}_s, \bar{v}_{s,t}(\bar{\xi}_s))\big]\mathrm{d}s. \qquad (6.37)$$

In view of (A.48) we can rewrite (6.37) as

$$\bar{v}_{r,t}(y) = \mathbf{P}_{y(r)}[f(y/r/\xi^{t-r})] - \int_r^t \mathbf{P}_{y(r)}\big[\phi(\xi(s-r), \bar{v}_{s,t}(y/r/\xi^{s-r}))\big]\mathrm{d}s.$$

A realization of the $(\bar{\xi}, \bar{\phi})$-superprocess is called a (ξ, ϕ)-*historical superprocess*. Let $p_t(y) = y(t)$ for $t \geq 0$ and $y \in D_E$. Then each p_t is a Borel mapping from D_E to E. Suppose that $X = (W, \mathscr{G}, \mathscr{G}_t, X_t, \mathbf{Q}_\mu)$ is a realization of the (ξ, ϕ)-superprocess and $\bar{X} = (\bar{W}, \bar{\mathscr{G}}, \bar{\mathscr{G}}_{r,t}, \bar{X}_t, \bar{\mathbf{Q}}_{r,\mu})$ is a realization of the (ξ, ϕ)-historical superprocess. If we identify D_E^0 with E, then for every $\mu \in M(E)$, the process $\{\bar{X}_t \circ p_t^{-1} : t \geq 0\}$ under $\bar{\mathbf{Q}}_{0,\mu}$ is distributed identically as $\{X_t : t \geq 0\}$ under $\mathbf{Q}_\mu$; see Dawson and Perkins (1991, p.29). The historical superprocess not only contains the information on the current distribution of the population but also the records the past histories of all the individuals. This feature makes it a very powerful tool in studying the structural properties of the superprocess.

Example 6.5 Let E be a complete separable metric space. Suppose that ξ is a càdlàg Borel right process with state space E satisfying Condition 4.7. In the setting of Example 4.1 we can use the ξ-Brownian snake $\{(\eta_s, \zeta_s) : s \geq 0\}$ to define a process $\{\bar{X}_t : t \geq 0\}$ taking values in $M(D_E)$ by

$$\bar{X}_t(F) = \int_0^{\sigma(u)} F(\eta_s)\mathrm{d}l_s(t), \qquad F \in B(D_E). \qquad (6.38)$$

Since $s \mapsto l_s(t)$ increases only when $\zeta_s = t$, the random measure $\bar{X}_t$ takes values in $M(D_E^t)$. In fact, the process $\{\bar{X}_t : t \geq 0\}$ is a realization of the (ξ, ϕ)-historical superprocess with local branching mechanism $\phi(z) = z^2$.

6.6 Notes and Comments

The transformation $\mu(\mathrm{d}x) \mapsto h(x)^{-1}\mu(\mathrm{d}x)$ in the construction of superprocesses in spaces of infinite measures was used in Schied (1999); see also El Karoui and Roelly (1991) and Li (1992b). A special form of the superprocess in Example 6.2 was first given in Fleischmann and Mueller (1997); see also Wang (2002).

The study of multitype superprocesses was initiated by Gorostiza and Lopez-Mimbela (1990); see also Gorostiza and Roelly (1991), Gorostiza et al. (1992) and Li (1992a). A special form of the two-type superprocess in Section 6.3 was studied in Hong and Li (1999), where $\{X_2(t) : t \geq 0\}$ was interpreted as a superprocess with immigration governed by the trajectory of $\{X_1(t) : t \geq 0\}$. Hong and Li (1999) proved a central limit theorem of $\{X_2(t) : t \geq 0\}$ for Brownian spatial motion and binary local branching. The corresponding moderate and large deviations were studied in Hong (2002, 2003) and the quenched mean limit theorems and moderate deviations were discussed in Hong (2005). A quenched central limit theorem was given in Hong and Zeitouni (2007). The multitype super-Brownian motion was studied in Ceci and Gerardi (2006) in the framework of marked trees.

Theorems 6.7 and 6.8 are modifications of the results of Roelly and Rouault (1989). Evans (1993) gave two representations of the conditioned superprocess with transition semigroup $(\tilde{Q}_t)_{t \geq 0}$ defined by (6.27) or (6.29). One of those involves an "immortal particle" that moves according to the underlying spatial motion and throw off pieces of mass which then proceed to evolve as the original superprocess; see also Etheridge and Williams (2003). This representation was used in Engländer and Kyprianou (2004) and Liu et al. (2009) to investigate the long-time growth rate of the process. A number of limit theorems of the conditioned superprocess were proved in Evans (1991) and Evans and Perkins (1990); see also Liu and Ren (2009), Overbeck (1993) and Zhao (1994, 1996).

The concepts of path process and historical superprocess were introduced by Dawson and Perkins (1991); see also Dynkin (1991a, 1991c). A nonstandard model containing the genealogical trees of the super-Brownian motion was introduced in Perkins (1988); see also Dawson et al. (1989b). The representation (6.38) of historical superprocesses using Brownian snakes was given in Le Gall (1993). A different approach to the genealogical structures was developed in Donnelly and Kurtz (1996, 1999a, 1999b) using lookdown processes. A super-Brownian motion with reflecting historical paths was constructed in Burdzy and Le Gall (2001) and Burdzy and Mytnik (2005) by discrete approximations.

The age-structured superprocess defined in Example 6.3 was first studied in Bose and Kaj (2000). A different age-structured branching model was introduced in Jagers (1995). The high-density limits of age-dependent branching particle systems was studied in Kaj and Sagitov (1998). Some other models of superprocesses that can be obtained by transformations were given in Dawson et al. (2002c).

Let $C_d := C([0, \infty), \mathbb{R}^d)$ be the set of continuous paths from $[0, \infty)$ to $\mathbb{R}^d$ furnished with the topology of locally uniform convergence. We consider the canonical realization $\xi = (C_d, \mathscr{F}, \mathscr{F}_t, \xi_t, \mathbf{P}_x)$ of the d-dimensional diffusion process generated by the differential operator A specified in Example 6.1. Let $(x, z) \mapsto \phi(x, z)$ be a subcritical local branching mechanism given by (2.45) which is jointly continuous in $(x, z) \in \mathbb{R}^d \times [0, \infty)$. Suppose that $D \subset \mathbb{R}^d$ is a bounded domain with smooth boundary ∂D. Let $\tau_D = \inf\{t \geq 0 : \xi_t \in D^c\}$ be the *exit time* of ξ from D and let $\xi_t^D = \xi_{t \wedge \tau_D}$ for $t \geq 0$. We consider the stopped diffusion process $\xi^D = (C_d, \mathscr{F}, \mathscr{F}_t, \xi_t^D, \mathbf{P}_x)$. Let $\phi^D(x, z) = 1_D(x)\phi(x, z)$ for $x \in \mathbb{R}^d$ and $z \geq 0$. Suppose that $X = (W, \mathscr{G}, \mathscr{G}_t, X_t^D, \mathbf{Q}_\mu)$ is a realization of the (ξ^D, ϕ^D)-superprocess. Then for any $\mu \in M(\mathbb{R}^d)$ we have

$$\mathbf{Q}_\mu \exp\{-X_t^D(f)\} = \exp\{-\mu(v_t^D)\}, \qquad t \geq 0, f \in C(\mathbb{R}^d)^+, \qquad (6.39)$$

where $(t, x) \mapsto v_t^D(x)$ is the unique bounded positive solution to the integral evolution equation

$$v_t^D(x) = \mathbf{P}_x[f(\xi_{t \wedge \tau_D})] - \mathbf{P}_x\left[\int_0^{t \wedge \tau_D} \phi(\xi_s, v_{t-s}^D(\xi_s))\mathrm{d}s\right]. \qquad (6.40)$$

In view of (6.39) and (6.40) one would expect there is a random measure $X^D \in M(D^c)$ defined on the probability space $(W, \mathscr{G}, \mathbf{Q}_\mu)$ so that

$$\mathbf{Q}_\mu \exp\{-X^D(f)\} = \exp\{-\mu(v^D)\}, \qquad f \in C(\mathbb{R}^d)^+,$$

where $x \mapsto v^D(x)$ is the unique bounded positive solution to the equation

$$v(x) = \mathbf{P}_x[f(\xi_{\tau_D})] - \mathbf{P}_x\left[\int_0^{\tau_D} \phi(\xi_s, v(\xi_s))\mathrm{d}s\right].$$

This observation was made rigorous by Dynkin (1991b), who showed that $x \mapsto v^D(x)$ can also be defined by the nonlinear partial differential equation

$$\begin{cases} Av(x) = \phi(x, v(x)), & x \in D, \\ v(x) = f(x), & x \in D^c. \end{cases} \qquad (6.41)$$

The random measure X^D is called the *exit measure* of the (ξ, ϕ)-superprocess from D. It can be obtained in a limit theorem of the type of Theorem 4.6 by freezing each particle at its exit time from D. Using the (ξ, ϕ)-historical superprocess $\bar{X}$, the exit measure can be represented formally as

$$X^D(f) = \lim_{\varepsilon \to 0+} \frac{1}{\varepsilon} \int_0^\infty \mathrm{d}t \int_{C_d} f(w(\tau_D(w)))1_{[t-\varepsilon,t]}(\tau_D(w))\bar{X}_t(\mathrm{d}w),$$

where $\tau_D(w) = \inf\{t \geq 0 : w(t) \in D^c\}$ for $w \in C_d$; see Dynkin (1991c). In the binary local branching case, it can also be represented in terms of the ξ-Brownian snake; see Le Gall (1999) for details. In Salisbury and Verzani (1999, 2000), some conditioned exit measures of the super-Brownian motion were defined and characterized. Mselati (2004) applied the Brownian snake to give classifications and probabilistic representations of positive solutions of the equation $\Delta v(x) = v(x)^2$ in a bounded smooth domain; see also Le Gall (1995). The interplay between superprocesses and nonlinear partial differential equations has led to many deep results. We refer the reader to Dynkin (2002, 2004) and Le Gall (1999, 2005) for the developments in this subject.

Chapter 7
Martingale Problems of Superprocesses

Martingale problems play a very important role in the study of Markov processes. In this chapter we investigate some martingale problems associated with Dawson–Watanabe superprocesses. In particular, we shall prove the equivalence of a number of martingale problems for the superprocesses. The martingale problems induce martingale measures which are not necessarily orthogonal, but still worthy in the sense of Walsh (1986). We give a representation for the superprocesses in terms of stochastic integrals with respect to the martingale measures. The Girsanov type transform of Dawson (1978) is used to derive superprocesses with interactive growth rates. For simplicity, we only consider locally compact underlying spaces and establish the results under Feller type assumptions.

7.1 The Differential Evolution Equation

Let E be a locally compact separable metric space and $(P_t)_{t \geq 0}$ be a transition semigroup on E. Suppose that $(P_t)_{t \geq 0}$ preserves $C_0(E)$ and $t \mapsto P_t f$ is continuous in the supremum norm for every $f \in C_0(E)$, but the semigroup is not necessarily conservative. Let A denote the strong generator of $(P_t)_{t \geq 0}$ defined by

$$Af(x) = \lim_{t \to 0} \frac{P_t f(x) - f(x)}{t}, \qquad x \in E, \tag{7.1}$$

where the limit is taken in the supremum norm. The domain $D_0(A)$ of A is the totality of functions $f \in C_0(E)$ for which the above limit exists. In general, this is much smaller than the domain of the weak generator of $(P_t)_{t \geq 0}$. Let ϕ be a branching mechanism given by (2.26) or (2.27). In this section, we always assume the following conditions:

Condition 7.1 $b \in C(E)$, $c \in C(E)^+$ and the operator $f \mapsto \gamma(\cdot, f)$ preserves $C_0(E)^+$.

Z. Li, *Measure-Valued Branching Markov Processes*,
Probability and Its Applications, DOI 10.1007/978-3-642-15004-3_7,
© Springer-Verlag Berlin Heidelberg 2011

Condition 7.2 $x \mapsto (\nu(1) \wedge \nu(1)^2)H(x, d\nu)$ *is continuous by weak convergence on* $M(E)^\circ$ *and* $C_0(E)^+$ *is preserved by the operator*

$$f \mapsto \int_{M(E)^\circ} (\nu(f) \wedge \nu(f)^2)H(x, d\nu).$$

Let $\phi_n(x, f)$ be defined by (2.29). Then the above conditions imply that both $f \mapsto \phi(\cdot, f)$ and $f \mapsto \phi_n(\cdot, f)$ map $C_0(E)^+$ into $C_0(E)$. Recall that $c_0 = \|b^-\| + \|\gamma(\cdot, 1)\|$.

Lemma 7.3 *Let* $f \in C_0(E)^+$ *and let* $t \mapsto \pi_t f$ *be defined by (2.35). Then as* $n \to \infty$ *we have* $\phi_n(x, \pi_t f) \to \phi(x, \pi_t f)$ *uniformly and increasingly on the set* $[0, T] \times E$ *for each* $T \geq 0$.

Proof. Clearly, $\phi_n(x, f) \to \phi(x, f)$ increasingly for $f \in C_0(E)^+$. Let $b^* = \|b^-\|$ and let $t \mapsto \pi_t^* f$ be defined by (2.35) with b replaced by $-b^*$. By Corollary 5.18 we have $\pi_t f \leq \pi_t^* f$ for $t \geq 0$. From (2.37) it is easy to see the operators $(\pi_t^*)_{t \geq 0}$ preserve $C_0(E)^+$ and $t \mapsto \pi_t^* f$ is strongly continuous for each $f \in C_0(E)^+$. By Proposition A.49 we have $\|\pi_t^* f\| \leq \|f\| e^{c_0 t}$ for $t \geq 0$. Observe that

$$\phi(x, f) - \phi_n(x, f) = c(x)f(x)^2 - 2n^2 c(x)\left[e^{-f(x)/n} - 1 + f(x)/n\right]$$
$$+ \int_{\{\nu(1) < 1/n\}} \left[e^{-\nu(f)} - 1 + \nu(f)\right][1 - n\nu(1)]H(x, d\nu). \qquad (7.2)$$

Then we have

$$0 \leq \phi(x, f) - \phi_n(x, f) \leq \varepsilon_n(x, f) + \eta_n(x, f),$$

where

$$\varepsilon_n(x, f) = c(x)f(x)^2 - 2n^2 c(x)\left[e^{-f(x)/n} - 1 + f(x)/n\right]$$

and

$$\eta_n(x, f) = \frac{1}{2} \int_{\{\nu(1) < 1/n\}} \nu(f)^2(1 - n\nu(1))H(x, d\nu).$$

By Taylor's expansion it is simple to see that

$$\varepsilon_n(x, \pi_t f) \leq \varepsilon_n(x, \pi_t^* f) \leq \frac{1}{3n} \|c\| \|\pi_t^* f\|^3 \leq \frac{1}{3n} e^{3c_0 t} \|c\| \|f\|^3.$$

Then $\varepsilon_n(x, \pi_t f) \to 0$ uniformly on the set $[0, T] \times E$ for each $T \geq 0$. By the assumptions on the kernel $H(x, d\nu)$, it is elementary to see that $f \mapsto \eta_n(\cdot, f)$ preserves $C_0(E)^+$ and $\eta_n(x, f) \to 0$ as $n \to \infty$. Moreover, we have

$$|\eta_n(x, \pi_t^* f) - \eta_n(x, f)|$$

$$\leq \frac{1}{2} \int_{\{\nu(1)<1/n\}} \nu(|\pi_t^* f - f|)\nu(|\pi_t^* f + f|)H(x, \mathrm{d}\nu)$$

$$\leq \frac{1}{2}\|\pi_t^* f - f\|\|\pi_t^* f + f\| \int_{\{\nu(1)<1/n\}} \nu(1)^2 H(x, \mathrm{d}\nu),$$

so $t \mapsto \eta_n(\cdot, \pi_t^* f)$ is strongly continuous on $C_0(E)^+$. Let $\bar{E} = E \cup \{\partial\}$ with ∂ being an isolated point if E is compact and with $\bar{E}$ being the one-point compactification of E otherwise. Then $(t, x) \mapsto \eta_n(x, \pi_t^* f)$ extends continuously onto $[0, \infty) \times \bar{E}$ with $\eta_n(\partial, \pi_t^* f) = 0$ for all $t \geq 0$. Given $\varepsilon > 0$ let $n_0(t, x) \geq 1$ be sufficiently large so that $\eta_{n_0(t,x)}(x, \pi_t^* f) < \varepsilon/2$. Consequently, there is a neighborhood $U(t, x)$ of (t, x) so that $\eta_{n_0(t,x)}(y, \pi_s^* f) < \varepsilon$ for $(s, y) \in U(t, x)$. By the compactness, for each $T \geq 0$ we can find a finite subset $\{(t_i, x_i) : i = 1, 2, \ldots, k\}$ of $[0, T] \times \bar{E}$ so that $\cup_{i=1}^k U(t_i, x_i) \supset [0, T] \times \bar{E}$. Then $\eta_n(x, \pi_s f) \leq \eta_n(x, \pi_s^* f) < \varepsilon$ for $(s, y) \in [0, T] \times \bar{E}$ and $n \geq n_0 := \max_{1 \leq i \leq k} n_0(t_i, x_i)$. It follows that $\eta_n(x, \pi_t f) \to 0$ uniformly on $[0, T] \times E$ for each $T \geq 0$. That proves the desired result. $\qquad \square$

Now let us prove a regularity property of the cumulant semigroup $(V_t)_{t \geq 0}$ of the (ξ, ϕ)-superprocess, which is defined by the nonlinear integral evolution equation

$$V_t f(x) + \int_0^t \mathrm{d}s \int_E \phi(y, V_s f) P_{t-s}(x, \mathrm{d}y) = P_t f(x), \quad t \geq 0, x \in E. \quad (7.3)$$

Recall that $(V_t)_{t \geq 0}$ has the canonical representation (2.5).

Theorem 7.4 *The operators* $(V_t)_{t \geq 0}$ *preserve* $C_0(E)^+$ *and* $t \mapsto V_t f$ *is continuous in the supremum norm for each* $f \in C_0(E)^+$.

Proof. For $n \geq 1$ and $f \in C_0(E)^+$ let $(t, x) \mapsto v_n(t, x, f)$ be the unique locally bounded positive solution of

$$v(t, x) + \int_0^t \mathrm{d}s \int_E \phi_n(y, v(s)) P_{t-s}(x, \mathrm{d}y) = P_t f(x), \quad t \geq 0, x \in E.$$

The above equation is a special case of (2.21). By Proposition 2.19 it is easy to infer that $f \mapsto v_n(t, \cdot, f)$ preserves $C_0(E)^+$ and $t \mapsto v_n(t) = v_n(t, \cdot, f)$ is continuous in the supremum norm for each $f \in C_0(E)^+$. By Proposition 2.20 we have $v_n(t, x, f) \to V_t f(x)$ decreasingly. Recall that $f \mapsto \phi_n(\cdot, f)$ maps $C_0(E)^+$ into $C_0(E)$. Now fix $T \geq 0$ and $f \in C_0(E)^+$ and let $a = \|f\|e^{c_0 T}$. Then

$$\|v_n(t, \cdot, f)\| \leq \|V_t f\| \leq \|\pi_t f\| \leq a, \quad 0 \leq t \leq T.$$

By Lemma 7.3 for $\varepsilon > 0$ there is an integer $N = N(\varepsilon, T) \geq 1$ so that $\phi(x, \pi_t f) - \phi_n(x, \pi_t f) \leq \varepsilon$ and hence $\phi(x, v_n(t)) - \phi_n(x, v_n(t)) \leq \varepsilon$ for $(t, x) \in [0, T] \times E$ and $n \geq N$. From (7.3) we have

$$\|v_n(t) - V_t f\| \leq \int_0^t \left(\varepsilon + \|\phi(\cdot, v_n(s)) - \phi(\cdot, V_s f)\|\right) \mathrm{d}s$$

$$\leq \varepsilon T + L_a \int_0^t \|v_n(s) - V_s f\| ds$$

for $0 \leq t \leq T$ and $n \geq N$, where $L_a \geq 0$ is a Lipschitz constant for the restriction of the operator $f \mapsto \phi(\cdot, f)$ on $C_0(E)^+ \cap B_a(E)^+$. By applying Gronwall's inequality we get

$$\|v_n(t) - V_t f\| \leq \varepsilon T \exp\{L_a t\}.$$

It follows that $v_n(t, x, f) \to V_t f(x)$ uniformly on $[0, T] \times E$. Then $(V_t)_{t \geq 0}$ preserves $C_0(E)^+$ and $t \mapsto V_t f$ is continuous in the supremum norm. $\qquad \square$

Corollary 7.5 *Let $f \in C_0(E)^+$ and let $t \mapsto V_t f$ be the unique locally bounded positive solution of (7.3). Then $t \mapsto V_t f$ is continuous in the supremum norm uniformly on each bounded interval.*

Corollary 7.6 *Let $f \in C_0(E)$ and let $t \mapsto \pi_t f$ be the unique locally bounded solution of (2.35). Then $t \mapsto \pi_t f$ is continuous in the supremum norm uniformly on each bounded interval.*

Proof. For $f \in C_0(E)^+$ the result follows from Corollary 7.5. The extension to $f \in C_0(E)$ is easy. $\qquad \square$

Let us introduce a differential form of the equation (7.3). Given $f \in D_0(A)^+$ we consider the nonlinear differential evolution equation

$$\begin{cases} \dfrac{d}{dt} V_t f(x) = A V_t f(x) - \phi(x, V_t f), & t \geq 0, x \in E, \\ V_0 f(x) = f(x), & x \in E. \end{cases} \qquad (7.4)$$

By a *positive solution* of (7.4) we mean a mapping $t \mapsto V_t f$ from $[0, \infty)$ to $D_0(A)^+$ that is continuously differentiable in $t \geq 0$ by the supremum norm and satisfies the equalities in (7.4). The main purpose of this section is to prove that (7.3) and (7.4) are equivalent for any $f \in D_0(A)^+$.

Theorem 7.7 *Let $f \in D_0(A)^+$. If $t \mapsto V_t f$ is a positive solution of the differential equation (7.4), it also solves the integral equation (7.3).*

Proof. Fix $t \geq 0$ and let $g(s) = P_{t-s} V_s f$ for $0 \leq s \leq t$. If $t \mapsto V_t f$ is a positive solution of the differential equation (7.4), it is easy to show that

$$\frac{d}{ds} g(s) = P_{t-s}\left(\frac{d}{ds} V_s f\right) - P_{t-s} A V_s f = -P_{t-s}\phi(V_s f), \quad 0 \leq s \leq t.$$

Then $s \mapsto (d/ds)g(s)$ is continuous by the supremum norm, and (7.3) follows by integrating both sides of the above equation over $[0, t]$. $\qquad \square$

Now we show that (7.3) also implies (7.4). Given $f \in D_0(A)^+$ we define $B_0 f(x) = A f(x) - \phi(x, f)$ and $B_t f(x) = V_t^{B_0 f} f(x)$, using the notation of Proposition 2.29. Let ψ be defined by (2.54). Then we have

$$B_t f(x) = P_t B_0(x) - \int_0^t ds \int_E \psi(y, V_s f, B_s f) P_{t-s}(x, dy) \qquad (7.5)$$

and

$$B_t f(x) = \lambda_t(x, B_0 f) + \int_{M(E)^\circ} e^{-\nu(f)} \nu(B_0 f) L_t(x, d\nu) \qquad (7.6)$$

for $t \geq 0$ and $x \in E$.

Lemma 7.8 *For every $t \geq 0$ and $f \in D_0(A)^+$ we have $B_t f \in C_0(E)$ and the mapping $t \mapsto B_t f$ is continuous in the supremum norm.*

Proof. Since $B_0 f \in C_0(E)$, by Proposition 2.32 we have immediately $B_t f \in C_0(E)$ for every $t \geq 0$. By (7.5) it is easy to show $\|B_t f - B_0 f\| \to 0$ as $t \to 0$. Moreover, for any $t \geq r \geq 0$ we have

$$\begin{aligned}
\|B_t f - B_r f\| &\leq \|P_t B_0 f - P_r B_0 f\| + \int_r^t \|P_{t-s}\psi(V_s f, B_s f)\| ds \\
&\quad + \int_0^r \|P_{t-s}\psi(V_s f, B_s f) - P_{r-s}\psi(V_s f, B_s f)\| ds \\
&\leq \|P_t B_0 f - P_r B_0 f\| + \int_r^t \|P_{t-s}\psi(V_s f, B_s f)\| ds \\
&\quad + \int_0^r \|P_{t-r}\psi(V_s f, B_s f) - \psi(V_s f, B_s f)\| ds.
\end{aligned}$$

Clearly, the right-hand side tends to zero as $t \to r$ or $r \to t$. Then $t \mapsto B_t f$ is continuous in the supremum norm. □

Lemma 7.9 *Let $f \in D_0(A)^+$ and let $t \mapsto V_t f$ be the unique locally bounded positive solution of the integral equation (7.3). Then*

$$\frac{d}{dt} V_t f(x) = B_t f(x), \qquad t \geq 0, x \in E, \qquad (7.7)$$

where the derivative is taken in the supremum norm.

Proof. By (7.3) and Corollary 7.5 it is easy to show that (7.7) holds at $t = 0$. For $t, s > 0$ we can use (2.5) and (7.6) to get

$$\begin{aligned}
\left| \frac{1}{s}[V_{t+s}f(x) - V_t f(x)] - B_t f(x) \right| \\
\leq \lambda_t \left(x, \left|\frac{1}{s}(V_s f - f) - B_0 f\right|\right) + \int_{M(E)^\circ} J_s(\nu, f) L_t(x, d\nu),
\end{aligned}$$

where

$$J_s(\nu, f) = \left| \frac{1}{s}(e^{-\nu(f)} - e^{-\nu(V_s f)}) - e^{-\nu(f)} \nu(B_0 f) \right|.$$

By the mean-value theorem we have

$$
\begin{aligned}
J_s(\nu, f) &= \left| e^{-\eta_s(f)} \frac{1}{s} \nu(V_s f - f) - e^{-\nu(f)} \nu(B_0 f) \right| \\
&\leq \left| \frac{1}{s} \nu(V_s f - f) - \nu(B_0 f) \right| + \left| e^{-\eta_s(f)} - e^{-\nu(f)} \right| \nu(|B_0 f|),
\end{aligned}
$$

where

$$
\nu(f) \wedge \nu(V_s f) \leq \eta_s(f) \leq \nu(f) \vee \nu(V_s f).
$$

Then we get

$$
\begin{aligned}
&\left| \frac{1}{s} [V_{t+s} f(x) - V_t f(x)] - B_t f(x) \right| \\
&\leq \left\| \frac{1}{s} (V_s f - f) - B_0 f \right\| \pi_t 1(x) \\
&\quad + \| B_0 f \| \int_{M(E)^\circ} \left| e^{-\eta_s(f)} - e^{-\nu(f)} \right| \nu(1) L_t(x, d\nu).
\end{aligned}
$$

Given $\varepsilon > 0$ we take $N = N_t(x, \varepsilon) \geq 1$ so that

$$
\int_{\{\nu(1) > N\}} \nu(1) L_t(x, d\nu) < \varepsilon.
$$

If $\nu(1) \leq N$, we have

$$
\left| e^{-\eta_s(f)} - e^{-\nu(f)} \right| \leq \left| \nu(V_s f) - \nu(f) \right| \leq N \| V_s f - f \|.
$$

It follows that

$$
\begin{aligned}
&\left| \frac{1}{s} [V_{t+s} f(x) - V_t f(x)] - B_t f(x) \right| \\
&\leq \left\| \frac{1}{s} (V_s f - f) - B_0 f \right\| \pi_t 1(x) + \varepsilon \| B_0 f \| \\
&\quad + N \| B_0 f \| \| V_s f - f \| \int_{M(E)^\circ} \nu(1) L_t(x, d\nu) \\
&\leq \left(\left\| \frac{1}{s} (V_s f - f) - B_0 f \right\| + N \| B_0 f \| \| V_s f - f \| \right) \pi_t 1(x) + \varepsilon \| B_0 f \|.
\end{aligned}
$$

Consequently,

$$
\lim_{s \to 0+} \left\| \frac{1}{s} (V_{t+s} f - V_t f) - B_t f \right\| = 0.
$$

In particular, for any $x \in E$ the function $t \mapsto V_t f(x)$ has continuous right derivative $t \mapsto B_t f(x)$, and thus $t \mapsto V_t f(x)$ is continuously differentiable. This implies

$$
V_t f(x) = f(x) + \int_0^t B_s f(x) ds, \qquad t \geq 0, x \in E.
$$

Then one can use the strong continuity of $t \mapsto B_t f$ to see (7.7) holds in the supremum norm.   □

Theorem 7.10 *For $f \in D_0(A)^+$ the unique locally bounded positive solution $t \mapsto V_t f$ of the integral equation (7.3) also solves the differential equation (7.4).*

Proof. Recall that $V_{t+r} f = V_r V_t f$ for $t, u \geq 0$. Then from (7.3) it follows that

$$P_r V_t f - V_t f = V_{t+r} f - V_t f + \int_0^r P_{r-s} \phi(V_{s+t} f) \, ds.$$

By Corollary 7.5 and Lemma 7.9 we see $V_t f \in D_0(A)^+$ and $A V_t f = B_t f + \phi(V_t f)$. That gives (7.4).   □

By a combination of Theorems 7.7 and 7.10 we obtain:

Theorem 7.11 *For any $f \in D_0(A)^+$, the integral equation (7.3) and the differential equation (7.4) for $t \mapsto V_t f$ are equivalent.*

By modifications of the arguments given above one can prove the following:

Theorem 7.12 *For any $f \in D_0(A)^+$ and $g \in C_0(E)^+$, the integral equation (5.32) is equivalent to the differential evolution equation*

$$\begin{cases} \dfrac{d}{dt} v_t(x) = A v_t(x) - \phi(x, v_t) + g(x), & t \geq 0, x \in E, \\ v_0(x) = f(x), & x \in E. \end{cases} \quad (7.8)$$

Suppose that there is a Hunt process ξ with transition semigroup $(P_t)_{t \geq 0}$. Let $X = (W, \mathscr{G}, \mathscr{G}_t, X_t, \mathbf{Q}_\mu)$ be a Hunt realization of the (ξ, ϕ)-superprocess. Since any function in $C(E)^+$ is the increasing limit of a sequence of functions from $C_0(E)^+$, we can define the transition semigroup $(Q_t)_{t \geq 0}$ of X by

$$\int_{M(E)} e^{-\nu(f)} Q_t(\mu, d\nu) = \exp\{-\mu(V_t f)\}, \qquad f \in C_0(E)^+, \quad (7.9)$$

where $t \mapsto V_t f$ is the unique locally bounded positive solution of (7.3). In fact, the operators $(V_t)_{t \geq 0}$ are uniquely determined by their restrictions on the smaller set $D_0(A)^+$, which is uniformly dense in $C_0(E)^+$. Then the transition semigroup $(Q_t)_{t \geq 0}$ can be defined by (7.9) for $f \in D_0(A)^+$ with $t \mapsto V_t f$ being the unique positive solution of the differential equation (7.4). Similarly, the joint distribution of X_t and $\int_0^t X_s ds$ can also be determined by (5.31) and (7.8). In applications we may also consider (7.4) in a smaller class of functions as shown in the following example.

Example 7.1 Let $C_0^2(\mathbb{R}^d)$ denote the set of twice continuously differentiable functions on $\mathbb{R}^d$ that together with all their partial derivatives up to the second order vanish at infinity. If ξ is a d-dimensional diffusion process with generator A specified in Example 6.1, then for any $f \in C_0^2(\mathbb{R}^d)^+$ we can also define $(t, x) \mapsto V_t f(x)$ by the nonlinear partial differential equation

$$\begin{cases} \dfrac{\mathrm{d}}{\mathrm{d}t} V_t f(x) = A V_t f(x) - \phi(x, V_t f), & t \geq 0, x \in \mathbb{R}^d, \\ V_0 f(x) = f(x), & x \in \mathbb{R}^d. \end{cases} \qquad (7.10)$$

The operators $(V_t)_{t\geq 0}$ are uniquely determined by their restrictions on $C_0^2(\mathbb{R}^d)^+$. This follows from the fact that any function in $C_0(\mathbb{R}^d)^+$ is the limit of a sequence of functions from $C_0^2(\mathbb{R}^d)^+$ in the supremum norm.

7.2 Generators and Martingale Problems

Suppose that E is a locally compact separable metric space. Let ξ be a Hunt process in E with transition semigroup $(P_t)_{t\geq 0}$ and let ϕ be a branching mechanism given by (2.26) or (2.27). We assume that $(P_t)_{t\geq 0}$ and ϕ satisfy the conditions specified at the beginning of the first section. Let $(Q_t)_{t\geq 0}$ and $(V_t)_{t\geq 0}$ denote respectively the transition semigroup and the cumulant semigroup of the (ξ, ϕ)-superprocess. Let $\mathscr{D}_0$ be the class of functions on $M(E)$ of the form

$$F(\mu) = G(\mu(f_1), \ldots, \mu(f_n)), \qquad (7.11)$$

where $G \in C^2(\mathbb{R}^n)$ and $\{f_1, \ldots, f_n\} \subset D_0(A)$. For $F \in \mathscr{D}_0$ define

$$\begin{aligned} L_0 F(\mu) = & \int_E \left[A F'(\mu; x) + \gamma(x, F'(\mu)) - b(x) F'(\mu; x) \right] \mu(\mathrm{d}x) \\ & + \int_E \mu(\mathrm{d}x) \int_{M(E)^\circ} \left[F(\mu + \nu) - F(\mu) - \nu(F'(\mu)) \right] H(x, \mathrm{d}\nu) \\ & + \int_E c(x) F''(\mu; x) \mu(\mathrm{d}x), \end{aligned} \qquad (7.12)$$

where

$$F'(\mu; x) = \lim_{\varepsilon \to 0+} \frac{1}{\varepsilon} \left[F(\mu + \varepsilon \delta_x) - F(\mu) \right] \qquad (7.13)$$

and $F''(\mu; x)$ is defined by the limit with $F(\cdot)$ replaced by $F'(\cdot; x)$.

Suppose that $(\Omega, \mathscr{G}, \mathscr{G}_t, \mathbf{P})$ is a filtered probability space satisfying the usual hypotheses and $\{X_t : t \geq 0\}$ is a càdlàg process in $M(E)$ that is adapted to $(\mathscr{G}_t)_{t\geq 0}$ and satisfies $\mathbf{P}[X_0(1)] < \infty$. Let us consider the following properties:

(1) For every $T \geq 0$ and $f \in C_0(E)^+$,

$$\exp\{-X_t(V_{T-t}f)\}, \qquad 0 \leq t \leq T,$$

is a martingale.

(2) For every $f \in D_0(A)^+$,

$$H_t(f) := \exp\left\{ -X_t(f) + \int_0^t X_s(Af - \phi(f))ds \right\}, \quad t \geq 0,$$

is a local martingale.

(3) (a) The process $\{X_t : t \geq 0\}$ has no negative jumps. Let $N(ds, d\nu)$ be the optional random measure on $[0, \infty) \times M(E)^\circ$ defined by

$$N(ds, d\nu) = \sum_{s>0} 1_{\{\Delta X_s \neq 0\}} \delta_{(s, \Delta X_s)}(ds, d\nu),$$

where $\Delta X_s = X_s - X_{s-}$, and let $\hat{N}(ds, d\nu)$ denote the predictable compensator of $N(ds, d\nu)$. Then $\hat{N}(ds, d\nu) = ds K(X_{s-}, d\nu)$ with

$$K(\mu, d\nu) = \int_E \mu(dx) H(x, d\nu).$$

(b) Let $\tilde{N}(ds, d\nu) = N(ds, d\nu) - \hat{N}(ds, d\nu)$. Then for any $f \in D_0(A)$ we have

$$X_t(f) = X_0(f) + M_t^c(f) + M_t^d(f) + \int_0^t X_s(Af + \gamma f - bf)ds,$$

where $t \mapsto M_t^c(f)$ is a continuous local martingale with quadratic variation $2X_t(cf^2)dt$ and

$$t \mapsto M_t^d(f) = \int_0^t \int_{M(E)^\circ} \nu(f)\tilde{N}(ds, d\nu)$$

is a purely discontinuous local martingale.

(4) For every $F \in \mathscr{D}_0$ we have

$$F(X_t) = F(X_0) + \int_0^t L_0 F(X_s)ds + \text{local mart.}$$

(5) For every $G \in C^2(\mathbb{R})$ and $f \in D_0(A)$ we have

$$\begin{aligned}
G(X_t(f)) = G(X_0(f)) + \int_0^t \Big\{ &\overset{\bullet}{G'}(X_s(f))X_s(Af + \gamma f - bf) \\
&+ G''(X_s(f))X_s(cf^2) + \int_E X_s(dx) \int_{M(E)^\circ} \Big[G(X_s(f) + \nu(f)) \\
&- G(X_s(f)) - \nu(f)G'(X_s(f)) \Big] H(x, d\nu) \Big\} ds + \text{local mart.}
\end{aligned}$$

Theorem 7.13 *The above properties* (1), (2), (3), (4) *and* (5) *are equivalent to each other. Those properties hold if and only if* $\{(X_t, \mathscr{G}_t) : t \geq 0\}$ *is a* (ξ, ϕ)-*superprocess with transition semigroup* $(Q_t)_{t\geq 0}$.

Proof. Clearly, (1) holds if and only if $\{X_t : t \geq 0\}$ is a Markov process relative to $(\mathcal{G}_t)_{t\geq 0}$ with transition semigroup $(Q_t)_{t\geq 0}$ defined by (7.9). Then we only need to prove the equivalence of the five properties.

(1)$\Rightarrow$(2): If (1) holds, then $\{X_t : t \geq 0\}$ is a (ξ, ϕ)-superprocess, so Corollary 2.28 implies

$$\mathbf{P}[X_t(f)] = \mathbf{P}[X_0(\pi_t f)], \quad t \geq 0, f \in B(E), \tag{7.14}$$

where $(\pi_t)_{t\geq 0}$ is defined by (2.35). Now we fix $r \geq 0$ and $B \in \mathcal{G}_r$ and define

$$J_t(f) = \mathbf{P}[1_B e^{-X_t(f)}] = \mathbf{P}[1_B e^{-X_r(V_{t-r}f)}]$$

for $t \geq r$ and $f \in D_0(A)^+$. In view of (7.14), we can use Theorem 7.10 and dominated convergence to show that $J_t(f)$ is continuously differentiable in $t \geq r$. By calculating the right derivative, we have

$$\frac{\mathrm{d}}{\mathrm{d}t} J_t(f) = \frac{\mathrm{d}}{\mathrm{d}s} \mathbf{P}[1_B e^{-X_t(V_s f)}]\Big|_{s=0} = -\mathbf{P}[1_B X_t(Af - \phi(f))e^{-X_t(f)}].$$

It follows that

$$Y_t(f) := e^{-X_t(f)} + \int_0^t X_s(Af - \phi(f))e^{-X_s(f)}\mathrm{d}s, \quad t \geq 0$$

is a martingale. By integration by parts applied to

$$Z_t(f) := e^{-X_t(f)} \quad \text{and} \quad W_t(f) := \exp\left\{ \int_0^t X_s(Af - \phi(f))\mathrm{d}s \right\} \tag{7.15}$$

we obtain

$$\mathrm{d}H_t(f) = e^{-X_{t-}(f)}\mathrm{d}W_t(f) + W_t(f)\mathrm{d}e^{-X_t(f)} = W_t(f)\mathrm{d}Y_t(f).$$

Then $\{H_t(f)\}$ is a local martingale.

(2)$\Rightarrow$(3): For $f \in D_0(A)^+$ define $Z_t(f)$ and $W_t(f)$ by (7.15). We have $Z_t(f) = H_t(f)W_t(f)^{-1}$ and so

$$\mathrm{d}Z_t(f) = W_t(f)^{-1}\mathrm{d}H_t(f) - Z_{t-}(f)X_{t-}(Af - \phi(f))\mathrm{d}t \tag{7.16}$$

by integration by parts. Then $\{Z_t(f)\}$ is a special semi-martingale; see, e.g., Dellacherie and Meyer (1982, p.213). By Itô's formula we find the $\{X_t(f)\}$ is also a special semi-martingale. Let $S(E)$ denote the space of finite Borel signed measures on E endowed with the σ-algebra generated by the mappings $\mu \mapsto \mu(B)$ for all $B \in \mathcal{B}(E)$. Let $S(E)^\circ = S(E) \setminus \{0\}$. We define the optional random measure $N(\mathrm{d}s, \mathrm{d}\nu)$ on $[0, \infty) \times S(E)^\circ$ by

$$N(\mathrm{d}s, \mathrm{d}\nu) = \sum_{s>0} 1_{\{\Delta X_s \neq 0\}} \delta_{(s,\Delta X_s)}(\mathrm{d}s, \mathrm{d}\nu),$$

where $\Delta X_s = X_s - X_{s-} \in S(E)$. Let $\hat{N}(ds, d\nu)$ denote the predictable compensator of $N(ds, d\nu)$ and let $\tilde{N}(ds, d\nu)$ denote the compensated random measure; see Dellacherie and Meyer (1982, pp.371–374). It follows that

$$X_t(f) = X_0(f) + U_t(f) + M_t^c(f) + M_t^d(f), \tag{7.17}$$

where $\{U_t(f)\}$ is a predictable process with locally bounded variations, $\{M_t^c(f)\}$ is a continuous local martingale and

$$M_t^d(f) = \int_0^t \int_{S(E)^\circ} \nu(f)\tilde{N}(ds, d\nu), \quad t \geq 0, \tag{7.18}$$

is a purely discontinuous local martingale; see Dellacherie and Meyer (1982, p.353 and p.376) or Jacod and Shiryaev (2003, p.84). Let $\{C_t(f)\}$ denote the quadratic variation process of $\{M_t^c(f)\}$. By Itô's formula,

$$Z_t(f) = Z_0(f) - \int_0^t Z_{s-}(f)dU_s(f) + \frac{1}{2}\int_0^t Z_{s-}(f)dC_s(f)$$
$$+ \int_0^t \int_{S(E)^\circ} Z_{s-}(f)\left[e^{-\nu(f)} - 1 + \nu(f)\right]\hat{N}(ds, d\nu)$$
$$+ \text{local mart.} \tag{7.19}$$

In view of (7.16) and (7.19) we get

$$dU_t(f) = \frac{1}{2}dC_t(f) + X_{t-}(Af - \phi(f))dt$$
$$+ \int_{S(E)^\circ} \left[e^{-\nu(f)} - 1 + \nu(f)\right]\hat{N}(dt, d\nu)$$

by the uniqueness of canonical decompositions of special semi-martingales; see Dellacherie and Meyer (1982, p.213). By substituting the representation (2.27) of ϕ into the above equation and comparing both sides it is easy to find that (3.a) holds and (3.b) holds for $f \in D_0(A)^+$. For an arbitrary $f \in D_0(A)$ set $f^+ = 0 \vee f$ and $f^- = 0 \vee (-f)$. For $n \geq 1$ define

$$f_n^+ = n\int_0^{\frac{1}{n}} P_s f^+ ds \quad \text{and} \quad f_n^- = n\int_0^{\frac{1}{n}} P_s f^- ds. \tag{7.20}$$

Then (3.b) holds for f_n^+ and $f_n^- \in D_0(A)^+$, so it holds for $f_n := f_n^+ - f_n^- \in D_0(A)$. It is easy to show that $f_n \to f$ and $Af_n \to Af$ in the supremum norm as $n \to \infty$. Therefore (3.b) is also satisfied for $f \in D_0(A)$.

(3)$\Rightarrow$(4): If $F \in \mathcal{D}_0$ is given by (7.11), it is easy to show that

$$F'(\mu; x) = \sum_{i=1}^n f_i(x)G_i'(\mu(f_1), \ldots, \mu(f_n))$$

and

$$F''(\mu; x) = \sum_{i,j=1}^{n} f_i(x) f_j(x) G''_{ij}(\mu(f_1), \ldots, \mu(f_n)).$$

Consequently, we have

$$\begin{aligned}
L_0 F(\mu) = {} & \sum_{i=1}^{n} G'_i(\mu(f_1), \ldots, \mu(f_n)) \mu(A f_i + \gamma f_i - b f_i) \\
& + \int_E \mu(\mathrm{d}x) \int_{M(E)^\circ} \Big[G(\mu(f_1) + \nu(f_1), \ldots, \mu(f_n) + \nu(f_n)) \\
& - G(\mu(f_1), \ldots, \mu(f_n)) - \sum_{i=1}^{n} \nu(f_i) G'_i(\mu(f_1), \ldots, \mu(f_n)) \Big] H(x, \mathrm{d}\nu) \\
& + \sum_{i,j=1}^{n} G''_{ij}(\mu(f_1), \ldots, \mu(f_n)) \mu(c f_i f_j).
\end{aligned} \tag{7.21}$$

Then (4) follows by (3) and Itô's formula.

(4)⇒(5): Let $F(\mu) = G(\mu(f))$ for $G \in C^2(\mathbb{R})$ and $f \in D_0(A)$. As a special case of (7.21), we have

$$\begin{aligned}
L_0 F(\mu) = {} & G'(\mu(f)) \mu(A f + \gamma f - b f) + G''(\mu(f)) \mu(c f^2) \\
& + \int_E \mu(\mathrm{d}x) \int_{M(E)^\circ} \Big[G(\mu(f) + \nu(f)) - G(\mu(f)) \\
& - \nu(f) G'(\mu(f)) \Big] H(x, \mathrm{d}\nu)
\end{aligned} \tag{7.22}$$

Then (5) follows from (4).

(5)⇒(1): Let $G \in C^2(\mathbb{R})$ and let $t \mapsto f_t$ be a mapping from $[0, T]$ to $D_0(A)^+$ such that $t \mapsto f_t$ is continuously differentiable and $t \mapsto A f_t$ is continuous by the supremum norm. For $0 \le t \le T$ and $k \ge 1$ we have

$$\begin{aligned}
G(X_t(f_t)) = {} & G(X_0(f_0)) + \sum_{j=0}^{\infty} \big[G(X_{t \wedge (j+1)/k}(f_{t \wedge j/k})) - G(X_{t \wedge j/k}(f_{t \wedge j/k})) \big] \\
& + \sum_{j=0}^{\infty} \big[G(X_{t \wedge (j+1)/k}(f_{t \wedge (j+1)/k})) - G(X_{t \wedge (j+1)/k}(f_{t \wedge j/k})) \big],
\end{aligned}$$

where the summations only consist of finitely many non-trivial terms. By applying (5) term by term we obtain

$$\begin{aligned}
G(X_t(f_t)) = {} & G(X_0(f_0)) + \sum_{j=0}^{\infty} \int_{t \wedge j/k}^{t \wedge (j+1)/k} \Big\{ G'(X_s(f_{t \wedge j/k})) X_s((A + \gamma) f_{t \wedge j/k}) \\
& - G'(X_s(f_{t \wedge j/k})) X_s(b f_{t \wedge j/k}) + G''(X_s(f_{t \wedge j/k})) X_s(c f_{t \wedge j/k}^2) \\
& + \int_E X_s(\mathrm{d}x) \int_{M(E)^\circ} \Big[G(X_s(f_{t \wedge j/k}) + \nu(f_{t \wedge j/k})) \\
& - G(X_s(f_{t \wedge j/k})) - \nu(f_{t \wedge j/k}) G'(X_s(f_{t \wedge j/k})) \Big] H(x, \mathrm{d}\nu) \Big\} \mathrm{d}s
\end{aligned}$$

$$+ \sum_{j=0}^{\infty} \int_{t \wedge j/k}^{t \wedge (j+1)/k} G'(X_{t \wedge (j+1)/k}(f_s)) X_{t \wedge (j+1)/k}(f'_s) \mathrm{d}s + M_k(t),$$

where $\{M_k(t)\}$ is a local martingale. Since $\{X_t\}$ is a càdlàg process, letting $k \to \infty$ in the equation above gives

$$\begin{aligned}
G(X_t(f_t)) = G(X_0(f_0)) + \int_0^t \Big\{ & G'(X_s(f_s)) X_s(Af_s + \gamma f_s - bf_s + f'_s) \\
& + G''(X_s(f_s)) X_s(cf_s^2) + \int_E X_s(\mathrm{d}x) \int_{M(E)^\circ} \Big[G(X_s(f_s) + \nu(f_s)) \\
& - G(X_s(f_s)) - \nu(f_s) G'(X_s(f_s)) \Big] H(x, \mathrm{d}\nu) \Big\} \mathrm{d}s + M(t),
\end{aligned}$$

where $\{M(t)\}$ is a local martingale. For any $f \in D_0(A)^+$ we may apply the above to $G(z) = \mathrm{e}^{-z}$ and $f_t = V_{T-t}f$ to see $t \mapsto \exp\{-X_t(V_{T-t}f)\}$ is a local martingale. Then the assertion of (1) follows by dominated convergence. □

Corollary 7.14 Let $\{(X_t, \mathscr{G}_t) : t \geq 0\}$ be a càdlàg (ξ, ϕ)-superprocess satisfying $\mathbf{P}[X_0(1)] < \infty$. Then for every $T \geq 0$ and $f \in D_0(A)$ there is a constant $C(T, f) \geq 0$ such that

$$\mathbf{P}\Big[\sup_{0 \leq t \leq T} |X_t(f)| \Big] \leq C(T, f) \Big\{ \mathbf{P}[X_0(1)] + \sqrt{\mathbf{P}[X_0(1)]} \Big\}.$$

Proof. By the above property (3.b) and Doob's martingale inequality we have

$$\begin{aligned}
\mathbf{P}\Big[\sup_{0 \leq t \leq T} |X_t(f)| \Big] \leq{} & \mathbf{P}[|X_0(f)|] + \mathbf{P}\Big[\sup_{0 \leq t \leq T} |M_t^c(f)| \Big] \\
& + \mathbf{P}\Big[\sup_{0 \leq t \leq T} \Big| \int_0^t \int_{M(E)^\circ} \nu(f) 1_{\{\nu(1) \leq 1\}} \tilde{N}(\mathrm{d}s, \mathrm{d}\nu) \Big| \Big] \\
& + \mathbf{P}\Big[\sup_{0 \leq t \leq T} \Big| \int_0^t \int_{M(E)^\circ} \nu(f) 1_{\{\nu(1) > 1\}} \tilde{N}(\mathrm{d}s, \mathrm{d}\nu) \Big| \Big] \\
& + \mathbf{P}\Big[\int_0^T |X_s(Af + \gamma f - bf)| \mathrm{d}s \Big] \\
\leq{} & \mathbf{P}[|X_0(f)|] + 2\Big\{ \mathbf{P}\Big[\int_0^T X_s(cf^2) \mathrm{d}s \Big] \Big\}^{1/2} \\
& + 2\Big\{ \mathbf{P}\Big[\int_0^T \mathrm{d}s \int_E X_s(\mathrm{d}x) \int_{\{\nu(1) \leq 1\}} \nu(f)^2 H(x, \mathrm{d}\nu) \Big] \Big\}^{1/2} \\
& + 2\mathbf{P}\Big[\int_0^T \mathrm{d}s \int_E X_s(\mathrm{d}x) \int_{\{\nu(1) > 1\}} \nu(|f|) H(x, \mathrm{d}\nu) \Big] \\
& + \mathbf{P}\Big[\int_0^T X_s(|Af + \gamma f - bf|) \mathrm{d}s \Big].
\end{aligned}$$

Then the desired inequality follows by simple estimates based on Corollary 2.28. □

Corollary 7.15 Suppose that $\nu(1)^2 H(x, \mathrm{d}\nu)$ is a bounded kernel from E to $M(E)^\circ$ and $\{(X_t, \mathscr{G}_t) : t \geq 0\}$ is a càdlàg (ξ, ϕ)-superprocess satisfying $\mathbf{P}[X_0(1)] < \infty$. Then for every $f \in D_0(A)$,

$$M_t(f) = X_t(f) - X_0(f) - \int_0^t X_s(Af + \gamma f - bf)\mathrm{d}s \qquad (7.23)$$

is a square integrable $(\mathscr{G}_t)$-martingale with increasing process

$$\langle M(f)\rangle_t = \int_0^t \mathrm{d}s \int_E q(x,f)X_s(\mathrm{d}x), \qquad (7.24)$$

where $q(x,f)$ is defined by (2.59).

Proof. Since $\{X_t : t \geq 0\}$ is a (ξ,ϕ)-superprocess, it satisfies the properties (1)–(5). In particular, from (3) one sees that (7.23) defines a local martingale with increasing process (7.24). From (7.14) we see that $t \mapsto \mathbf{P}[X_t(1)]$ is locally bounded, so $\{M_t(f)\}$ is actually a square integral martingale.   □

Corollary 7.16 *Suppose that the branching mechanism ϕ has the special form with $H(x, M(E)^\circ) = 0$ for all $x \in E$. Then any càdlàg (ξ,ϕ)-superprocess $\{(X_t, \mathscr{G}_t) : t \geq 0\}$ is a.s. continuous. Conversely, if $\{(X_t, \mathscr{G}_t) : t \geq 0\}$ is a continuous process in $M(E)$ and if for every $f \in D_0(A)$ the process $\{M_t(f) : t \geq 0\}$ defined by (7.23) is a $(\mathscr{G}_t)$-local martingale with increasing process*

$$\langle M(f)\rangle_t = 2\int_0^t \mathrm{d}s \int_E c(x)f(x)^2 X_s(\mathrm{d}x), \qquad (7.25)$$

then $\{(X_t, \mathscr{G}_t) : t \geq 0\}$ is a (ξ,ϕ)-superprocess.

Proof. If $\{(X_t, \mathscr{G}_t) : t \geq 0\}$ is a càdlàg (ξ,ϕ)-superprocess, by Theorem 7.13 it has property (3) with $K(\mu, M(E)^\circ) = 0$. Then $\{X_t : t \geq 0\}$ is a.s. continuous. That gives the first assertion. Conversely, if $\{X_t : t \geq 0\}$ is continuous in $M(E)$ and if for every $f \in D_0(A)$ the process in (7.23) is a local martingale with increasing process given by (7.25), we can use Itô's formula to obtain

$$G(X_t(f)) = G(X_0(f)) + \int_0^t G'(X_s(f))X_s(Af + \gamma f - bf)\mathrm{d}s$$
$$+ \int_0^t G''(X_s(f))X_s(cf^2)\mathrm{d}s + \text{local mart.}$$

Then another application of Theorem 7.13 gives the second assertion.   □

The above property (4) implies that the generator of the (ξ,ϕ)-superprocess is the closure of $(L_0, \mathscr{D}_0)$ in the sense of Ethier and Kurtz (1986). Under suitable assumptions, we can replace $D_0(A)$ by a larger function class in Theorem 7.13 and its corollaries. In particular, if $P_t 1 \in C(E)$ for every $t \geq 0$ and there exists a function $A1 \in C(E)$ such that

$$\lim_{t\to 0}\frac{1}{t}\big[P_t 1(x) - 1\big] = A1(x), \qquad x \in E \qquad (7.26)$$

by the uniform convergence, we can extend the operator A to the linear span $D(A)$ of $D_0(A)$ and the constant functions. In this case, the results of Theorem 7.13 and its corollaries remain true with $D_0(A)$ replaced by $D(A)$. Of course, we have $D(A) = D_0(A)$ if E is a compact metric space. Recall that $C^2(\mathbb{R}_+)$ denotes the set of bounded continuous real functions on $\mathbb{R}_+$ with bounded continuous derivatives up to the second order. From Theorem 7.13 we derive immediately the following characterization of a CB-process.

Theorem 7.17 *Suppose that $\{(x(t), \mathscr{G}_t) : t \geq 0\}$ is a positive càdlàg process such that $\mathbf{P}[x(0)] < \infty$. Then $\{(x(t), \mathscr{G}_t) : t \geq 0\}$ is a CB-process with branching mechanism given by (3.1) if and only if for every $f \in C^2(\mathbb{R}_+)$ we have*

$$f(x(t)) = f(x(0)) + \int_0^t L_0 f(x(s)) \mathrm{d}s + local\ mart.,$$

where

$$L_0 f(x) = cx f''(x) - bx f'(x) + \int_0^\infty x[f(x+z) - f(x) - zf'(x)]m(\mathrm{d}z).$$

The martingale problems for the (ξ, ϕ)-superprocess can also be reformulated on the state space of tempered measures. In the next two theorems we assume $h \in D_0(A)$ is a strictly positive function satisfying $Ah \leq \alpha h$ for some constant $\alpha > 0$. From (6.12) we see that h is an α-excessive function for $(P_t)_{t \geq 0}$. Recall that $M_h(E)$ is the space of measures μ on E satisfying $\mu(h) < \infty$ and $C_h(E)$ is the set of continuous functions f on E satisfying $|f| \leq \mathrm{const} \cdot h$. Let $D_h(A) = \{f \in D_0(A) \cap C_h(E) : Af \in C_h(E)\}$. Let $f \mapsto \phi(\cdot, f)$ be a branching mechanism given as in Section 6.1 with $\rho = h$ and let $(Q_t)_{t \geq 0}$ be the transition semigroup on $M_h(E)$ defined by (6.6) and (6.11). Suppose that $f \mapsto h^{-1}\phi(\cdot, hf) - \alpha f$ satisfies the conditions for the branching mechanism specified at the beginning of the first section. The proof of the following theorem is similar to that of Theorem 7.13.

Theorem 7.18 *Let $(\Omega, \mathscr{G}, \mathscr{G}_t, \mathbf{P})$ be a filtered probability space satisfying the usual hypotheses and let $\{X_t : t \geq 0\}$ be a càdlàg process in $M_h(E)$ that is adapted to $(\mathscr{G}_t)_{t \geq 0}$ and satisfies $\mathbf{P}[X_0(h)] < \infty$. Then Theorem 7.13 still holds when $M(E)$, $C_0(E)$ and $D_0(A)$ are replaced by $M_h(E)$, $C_h(E)$ and $D_h(A)$, respectively.*

If the semigroup $(\tilde{P}_t)_{t \geq 0}$ given by (6.10) has a Hunt realization, the (ξ, ϕ)-superprocess has a càdlàg realization in $M_h(E)$ by Theorem 6.3. The following theorem describes another situation where a càdlàg realization in $M_h(E)$ of the (ξ, ϕ)-superprocess exists.

Theorem 7.19 *Suppose that ξ is a Hunt process. Then for every $\mu \in M_h(E)$ there is a càdlàg realization $\{X_t : t \geq 0\}$ in $M_h(E)$ of the (ξ, ϕ)-superprocess with initial value $X_0 = \mu$.*

Proof. Given $\mu \in M_h(E)$ we write $\mu = \sum_{i=1}^\infty \mu_i$ for a sequence of finite measures $\{\mu_i : i = 1, 2, \ldots\} \subset M(E)$. Let $\{X_i(t) : t \geq 0\}$, $i = 1, 2, \ldots$ be a sequence of

independent càdlàg (ξ, ϕ)-superprocesses in $M(E)$ with $X_i(0) = \mu_i$, $i = 1, 2, \ldots$.
For $n \geq k \geq 1$ it is easy to see that

$$Z_{k,n}(t) = \sum_{i=k}^{n} X_i(t), \qquad t \geq 0$$

is a càdlàg realization of the (ξ, ϕ)-superprocess in $M(E)$ with initial state $\mu_{k,n} :=$ $\sum_{i=k}^{n} \mu_i$. By the assumptions on the branching mechanism one can see as in the proof of Corollary 7.14 that

$$\mathbf{P}\left[\sup_{0 \leq s \leq t} \langle Z_{k,n}(s), h \rangle \right] \leq C(t, h)\left[\langle \mu_{k,n}, h \rangle + \langle \mu_{k,n}, h \rangle^{1/2} \right],$$

where $t \mapsto C(t, h)$ is a locally bounded function. The right-hand side tends to zero as $k, n \to \infty$. Then

$$X_t = \sum_{i=1}^{\infty} X_i(t), \qquad t \geq 0$$

defines a càdlàg process in $M_h(E)$. This process is clearly a realization of the (ξ, ϕ)-superprocess with $X_0 = \mu$. $\square$

7.3 Worthy Martingale Measures

In this section, we assume E is a Lusin topological space. However, the results obtained here can obviously be modified to the case of a Lusin measurable space. Given a signed measure $K(\mathrm{d}s, \mathrm{d}x, \mathrm{d}y)$ on $\mathscr{B}((0, \infty) \times E^2)$ with the total variation $|K|(\mathrm{d}s, \mathrm{d}x, \mathrm{d}y)$ satisfying $|K|((0, T] \times E^2) < \infty$ for all $T \geq 0$, we define the bilinear form

$$(f, g)_{K,T} = \int_0^T \int_{E^2} f(s, x)g(s, y)K(\mathrm{d}s, \mathrm{d}x, \mathrm{d}y), \qquad (7.27)$$

where $f, g \in B((0, T] \times E)$. We say the signed measure is *symmetric* if $(f, g)_{K,T} = (g, f)_{K,T}$ for all $T \geq 0$ and $f, g \in B((0, T] \times E)$, and say $K(\mathrm{d}s, \mathrm{d}x, \mathrm{d}y)$ is *positive definite* if $(f, f)_{K,T} \geq 0$ for all $T \geq 0$ and $f \in B((0, T] \times E)$. For a symmetric and positive definite signed measure $K(\mathrm{d}s, \mathrm{d}x, \mathrm{d}y)$, one shows by a standard argument the following *Schwarz's inequality*

$$(f, g)_{K,T}^2 \leq (f, f)_{K,T}(g, g)_{K,T}. \qquad (7.28)$$

In particular, those apply to random (positive) measures $K(\mathrm{d}s, \mathrm{d}x, \mathrm{d}y)$.

Now let $(\Omega, \mathscr{G}, \mathscr{G}_t, \mathbf{P})$ be a filtered probability space satisfying the usual hypotheses. Suppose that for each $B \in \mathscr{B}(E)$ there exists a square-integrable càdlàg

$(\mathscr{G}_t)$-martingale $\{M_t(B) : t \geq 0\}$ satisfying $M_0(B) = 0$. The system $\{M_t(B) : t \geq 0; B \in \mathscr{B}(E)\}$ is called a *martingale measure* on E if for every $t \geq 0$ and every disjoint sequence $\{B_1, B_2, \ldots\} \subset \mathscr{B}(E)$ we have

$$M_t\left(\bigcup_{k=1}^{\infty} B_k\right) = \sum_{k=1}^{\infty} M_t(B_k)$$

by the convergence in $L^2(\Omega, \mathbf{P})$. A martingale measure $\{M_t(B) : t \geq 0; B \in \mathscr{B}(E)\}$ is said to be *worthy* if there is a random measure $K(\mathrm{d}s, \mathrm{d}x, \mathrm{d}y)$ on $\mathscr{B}((0, \infty) \times E^2)$ such that:

(1) $K(\mathrm{d}s, \mathrm{d}x, \mathrm{d}y)$ is symmetric and positive definite;
(2) $t \mapsto K((0, t] \times A \times B)$ is predictable for all $A, B \in \mathscr{B}(E)$ and

$$\mathbf{P}\{K((0, t] \times E^2)\} < \infty, \qquad t \geq 0; \tag{7.29}$$

(3) for every $t \geq s \geq 0$ and $A, B \in \mathscr{B}(E)$ we have

$$|\langle M(A), M(B)\rangle_t - \langle M(A), M(B)\rangle_s| \leq K((s, t] \times A \times B). \tag{7.30}$$

In this case, we call $K(\mathrm{d}s, \mathrm{d}x, \mathrm{d}y)$ the *dominating measure* of $\{M_t(B) : t \geq 0; B \in \mathscr{B}(E)\}$.

Let $\mathscr{R}$ denote the semi-algebra consisting of rectangles on $(0, \infty) \times E^2$ of the form $(s, t] \times A \times B$ for $t \geq s \geq 0$ and $A, B \in \mathscr{B}(E)$. Given a worthy martingale measure $\{M_t(B) : t \geq 0; B \in \mathscr{B}(E)\}$ with dominating measure $K(\mathrm{d}s, \mathrm{d}x, \mathrm{d}y)$, we define a random set function $\eta(\omega, \cdot)$ on $\mathscr{R}$ by

$$\eta((s, t] \times A \times B) = \langle M(A), M(B)\rangle_t - \langle M(A), M(B)\rangle_s.$$

Since the σ-algebra $\mathscr{B}(E)$ is separable, $\eta(\omega, \cdot)$ can be extended to a random signed measure on $\mathscr{B}((0, \infty) \times E^2)$ with total variation dominated by $K(\omega, \cdot)$. It is simple to see that $\eta(\omega, \mathrm{d}s, \mathrm{d}x, \mathrm{d}y)$ is symmetric and positive definite. We refer to $\eta(\omega, \mathrm{d}s, \mathrm{d}x, \mathrm{d}y)$ as the *covariance measure* of $\{M_t(B) : t \geq 0; B \in \mathscr{B}(E)\}$. We say the martingale measure is *orthogonal* if $\eta(\omega, \mathrm{d}s, \mathrm{d}x, \mathrm{d}y)$ is a.s. carried by $[0, \infty) \times \Delta(E)$, where $\Delta(E) = \{(x, x) : x \in E\}$. An orthogonal martingale measure is called a *time–space white noise* if $\{M_t(B) : t \geq 0\}$ is a one-dimensional Brownian motion for every $B \in \mathscr{B}(E)$.

Proposition 7.20 *A worthy martingale measure* $\{M_t(B) : t \geq 0; B \in \mathscr{B}(E)\}$ *is orthogonal if and only if* $\{M_t(A) : t \geq 0\}$ *and* $\{M_t(B) : t \geq 0\}$ *are orthogonal martingales whenever A and $B \in \mathscr{B}(E)$ are disjoint.*

Proof. Suppose that $\{M_t(B) : t \geq 0; B \in \mathscr{B}(E)\}$ is orthogonal and $A, B \in \mathscr{B}(E)$ are disjoint sets. Then $\langle M(A), M(B)\rangle_t = \eta((0, t] \times A \times B)$ vanishes, so $\{M_t(A) : t \geq 0\}$ and $\{M_t(B) : t \geq 0\}$ are orthogonal martingales. Conversely, suppose that $\{M_t(A) : t \geq 0\}$ and $\{M_t(B) : t \geq 0\}$ are orthogonal whenever A and $B \in \mathscr{B}(E)$ are disjoint. Then $\eta((0, t] \times A \times B) = \langle M(A), M(B)\rangle_t$ vanishes when A and

$B \in \mathscr{B}(E)$ are disjoint, and hence $\eta(\omega, \mathrm{d}s, \mathrm{d}x, \mathrm{d}y)$ is carried by $[0, \infty) \times \Delta(E)$.

□

Let $\mathscr{L}$ be a linear space of Borel functions on E. Suppose that for each $f \in \mathscr{L}$ there is a square-integrable càdlàg $(\mathscr{G}_t)$-martingale $\{M_t(f) : t \geq 0\}$ satisfying $M_0(f) = 0$. The family $\{M_t(f) : t \geq 0; f \in \mathscr{L}\}$ is called a *martingale functional* if for every $t \geq 0$ the following properties hold:

(1) For each $c \in \mathbb{R}$ and each $f \in \mathscr{L}$ we have a.s. $M_t(cf) = cM_t(f)$.
(2) If $f, f_1, f_2, \ldots \in \mathscr{L}$ and $f = \sum_{k=1}^{\infty} f_k$ by bounded pointwise convergence, then

$$M_t(f) = \sum_{k=1}^{\infty} M_t(f_k)$$

by the convergence in $L^2(\Omega, \mathbf{P})$.

Proposition 7.21 *For each worthy martingale measure $\{M_t(B) : t \geq 0; B \in \mathscr{B}(E)\}$ there is a martingale functional $\{M_t(f) : t \geq 0; f \in B(E)\}$ so that $M_t(1_B) = M_t(B)$ a.s. for every $t \geq 0$ and every $B \in \mathscr{B}(E)$. Moreover, for any $f \in B(E)$ the $(\mathscr{G}_t)$-martingale $\{M_t(f) : t \geq 0\}$ has increasing process*

$$(f, f)_{\eta, t} = \int_0^t \int_{E^2} f(x)f(y)\eta(\mathrm{d}s, \mathrm{d}x, \mathrm{d}y). \tag{7.31}$$

Proof. We shall give an explicit construction of the martingale functional $\{M_t(f) : t \geq 0; f \in B(E)\}$. If $f \in B(E)$ is a simple function given by

$$f(x) = \sum_{i=1}^{n} b_i 1_{B_i}(x), \qquad x \in E,$$

where $b_i \in \mathbb{R}$ and $B_i \in \mathscr{B}(E)$ for $i = 1, \ldots, n$, we define

$$M_t(f) = \sum_{i=1}^{n} b_i M_t(B_i), \qquad t \geq 0.$$

It is easy to see that $\{M_t(f) : t \geq 0\}$ is a càdlàg martingale with increasing process given by (7.31). For a general function $f \in B(E)$ let $\{f_k\}$ be a sequence of simple functions on E so that $\|f_k - f\| \to 0$ as $k \to \infty$. By Doob's martingale inequality and the definition of the worthy martingale measure,

$$\mathbf{P}\Big[\sup_{0 \leq s \leq t} |M_s(f_k) - M_s(f_j)|^2 \Big] \leq 4\mathbf{P}\big[(f_k - f_j, f_k - f_j)_{\eta, t} \big]$$
$$\leq 4\|f_k - f_j\|^2 \mathbf{P}\big[K([0, t] \times E^2) \big]$$

for any $k \geq j \geq 1$. Then there is a square-integrable càdlàg $(\mathscr{G}_t)$-martingale $\{M_t(f) : t \geq 0\}$ independent of the choice of $\{f_k\}$ so that

$$\lim_{k \to \infty} \mathbf{P} \left[\sup_{0 \le s \le t} |M_s(f_k) - M_s(f)|^2 \right] = 0, \qquad t \ge 0.$$

Since (7.31) holds when f is replaced by f_k, for any $t \ge r \ge 0$ we have

$$\mathbf{E} \left[M_t(f_k)^2 - M_r(f_k)^2 - (f_k, f_k)_{\eta,t} + (f_k, f_k)_{\eta,r} \right] = 0.$$

Then we can let $k \to \infty$ to see $\{M_t(f) : t \ge 0\}$ has increasing process (7.31). It is easy to show that $\{M_t(f) : t \ge 0; f \in B(E)\}$ satisfies the two properties in the definition of a martingale functional. $\qquad\square$

The worthy martingale measure defined above is *finite*. Let $h \in p\mathscr{B}(E)$ be a strictly positive function. Recall that $B_h(E)$ denotes the set of functions $f \in \mathscr{B}(E)$ such that $|f| \le \text{const} \cdot h$. Instead of (7.29) one can also define a worthy martingale measure with dominating measure $K(ds, dx, dy)$ satisfying

$$\mathbf{P} \left[\int_0^t \int_{E^2} h(x)h(y)K(ds, dx, dy) \right] < \infty, \qquad t \ge 0.$$

Let $E_n = \{x \in E : h(x) \ge 1/n\}$ and let $\mathscr{B}_0(E) = \{B \in \mathscr{B}(E) : B \subset E_n$ for some $n \ge 1\}$. A family of square-integrable càdlàg $(\mathscr{G}_t)$-martingales $\{M_t(B) : t \ge 0; B \in \mathscr{B}_0(E)\}$ is called a *σ-finite worthy martingale measure* on E if (7.30) holds for all $t \ge s \ge 0$ and $A, B \in \mathscr{B}_0(E)$. Following the arguments in the proof above, we can extend the σ-finite worthy martingale measure to a martingale functional $\{M_t(f) : t \ge 0; f \in B_h(E)\}$. For simplicity we shall only discuss finite worthy martingale measures in this section, but all the results can be reformulated for σ-finite worthy martingale measures.

Now suppose we are given a worthy martingale measure $\{M_t(B) : t \ge 0; B \in \mathscr{B}(E)\}$ on E with covariance and dominating measures $\eta(ds, dx, dy)$ and $K(ds, dx, dy)$, respectively. The martingale functional $\{M_t(f) : t \ge 0; f \in B(E)\}$ given in Proposition 7.21 is clearly unique in the following sense: If $\{Z_t(f) : t \ge 0; f \in B(E)\}$ is also a martingale functional so that $Z_t(1_B) = M_t(B)$ a.s. for every $t \ge 0$ and every $B \in \mathscr{B}(E)$, then $M_t(f) = Z_t(f)$ a.s. for every $t \ge 0$ and every $f \in B(E)$. A real-valued two-parameter process $\{h_s(x) : s \ge 0, x \in E\}$ is said to be *progressive* if for every $t \ge 0$ the mapping $(\omega, s, x) \mapsto h_s(\omega, x)$ restricted to $\Omega \times [0, t] \times E$ is measurable relative to $\mathscr{G}_t \times \mathscr{B}([0, t] \times E)$. Let $\mathscr{P} = \mathscr{P}(\mathscr{G}_t)$ denote the σ-algebra on $\Omega \times [0, \infty)$ generated by all real-valued left continuous processes adapted to $(\mathscr{G}_t)$. A progressive process $\{h_s(x) : s \ge 0, x \in E\}$ is said to be *predictable* if the mapping $(\omega, s, x) \mapsto h_s(\omega, x)$ is $(\mathscr{P} \times \mathscr{B}(E))$-measurable. Let $\mathscr{L}_K^2(E)$ be the space of two-parameter predictable processes $h = \{h_s(x) : s \ge 0, x \in E\}$ satisfying

$$\|h\|_{K,T} := \left\{ \mathbf{P} \left[(|h|, |h|)_{K,T} \right] \right\}^{1/2} < \infty, \qquad T \ge 0. \tag{7.32}$$

It is easy to show that each $\| \cdot \|_{K,T}$ is a seminorm on $\mathscr{L}_K^2(E)$. We identify h_1 and $h_2 \in \mathscr{L}_K^2(E)$ if $\|h_1 - h_2\|_{K,T} = 0$ for every $T \ge 1$. Then

$$d_2(h_1, h_2) = \sum_{n=1}^{\infty} \frac{1}{2^n} (1 \wedge \|h_1 - h_2\|_{K,n}) \tag{7.33}$$

defines a metric on $\mathscr{L}_K^2(E)$. We call $\{q_s(x) : s \geq 0, x \in E\}$ a *step process* if it is of the form

$$q_s(x) = g_0(x) 1_{\{0\}}(s) + \sum_{i=0}^{\infty} g_i(x) 1_{(r_i, r_{i+1}]}(s), \tag{7.34}$$

where each $(\omega, x) \mapsto g_i(\omega, x)$ is a $(\mathscr{G}_{r_i} \times \mathscr{B}(E))$-measurable function and $\{0 = r_0 < r_1 < r_2 < \cdots\}$ is a sequence increasing to infinity. Clearly, a step process is predictable. Let $\mathscr{L}_K^0(E)$ be the set of step processes in $\mathscr{L}_K^2(E)$.

Proposition 7.22 *The metric space $(\mathscr{L}_K^2(E), d_2)$ is complete and $\mathscr{L}_K^0(E)$ is a dense subset of $\mathscr{L}_K^2(E)$.*

Proof. Suppose that $\{h_k\}$ is a Cauchy sequence in $\mathscr{L}_K^2(E)$. Then for any fixed $n \geq 1$ the restrictions of $\{h_k\}$ to $\Omega \times [0, n] \times E$ form a Cauchy sequence with respect to the seminorm $\|\cdot\|_{K,n}$ defined by (7.32). It is easily seen that

$$\mathbf{Q}_n(d\omega, ds, dx, dy) = \mathbf{P}(d\omega) K(\omega, ds, dx, dy)$$

defines a finite measure on $\mathscr{G} \times \mathscr{B}((0, n] \times E^2)$. For any $\varepsilon > 0$ and $j, k \geq 1$ we have

$$\mathbf{Q}_n\{(\omega, s, x, y) : |h_j(\omega, s, x) - h_k(\omega, s, x)| \geq \varepsilon\}$$
$$\leq \frac{1}{\varepsilon} \mathbf{P}\left[\int_0^n \int_{E^2} |h_j(s, x) - h_k(s, x)| K(ds, dx, dy)\right]$$
$$\leq \frac{1}{\varepsilon} \mathbf{P}\left[(|h_j - h_k|, |h_j - h_k|)_{K,n}^{1/2} K((0, n] \times E^2)^{1/2}\right]$$
$$\leq \frac{1}{\varepsilon} \|h_j - h_k\|_{K,n} \mathbf{P}\left[K((0, n] \times E^2)\right]^{1/2},$$

where the second inequality follows from (7.28). Then for each $i \geq 1$ we can choose $l_i \geq 1$ so that

$$\mathbf{Q}_n\{(\omega, s, x, y) : |h_j(\omega, s, x) - h_k(\omega, s, x)| \geq 1/2^i\} < 1/2^i$$

for all $j, k \geq l_i$. In addition, we can assume $l_i \to \infty$ increasingly as $i \to \infty$. Let

$$F_i = \{(\omega, s, x, y) : |h_{l_i}(\omega, s, x) - h_{l_{i+1}}(\omega, s, x)| \geq 1/2^i\}$$

and $N_1 = \cap_{m=1}^{\infty} \cup_{i=m}^{\infty} F_i$. We have

$$\mathbf{Q}_n\left(\bigcup_{i=m}^{\infty} F_i\right) \leq \sum_{i=m}^{\infty} \frac{1}{2^i} = \frac{1}{2^{m-1}}$$

and hence $\mathbf{Q}_n(N_1) = 0$. For any $(\omega, s, x, y) \in N_1^c$ there is some $m \geq 1$ so that $|h_{l_i}(\omega, s, x) - h_{l_{i+1}}(\omega, s, x)| < 1/2^i$ for all $i \geq m$, and hence $\{h_{l_i}(\omega, s, x)\}$ is a Cauchy sequence. Now define the predictable process

$$h(\omega, s, x) = \limsup_{i \to \infty} h_{l_i}(\omega, s, x), \qquad \omega \in \Omega, s \geq 0, x \in E.$$

We have $h_{l_i}(\omega, s, x) \to h(\omega, s, x)$ for all $(\omega, s, x, y) \in N_1^c$. Let

$$N = \{(\omega, s, x, y) \in \Omega \times (0, n] \times E^2 : (\omega, s, x, y) \text{ or } (\omega, s, y, x) \in N_1\}.$$

Then $\mathbf{Q}_n(N) = 0$ by the symmetry of $K(\mathrm{d}s, \mathrm{d}x, \mathrm{d}y)$. Moreover, $h_{l_i}(\omega, s, x) \to h(\omega, s, x)$ and $h_{l_i}(\omega, s, y) \to h(\omega, s, y)$ for all $(\omega, s, x, y) \in N^c$. For any $\varepsilon > 0$ let $m(\varepsilon) \geq 1$ be such that $\|h_k - h_j\|_{K,n} \leq \varepsilon$ for $j, k \geq m(\varepsilon)$. Letting $j \to \infty$ along the sequence $\{l_i\}$ and applying Fatou's lemma we see $\|h_k - h\|_{K,n} \leq \varepsilon$ for $k \geq m(\varepsilon)$. Thus $\|h_k - h\|_{K,n} \to 0$ as $k \to \infty$. Since $n \geq 1$ was arbitrary, it is easy to define a process $h \in \mathscr{L}_K^2(E)$ so that $h_k \to h$ relative to the metric defined by (7.33). That gives the first assertion of the proposition. To prove the second assertion, note that the σ-algebra $\mathscr{P} \times \mathscr{B}(E)$ can be generated by bounded two-parameter step processes. Let $\mathrm{b}\mathscr{L}_K^2(E)$ and $\mathrm{b}\mathscr{L}_K^0(E)$ denote respectively the sets of bounded elements of $\mathscr{L}_K^2(E)$ and $\mathscr{L}_K^0(E)$. By Proposition A.1 one sees that $\mathrm{b}\mathscr{L}_K^0(E)$ is dense in $\mathrm{b}\mathscr{L}_K^2(E)$. Now let $h \in \mathscr{L}_K^2(E)$. For any $k \geq 1$ define $h_k \in \mathrm{b}\mathscr{L}_K^2(E, k)$ by $h_k(\omega, s, x) = h(\omega, s, x)1_{\{|h(\omega, s, x)| \leq k\}}$. Then we have

$$\|h_k - h\|_{K,n}^2 = \mathbf{P}\left[\int_0^n \int_{E^2} |h(s, x)h(s, y)|1_{\{|h(s,x)|>k, |h(s,y)|>k\}} K(\mathrm{d}s, \mathrm{d}x, \mathrm{d}y)\right],$$

which tends to zero as $k \to \infty$. Then $\mathrm{b}\mathscr{L}_K^2(E)$ is dense in $\mathscr{L}_K^2(E)$, so $\mathrm{b}\mathscr{L}_K^0(E)$ is also dense in $\mathscr{L}_K^2(E)$. □

We are now ready to define the stochastic integrals of processes in $\mathscr{L}_K^2(E)$ with respect to the martingale measure $\{M_t(B) : t \geq 0; B \in \mathscr{B}(E)\}$. For a step process $q \in \mathscr{L}_K^0(E)$ given by (7.34), each g_i is a deterministic Borel function on E under the conditional probability $\mathbf{P}\{\cdot|\mathscr{G}_{r_i}\}$. Then we can use the martingale functional induced by $\{M_t(B) : t \geq 0; B \in \mathscr{B}(E)\}$ to define the process $\{M_{r_{i+1} \wedge t}(g_i) - M_{r_i \wedge t}(g_i) : t \geq 0\}$, which is a square-integrable càdlàg $(\mathscr{G}_t)$-martingale first under $\mathbf{P}\{\cdot|\mathscr{G}_{r_i}\}$ and then under $\mathbf{P}$. It follows that

$$M_t(q_t) = \sum_{i=0}^{\infty} [M_{r_{i+1} \wedge t}(g_i) - M_{r_i \wedge t}(g_i)], \qquad t \geq 0$$

is a square-integrable càdlàg $(\mathscr{G}_t)$-martingale. The increasing process of $\{M_t(q_t) : t \geq 0\}$ is clearly given by

$$\langle M(q) \rangle_t = \int_0^t \int_{E^2} q_s(x)q_s(y)\eta(\mathrm{d}s, \mathrm{d}x, \mathrm{d}y).$$

For a general process $h \in \mathscr{L}_K^2(E)$, choose a sequence $\{q_k\} \subset \mathscr{L}_K^0(E)$ so that $d_2(q_k, h) \to 0$ as $k \to \infty$. By Doob's martingale inequality,

$$\mathbf{P}\left[\sup_{0\le s\le n} M_s(q_k(s)-q_j(s))^2\right] \le 4\mathbf{P}\left[(q_k-q_j,q_k-q_j)_{\eta,n}\right]$$
$$\le 4\|q_k-q_j\|_{K,n}^2,$$

which tends to zero as $j,k \to \infty$. Then there is a square-integrable càdlàg $(\mathscr{G}_t)$-martingale $\{M_t(h_t) : t \ge 0\}$ so that

$$\lim_{k\to\infty} \mathbf{P}\left[\sup_{0\le s\le n} |M_s(q_k(s)) - M_s(h_s)|^2\right] = 0, \qquad n \ge 1.$$

It is easy to see that $\{M_t(h_t) : t \ge 0\}$ has increasing process

$$(h,h)_{\eta,t} = \int_0^t \int_{E^2} h_s(x)h_s(y)\eta(\mathrm{d}s,\mathrm{d}x,\mathrm{d}y). \tag{7.35}$$

We shall write

$$M_t(h_t) = \int_0^t \int_E h_s(x)M(\mathrm{d}s,\mathrm{d}x)$$

and call it the *stochastic integral* of $h \in \mathscr{L}_K^2(E)$ with respect to $\{M_t(B) : t \ge 0; B \in \mathscr{B}(E)\}$.

We next prove an important property of stochastic integrals with respect to the martingale measure. Suppose that F is another Lusin topological space and λ is a finite Borel measure on F. Let $\{h(s,x,z) : s \ge 0, x \in E, z \in F\}$ be a predictable process satisfying

$$\mathbf{P}\left[\int_F \lambda(\mathrm{d}z) \int_0^n \int_{E^2} |h(s,x,z)h(s,y,z)|K(\mathrm{d}s,\mathrm{d}x,\mathrm{d}y)\right] < \infty \tag{7.36}$$

for every $n \ge 1$. The above condition implies

$$\mathbf{P}\left[\int_0^n \int_{E^2} |h(s,x,z)h(s,y,z)|K(\mathrm{d}s,\mathrm{d}x,\mathrm{d}y)\right] < \infty$$

for λ-a.e. $z \in F$. Then the stochastic integral

$$M_t(z) := \int_0^t \int_E h(s,x,z)M(\mathrm{d}s,\mathrm{d}x) \tag{7.37}$$

is well-defined for λ-a.e. $z \in F$. On the other hand, using (7.28) we have

$$\mathbf{P}\left[\int_{F^2} (|h(z_1)|,|h(z_2)|)_{K,n}\lambda(\mathrm{d}z_1)\lambda(\mathrm{d}z_2)\right]$$
$$\le \int_{F^2} \mathbf{P}\left[(|h(z_1)|,|h(z_1)|)_{K,n}^{1/2}(|h(z_2)|,|h(z_2)|)_{K,n}^{1/2}\right]\lambda(\mathrm{d}z_1)\lambda(\mathrm{d}z_2)$$
$$\le \int_F \left\{\mathbf{P}\left[(|h(z_1)|,|h(z_1)|)_{K,n}\right]\right\}^{1/2}\lambda(\mathrm{d}z_1)$$

$$\cdot \int_F \left\{ \mathbf{P} \left[\left(|h(z_2)|, h(z_2)| \right)_{K,n} \right] \right\}^{1/2} \lambda(dz_2)$$

$$\leq \lambda(1) \int_F \mathbf{P} \left[\left(|h(z)|, |h(z)| \right)_{K,n} \right] \lambda(dz). \tag{7.38}$$

The right-hand side is finite by (7.36). It follows that

$$H(s, x) := \int_F h(s, x, z) \lambda(dz), \qquad s \geq 0, x \in E \tag{7.39}$$

defines a predictable process $H \in \mathscr{L}_K^2(E)$. Therefore

$$t \mapsto \int_0^t \int_E H(s, x) M(ds, dx)$$

is well-defined as a square-integrable càdlàg martingale. From (7.39) and (7.37) it is natural to expect

$$\int_F M_t(z) \lambda(dz) = \int_0^t \int_E H(s, x) M(ds, dx). \tag{7.40}$$

The above formula is called a *stochastic Fubini's theorem* for the martingale measure, which means that $(\omega, z) \mapsto M_t(\omega, z)$ has a $(\mathscr{G}_t \times \mathscr{B}(F))$-measurable version and the equality holds with probability one. To establish the formula rigorously we first prove the following:

Lemma 7.23 *Let h and h_k be predictable processes satisfying condition (7.36). Suppose that (7.40) holds for every h_k and*

$$\int_F \mathbf{P} \left[\left(|h_k(z) - h(z)|, |h_k(z) - h(z)| \right)_{K,n} \right] \lambda(dz) \to 0 \tag{7.41}$$

as $k \to \infty$ for every $n \geq 1$. Then (7.40) also holds for the process h.

Proof. Step 1. Let $M_k(t, z)$ be defined by the right-hand side of (7.37) with h replaced by h_k. Since (7.40) holds for h_k, the function $(\omega, z) \mapsto M_k(\omega, t, z)$ has a $(\mathscr{G}_t \times \mathscr{B}(F))$-measurable version. By (7.41) it is easy to show that

$$\mathbf{P} \left[\int_F |M_k(t, z) - M_j(t, z)|^2 \lambda(dz) \right] \to 0$$

as $j, k \to \infty$. Then there is a $(\mathscr{G}_t \times \mathscr{B}(F))$-measurable function $(\omega, z) \mapsto N_t(\omega, z)$ so that

$$\mathbf{P} \left[\int_F |M_k(t, z) - N_t(z)|^2 \lambda(dz) \right] \to 0, \tag{7.42}$$

and hence

$$\int_F M_k(t, z)\lambda(\mathrm{d}z) \to \int_F N_t(z)\lambda(\mathrm{d}z) \tag{7.43}$$

in $L^2(\Omega, \mathbf{P})$ because λ is a finite measure. By (7.42) we can choose a sequence $\{k_i\}$ so that $M_{k_i}(t, z) \to N_t(z)$ in $L^2(\Omega, \mathbf{P})$ for λ-a.e. $z \in F$. On the other hand, by (7.41) there is a subsequence $\{k_i'\} \subset \{k_i\}$ so that $\|h_{k_i'}(\cdot, \cdot, z) - h(\cdot, \cdot, z)\|_{K,n} \to 0$ for every $n \geq 1$ and λ-a.e. $z \in F$. Then $M_{k_i'}(t, z) \to M_t(z)$ in $L^2(\Omega, \mathbf{P})$ for λ-a.e. $z \in F$. It follows that $M_t(z) = N_t(z)$ a.s. for λ-a.e. $z \in F$.

Step 2. Let H_k be defined by the right-hand side of (7.39) with h replaced by h_k. From (7.41) and the calculations in (7.38) we have $\|H_k - H\|_{K,n} \to 0$ as $k \to \infty$ for every $n \geq 1$. It follows that

$$\int_0^t \int_E H_k(s, x)M(\mathrm{d}s, \mathrm{d}x) \to \int_0^t \int_E H(s, x)M(\mathrm{d}s, \mathrm{d}x) \tag{7.44}$$

in $L^2(\Omega, \mathbf{P})$ as $k \to \infty$. By the assumption of the lemma,

$$\int_F M_k(t, z)\lambda(\mathrm{d}z) = \int_0^t \int_E H_k(s, x)M(\mathrm{d}s, \mathrm{d}x).$$

This together with (7.43) and (7.44) yields a.s.

$$\int_F N_t(z)\lambda(\mathrm{d}z) = \int_0^t \int_E H(s, x)M(\mathrm{d}s, \mathrm{d}x),$$

which is just what (7.40) means.   □

Theorem 7.24 *Let $\{h(s, x, z) : s \geq 0, x \in E, z \in F\}$ be a predictable process satisfying (7.36). Let $\{M_t(z) : t \geq 0, z \in F\}$ and $\{H(s, x) : s \geq 0, x \in E\}$ be defined by (7.37) and (7.39), respectively. Then (7.40) holds a.s. for every $t \geq 0$.*

Proof. Let $\mathscr{H}$ be the class of bounded predictable process $h = \{h(s, x, z) : s \geq 0, x \in E, z \in F\}$ for which the theorem holds. Then Lemma 7.23 implies that $\mathscr{H}$ is closed under bounded pointwise convergence. If $h(\omega, s, x, z) = q(\omega, s, x)f(z)$ for a bounded $(\mathscr{P} \times \mathscr{B}(E))$-measurable function q on $\Omega \times [0, \infty) \times E$ and a bounded $\mathscr{B}(F)$-measurable function f on F, then (7.40) holds clearly. By Proposition A.1, the class $\mathscr{H}$ contains all bounded $(\mathscr{P} \times \mathscr{B}(E) \times \mathscr{B}(F))$-measurable processes. For a general predictable process $h = \{h(s, x, z) : s \geq 0, x \in E, z \in F\}$ satisfying (7.36), the result follows from Lemma 7.23 by an approximation using the sequence defined by $h_k := h1_{\{|h| \leq k\}}$.   □

7.4 A Representation for Superprocesses

Suppose that E is a locally compact separable metric space. Let ξ be a Hunt process in E with transition semigroup $(P_t)_{t \geq 0}$ and ϕ a branching mechanism given by

(2.26) or (2.27). We assume that $(P_t)_{t\geq 0}$ and ϕ satisfy the conditions specified at the beginning of the first section and $\nu(1)^2 H(x, d\nu)$ is a bounded kernel from E to $M(E)^\circ$. Suppose that $\{X_t : t \geq 0\}$ is a càdlàg (ξ, ϕ)-superprocess relative to the filtration $(\mathscr{G}_t)_{t\geq 0}$ satisfying $\mathbf{P}[X_0(1)] < \infty$.

Theorem 7.25 *The martingales defined in (7.23) induce a worthy $(\mathscr{G}_t)$-martingale measure $\{M_t(B) : t \geq 0; B \in \mathscr{B}(E)\}$ satisfying*

$$M_t(f) = \int_0^t \int_E f(x) M(ds, dx), \qquad t \geq 0, f \in D_0(A) \tag{7.45}$$

and having covariance measure defined by

$$\eta(ds, dx, dy) = ds \int_E 2c(z)\delta_z(dx)\delta_z(dy)X_s(dz)$$

$$+ ds \int_E X_s(dz) \int_{M(E)^\circ} \nu(dx)\nu(dy)H(z, d\nu). \tag{7.46}$$

Proof. We first note that (7.23) and (7.24) define a martingale functional $\{M_t(f) : t \geq 0; f \in D_0(A)\}$. For each $n \geq 1$ define the measure $\mu_n \in M(E)$ by

$$\mu_n(f) = \mathbf{P}\left\{ \int_0^n ds \int_E \left[2c(z)f(z) + \int_{M(E)^\circ} \nu(f)\nu(1)H(z, d\nu) \right] X_s(dz) \right\},$$

where $f \in B(E)$. It is well-known that $C_0(E)$ is dense in $L^2(\mu_n)$; see, e.g., Hewitt and Stromberg (1965, p.197). Since $D_0(A)$ is dense in $C_0(E)$ by the supremum norm, it is also dense in $L^2(\mu_n)$. Consequently, for any $f \in B(E)$ there is a sequence $\{f_k\} \subset D_0(A)$ so that $\lim_{k\to\infty} \mu_n(|f_k - f|^2) = 0$ for every $n \geq 1$. By Doob's martingale inequality and Corollary 7.15,

$$\mathbf{P}\left[\sup_{0\leq s\leq n} M_s(f_k - f_j)^2 \right] \leq 4\mathbf{P}\left\{ \int_0^n ds \int_E \left[2c(z)|f_k(z) - f_j(z)|^2 \right. \right.$$

$$\left. + \int_{M(E)^\circ} \nu(|f_k - f_j|)^2 H(z, d\nu) \right] X_s(dz) \right\}$$

$$\leq 4\mathbf{P}\left\{ \int_0^n ds \int_E \left[2c(z)|f_k(z) - f_j(z)|^2 \right. \right.$$

$$\left. + \int_{M(E)^\circ} \nu(|f_k - f_j|^2)\nu(1)H(z, d\nu) \right] X_s(dz) \right\}$$

$$\leq 4\mu_n(|f_k - f_j|^2).$$

The right-hand side goes to zero as $j, k \to \infty$. Then there is a square-integrable càdlàg $(\mathscr{G}_t)$-martingale $\{M_t(f) : t \geq 0\}$ so that

$$\lim_{k\to\infty} \mathbf{P}\left[\sup_{0\leq s\leq t} |M_s(f_k) - M_s(f)|^2 \right] = 0, \qquad t \geq 0.$$

It is easy to see that $\{M_t(f) : t \geq 0; f \in B(E)\}$ is a martingale functional. Let $M_t(B) = M_t(1_B)$ for $t \geq 0$ and $B \in \mathscr{B}(E)$. Then $\{M_t(B) : t \geq 0; B \in \mathscr{B}(E)\}$ is a worthy martingale measure with covariance measure $\eta(ds, dx, dy)$.   □

It is easy to see that the martingale measure defined by (7.45) is orthogonal if and only if the branching mechanism ϕ can be chosen in a way so that $H(x, d\nu)$ is concentrated on $\{u\delta_x : u > 0\}$ for all $x \in E$. The following theorem gives a representation of the superprocess in terms of a stochastic integral with respect to the martingale measure.

Theorem 7.26 *Let $\{M_t(B) : t \geq 0; B \in \mathscr{B}(E)\}$ be the worthy $(\mathscr{G}_t)$-martingale measure defined by (7.45) and (7.46). Then for any $t \geq 0$ and $f \in B(E)$ we have a.s.*

$$X_t(f) = X_0(\pi_t f) + \int_0^t \int_E \pi_{t-s} f(x) M(ds, dx), \qquad (7.47)$$

where $t \mapsto \pi_t f$ is defined by (2.35).

Proof. Since $(s, x) \mapsto 1_{\{s \leq t\}} \pi_{t-s} f(x)$ is a deterministic measurable function, the stochastic integral on the right-hand side of (7.47) is well-defined. Let us consider a partition $\Delta = \{0 = t_0 < t_1 < \cdots < t_n = t\}$ of $[0, t]$. Let $|\Delta| = \max_{1 \leq i \leq n} |t_i - t_{i-1}|$. For any $f \in D_0(A)^+$ one can use Theorem 7.10 to see

$$\frac{d}{dt} \pi_t f(x) = (A + \gamma - b) \pi_t f(x), \qquad t \geq 0, x \in E.$$

It follows that

$$X_t(f) = X_0(\pi_t f) + \sum_{i=1}^n X_{t_i}(\pi_{t-t_i} f - \pi_{t-t_{i-1}} f)$$

$$+ \sum_{i=1}^n \left[X_{t_i}(\pi_{t-t_{i-1}} f) - X_{t_{i-1}}(\pi_{t-t_{i-1}} f) \right]$$

$$= X_0(\pi_t f) - \sum_{i=1}^n \int_{t-t_i}^{t-t_{i-1}} X_{t_i}((A + \gamma - b)\pi_s f) ds$$

$$+ \sum_{i=1}^n \left[M_{t_i}(\pi_{t-t_{i-1}} f) - M_{t_{i-1}}(\pi_{t-t_{i-1}} f) \right]$$

$$+ \sum_{i=1}^n \int_{t_{i-1}}^{t_i} X_s((A + \gamma - b)\pi_{t-t_{i-1}} f) ds.$$

By letting $|\Delta| \to 0$ and using the right continuity of $s \mapsto X_s$ and the strong continuity of $s \mapsto \pi_s f$ we obtain (7.47) for $f \in D_0(A)^+$. For any $f \in C_0(E)$ we let $f_n = f_n^+ - f_n^-$ for the functions f_n^+ and f_n^- defined by (7.20). Note that $f_n \to f$ in the supremum norm as $n \to \infty$. Using this approximation we get (7.47) for $f \in C_0(E)$. The result for a general function $f \in B(E)$ follows by Proposition A.1.   □

To conclude this section, we give an application of the results above by proving a structural property of the (ξ, ϕ)-superprocess. For this purpose we consider the following condition:

Condition 7.27 *There exists a σ-finite measure λ on E and a Borel function $(t, x, y) \mapsto p_t(x, y)$ on $(0, \infty) \times E^2$ so that*

$$P_t(x, \mathrm{d}y) = p_t(x, y)\lambda(\mathrm{d}y), \qquad t > 0, \; x, y \in E,$$

and there exists a constant $0 < \alpha < 1$ and a locally bounded function $t \mapsto C(t)$ on $[0, \infty)$ so that

$$p_t(x, y) \leq t^{-\alpha}C(t), \qquad t > 0, \; x, y \in E.$$

Theorem 7.28 *Suppose that Condition 7.27 holds. Then for every $t > 0$ we have $\mathbf{P}\{X_t$ is absolutely continuous with respect to $\lambda\} = 1$.*

Proof. By the expression (2.37) one can show that for any $t > 0$ and $x \in E$ the finite measure $\pi_t(x, \mathrm{d}y)$ is absolutely continuous with respect to $\lambda(\mathrm{d}y)$ with density $q_t(x, y)$ satisfying

$$q_t(x, y) \leq g(t) := t^{-\alpha}e^{\|b\|t}C(t) + K(t), \qquad t > 0, \; x, y \in E,$$

where $t \mapsto K(t)$ is a locally bounded function on $[0, \infty)$. Let $f \in C_0(E)^+$ be a strictly positive function satisfying $\langle \lambda, f \rangle < \infty$ and define $\lambda_f(\mathrm{d}z) = f(z)\lambda(\mathrm{d}z)$. Let $\{M_t(B) : t \geq 0; B \in \mathscr{B}(\mathbb{R})\}$ be the martingale measure defined by (7.45) and (7.46). Recall that $c_0 = \|b^-\| + \|\gamma(\cdot, 1)\|$ and set $q_s(x, z) = 0$ for $s \leq 0$. For any $n \geq 1$ we have

$$\mathbf{P}\left[\int_E \lambda_f(\mathrm{d}z) \int_0^n \int_{E^2} q_{t-s}(x, z)q_{t-s}(y, z)\eta(\mathrm{d}s, \mathrm{d}x, \mathrm{d}y)\right]$$
$$\leq \mathbf{P}\left[\int_0^t g(t - s)\mathrm{d}s \int_E 2c(x)\pi_{t-s}f(x)X_s(\mathrm{d}x)\right]$$
$$+ \mathbf{P}\left[\int_0^t g(t - s)\mathrm{d}s \int_E X_s(\mathrm{d}x) \int_{M(E)^\circ} \nu(\pi_{t-s}f)\nu(1)H(x, \mathrm{d}\nu)\right]$$
$$\leq e^{c_0t}\|f\|\mathbf{P}\left[\int_0^t g(t - s)\mathrm{d}s \int_E q(x, 1)X_s(\mathrm{d}x)\right] < \infty,$$

where $q(x, 1)$ is defined by (2.59) with $f = 1$. Then using Theorems 7.24 and 7.26 we get

$$X_t(f) = \int_E Y_t(z)\lambda_f(\mathrm{d}z) = \int_E f(z)Y_t(z)\lambda(\mathrm{d}z),$$

where

$$Y_t(z) = \int_E q_t(x, z)X_0(\mathrm{d}x) + \int_0^t \int_E q_{t-s}(x, z)M(\mathrm{d}s, \mathrm{d}x).$$

By considering a sequence $\{f_n\}$ dense in $C_0(E)^+$ we obtain the desired result. □

Example 7.2 If ξ is a Brownian motion in $\mathbb{R}$, then Condition 7.27 holds with λ being the Lebesgue measure. Thus for the super-Brownian motion the random measures $\{X_t : t > 0\}$ are absolutely continuous with respect to the Lebesgue measure.

7.5 Transforms by Martingales

In this section, we assume E is a locally compact separable metric space and ξ is a Hunt process with Feller transition semigroup $(P_t)_{t\geq0}$. Let us consider a local branching mechanism ϕ given by (2.45) with constant function $b(x) \equiv b \geq 0$. Suppose in addition that $c \in C(E)^+$ and $x \mapsto (u \wedge u^2)m(x, du)$ is continuous by weak convergence on $(0, \infty)$. Let $X = (W_0, \mathscr{G}, \mathscr{G}_t, X_t, \tilde{\mathbf{Q}}_\mu)$ be the subprocess of the (ξ, ϕ)-superprocess generated by the multiplicative functional $\{m_t : t \geq 0\}$ given by (6.25). Recall that $\hat{\mu} = \mu(1)^{-1}\mu$ for $\mu \in M(E)^\circ$. Let L_0 be defined by (7.12) for the local branching mechanism.

Theorem 7.29 *Under $\tilde{\mathbf{Q}}_\mu$ the process $\{X_t : t \geq 0\}$ solves the martingale problem: For any $F \in \mathscr{D}_0$ given by (7.11),*

$$F(X_t) = \int_0^t L_0F(X_s)ds + 2 \int_0^t ds \int_E c(x)F'(X_s; x)\hat{X}_s(dx)$$
$$+ \int_0^t ds \int_E \hat{X}_s(dx) \int_0^\infty [F(X_s + u\delta_x) - F(X_s)]um(x, du)$$
$$+ local\ mart. \tag{7.48}$$

Proof. Let $H(\mu) = \mu(1)F(\mu)$ for $\mu \in M(E)$. The operator L_0 can still be applied to H although the function is not necessarily in $\mathscr{D}_0$. In fact, it is easy to see that

$$H'(\mu; x) = F(\mu) + \mu(1)F'(\mu; x)$$

and

$$H''(\mu; x) = 2F'(\mu; x) + \mu(1)F''(\mu; x).$$

Then we have

$$L_0H(\mu) = \mu(1)(L_0 - b)F(\mu) + 2 \int_E c(x)F'(\mu; x)\mu(dx)$$
$$+ \int_E \mu(dx) \int_0^\infty [F(\mu + u\delta_x) - F(\mu)]um(x, du).$$

By extending Theorem 7.13 slightly and applying it to suitable truncations of H we get

$$H(X_t) = \int_0^t X_s(1)(L_0 - b)F(X_s)\mathrm{d}s + 2\int_0^t \mathrm{d}s \int_E c(x)F'(X_s;x)X_s(\mathrm{d}x)$$
$$+ \int_0^t \mathrm{d}s \int_E X_s(\mathrm{d}x) \int_0^\infty [F(X_s + u\delta_x) - F(X_s)]um(x,\mathrm{d}u)$$
$$+ \text{local mart.}$$

under $\mathbf{Q}_\mu$. By integration by parts,

$$e^{bt}H(X_t) = \int_0^t e^{bs}X_s(1)L_0F(X_s)\mathrm{d}s + 2\int_0^t e^{bs}\mathrm{d}s \int_E c(x)F'(X_s;x)X_s(\mathrm{d}x)$$
$$+ \int_0^t e^{bs}\mathrm{d}s \int_E X_s(\mathrm{d}x) \int_0^\infty [F(X_s + u\delta_x) - F(X_s)]um(x,\mathrm{d}u)$$
$$+ \text{local mart.}$$
$$= \int_0^t e^{bs}X_s(1)J(X_s)\mathrm{d}s + \text{local mart.}, \tag{7.49}$$

where

$$J(X_s) = L_0F(X_s) + 2\int_E c(x)F'(X_s;x)\hat{X}_s(\mathrm{d}x)$$
$$+ \int_E \hat{X}_s(\mathrm{d}x) \int_0^\infty [F(X_s + u\delta_x) - F(X_s)]um(x,\mathrm{d}u).$$

Since $t \mapsto e^{bt}X_t(1)$ is a martingale under $\mathbf{Q}_\mu$, we can use integration by parts again to the right-hand side of (7.49) to see

$$e^{bt}X_t(1)F(X_t) = e^{bt}H(X_t) = e^{bt}X_t(1)\int_0^t J(X_s)\mathrm{d}s + \text{local mart.}$$

Then we have (7.48) under $\tilde{\mathbf{Q}}_\mu$ by a simple calculation. $\qquad\square$

Now suppose that $(\Omega, \mathscr{G}, \mathscr{G}_t, \mathbf{P})$ is a probability space satisfying the usual hypothesis. Let $\{X_t : t \geq 0\}$ be a continuous $M(E)$-valued adapted process satisfying $\mathbf{P}[\langle X_0, 1\rangle] < \infty$. For $b \in C(M(E) \times E)$ and $c \in C(E)^+$ we consider the following martingale problem: For every $f \in D_0(A)$ the process

$$M_t(f) = \langle X_t, f\rangle - \langle X_0, f\rangle - \int_0^t \langle X_s, Af - b(X_s)f\rangle \mathrm{d}s \tag{7.50}$$

is a square-integrable $(\mathscr{G}_t)$-martingale with increasing process

$$\langle M(f)\rangle_t = \int_0^t \mathrm{d}s \int_E 2c(x)f(x)^2 X_s(\mathrm{d}x). \tag{7.51}$$

This should be compared with the martingale problem given by (7.23) and (7.25). If $\{X_t : t \geq 0\}$ is a solution of the martingale problem above, we can follow the arguments in Section 7.3 to show that there is a continuous $(\mathscr{G}_t)$-martingale measure $\{M_t(B) : t \geq 0; B \in \mathscr{B}(E)\}$ satisfying

$$M_t(f) = \int_0^t \int_E f(x) M(\mathrm{d}s, \mathrm{d}x), \qquad t \geq 0, f \in D_0(A) \tag{7.52}$$

and having covariance measure

$$\eta(\mathrm{d}s, \mathrm{d}x, \mathrm{d}y) = \mathrm{d}s \int_E 2c(z) \delta_z(\mathrm{d}x) \delta_z(\mathrm{d}y) X_s(\mathrm{d}z). \tag{7.53}$$

Then for any function $\beta \in C(M(E) \times E)$ we can define the continuous and strictly positive local martingale $\{Z_t : t \geq 0\}$ by

$$Z_t = \exp\left\{ \int_0^t \int_E \beta(X_s, x) M(\mathrm{d}s, \mathrm{d}x) - \int_0^t \mathrm{d}s \int_E c(x) \beta(X_s, x)^2 X_s(\mathrm{d}x) \right\}.$$

Lemma 7.30 *Suppose that* $\{X_t : t \geq 0\}$ *is a solution of the martingale problem given by* (7.50) *and* (7.51). *Then* $\{Z_t : t \geq 0\}$ *is actually a* $(\mathscr{G}_t)$-*martingale.*

Proof. It suffices to prove $\mathbf{P}[Z_t] = 1$ for every $t \geq 0$; see, e.g., Ikeda and Watanabe (1989, p.152). For each $n \geq 1$ let $\tau_n = \inf\{t \geq 0 : \langle X_t, 1 \rangle \geq n\}$. It is easy to see that $\tau_n \to \infty$ as $n \to \infty$. Observe also that

$$\mathbf{P}\left[\exp\left\{ \int_0^{t \wedge \tau_n} \mathrm{d}s \int_E c(x) \beta(X_s, x)^2 X_s(\mathrm{d}x) \right\} \right] < \infty.$$

Then $\{Z_{t \wedge \tau_n} : t \geq 0\}$ is a continuous and strictly positive $(\mathscr{G}_t)$-martingale. It follows that

$$1 = \mathbf{P}[Z_{t \wedge \tau_n}] = \mathbf{P}[Z_{t \wedge \tau_n} 1_{\{\tau_n \leq t\}}] + \mathbf{P}[Z_{t \wedge \tau_n} 1_{\{\tau_n > t\}}]. \tag{7.54}$$

For fixed $n \geq 1$ and $t \geq 0$ we define the new probability measure $\mathbf{P}^t$ on $\mathscr{G}_t$ by $\mathbf{P}^t(\mathrm{d}\omega) = Z_{t \wedge \tau_n}(\omega) \mathbf{P}(\mathrm{d}\omega)$. Under the measure $\mathbf{P}^t$, for each $f \in D_0(A)$,

$$N_u(f) := M_{u \wedge \tau_n}(f) - 2 \int_0^{u \wedge \tau_n} \langle X_s, c\beta(X_s) f \rangle \mathrm{d}s, \quad 0 \leq u \leq t, \tag{7.55}$$

is a square-integrable martingale with increasing process given by

$$\langle N(f) \rangle_u = 2 \int_0^{u \wedge \tau_n} \mathrm{d}s \int_E c(x) f(x)^2 X_s(\mathrm{d}x);$$

see, e.g., Ikeda and Watanabe (1989, p.191). Since $(P_t)_{t \geq 0}$ is conservative, it is easy to extend the martingale problems to the constant function $f = 1$ with $A1 = 0$. Consequently, for any $0 \leq u \leq t$ we have

$$\mathbf{P}^t[\langle X_{u \wedge \tau_n}, 1 \rangle] \leq \mathbf{P}^t[\langle X_0, 1 \rangle] + \|2c\beta - b\| \int_0^u \mathbf{P}^t[\langle X_{s \wedge \tau_n}, 1 \rangle] \mathrm{d}s,$$

where

$$\mathbf{P}^t[\langle X_0, 1 \rangle] = \mathbf{P}[\langle X_0, 1 \rangle Z_{t \wedge \tau_n}] = \mathbf{P}[\langle X_0, 1 \rangle].$$

Then Gronwall's inequality implies

$$\mathbf{P}^t[\langle X_{u \wedge \tau_n}, 1 \rangle] \leq \mathbf{P}[\langle X_0, 1 \rangle] \exp\{\|2c\beta - b\|u\}. \tag{7.56}$$

From (7.50) and (7.55) it follows that

$$\mathbf{P}^t\left\{ \sup_{0 \leq s \leq t} \langle X_{s \wedge \tau_n}, 1 \rangle \geq n \right\}$$

$$\leq \mathbf{P}^t\{\langle X_0, 1 \rangle \geq n/3\} + \mathbf{P}^t\left\{ \sup_{0 \leq s \leq t} |N_s(1)| \geq n/3 \right\}$$

$$+ \|2c\beta - b\| \mathbf{P}^t\left\{ \int_0^t \langle X_{s \wedge \tau_n}, 1 \rangle ds \geq n/3 \right\}.$$

By Doob's martingale inequality,

$$\mathbf{P}^t\left\{ \sup_{0 \leq s \leq t} |N_s(1)| \geq n/3 \right\} \leq \frac{18\|c\|}{n^2} \int_0^t \mathbf{P}^t[\langle X_{s \wedge \tau_n}, 1 \rangle] ds.$$

In view of (7.56), we can use Chebyshev's inequality to see

$$\lim_{n \to \infty} \mathbf{P}[Z_{t \wedge \tau_n} 1_{\{\tau_n \leq t\}}] = \lim_{n \to \infty} \mathbf{P}^t\left\{ \sup_{0 \leq s \leq t} \langle X_{s \wedge \tau_n}, 1 \rangle \geq n \right\} = 0.$$

Then letting $n \to \infty$ in (7.54) we get $\mathbf{P}[Z_t] = 1$. □

Theorem 7.31 *Suppose that $c \in C(E)^+$ is bounded away from zero. Then there is a unique solution to the martingale problem given by (7.50) and (7.51).*

Proof. If $\{X_t : t \geq 0\}$ is a solution of the martingale problem given by (7.50) and (7.51) under $\mathbf{P}$ and if $\mathbf{P}^Z$ is the probability measure on $(\Omega, \mathscr{G})$ such that $\mathbf{P}^Z(d\omega) = Z_t(\omega)\mathbf{P}(d\omega)$ on $\mathscr{G}_t$ for every $t \geq 0$, then for each $f \in D_0(A)$,

$$M_t(f) = \langle X_t, f \rangle - \langle X_0, f \rangle - \int_0^t \langle X_s, Af - b(X_s)f + 2c\beta(X_s)f \rangle ds$$

is a square-integrable $(\mathscr{G}_t)$-martingale with increasing process (7.51) under $\mathbf{P}^Z$. Here we may assume $(\Omega, \mathscr{G}, \mathscr{G}_t)$ is the $\mathbf{P}$-augmentation of the canonical space consisting of continuous paths from $[0, \infty)$ to $M(E)$, which is a standard measurable space, so that the measure $\mathbf{P}^Z$ described as above is well-defined; see, e.g., Ikeda and Watanabe (1989, p.190). By Corollary 7.16 the existence and uniqueness of the martingale problem given by (7.50) and (7.51) hold for $b(x) \equiv 0$. Since $c \in C(E)^+$ is bounded away from zero, using changes of the probability measures as the above one can see the existence and uniqueness also hold for a general $b \in C(M(E) \times E)$. □

Here we may interpret $\{X_t : t \geq 0\}$ as a superprocess with *interactive growth rate* given by the function $b(\mu, x)$. The transformation based on the strictly positive martingale $\{Z_t : t \geq 0\}$ used in the above proof is known as *Dawson's Girsanov transform*.

7.6 Notes and Comments

A systematic treatment of martingale problems for diffusions was given in Stroock and Varadhan (1979). Those for Markov processes with abstract state spaces were discussed in Ethier and Kurtz (1986). Nonlinear functional integral and differential evolution equations were discussed in Pazy (1983). Our approach in Section 7.1 is different from that of Pazy (1983) and uses heavily the special structures of the cumulant semigroup.

The approach of martingale problems plays an important role in the study of measure-valued processes. Martingale problems for Dawson–Watanabe superprocesses with Feller spatial motion and binary branching mechanism were studied in Roelly (1986). The treatment in Section 7.2 follows El Karoui and Roelly (1991). Fitzsimmons (1988, 1992) studied martingale problems of superprocesses in the Borel right setting. Our main references for worthy martingale measures are El Karoui and Méléard (1990) and Walsh (1986). Dawson (1978) first used the Girsanov type transform to derive superprocesses with interactive branching structures. Martingale problems of the type given by (7.50) and (7.51) were considered in Etheridge (2004) and Fournier and Méléard (2004) in the study of locally regulated population models; see also Méléard and Roelly (1993). Martingale problems for superprocesses with general killing rates were studied in Leduc (2006). Champagnat and Roelly (2008) gave a martingale problem characterization for a continuous multitype superprocess conditioned on non-extinction, and proved several results on the long-time behavior of the conditioned superprocess.

Let $b \in C(\mathbb{R})$ and $c \in C(\mathbb{R})^+$. Then the super-Brownian motion $\{X_t : t \geq 0\}$ on $\mathbb{R}$ with local branching mechanism $\phi(x, z) \equiv b(x)z + c(x)z^2$ has a continuous realization. It was proved in Konno and Shiga (1988) that $\{X_t : t > 0\}$ has a continuous density field $\{X_t(x) : t > 0, x \in \mathbb{R}\}$ with respect to the Lebesgue measure. The density field solves the following stochastic partial differential equation:

$$\frac{\partial}{\partial t} X_t(x) = \sqrt{2c(x)X_t(x)} \dot{W}(t, x) + \frac{1}{2} \Delta X_t(x) - b(x)X_t(x), \qquad (7.57)$$

where $\{W(t, x) : t \geq 0, x \in \mathbb{R}\}$ is a time–space white noise based on the Lebesgue measure and the dot denotes the derivative in distribution sense. The above equation should be understood in the weak sense, that is, for any $f \in C^2(\mathbb{R})$ we have

$$\int_{\mathbb{R}} f(x)X_t(x)\mathrm{d}x = \int_{\mathbb{R}} f(x)X_0(\mathrm{d}x) + \int_0^t \int_{\mathbb{R}} f(x)\sqrt{2c(x)X_s(x)}W(\mathrm{d}s, \mathrm{d}x)$$
$$+ \int_0^t \mathrm{d}s \int_{\mathbb{R}} \left[\frac{1}{2}f''(x) - b(x)f(x)\right] X_s(x)\mathrm{d}x.$$

A special case of (7.57) was established independently in Reimers (1989). Similar stochastic partial differential equations driven by stable noises were studied in Mueller (1998) and Mytnik (1998a, 2002). In the past years, a lot of attention has been paid to the study of stochastic partial differential equations. A theory of the subject was developed in Walsh (1986) on the basis of martingale measures. Mueller (2009) gave a recent survey of the tools and results for stochastic parabolic equations with emphasis on the techniques from Dawson–Watanabe superprocesses and interacting particle systems; see also Krylov (1997). The approaches of Hilbert spaces and Sobolev spaces for stochastic partial differential equations were developed in Da Prato and Zabczyk (1992) and Krylov (1996).

A mutually catalytic super-Brownian motion on the real line was constructed in Dawson and Perkins (1998) as the solution of a system of stochastic partial differential equations. The uniqueness in law of the solution was proved in Mytnik (1998b) by a duality method. The construction of the mutually catalytic super-Brownian motion on the plane is a hard problem. This was settled by Dawson et al. (2002a, 2002b, 2003). See Dawson and Fleischmann (2002) and Klenke (2000) for reviews of the study of catalytic and mutually catalytic branching models. Stochastic differential equations driven by the path processes of Brownian motions were introduced in Perkins (1995, 2002) in the construction of superprocesses with interaction. A super-Brownian motion with interaction was constructed in Delmas and Dhersin (2003) using the technique of Brownian snake. Athreya et al. (2002) constructed some classes of super-Markov chains with state-dependent branching rates and spatial motions.

Let $\{X_t : t \geq 0\}$ be a super-Brownian motion on $\mathbb{R}^d$ with binary local branching mechanism. Then for $d \geq 2$ and $t > 0$ the random measure X_t has support with Hausdorff dimension two and distributes its mass over the support in a deterministic manner; see, e.g., Perkins (2002, p.209 and p.212). For $d \geq 2$ it was proved in Tribe (1994) that X_t can be approximated by suitably normalized restrictions of the Lebesgue measure to the ε-neighborhoods of support of the random measure. The analogous result for the more difficult case $d = 2$ was established in Kallenberg (2008), which leads to a simple derivation of the property of deterministic mass distribution.

The key assumption of a Dawson–Watanabe superprocess is the independence of different particles in the approximating system. When dependence is introduced into the branching or migrating mechanisms, the characterization of the limiting measure-valued process usually becomes very difficult. The method of dual processes plays an important role in the analysis of the uniqueness of martingale problems for measure-valued processes. A general theory of duality was developed in Ethier and Kurtz (1986). The reader may refer to Dawson (1993) for systematic applications of this method to measure-valued processes. The approach of filtered martingale problems introduced by Kurtz (1998) and Kurtz and Ocone (1988) is another important tool in handling the uniqueness of martingale problems.

A *superprocess with dependent spatial motion* over the real line $\mathbb{R}$ was constructed in Dawson et al. (2001), generalizing the model of Wang (1997a, 1998a). Let $c \in C^2(\mathbb{R})$ and $\sigma \in C^2(\mathbb{R})^+$. Let $h \in C^1(\mathbb{R})$ and assume both h and h' are square-integrable. Let

$$\rho(x) = \int_{\mathbb{R}} h(y - x)h(y)\mathrm{d}y \quad \text{and} \quad a(x) = c(x)^2 + \rho(0), \quad x \in \mathbb{R}. \quad (7.58)$$

The superprocess with dependent spatial motion is a diffusion process $\{X_t : t \geq 0\}$ in $M(\mathbb{R})$ characterized by the following martingale problem: For each $f \in C^2(\mathbb{R})$,

$$M_t(f) = \langle X_t, f \rangle - \langle X_0, f \rangle - \frac{1}{2} \int_0^t \langle X_s, af'' \rangle \mathrm{d}s, \quad (7.59)$$

is a continuous martingale with quadratic variation process

$$\langle M(f) \rangle_t = \int_0^t \langle X_s, \sigma f^2 \rangle \mathrm{d}s + \int_0^t \mathrm{d}s \int_{\mathbb{R}} \langle X_s, h(z - \cdot)f' \rangle^2 \mathrm{d}z. \quad (7.60)$$

The process $\{X_t : t \geq 0\}$ arises as a weak limit of critical branching particle systems with dependent spatial motion. Consider a family of independent Brownian motions $\{B_i(t) : t \geq 0, i = 1, 2, \ldots\}$ and a time–space white noise $\{W(\mathrm{d}t, \mathrm{d}y) : t \geq 0, y \in \mathbb{R}\}$. Suppose that $\{B_i(t) : t \geq 0, i = 1, 2, \ldots\}$ and $\{W(\mathrm{d}t, \mathrm{d}y) : t \geq 0, y \in \mathbb{R}\}$ are independent. The migration of the particle with label $i \geq 1$ in the approximating system is defined by the stochastic differential equation

$$\mathrm{d}x_i(t) = c(x_i(t))\mathrm{d}B_i(t) + \int_{\mathbb{R}} h(y - x_i(t))W(\mathrm{d}t, \mathrm{d}y).$$

The uniqueness of solution of the martingale problem given by (7.59) and (7.60) was established in Dawson et al. (2001) by considering a function-valued dual process. Clearly, the superprocess with dependent spatial motion reduces to a usual critical branching Dawson–Watanabe superprocess if $h \equiv 0$. On the other hand, when $\sigma \equiv 0$, branching does not occur and the total mass of the process remains unchanged as time passes. By considering a stochastic equation driven by a time–space white noise and the path process of a Brownian motion, Gill (2009) unified the approaches of Dawson et al. (2001) and Perkins (1995, 2002) and gave a new class of measure-valued diffusions. Ren et al. (2009) introduced a superprocess with dependent spatial motion in a bounded domain in $\mathbb{R}^d$ with killing boundary. A discontinuous superprocess with dependent spatial motion and general branching mechanism was constructed in He (2009). Some probability-valued Markov processes arising from consistent particle systems were studied in Ma and Xiang (2001) and Xiang (2009).

Chapter 8
Entrance Laws and Excursion Laws

The main purpose of this chapter is to investigate the structures of entrance laws for Dawson–Watanabe superprocesses. In particular, we establish a one-to-one correspondence between minimal probability entrance laws for a superprocess and entrance laws for its spatial motion. Based on this result, a complete characterization is given for infinitely divisible probability entrance laws of the superprocess. We also prove some supporting properties of Kuznetsov measures determined by entrance laws. Finally we discuss briefly the special case where the underlying process is an absorbing-barrier Brownian motion in a domain. The results presented here will be used in the study of immigration superprocesses.

8.1 Some Simple Properties

Suppose that E is a Lusin topological space. Let $(Q_t)_{t\geq 0}$ and $(V_t)_{t\geq 0}$ denote respectively the transition semigroup and the cumulant semigroup of an MB-process with state space $M(E)$. Recall that $(V_t)_{t\geq 0}$ always has the representation (2.5) and E° is the set of points $x \in E$ so that (2.9) holds. Let $(Q_t^\circ)_{t\geq 0}$ denote the restriction of $(Q_t)_{t\geq 0}$ to $M(E)^\circ$.

Theorem 8.1 *Given a bounded entrance law $(K_t^\circ)_{t>0}$ for $(Q_t^\circ)_{t\geq 0}$, we can define a bounded entrance law $(K_t)_{t>0}$ for $(Q_t)_{t\geq 0}$ by*

$$K_t = \lim_{s\to 0} \int_{M(E)^\circ} K_s^\circ(\mathrm{d}\mu)Q_{t-s}(\mu, \cdot), \qquad t > 0. \tag{8.1}$$

Moreover, the above relation determines a one-to-one correspondence of bounded entrance laws $(K_t^\circ)_{t>0}$ for $(Q_t^\circ)_{t\geq 0}$ with bounded entrance laws $(K_t)_{t>0}$ for $(Q_t)_{t\geq 0}$ satisfying $\lim_{t\to 0} K_t(\{0\}) = 0$.

Proof. If $(K_t^\circ)_{t>0}$ is a bounded entrance law for $(Q_t^\circ)_{t\geq 0}$, the limit (8.1) clearly exists and defines a bounded entrance law $(K_t)_{t>0}$ for $(Q_t)_{t\geq 0}$. In fact, K_t is the

Z. Li, *Measure-Valued Branching Markov Processes*,
Probability and Its Applications, DOI 10.1007/978-3-642-15004-3_8,
© Springer-Verlag Berlin Heidelberg 2011

extension of K_t° to $M(E)$ so that

$$K_t(\{0\}) = \lim_{s \to 0} K_s^\circ(M(E)^\circ) - K_t^\circ(M(E)^\circ),$$

which implies $\lim_{t \to 0} K_t(\{0\}) = 0$. Conversely, if $(K_t)_{t>0}$ is a bounded entrance law for $(Q_t)_{t \geq 0}$ satisfying $\lim_{t \to 0} K_t(\{0\}) = 0$, we let K_t° be the restriction of K_t to $M(E)^\circ$. It is easy to see that $(K_t^\circ)_{t>0}$ is a bounded entrance law for $(Q_t^\circ)_{t \geq 0}$ and (8.1) holds. Then we have the desired one-to-one correspondence.   □

Theorem 8.2 *Let $K = (K_t)_{t>0}$ be a family of infinitely divisible probability measures on $M(E)$ given by*

$$\int_{M(E)} e^{-\nu(f)} K_t(d\nu)$$
$$= \exp\left\{ -\eta_t(f) - \int_{M(E)^\circ} (1 - e^{-\nu(f)}) H_t(d\nu) \right\}, \qquad (8.2)$$

where $\eta_t \in M(E)$ and $[1 \wedge \nu(1)] H_t(d\nu)$ is a finite measure on $M(E)^\circ$. Then K is an entrance law for $(Q_t)_{t \geq 0}$ if and only if

$$\eta_{r+t} = \int_E \eta_r(dy) \lambda_t(y, \cdot) \quad and \quad H_{r+t} = \int_E \eta_r(dy) L_t(y, \cdot) + H_r Q_t^\circ, \quad (8.3)$$

for all $r, t > 0$.

Proof. By Theorem 1.35 the family of infinitely divisible probability measures $(K_t)_{t>0}$ on $M(E)$ can be represented by (8.2). By Proposition 2.6 one can see (8.3) gives an alternative expression for the relation $K_{r+t} = K_r Q_t$.   □

Corollary 8.3 *If $H = (H_t)_{t>0}$ is a σ-finite entrance law for the restricted semigroup $(Q_t^\circ)_{t \geq 0}$ satisfying*

$$\int_{M(E)^\circ} [1 \wedge \nu(1)] H_t(d\nu) < \infty, \qquad t > 0, \qquad (8.4)$$

then

$$\int_{M(E)} e^{-\nu(f)} K_t(d\nu) = \exp\left\{ -\int_{M(E)^\circ} (1 - e^{-\nu(f)}) H_t(d\nu) \right\} \qquad (8.5)$$

defines an infinitely divisible probability entrance law $K = (K_t)_{t>0}$ for $(Q_t)_{t \geq 0}$.

Corollary 8.4 *If $E^\circ = E$, then (8.5) establishes a one-to-one correspondence between infinitely divisible probability entrance laws K for $(Q_t)_{t \geq 0}$ and σ-finite entrance laws H for $(Q_t^\circ)_{t \geq 0}$ satisfying (8.4).*

We next turn to the special case of a (ξ, ϕ)-superprocess X. Here we assume ξ is a Borel right process in E with transition semigroup $(P_t)_{t \geq 0}$ and ϕ is a branching mechanism given by (2.26) or (2.27). A sufficient condition for the cumulant semi-group of the (ξ, ϕ)-superprocess to admit the representation (2.9) for all $x \in E$ is the following:

Condition 8.5 *There is a spatially constant local branching mechanism $z \mapsto \phi_*(z)$ so that $\phi_*'(z) \to \infty$ as $z \to \infty$ and ϕ is bounded below by ϕ_* in the sense*

$$\phi(x, f) \geq \phi_*(f(x)), \qquad x \in E, f \in B(E)^+. \tag{8.6}$$

Theorem 8.6 *Suppose that Condition 8.5 is satisfied. Then we have $E^\circ = E$.*

Proof. Let $V_t : f \mapsto v_t(\cdot, f)$ and $V_t^* : f \mapsto v_t^*(\cdot, f)$ denote the cumulant semi-groups of the (ξ, ϕ)- and the (ξ, ϕ_*)-superprocesses, respectively. Let $\lambda \mapsto v_t^*(\lambda)$ denote the cumulant semigroup of the CB-process with branching mechanism ϕ_*. Then for any constant $\lambda \geq 0$ we have $v_t^*(x, \lambda) \leq v_t^*(\lambda)$ with equality if $(P_t)_{t \geq 0}$ is conservative. On the other hand, by Corollary 5.18 we have $v_t(x, \lambda) \leq v_t^*(x, \lambda)$. Now suppose there exist $t > 0$ and $x \in E$ so that $v_t(x, f)$ is represented by the right-hand side of (2.5) with $\lambda_t(x, 1) > 0$. By Theorem 3.10, $v_t^*(\lambda)$ has the representation (3.15) for $t > 0$, so $(\partial/\partial\lambda)v_t^*(\lambda) \to 0$ as $\lambda \to \infty$. Then $v_t(x, \lambda) \geq \lambda_t(x, 1)\lambda \geq v_t^*(\lambda)$ for sufficiently large $\lambda \geq 0$, yielding a contradiction. That proves $\lambda_t(x, 1) = 0$ for all $t > 0$ and $x \in E$. $\square$

Let $\gamma(x, dy)$ be the kernel on E defined by (2.28) and let $(\pi_t)_{t \geq 0}$ be the semigroup of kernels defined by (2.35). By Proposition A.49 we have $\|\pi_t\| \leq e^{c_0 t}$ for all $t \geq 0$, where $c_0 = \|b^-\| + \|\gamma(\cdot, 1)\|$. To study the structures of entrance laws for the (ξ, ϕ)-superprocess, we need to clarify some connections between entrance laws for the underlying semigroup $(P_t)_{t \geq 0}$ and those for $(\pi_t)_{t \geq 0}$. Let $\mathscr{K}(P)$ be the set of entrance laws $\kappa = (\kappa_t)_{t > 0}$ for $(P_t)_{t \geq 0}$ satisfying

$$\int_0^1 \kappa_s(1)\mathrm{d}s < \infty \tag{8.7}$$

and let $\mathscr{K}(\pi)$ be the set of entrance laws for $(\pi_t)_{t \geq 0}$ satisfying the above integral condition. In particular, if $(P_t)_{t \geq 0}$ is a conservative semigroup, then $\mathscr{K}(P)$ coincides with the space of bounded entrance laws for $(P_t)_{t \geq 0}$.

Proposition 8.7 *There is a one-to-one correspondence between $\kappa \in \mathscr{K}(P)$ and $\eta \in \mathscr{K}(\pi)$ given by*

$$\eta_t(f) = \lim_{r \to 0} \kappa_r(\pi_{t-r}f) \quad and \quad \kappa_t(f) = \lim_{r \to 0} \eta_r(P_{t-r}f), \tag{8.8}$$

where $t > 0$ and $f \in B(E)$. Moreover, if the two entrance laws are related by (8.8), we have

$$\eta_t(f) = \kappa_t(f) + \int_0^t \kappa_{t-s}((\gamma - b)\pi_s f)\mathrm{d}s \tag{8.9}$$

for $t > 0$ and $f \in B(E)$.

Proof. Suppose that $\kappa \in \mathscr{K}(P)$. For $t > r > 0$ and $f \in B(E)$ we can use (2.35) and the entrance law property of $\kappa = (\kappa_t)_{t>0}$ to see

$$\kappa_r(\pi_{t-r}f) = \kappa_t(f) + \int_0^{t-r} \kappa_{t-s}((\gamma - b)\pi_s f)ds. \qquad (8.10)$$

Then the first limit in (8.8) exists and is given by (8.9). Clearly, the family $\eta = (\eta_t)_{t>0}$ constitute an entrance law for $(\pi_t)_{t\geq0}$. Moreover, we have

$$\eta_t(1) \leq \kappa_t(1) + c_0 \int_0^t \kappa_{t-s}(\pi_s 1)ds \leq \kappa_t(1) + c_0 e^{c_0 t} \int_0^t \kappa_s(1)ds,$$

and hence $\eta \in \mathscr{K}(\pi)$. From (8.9) and the entrance law property of $(\kappa_t)_{t>0}$ it follows that

$$\eta_r(P_{t-r}f) = \kappa_t(f) + \int_0^r \kappa_{r-s}((\gamma - b)\pi_s P_{t-r}f)ds.$$

By letting $r \to 0$ we obtain the second equality in (8.8). Conversely, suppose that $\eta \in \mathscr{K}(\pi)$. For $f \in B(E)^+$ we get from (2.34) and (2.36) that

$$e^{-\|b^+\|t}P_t f(x) \leq P_t^b f(x) \leq \pi_t f(x). \qquad (8.11)$$

Then for any $t > s > r > 0$ we have

$$e^{\|b^+\|r}\eta_r(P_{t-r}f) \leq e^{\|b^+\|s}\eta_r(\pi_{s-r}P_{t-s}f) = e^{\|b^+\|s}\eta_s(P_{t-s}f).$$

Consequently, we can define an entrance law $\kappa = (\kappa_t)_{t>0}$ for $(P_t)_{t\geq0}$ by

$$\kappa_t(f) = \lim_{r\to0} e^{\|b^+\|r}\eta_r(P_{t-r}f) = \lim_{r\to0} \eta_r(P_{t-r}f). \qquad (8.12)$$

Clearly, the above relation also holds for all $f \in B(E)$. In view of (8.11) and (8.12), we have

$$\kappa_t(1) = \lim_{r\to0} \eta_r(P_{t-r}1) \leq \lim_{r\to0} e^{\|b^+\|t}\eta_r(\pi_{t-r}1) \leq e^{\|b^+\|t}\eta_t(1),$$

and hence $\kappa \in \mathscr{K}(P)$. Then we use (2.35) and the entrance law property of $(\eta_t)_{t>0}$ to see

$$\eta_t(f) = \eta_r(P_{t-r}f) + \int_0^{t-r} \eta_r(P_{t-r-s}(\gamma - b)\pi_s f)ds.$$

By letting $r \to 0$ in both sides we get (8.9). The first equality in (8.8) follows from (8.10). $\qquad\square$

The above proof also gives the following:

Corollary 8.8 *If $\kappa \in \mathscr{K}(P)$ and $\eta \in \mathscr{K}(\pi)$ are related by (8.8) and (8.9), then for every $t > 0$ we have*

$$
e^{-\|b^+\|t}\kappa_t(1) \le \eta_t(1) \le \kappa_t(1) + c_0 e^{c_0 t} \int_0^t \kappa_s(1)\mathrm{d}s. \tag{8.13}
$$

Let $(Q_t)_{t\geq0}$ denote the transition semigroup of the (ξ, ϕ)-superprocess defined by (2.32) and (2.33). Let $\mathscr{K}(Q)$ be the set of σ-finite entrance laws $K = (K_t)_{t>0}$ for the semigroup $(Q_t)_{t\geq0}$ satisfying

$$
\int_0^1 \mathrm{d}s \int_{M(E)^\circ} \nu(1)K_s(\mathrm{d}\nu) < \infty \tag{8.14}
$$

and let $\mathscr{K}(Q^\circ)$ be the set of entrance laws for the restricted semigroup $(Q_t^\circ)_{t\geq0}$ satisfying the above integral condition. By Corollary 2.28 we have

$$
\int_{M(E)} \nu(f)Q_t(\mu, \mathrm{d}\nu) = \mu(\pi_t f), \qquad t \ge 0, f \in B(E)^+, \tag{8.15}
$$

where $(\pi_t)_{t\geq0}$ is the semigroup defined by (2.35). Based on (8.15) and Corollary 8.8 it is simple to check that for any $K \in \mathscr{K}(Q)$ or $\mathscr{K}(Q^\circ)$ we can define $\eta := \pi K \in \mathscr{K}(\pi)$ and $\kappa := pK \in \mathscr{K}(P)$ by

$$
\eta_t(f) = \int_{M(E)^\circ} \nu(f)K_t(\mathrm{d}\nu) \tag{8.16}
$$

and

$$
\kappa_t(f) = \lim_{r\to0} \int_{M(E)^\circ} \nu(P_{t-r}f)K_r(\mathrm{d}\nu), \tag{8.17}
$$

where $t > 0$ and $f \in B(E)$.

8.2 Minimal Probability Entrance Laws

Suppose that ξ is a Borel right process in the Lusin topological space E with transition semigroup $(P_t)_{t\geq0}$ and ϕ is a branching mechanism given by (2.26) or (2.27). Let $(Q_t)_{t\geq0}$ be the transition semigroup of the (ξ, ϕ)-superprocess defined by (2.32) and (2.33). Given $\kappa \in \mathscr{K}(P)$ we set

$$
S_t(\kappa, f) = \kappa_t(f) - \int_0^t \mathrm{d}s \int_E \phi(y, V_s f)\kappa_{t-s}(\mathrm{d}y) \tag{8.18}
$$

for $t > 0$ and $f \in B(E)^+$. In particular, if $\kappa \in \mathscr{K}(P)$ is closed by $\mu \in M(E)$, we have $S_t(\kappa, f) = \mu(V_t f)$.

Lemma 8.9 *If $\kappa \in \mathscr{K}(P)$ and $\eta \in \mathscr{K}(\pi)$ are related by (8.8), then for any $t > 0$ and $f \in B(E)^+$ we have*

$$S_t(\kappa, f) = \lim_{r \to 0} \kappa_r(V_{t-r}f) = \downarrow \lim_{r \to 0} \eta_r(V_{t-r}f). \tag{8.19}$$

Proof. By (2.33) for $t > r > 0$ and $f \in B(E)^+$ we have

$$\kappa_r(V_{t-r}f) = \kappa_t(f) - \int_0^{t-r} ds \int_E \phi(y, V_s f) \kappa_{t-s}(dy).$$

Then the first equality in (8.19) holds. The second equality follows similarly from (2.33) and (8.9).   □

Lemma 8.10 *The entrance law $\kappa \in \mathscr{K}(P)$ is non-trivial if and only if we have $\lim_{t \to 0} \lim_{\theta \to \infty} S_t(\kappa, \theta) = \infty$.*

Proof. From (8.19) we see $f \mapsto S_t(\kappa, f)$ is an increasing functional, so the limit $\lim_{\theta \to \infty} S_t(\kappa, \theta)$ exists in $[0, \infty]$. By (8.19) and (8.18) for any $\theta_0 \geq 0$ we have

$$\liminf_{t \to 0} \lim_{\theta \to \infty} S_t(\kappa, \theta) \geq \lim_{t \to 0} S_t(\kappa, \theta_0) = \lim_{t \to 0} \kappa_t(\theta_0) = \theta_0 \lim_{t \to 0} \kappa_t(1).$$

If $\kappa \in \mathscr{K}(P)$ is non-trivial, then $\lim_{t \to 0} \kappa_t(1) > 0$ and hence

$$\lim_{t \to 0} \lim_{\theta \to \infty} S_t(\kappa, \theta) = \infty.$$

If $\kappa \in \mathscr{K}(P)$ is trivial, then $S_t(\kappa, \theta) = 0$ for all $t > 0$ and $\theta \geq 0$.   □

For $0 < a \leq \infty$ write $K \in \mathscr{K}^a(Q)$ if $K \in \mathscr{K}(Q)$ and $K_t(1) = a$ for all $t > 0$. Similarly, we write $K \in \mathscr{K}^a(Q^\circ)$ if $K \in \mathscr{K}(Q^\circ)$ and $\lim_{t \to 0} K_t(1) = a$. Let $\mathscr{K}_m^a(Q)$ and $\mathscr{K}_m^a(Q^\circ)$ denote the sets of minimal elements of $\mathscr{K}^a(Q)$ and $\mathscr{K}^a(Q^\circ)$, respectively. Let $\mathscr{K}(P)^\circ = \mathscr{K}(P) \setminus \{0\}$, where 0 is the trivial entrance law of $(P_t)_{t \geq 0}$.

Theorem 8.11 *To each $\kappa \in \mathscr{K}(P)$ there corresponds an entrance law $K := l\kappa \in \mathscr{K}_m^1(Q)$ given by*

$$\int_{M(E)} e^{-\nu(f)} K_t(d\nu) = \exp\{-S_t(\kappa, f)\}, \quad t > 0, f \in B(E)^+. \tag{8.20}$$

Moreover, (8.20) and (8.17) give a one-to-one correspondence between $\mathscr{K}_m^1(Q)$ and $\mathscr{K}(P)$.

Proof. Step 1. Suppose that $\kappa \in \mathscr{K}(P)$ and $\eta \in \mathscr{K}(\pi)$ are related by (8.8). By Lemma 8.9 we have

$$\int_{M(E)} e^{-\nu(f)} Q_{t-r}(\eta_r, d\nu) = \exp\{-\eta_r(V_{t-r}f)\} \to \exp\{-S_t(\kappa, f)\}$$

increasingly as $r \to 0$. Then an application of Theorem 1.20 shows that (8.20) really defines a family of probability measures $K = (K_t)_{t>0}$ on $M(E)$. By (2.33) and (8.18) it is easy to show that $S_{r+t}(\kappa, f) = S_r(\kappa, V_t f)$, so K is an entrance law for $(Q)_{t\geq0}$. In view of (8.18) and (8.9) we have $(d/d\theta)S_t(\kappa, \theta f)|_{\theta=0+} = \eta_t(f)$ and hence (8.16) holds. In particular, we have $K \in \mathscr{K}^1(Q)$. Write $K = l\kappa = \lambda\eta$. By Proposition 8.7 we have $\pi K = \eta$ and $pK = \kappa$. Therefore $pl\kappa = \kappa$ for $\kappa \in \mathscr{K}(P)$ and $\pi\lambda\eta = \eta$ for $\eta \in \mathscr{K}(\pi)$.

Step 2. We claim $K = \lambda\pi K = lpK$ for every $K \in \mathscr{K}_m^1(Q)$. To see this let $\mathbf{Q}_K$ be the probability measure on $M(E)^{(0,\infty)}$ under which the coordinate process $\{w_t : t > 0\}$ is a Markov process with one-dimensional distributions $(K_t)_{t>0}$ and semigroup $(Q_t)_{t\geq0}$. Since K is minimal, by Dynkin (1978, p.724) we have $\mathbf{Q}_K$-a.s.

$$\int_{M(E)} e^{-\nu(f)} K_t(d\nu) = \lim_{n\to\infty} \exp\{-w_{r_n}(V_{t-r_n} f)\} \qquad (8.21)$$

for any sequence $r_n \to 0$. By (8.15),

$$w_{r_n}(V_{t-r_n} f) \leq w_{r_n}(\pi_{t-r_n} f) = \mathbf{Q}_K[w_t(f)|w_s : 0 < s \leq r_n].$$

Then the family of random variables $\{w_{r_n}(V_{t-r_n} f) : 0 < r_n \leq t\}$ is uniformly $\mathbf{Q}_K$-integrable. By (8.21) and dominated convergence we have

$$-\log \int_{M(E)} e^{-\nu(f)} K_t(d\nu) = \lim_{n\to\infty} \mathbf{Q}_K[w_{r_n}(V_{t-r_n} f)]$$
$$= \lim_{n\to\infty} \pi K_{r_n}(V_{t-r_n} f) = S_t(pK, f),$$

where the last equality follows by Lemma 8.9. That proves $K = lpK$. Then the results in the first step imply $K = \lambda\pi K$.

Step 3. Now it suffices to show $l\kappa \in \mathscr{K}_m^1(Q)$ for all $\kappa \in \mathscr{K}(P)$. By Dynkin (1978, p.723) there is a probability measure F on $\mathscr{K}_m^1(Q)$ such that

$$l\kappa_t = \int_{\mathscr{K}_m^1(Q)} H_t F(dH).$$

Let G be the image of F under the mapping $p : \mathscr{K}_m^1(Q) \to \mathscr{K}(P)$. By the results proved in the first two steps it follows that

$$\exp\{-S_t(\kappa, f)\} = \int_{\mathscr{K}(P)} \exp\{-S_t(\mu, f)\} G(d\mu).$$

Since $u \mapsto e^{-u}$ is a strictly convex function, G must be the unit mass concentrated at κ. Then F is the unit mass at $l\kappa$, yielding $l\kappa \in \mathscr{K}_m^1(Q)$. $\qquad\square$

Corollary 8.12 *There is a one-to-one correspondence between $K^\circ \in \mathscr{K}_m^1(Q^\circ)$ and $\kappa \in \mathscr{K}(P)^\circ$ given by*

$$\int_{M(E)^\circ} (1 - e^{-\nu(f)}) K_t^\circ(d\nu) = 1 - \exp\{-S_t(\kappa, f)\}, \qquad (8.22)$$

where $t > 0$ and $f \in B(E)^+$.

Proof. For the entrance laws $K \in \mathcal{K}^1(Q)$ and $K^\circ \in \mathcal{K}^1(Q^\circ)$ related by (8.1) one can see that $K \in \mathcal{K}_m^1(Q)$ if and only if $K^\circ \in \mathcal{K}_m^1(Q^\circ)$. On the other hand, for the entrance laws $\kappa \in \mathcal{K}(P)$ and $K \in \mathcal{K}_m^1(Q)$ related by (8.20) we have

$$K_t(\{0\}) = \lim_{\theta \to \infty} \int_{M(E)} e^{-\nu(\theta)} K_t(d\nu) = \lim_{\theta \to \infty} \exp\{-S_t(\kappa, \theta)\}.$$

By Lemma 8.10, we have $\lim_{t \to 0} K_t(\{0\}) = 0$ if and only if $\kappa \in \mathcal{K}(P)$ is non-trivial. Then the result follows from Theorem 8.11. □

In the special case where $(P_t)_{t \geq 0}$ is a conservative Borel right semigroup, let $\bar{E}$ be a Ray–Knight completion of E with respect to this semigroup. Let $(\bar{P}_t)_{t \geq 0}$ be the Ray extension of $(P_t)_{t \geq 0}$ to $\bar{E}$. Let $E_D \subset \bar{E}$ be the entrance space of $(P_t)_{t \geq 0}$. In this chapter, we only need the restriction of $(\bar{P}_t)_{t \geq 0}$ to E_D, which is also a Borel right semigroup. We extend $f \mapsto \phi(\cdot, f)$ to an operator $\bar{f} \mapsto \bar{\phi}(\cdot, \bar{f})$ from $B(E_D)^+$ to $B(E_D)$ by setting $\bar{\phi}(x, \bar{f}) = \phi(x, f)$ for $x \in E$ and $\bar{\phi}(x, \bar{f}) = 0$ for $x \in E_D \setminus E$, where $f = \bar{f}|_E$ is the restriction to E of $\bar{f} \in B(E_D)^+$. Then for every $\bar{f} \in B(E_D)^+$ there is a unique locally bounded positive solution $t \mapsto \bar{V}_t \bar{f}$ to the equation

$$\bar{V}_t \bar{f}(x) = \bar{P}_t \bar{f}(x) - \int_0^t ds \int_{E_D} \bar{\phi}(y, \bar{V}_s \bar{f}(y)) \bar{P}_{t-s}(x, dy), \qquad (8.23)$$

where $t \geq 0$ and $x \in E_D$. That defines a cumulant semigroup $(\bar{V}_t)_{t \geq 0}$ with underlying space E_D. By Proposition A.36 for $t > 0$ and $x \in E_D$ the probability measure $\bar{P}_t(x, \cdot)$ is carried by E. Then we can also regard $(\bar{V}_t)_{t > 0}$ as operators from $B(E)^+$ to $B(E_D)^+$. Indeed, for $f \in B(E)^+$ we have

$$\bar{V}_t f(x) = \bar{P}_t f(x) - \int_0^t ds \int_E \phi(y, V_s f) \bar{P}_{t-s}(x, dy), \qquad (8.24)$$

where $t > 0$ and $x \in E_D$.

Theorem 8.13 *If $(P_t)_{t \geq 0}$ is a conservative semigroup, there is a one-to-one correspondence between $K \in \mathcal{K}_m^1(Q)$ and $\mu \in M(E_D)$ given by*

$$\int_{M(E)} e^{-\nu(f)} K_t(d\nu) = \exp\{-\mu(\bar{V}_t f)\}, \quad t > 0, f \in B(E)^+. \qquad (8.25)$$

Proof. Since $(P_t)_{t \geq 0}$ is conservative, every $\kappa \in \mathcal{K}(P)$ is finite. By Theorem A.37 the relation $\kappa_t = \mu \bar{P}_t$ gives a one-to-one correspondence between $\kappa \in \mathcal{K}(P)$ and $\mu \in M(E_D)$. Using Lemma 8.9 it is easy to show that $S_t(\kappa, f) = \mu(\bar{V}_t f)$ for $t > 0$ and $f \in B(E)^+$. Then (8.25) follows from (8.20). □

Corollary 8.14 *If $(P_t)_{t\geq 0}$ is a conservative semigroup, there is a one-to-one correspondence between $K^\circ \in \mathscr{K}_m^1(Q^\circ)$ and $\mu \in M(E_D)^\circ$ given by*

$$\int_{M(E)^\circ} (1 - e^{-\nu(f)}) K_t^\circ(d\nu) = 1 - \exp\{-\mu(\bar{V}_t f)\}, \quad t > 0, f \in B(E)^+.$$

8.3 Infinitely Divisible Probability Entrance Laws

In this section, we study the structures of infinitely divisible probability entrance laws for the (ξ, ϕ)-superprocess. Suppose that ξ is a Borel right process in E with transition semigroup $(P_t)_{t\geq 0}$ and ϕ is a branching mechanism given by (2.26) or (2.27). Let $b^+ = 0 \vee b$ and $b^- = 0 \vee (-b)$. Let $\gamma(x, dy)$ be the kernel on E defined by (2.28) and let $c_0 = \|b^-\| + \|\gamma(\cdot, 1)\|$. Let $(Q_t)_{t\geq 0}$ denote the transition semigroup of the (ξ, ϕ)-superprocess defined by (2.32) and (2.33).

We first consider the case where $(P_t)_{t\geq 0}$ is conservative. Let E_D be the entrance space of ξ and let $(\bar{V}_t)_{t\geq 0}$ be the extension of $(V_t)_{t\geq 0}$ on $B(E_D)^+$ defined by (8.23). Let $\bar{\gamma}(x, dy)$ be the extension of $\gamma(x, dy)$ to E_D so that $\bar{\gamma}(x, E_D \setminus E) = 0$ for $x \in E$ and $\bar{\gamma}(x, E_D) = 0$ for $x \in E_D \setminus E$. Let $(\bar{\pi}_t)_{t\geq 0}$ be the semigroup of kernels on E_D defined by (2.35) from $(\bar{P}_t)_{t\geq 0}$ and $\bar{\gamma}(x, dy)$. Since $(\bar{V}_t)_{t\geq 0}$ is a cumulant semigroup, it can be represented in the form of (2.5). However, in view of (8.24), for $t > 0$ and $x \in E_D$ we can write

$$\bar{V}_t f(x) = \lambda_t(x, f) + \int_{M(E)^\circ} (1 - e^{-\nu(f)}) L_t(x, d\nu), \quad f \in B(E)^+, \quad (8.26)$$

where $\lambda_t(x, dy)$ is a bounded kernel from E_D to E and $\nu(1)L_t(x, d\nu)$ is a bounded kernel from E_D to $M(E)^\circ$. Let E_D° be the set of points $x \in E_D$ so that $\lambda_t(x, E) = 0$ for all $t > 0$.

Theorem 8.15 *Suppose that $(P_t)_{t\geq 0}$ is a conservative semigroup. Then an entrance law $K \in \mathscr{K}^1(Q)$ is infinitely divisible if and only if it has the representation*

$$\int_{M(E)} e^{-\nu(f)} K_t(d\nu)$$
$$= \exp\left\{ -\gamma_D(\bar{V}_t f) - \int_{M(E_D)^\circ} (1 - e^{-\nu(\bar{V}_t f)}) G_D(d\nu) \right\}, \quad (8.27)$$

where $\gamma_D \in M(E_D)$ and $G_D(d\nu)$ is a σ-finite measure on $M(E_D)^\circ$ satisfying

$$\int_{M(E_D)^\circ} \nu(1) G_D(d\nu) < \infty. \quad (8.28)$$

Proof. By Theorem 8.13 it is easy to see that (8.27) defines an infinitely divisible entrance law $K \in \mathscr{K}^1(Q)$. By letting $f \equiv \theta \geq 0$ and differentiating both sides at $\theta = 0$ we obtain

$$\int_{M(E)^\circ} \nu(1)K_t(\mathrm{d}\nu) = \gamma_D(\bar{\pi}_t 1) - \int_{M(E_D)^\circ} \nu(\bar{\pi}_t 1)G_D(\mathrm{d}\nu).$$

Applying (8.11) and Proposition A.49 to $(\bar{\pi}_t)_{t\geq 0}$ gives

$$\mathrm{e}^{-\|b^+\|t} \leq \bar{\pi}_t 1(x) \leq \mathrm{e}^{c_0 t}, \qquad t \geq 0, x \in E_D.$$

Then (8.14) is equivalent to (8.28). On the other hand, since $\mathscr{K}^1(Q)$ is a simplex, if $K \in \mathscr{K}^1(Q)$ is an infinitely divisible entrance law, by Theorem 8.13 there is a probability measure $F_D(\mathrm{d}\nu)$ on $M(E_D)$ so that

$$\int_{M(E)} \mathrm{e}^{-\nu(f)}K_t(\mathrm{d}\nu) = \int_{M(E_D)} \mathrm{e}^{-\mu(\bar{V}_t f)}F_D(\mathrm{d}\mu), \quad t > 0, f \in B(E)^+.$$

Since $(\bar{V}_t)_{t\geq 0}$ corresponds to a Borel right semigroup $(\bar{Q}_t)_{t\geq 0}$ on $M(E_D)$, we have $F_D = \lim_{t\to 0} K_t$ by the weak convergence of probability measures on $M(E_D)$. Then F_D is infinitely divisible and the representation (8.27) follows. □

Corollary 8.16 *Suppose $(P_t)_{t\geq 0}$ is a conservative semigroup. Then $H \in \mathscr{K}(Q^\circ)$ if and only if it is given by*

$$\int_{M(E)^\circ} (1 - \mathrm{e}^{-\nu(f)})H_t(\mathrm{d}\nu)$$
$$= \gamma_D(\bar{V}_t f) + \int_{M(E_D)^\circ} (1 - \mathrm{e}^{-\nu(\bar{V}_t f)})G_D(\mathrm{d}\nu), \qquad (8.29)$$

where $\gamma_D \in M(E_D^\circ)$ and $G_D(\mathrm{d}\nu)$ is a σ-finite measure on $M(E_D)^\circ$ satisfying (8.28).

Proof. It is easy to see that (8.29) defines an entrance law $H \in \mathscr{K}(Q^\circ)$. Conversely, by Corollary 8.3 and Theorem 8.15 any $H \in \mathscr{K}(Q^\circ)$ can be represented by (8.29) for $\gamma_D \in M(E_D)$ and a σ-finite measure $G_D(\mathrm{d}\nu)$ on $M(E_D)^\circ$ satisfying (8.28). Since the formula defines a family of σ-finite measures $(H_t)_{t>0}$ on $M(E)^\circ$, the measure $\gamma_D \in M(E_D)$ must be carried by E_D°. □

Corollary 8.17 *Suppose $(P_t)_{t\geq 0}$ is a conservative semigroup. Then $H \in \mathscr{K}_m^\infty(Q^\circ)$ if and only if there exist $q > 0$ and $x \in E_D^\circ$ so that*

$$H_t(\mathrm{d}\nu) = qL_t(x, \mathrm{d}\nu), \qquad t > 0, \nu \in M(E)^\circ.$$

Now let us turn to a general underlying semigroup $(P_t)_{t\geq 0}$ not necessarily conservative. Let $(\pi_t)_{t\geq 0}$ be the semigroup defined by (2.35). We can define a strictly positive function $h \in B(E)^+$ by

$$h(x) = \int_0^1 \pi_s 1(x)\mathrm{d}s, \qquad x \in E. \qquad (8.30)$$

Proposition 8.18 *Let $b_0 = c_0 + \|b^+\|$. Then for $t \geq 0$ and $x \in E$ we have*

$$e^{-b_0 t} P_t h(x) \le e^{-c_0 t} \pi_t h(x) \le h(x). \tag{8.31}$$

Moreover, the function h is b_0-excessive for $(P_t)_{t \ge 0}$.

Proof. By Proposition A.49 it is simple to see that $(e^{-c_0 t} \pi_t)_{t \ge 0}$ is a contraction semigroup. Then by (8.11) we have

$$e^{-b_0 t} P_t h(x) \le e^{-c_0 t} \pi_t h(x) = \int_0^1 e^{-c_0 t} \pi_s \pi_t 1(x) ds \le h(x).$$

That proves (8.31). On the other hand, from (8.30) we get

$$\pi_t h(x) = \int_0^{1+t} \pi_s 1(x) ds - \int_0^t \pi_s 1(x) ds.$$

Then $t \mapsto \pi_t h(x)$ is right continuous. By Proposition A.42 we see $t \mapsto P_t h(x)$ is also right continuous. Therefore h is a b_0-excessive function for $(P_t)_{t \ge 0}$. $\square$

To investigate the structures of infinitely divisible entrance laws $K \in \mathscr{K}^1(Q)$ for a general underlying semigroup $(P_t)_{t \ge 0}$, we introduce some transformations based on the result of Proposition 8.18. We first define a Borel right semigroup $(T_t)_{t \ge 0}$ on E by

$$T_t f(x) = h(x)^{-1} P_t^{bo}(hf)(x), \qquad t \ge 0, x \in E, f \in B(E); \tag{8.32}$$

see, e.g., Sharpe (1988, pp.298–299). Moreover, by (2.5) it is easy to show that

$$U_t f(x) = h(x)^{-1} V_t(hf)(x), \qquad t \ge 0, x \in E, f \in B(E)^+$$

defines a cumulant semigroup on E. Indeed, by Proposition 8.18 we have

$$U_t f(x) \le \pi_t^h f(x) := h(x)^{-1} \pi_t(hf)(x) \le \|f\| e^{c_0 t}. \tag{8.33}$$

By Proposition 2.9 we can rewrite (2.33) into

$$V_t f(x) = P_t^{bo} f(x) + \int_0^t ds \int_E \left[b_0 V_s f(y) - \phi(y, V_s f) \right] P_{t-s}^{bo}(x, dy).$$

Then $(t, x) \mapsto U_t f(x)$ satisfies

$$\begin{aligned}
U_t f(x) = T_t f(x) &+ \int_0^t ds \int_E \gamma_0(y, U_s f) T_{t-s}(x, dy) \\
&+ \int_0^t ds \int_E [b_0 - b(y)] U_s f(y) T_{t-s}(x, dy) \\
&- \int_0^t ds \int_E \psi_0(y, U_s f) T_{t-s}(x, dy),
\end{aligned} \tag{8.34}$$

where $\gamma_0(y, f) = h(y)^{-1} \gamma(y, hf)$ and

$$\psi_0(y, f) = c(y)h(y)f(y)^2 + h(y)^{-1} \int_{M(E)^\circ} \left[e^{-\nu(hf)} - 1 + \nu(hf) \right] H(y, d\nu).$$

Note that although $f \mapsto \gamma_0(\cdot, f)$ and $f \mapsto \psi_0(\cdot, f)$ are not necessarily bounded operators on $B(E)^+$, the second and last terms on the right-hand side of (8.34) are bounded. Indeed, by (2.36) and Proposition 2.9 we have

$$\pi_t f(x) = P_t^{bo} f(x) + \int_0^t ds \int_E \gamma(y, \pi_s f) P_{t-s}^{bo}(x, dy)$$
$$+ \int_0^t ds \int_E [b_0 - b(y)]\pi_s f(y) P_{t-s}^{bo}(x, dy).$$

Then (8.33) yields

$$\pi_t^h f(x) = T_t f(x) + \int_0^t ds \int_E \gamma_0(y, \pi_s^h f) T_{t-s}(x, dy)$$
$$+ \int_0^t ds \int_E [b_0 - b(y)]\pi_s^h f(y) T_{t-s}(x, dy)$$
$$\geq T_t f(x) + \int_0^t ds \int_E \gamma_0(y, U_s f) T_{t-s}(x, dy)$$
$$+ \int_0^t ds \int_E [b_0 - b(y)]U_s f(y) T_{t-s}(x, dy).$$

Since $U_t f(x)$ is positive, each term in (8.34) is bounded by $\|f\| e^{c_0 t}$.

Now let $(T_t^\partial)_{t \geq 0}$ be the conservative extension of $(T_t)_{t \geq 0}$ to $E^\partial := E \cup \{\partial\}$ with ∂ being an isolated point. Let $(\bar{T}_t^\partial)_{t \geq 0}$ be the Ray extension of $(T_t^\partial)_{t \geq 0}$ to its entrance space $E_D^{\partial, T}$ with the Ray topology. Let $E_D^T = E_D^{\partial, T} \setminus \{\partial\}$ and let $(\bar{T}_t)_{t \geq 0}$ be the restriction of $(\bar{T}_t^\partial)_{t \geq 0}$ to E_D^T. Then E_D^T is Lusin and $(\bar{T}_t)_{t \geq 0}$ is a Borel right semigroup. It is known that for any $t > 0$ and $x \in E_D^T$ the measure $\bar{T}_t(x, \cdot)$ is supported by E; see Proposition A.36. Given $\bar{f} \in B(E_D^T)^+$ let $f = \bar{f}|_E$. By (8.34) it is easy to show that the limit $\bar{U}_t \bar{f}(x) := \lim_{r \to 0} \bar{T}_r U_{t-r} f(x)$ exists for all $t > 0$ and $x \in E_D^T$. Let $\bar{U}_0 \bar{f}(x) = \bar{f}(x)$ for $x \in E_D^T$. Then $(\bar{U}_t)_{t \geq 0}$ constitute a cumulant semigroup on E_D^T. Moreover, we have

$$\bar{U}_t \bar{f}(x) = \bar{T}_t \bar{f}(x) + \int_0^t ds \int_E \gamma_0(y, U_s f) \bar{T}_{t-s}(x, dy)$$
$$+ \int_0^t ds \int_E [b_0 - b(y)]U_s f(y) \bar{T}_{t-s}(x, dy)$$
$$- \int_0^t ds \int_E \psi_0(y, U_s f) \bar{T}_{t-s}(x, dy) \tag{8.35}$$

for $t \geq 0$ and $x \in E_D^T$. By the observations in the last paragraph, each term in (8.35) is bounded by $\|\bar{f}\| e^{c_0 t}$. Obviously, we can also regard $(\bar{U}_t)_{t > 0}$ as operators from $B(E)^+$ to $B(E_D^T)^+$.

Lemma 8.19 *There is a one-to-one correspondence between $\mu \in M(E_D^T)$ and $\kappa \in \mathscr{K}(P)$ determined by*

$$\kappa_t(f) = e^{b_0 t}\mu(\bar{T}_t(h^{-1}f)), \qquad t > 0, f \in B(E)^+. \tag{8.36}$$

Moreover, if κ and μ are related by (8.36), we have

$$S_t(\kappa, f) = \mu(\bar{U}_t(h^{-1}f)), \qquad t > 0, f \in B(E)^+. \tag{8.37}$$

Proof. Let $\mu \in M(E_D^T)$ and define the family of measures $\kappa = (\kappa_t)_{t>0}$ by (8.36). Observe that

$$\kappa_r(P_t f) = e^{b_0 r}\mu\bar{T}_r(h^{-1}P_t f) = e^{b_0(r+t)}\mu\bar{T}_r T_t(h^{-1}f) = \kappa_{r+t}(f)$$

for all $r, t > 0$ and $f \in B(E)$. Moreover, by (8.32) and (8.36) it is easy to see that

$$\int_0^1 \kappa_s(1)\mathrm{d}s = \lim_{r \to 0}\int_r^1 e^{b_0 s}\mu\bar{T}_s(h^{-1})\mathrm{d}s = \lim_{r \to 0}\int_r^1 \mu\bar{T}_r(h^{-1}P_{s-r}1)\mathrm{d}s$$

$$\leq \lim_{r \to 0}\int_r^1 e^{\|b^+\|(s-r)}\mu\bar{T}_r(h^{-1}\pi_{s-r}1)\mathrm{d}s$$

$$\leq \lim_{r \to 0} e^{\|b^+\|}\mu\bar{T}_r(1) = e^{\|b^+\|}\mu(1),$$

where we also used (8.11) for the first inequality. Then we have $\kappa \in \mathscr{K}(P)$. Conversely, given $\kappa \in \mathscr{K}(P)$, we first define an entrance law $\nu = (\nu_t)_{t>0}$ for the semigroup $(T_t)_{t\geq0}$ by $\nu_t(f) = e^{-b_0 t}\kappa_t(hf)$. Observe that

$$\nu_{0+}(1) := \uparrow\lim_{t \to 0}\nu_t(1) = \uparrow\lim_{t \to 0} e^{-b_0 t}\int_0^1 \kappa_t(\pi_s 1)\mathrm{d}s = \int_0^1 \eta_s(1)\mathrm{d}s < \infty,$$

where $\eta \in \mathscr{K}(\pi)$ is defined by (8.9). For $t > 0$ define $\tilde{\nu}_t \in M(E^\partial)$ by $\tilde{\nu}_t|_E = \nu_t$ and $\tilde{\nu}_t(\{\partial\}) = \nu_{0+}(1) - \nu_t(1)$. It is simple to see that $(\tilde{\nu}_t)_{t>0}$ is a finite entrance law for the conservative Borel right semigroup $(T_t^\partial)_{t\geq0}$. By Theorem A.37 there exists a measure $\tilde{\nu}_0 \in M(E_D^{\partial,T})$ such that $\tilde{\nu}_t = \tilde{\nu}_0\bar{T}_t^\partial$ for $t > 0$. Then

$$\kappa_t(f) = e^{b_0 t}\nu_t(h^{-1}f) = e^{b_0 t}\mu\bar{T}_t(h^{-1}f)$$

with μ being the restriction of $\tilde{\nu}_0$ to E_D^T. Finally, assume that κ and μ are related by (8.36). If $f \in B(E)^+$ is bounded by const $\cdot h$ we can use Lemma 8.9 to see

$$S_t(\kappa, f) = \lim_{r \to 0}\kappa_r(V_{t-r}f) = \lim_{r \to 0} e^{b_0 r}\mu\bar{T}_r(h^{-1}V_{t-r}f)$$

$$= \lim_{r \to 0}\mu\bar{T}_r(U_{t-r}(h^{-1}f)) = \mu(\bar{U}_t(h^{-1}f)).$$

Then we obtain (8.37) for all $f \in B(E)^+$ by taking increasing limits. $\square$

Theorem 8.20 *The entrance law $K \in \mathscr{K}^1(Q)$ is infinitely divisible if and only if it has the representation*

$$\int_{M(E)} e^{-\nu(f)} K_t(d\nu)$$

$$= \exp\left\{ -S_t(\kappa, f) - \int_{\mathscr{K}(P)^\circ} (1 - e^{-S_t(\mu,f)}) F(d\mu) \right\}, \quad (8.38)$$

where $\kappa \in \mathscr{K}(P)$ and $F(d\mu)$ is a σ-finite measure on $\mathscr{K}(P)^\circ$ satisfying

$$\int_0^1 ds \int_{\mathscr{K}(P)^\circ} \mu_s(1) F(d\mu) < \infty. \quad (8.39)$$

Proof. By Theorem 8.11 any entrance law $K \in \mathscr{K}^1(Q)$ corresponds to a probability measure J on $\mathscr{K}(P)$ such that

$$\int_{M(E)} e^{-\nu(f)} K_t(d\nu) = \int_{\mathscr{K}(P)} \exp\{-S_t(\mu, f)\} J(d\mu)$$

for every $f \in B(E)^+$. Then by Lemma 8.19 there is a probability measure H on $M(E_D^T)$ such that

$$\int_{M(E)} e^{-\nu(f)} K_t(d\nu) = \int_{M(E_D^T)} \exp\{-\mu(\bar{U}_t(h^{-1}f))\} H(d\mu). \quad (8.40)$$

For any $\bar{f} \in C(E_D^T)^+$ we can use the above equality to see

$$\int_{M(E_D^T)} e^{-\mu(\bar{f})} H(d\mu) = \lim_{t \to 0} \int_{M(E)} e^{-\nu(h\bar{f})} K_t(d\nu). \quad (8.41)$$

Using (8.40) and (8.41) one can see K is an infinitely divisible probability entrance law if and only if H is an infinitely divisible probability measure. In this case, we have the representation

$$\int_{M(E_D^T)} e^{-\mu(\bar{f})} H(d\mu) = \exp\left\{ -\gamma_D^T(\bar{f}) - \int_{M(E_D^T)^\circ} (1 - e^{-\nu(\bar{f})}) G_D^T(d\nu) \right\},$$

where $\gamma_D^T \in M(E_D^T)$ and $[1 \wedge \nu(1)] G_D^T(d\nu)$ is a finite measure on $M(E_D^T)^\circ$. Then (8.38) follows by (8.40) and another application of Lemma 8.19. Using the notation introduced in the proof of Theorem 8.11, we can differentiate both sides of (8.38) to get

$$\int_{M(E)} \nu(f) K_t(d\nu) = \pi l \kappa_t(f) + \int_{\mathscr{K}(P)^\circ} \pi l \mu_t(f) F(d\mu), \quad (8.42)$$

where $\eta_t = \pi l \kappa_t$ is defined by (8.9) and $\pi l \mu_t$ is defined similarly. By (8.42) and Corollary 8.8 one can show (8.39) is equivalent to (8.14). $\qquad\square$

The theorem above gives a complete characterization of infinitely divisible entrance laws in $\mathscr{K}^1(Q)$. This result also yields a representation for the entrance laws for the restricted semigroup $(Q_t^\circ)_{t\geq 0}$. Indeed, by Corollary 8.3 and Theorem 8.20 an entrance law $H \in \mathscr{K}(Q^\circ)$ can always be represented as

$$\int_{M(E)^\circ} (1 - e^{-\nu(f)})H_t(d\nu)$$
$$= S_t(\kappa, f) + \int_{\mathscr{K}(P)^\circ} (1 - \exp\{-S_t(\mu, f)\})F(d\mu), \qquad (8.43)$$

where $\kappa \in \mathscr{K}(P)$ and $F(d\mu)$ is a σ-finite measure on $\mathscr{K}(P)^\circ$ satisfying (8.39). Then an application of Corollary 8.4 gives the following:

Corollary 8.21 *Suppose that Condition 8.5 is satisfied. Then $H \in \mathscr{K}(Q^\circ)$ if and only if it is given by (8.43) for $\kappa \in \mathscr{K}(P)$ and for a σ-finite measure $F(d\mu)$ on $\mathscr{K}(P)^\circ$ satisfying (8.39).*

8.4 Kuznetsov Measures and Excursion Laws

In this section, we prove some supporting properties of the Kuznetsov measures corresponding to unbounded entrance laws for the (ξ, ϕ)-superprocess, which typically lead to excursion laws. We assume ξ is a Borel right process in E and ϕ is a branching mechanism given by (2.26) or (2.27). Let $(Q_t)_{t\geq 0}$ and $(V_t)_{t\geq 0}$ be respectively the transition and cumulant semigroups of the (ξ, ϕ)-superprocess. Recall that $\mathscr{K}(Q^\circ)$ is the set of entrance laws for the restricted semigroup $(Q_t^\circ)_{t\geq 0}$ satisfying (8.14). Let W_0^+ be the space of right continuous paths $t \mapsto w_t$ from $(0, \infty)$ to $M(E)$ having zero as a trap. Let $(\mathscr{A}^0, \mathscr{A}_t^0)$ be the natural σ-algebras on W_0^+ generated by the coordinate process. By Theorem A.40 each entrance law $H \in \mathscr{K}(Q^\circ)$ determines a *Kuznetsov measure* $\mathbf{Q}_H(dw)$, which is the unique σ-finite measure on $(W_0^+, \mathscr{A}^0)$ such that $\mathbf{Q}_H(\{0\}) = 0$ and

$$\mathbf{Q}_H(w_{t_1} \in d\nu_1, w_{t_2} \in d\nu_2, \ldots, w_{t_n} \in d\nu_n)$$
$$= H_{t_1}(d\nu_1)Q_{t_2-t_1}^\circ(\nu_1, d\nu_2) \cdots Q_{t_n-t_{n-1}}^\circ(\nu_{n-1}, d\nu_n) \qquad (8.44)$$

for every $\{t_1 < \cdots < t_n\} \subset (0, \infty)$ and $\{\nu_1, \ldots, \nu_n\} \subset M(E)^\circ$. Roughly speaking, the above formula means that $\{w_t : t > 0\}$ under $\mathbf{Q}_H$ is a Markov process in $M(E)^\circ$ with transition semigroup $(Q_t^\circ)_{t\geq 0}$ and one-dimensional distributions $(H_t)_{t>0}$.

Let $\gamma(x, dy)$ be the kernel on E defined by (2.28) and let $(\pi_t)_{t\geq 0}$ be the semigroup defined by (2.35). If the underlying semigroup $(P_t)_{t\geq 0}$ is conservative, the function $(s, x) \mapsto \pi_s 1(x)$ is bounded away from zero on $[0, u] \times E$ for every $u > 0$. In this case, we fix a constant $u > 0$ and define the *conservative* inhomogeneous transition semigroup $(Q_{r,t}^u : 0 \leq r \leq t \leq u)$ on $M(E)^\circ$ by

$$Q_{r,t}^u(\mu, d\nu) = \mu(\pi_{u-r}1)^{-1}\nu(\pi_{u-t}1)Q_{t-r}^\circ(\mu, d\nu).$$

For any $a > 0$ let $\mathscr{K}^a(Q^u)$ denote the class of finite entrance laws $H := (H_t : 0 < t \leq u)$ for $(Q^u_{r,t} : 0 \leq r \leq t \leq u)$ satisfying $H_t(M(E)^\circ) = a$ for $0 < t \leq u$. Given any $K \in \mathscr{K}(Q^\circ)$ we can let

$$a = \int_{M(E)^\circ} \nu(1) K_u(d\nu)$$

and define $H^u \in \mathscr{K}^a(Q^u)$ by

$$H^u_t(d\nu) = \nu(\pi_{u-t}1) K_t(d\nu), \qquad 0 < t \leq u. \tag{8.45}$$

Recall that $\|\pi_t\| \leq e^{c_0 t}$ for all $t \geq 0$, where $c_0 = \|b^-\| + \|\gamma(\cdot, 1)\|$.

Theorem 8.22 *Let $x \in E^\circ$ and let $\mathbf{Q}_x$ be the Kuznetsov measure on W^+_0 corresponding to the entrance law $L(x) := \{L_t(x, \cdot) : t > 0\}$ defined by (2.9). Then for $\mathbf{Q}_x$-a.e. $w \in W^+_0$ we have $w_t \to 0$ and $w_t(1)^{-1} w_t \to \delta_x$ in $M(E)$ as $t \to 0$.*

Proof. Step 1. We first assume the underlying semigroup $(P_t)_{t \geq 0}$ is conservative. For fixed $u > 0$ let $a = \pi_u 1(x)$ and define $H^u \in \mathscr{K}^a(Q^u)$ by (8.45) with K replaced by $L(x)$. Then

$$\mathbf{Q}^u_x(dw) := a^{-1} w_u(1) \mathbf{Q}_x(dw) = \pi_u 1(x)^{-1} w_u(1) \mathbf{Q}_x(dw)$$

defines a probability measure on W^+_0. Under this measure, the coordinate process $\{w_t : 0 < t \leq u\}$ is Markovian with transition semigroup $(Q^u_{r,t} : 0 \leq r \leq t \leq u)$ and one-dimensional distributions $(a^{-1} H^u_t : 0 < t \leq u)$. By Corollary 8.17, $L(x) \in \mathscr{K}^\infty(Q^\circ)$ is a minimal entrance law. Then $H^u \in \mathscr{K}^a(Q^u)$ is a minimal entrance law. It follows that $\mathbf{Q}^u_x$ is trivial on $\mathscr{A}^0_{0+}$; see Dynkin (1978, p.724) or Sharpe (1988, p.199). For any $f \in B(E)^+$ we can use the martingale convergence theorem to get $\mathbf{Q}^u_x$-a.s.

$$\begin{aligned}
V_u f(x) &= \int_{M(E)^\circ} (1 - e^{-\nu(f)}) \nu(1)^{-1} H^u_u(d\nu) \\
&= \pi_u 1(x) \mathbf{Q}^u_x \Big[(1 - e^{-w_u(f)}) w_u(1)^{-1} \Big] \\
&= \lim_{s \to 0} \pi_u 1(x) \mathbf{Q}^u_x \Big[(1 - e^{-w_u(f)}) w_u(1)^{-1} \big| \mathscr{A}^0_s \Big] \\
&= \lim_{s \to 0} \pi_u 1(x) \int_{M(E)^\circ} (1 - e^{-\nu(f)}) \nu(1)^{-1} Q^u_{s,u}(w_s, d\nu) \\
&= \lim_{s \to 0} \pi_u 1(x) w_s(\pi_{u-s}1)^{-1} (1 - e^{-w_s(V_{u-s}f)}). \tag{8.46}
\end{aligned}$$

Let $W^+_u = \{w \in W^+_0 : w_u(1) > 0\}$. Then $\mathbf{Q}^u_x(dw)$ and $\mathbf{Q}_x(dw)$ are absolutely continuous with respect to each other on W^+_u. Observe also that $W^+_v \subset W^+_u$ for any $v \geq u$. Then (8.46) holds for $\mathbf{Q}_x$-a.e. $w \in W^+_v$. Since $(P_t)_{t \geq 0}$ is conservative, (8.11) and (8.46) imply

$$V_u f(x) \leq \liminf_{s \to 0} e^{c_0 u} w_s(\pi_{u-s}1)^{-1} \leq \liminf_{s \to 0} e^{(c_0 + \|b^+\|)u} w_s(1)^{-1}.$$

Then letting $u \to 0$ and $f \to \infty$ we see $w_s(1) \to 0$ as $s \to 0$ for $\mathbf{Q}_x$-a.e. $w \in W_v^+$. Since $W_0^+ = \{0\} \cup (\cup_{v>0} W_v^+)$ and $\mathbf{Q}_x(\{0\}) = 0$, we have $w_s(1) \to 0$ as $s \to 0$ for $\mathbf{Q}_x$-a.e. $w \in W_0^+$.

Step 2. Let $\mathscr{R}$ be a countable Ray cone for $(P_t)_{t\geq 0}$ that generates the Ray–Knight completion $\bar{E}$ of E. Recall that each $f \in \mathscr{R}$ can be extended uniquely to a function $\bar{f} \in C(\bar{E})^+$. By a similar reasoning as in the first step, for any $v \geq u$ we have

$$
\begin{aligned}
\pi_u f(x) &= \int_{M(E)^\circ} \nu(f) L_u(x, d\nu) \\
&= \lim_{s \to 0} \pi_u 1(x) w_s(\pi_{u-s} 1)^{-1} w_s(\pi_{u-s} f)
\end{aligned}
\tag{8.47}
$$

for $\mathbf{Q}_x$-a.e. $w \in W_v^+$. Take any $w \in W_v^+$ along which the above relation holds for all $f \in \mathscr{R}$ and all rational $u \in (0, v]$. Let $\alpha = \alpha(f) \geq 0$ be a constant so that $f \in \mathscr{R}$ is α-excessive for $(P_t)_{t\geq 0}$. By the proof of Corollary 2.34, there is a constant $\beta \geq \alpha$ so that $\pi_t f \leq e^{2\beta t} f$ for all $t \geq 0$. Then by (8.11) and (8.47) we can see

$$
\pi_u f(x) \leq \liminf_{s \to 0} e^{(c_0 - \|b^+\| + 2\beta)u} w_s(1)^{-1} w_s(f)
\tag{8.48}
$$

and

$$
e^{2\beta u} f(x) \geq \limsup_{s \to 0} e^{-(2\|b^+\| + c_0)u} w_s(1)^{-1} w_s(P_u f),
\tag{8.49}
$$

where we have also used the relation $P_{u-s} f \geq e^{-\alpha s} P_u f$ for the second inequality.

Step 3. Let $s_k = s_k(w) > 0$ be any sequence so that $s_k \to 0$ and $w_{s_k}(1)^{-1} w_{s_k} \to \hat{w}_0$ in $M(\bar{E})$ as $k \to \infty$, where $\hat{w}_0$ is a probability measure on $\bar{E}$. By (8.48) we have

$$
\pi_u f(x) \leq e^{(c_0 - \|b^+\| + 2\beta)u} \hat{w}_0(\bar{f}).
$$

Then letting $u \to 0$ along rationals gives $f(x) \leq \hat{w}_0(\bar{f})$. Let $\theta > 2\beta$. Since $u \mapsto e^{-(\theta + 2\|b^+\| + c_0)u} P_u f$ is decreasing, by (8.49) for any integer $n \geq 1$ we have

$$
\sum_{i=0}^\infty \frac{1}{n} e^{-(\theta - 2\beta)i/n} f(x) \geq \limsup_{s \to 0} w_s(1)^{-1} w_s(f_v),
$$

where

$$
f_v = \int_0^v e^{-(\theta + 2\|b^+\| + c_0)u} P_u f \, du.
$$

Then letting $n \to \infty$ gives

$$
\int_0^\infty e^{-(\theta - 2\beta)u} f(x) du \geq \limsup_{s \to 0} w_s(1)^{-1} w_s(f_v).
\tag{8.50}
$$

Let $(U^\alpha)_{\alpha>0}$ denote the resolvent of $(P_t)_{t\geq 0}$ and let $(\bar{U}^\alpha)_{\alpha>0}$ denote its Ray extension. It is elementary to see

$$U^{\theta+2\|b^+\|+c_0}f = \sum_{k=0}^{\infty} \int_0^v e^{-(\theta+2\|b^+\|+c_0)(kv+u)} P_{kv+u}f du$$

$$\leq \sum_{k=0}^{\infty} \int_0^v e^{-(\theta+2\|b^+\|+c_0)(kv+u)} e^{k\alpha v} P_u f du$$

$$= \sum_{k=0}^{\infty} e^{-k(\theta+2\|b^+\|+c_0-\alpha)v} f_v$$

$$\leq (1 - e^{-(\theta-\alpha)v})^{-1} f_v. \tag{8.51}$$

From (8.50) and (8.51) we get

$$\int_0^\infty e^{-(\theta-2\beta)u} f(x) du \geq (1 - e^{-(\theta-\alpha)v}) \hat{w}_0 (\bar{U}^{\theta+2\|b^+\|+c_0} \bar{f}).$$

Multiplying both sides of the above equality by $\theta > 2\beta$ and letting $\theta \to \infty$ we obtain $f(x) \geq \hat{w}_0(\bar{f})$. It follows that $f(x) = \hat{w}_0(\bar{f})$ and hence $\hat{w}_0 = \delta_x$ because $\{\bar{f}_1 - \bar{f}_2 : f_1, f_2 \in \mathcal{R}\}$ is uniformly dense in $C(\bar{E})$. That implies $w_s(1)^{-1} w_s \to \delta_x$ for $\mathbf{Q}_x$-a.e. $w \in W_v^+$ first in $M(\bar{E})$ and then in $M(E)$. Since $u > 0$ was arbitrary, the theorem follows when $(P_t)_{t\geq 0}$ is conservative.

Step 4. For a non-conservative underlying semigroup $(P_t)_{t\geq 0}$, we extend it to a conservative semigroup $(\tilde{P}_t)_{t\geq 0}$ on the Lusin topological space $\tilde{E} := E \cup \{\partial\}$ with ∂ being an isolated cemetery. Take a spatially constant local branching mechanism $z \mapsto \phi_*(z)$ satisfying $\phi_*'(z) \to \infty$ as $z \to \infty$. For $\tilde{f} \in B(\tilde{E})^+$ let $\tilde{\phi}(\partial, \tilde{f}) = \phi_*(\tilde{f}(\partial))$ and let $\tilde{\phi}(x, \tilde{f}) = \phi(x, \tilde{f}|_E)$ if $x \in E$. From those extensions we can use an analogue of (2.33) to define the cumulant semigroup $(\tilde{V}_t)_{t\geq 0}$, which can be represented by (2.5) with (λ_t, L_t) replaced by $(\tilde{\lambda}_t, \tilde{L}_t)$. If we extend the definition of $f \in B(E)^+$ to $\tilde{E}$ by setting $f(\partial) = 0$, it is not hard to show $\tilde{V}_t f(x) = V_t f(x)$ for $t \geq 0$ and $x \in E$. Then $\lambda_t(x, \cdot) = \tilde{\lambda}_t(x, \cdot)|_E$ for $t \geq 0$ and $x \in E$. In particular, we have $\tilde{\lambda}_t(x, E) = 0$ for $t > 0$ and $x \in E^\circ$. Let $(v_t)_{t\geq 0}$ be the cumulant semigroup of the CB-process with branching mechanism $z \mapsto \phi_*(z)$, which admits the representation (3.4). Since ∂ is a cemetery for $(\tilde{P}_t)_{t\geq 0}$, we have

$$\tilde{V}_t \tilde{f}(\partial) = \int_0^\infty \left(1 - e^{-u\tilde{f}(\partial)}\right) l_t(du), \quad t > 0, \tilde{f} \in B(\tilde{E})^+.$$

Thus $\tilde{\lambda}_t(\partial, \tilde{E}) = 0$ for $t > 0$. For any $t > 0$ and $x \in E^\circ$ one can choose sufficiently small $r > 0$ and use Corollary 2.7 to see

$$\tilde{\lambda}_t(x, \tilde{E}) = \int_{\tilde{E}} \tilde{\lambda}_r(x, dy) \tilde{\lambda}_{t-r}(y, \tilde{E}) = 0.$$

It follows that

$$\tilde{V}_t \tilde{f}(x) = \int_{M(\tilde{E})^\circ} \left(1 - e^{-\nu(\tilde{f})}\right) \tilde{L}_t(x, d\nu), \quad \tilde{f} \in B(\tilde{E})^+. \tag{8.52}$$

Let $\tilde{W}_0^+$ be the space of right continuous paths $t \mapsto \tilde{w}_t$ from $(0, \infty)$ to $M(\tilde{E})$ having zero as a trap and let $\tilde{\mathbf{Q}}_x$ be the Kuznetsov measure on $\tilde{W}_0^+$ corresponding to the entrance law $\tilde{L}(x) := \{\tilde{L}_t(x, \cdot) : t > 0\}$ defined by (8.52). Then $\{\tilde{w}_t|_E : t > 0\}$ under $\tilde{\mathbf{Q}}_x$ is equivalent to $\{w_t : t > 0\}$ under $\mathbf{Q}_x$. By Steps 1 and 3, for $\tilde{\mathbf{Q}}_x$-a.e. $\tilde{w} \in \tilde{W}_0^+$ we have $\lim_{t \to 0} \tilde{w}_t(E) \le \lim_{t \to 0} \tilde{w}_t(\tilde{E}) = 0$ and

$$\lim_{t \to 0} \tilde{w}_t(E)^{-1} \tilde{w}_t(f) = \lim_{t \to 0} \frac{\tilde{w}_t(\tilde{E})^{-1} \tilde{w}_t(f)}{\tilde{w}_t(\tilde{E})^{-1} \tilde{w}_t(E)} = f(x)$$

for any $f \in C(E)$. That proves the theorem. $\qquad\square$

By Theorem 8.22, for any $x \in E^\circ$ the Kuznetsov measure $\mathbf{Q}_x$ is actually carried by the paths $w \in W_0^+$ satisfying $w_t \to 0$ and $w_t(1)^{-1} w_t \to \delta_x$ in $M(E)$ as $t \to 0$. We call those paths *excursions* starting at $x \in E^\circ$ and call $\mathbf{Q}_x$ an *excursion law* for the (ξ, ϕ)-superprocess. The following theorem gives an alternate characterization of the excursion law.

Theorem 8.23 *Let $t > 0$ and assume $\lambda \in M([0, t])$ satisfies $\lambda(\{0\}) = 0$. For a bounded positive Borel function $(s, x) \mapsto f_s(x)$ on $[0, t] \times E$ let $(r, x) \mapsto u_r(x)$ be the unique bounded positive solution of (5.26). Then for any $x \in E^\circ$ we have*

$$\mathbf{Q}_x\left[1 - \exp\left\{-\int_{(0,t]} w_s(f_s)\lambda(\mathrm{d}s)\right\}\right] = u_0(x). \tag{8.53}$$

Proof. Let $\xi = (\Omega, \mathscr{F}, \mathscr{F}_{r,t}, \xi_t, \mathbf{P}_{r,x})$ and $X = (W, \mathscr{G}, \mathscr{G}_{r,t}, X_t, \mathbf{Q}_{r,\mu})$ be realizations of the underlying spatial motion and the (ξ, ϕ)-superprocess, respectively, started from the arbitrary initial time $r \ge 0$. By the Markov property of $\mathbf{Q}_x$ and Theorem 5.15 we have

$$\begin{aligned}
\mathbf{Q}_x&\left[1 - \exp\left\{-\int_{(0,t]} w_s(f_s)\lambda(\mathrm{d}s)\right\}\right] \\
&= \lim_{\varepsilon \to 0} \mathbf{Q}_x\left[1 - \exp\left\{-\int_{[\varepsilon,t]} w_s(f_s)\lambda(\mathrm{d}s)\right\}\right] \\
&= \lim_{\varepsilon \to 0} \mathbf{Q}_x \mathbf{Q}_{\varepsilon, w_\varepsilon}\left[1 - \exp\left\{-\int_{[\varepsilon,t]} X_s(f_s)\lambda(\mathrm{d}s)\right\}\right] \\
&= \lim_{\varepsilon \to 0} \mathbf{Q}_x\left[1 - \exp\{-w_\varepsilon(u_\varepsilon)\}\right] = \lim_{\varepsilon \to 0} v_0(x, u_\varepsilon), \tag{8.54}
\end{aligned}$$

where $(r, x) \mapsto u_r(x)$ is defined by (5.26) and $(r, x) \mapsto v_r(x) = v_r(x, u_\varepsilon)$ is defined by

$$v_r(x) + \int_r^\varepsilon \mathbf{P}_{r,x}[\phi(\xi_s, v_s)]\mathrm{d}s = \mathbf{P}_{r,x} u_\varepsilon(\xi_\varepsilon). \tag{8.55}$$

Combining (5.26) and (8.55) gives

$$v_0(x, u_\varepsilon) = \mathbf{P}_{0,x} u_\varepsilon(\xi_\varepsilon) - \int_0^\varepsilon \mathbf{P}_{0,x}[\phi(\xi_s, v_s(\cdot, u_\varepsilon))]\mathrm{d}s$$

$$= \mathbf{P}_{0,x} \mathbf{P}_{\varepsilon,\xi_\varepsilon} \left[\int_{[\varepsilon,t]} f_s(\xi_s) \lambda(\mathrm{d}s) \right] - \mathbf{P}_{0,x} \mathbf{P}_{\varepsilon,\xi_\varepsilon} \left[\int_\varepsilon^t \phi(\xi_s, u_s) \mathrm{d}s \right]$$

$$- \int_0^\varepsilon \mathbf{P}_{0,x} [\phi(\xi_s, v_s(\cdot, u_\varepsilon))] \mathrm{d}s$$

$$= \mathbf{P}_{0,x} \left[\int_{[\varepsilon,t]} f_s(\xi_s) \lambda(\mathrm{d}s) \right] - \mathbf{P}_{0,x} \left[\int_\varepsilon^t \phi(\xi_s, u_s) \mathrm{d}s \right]$$

$$- \int_0^\varepsilon \mathbf{P}_{0,x} [\phi(\xi_s, v_s(\cdot, u_\varepsilon))] \mathrm{d}s,$$

which converges to $u_0(x)$ as $\varepsilon \to 0$. Then (8.53) follows from (8.54).   □

Using the excursion law, we can give a reconstruction of the trajectory of the (ξ, ϕ)-superprocess. Suppose that $\mu \in M(E)$ is supported by E° and $N(\mathrm{d}w)$ is a Poisson random measure on W_0^+ with intensity

$$\int_{E^\circ} \mu(\mathrm{d}x) \mathbf{Q}_x(\mathrm{d}w), \qquad w \in W_0^+.$$

We define the measure-valued process $\{X_t : t \geq 0\}$ by $X_0 = \mu$ and

$$X_t = \int_{W_0^+} w_t N(\mathrm{d}w), \qquad t > 0. \tag{8.56}$$

Theorem 8.24 *For $t \geq 0$ let $\mathcal{G}_t$ be the σ-algebra generated by the collection of random variables $\{N(A) : A \in \mathcal{A}_t^0\}$. Then $\{(X_t, \mathcal{G}_t) : t \geq 0\}$ is a realization of the (ξ, ϕ)-superprocess.*

Proof. For any $t > 0$ and $f \in B(E)^+$ we can use Theorem 1.25 to see

$$\mathbf{P} \left[\exp\{-X_t(f)\} \right] = \exp \left\{ - \int_{E^\circ} \mathbf{Q}_x \left(1 - \exp\{-w_t(f)\} \right) \mu(\mathrm{d}x) \right\}$$

$$= \exp \left\{ - \int_{E^\circ} \mu(\mathrm{d}x) \int_{M(E)^\circ} \left(1 - e^{-\nu(f)} \right) L_t(x, \mathrm{d}\nu) \right\}$$

$$= \exp \left\{ - \int_E V_t f(x) \mu(\mathrm{d}x) \right\}.$$

Then the random measure X_t has distribution $Q_t(\mu, \cdot)$ on $M(E)$. Let $t > r > 0$ and let h be a bounded positive function on W_0^+ measurable relative to $\mathcal{A}_r^0$. For $f \in B(E)^+$,

$$\mathbf{P} \left[\exp \left\{ - \int_{W_0^+} h(w) N(\mathrm{d}w) - X_t(f) \right\} \right]$$

$$= \exp \left\{ - \int_{E^\circ} \mathbf{Q}_x \left(1 - \exp\{-h(w) - w_t(f)\} \right) \mu(\mathrm{d}x) \right\}$$

$$= \exp \left\{ - \int_{E^\circ} \mathbf{Q}_x \left(1 - e^{-h(w)} \right) \mu(\mathrm{d}x) \right\}$$

$$\cdot \exp\left\{ - \int_{E^\circ} \mathbf{Q}_x\left[\mathrm{e}^{-h(w)}\left(1 - \mathrm{e}^{-w_t(f)}\right)\right]\mu(\mathrm{d}x)\right\},$$

where we made the convention $\mathrm{e}^{-\infty} = 0$. By the Markov property of $\mathbf{Q}_x$ we have

$$\mathbf{Q}_x\left[\mathrm{e}^{-h(w)}\left(1 - \mathrm{e}^{-w_t(f)}\right)\right]$$
$$= \mathbf{Q}_x\left[\mathrm{e}^{-h(w)}\int_{M(E)^\circ}\left(1 - \mathrm{e}^{-\nu(f)}\right)Q^\circ_{t-r}(w_r, \mathrm{d}\nu)\right]$$
$$= \mathbf{Q}_x\left[\mathrm{e}^{-h(w)}\left(1 - \mathrm{e}^{-w_r(V_{t-r}f)}\right)\right].$$

It follows that

$$\mathbf{P}\left[\exp\left\{ - \int_{W_0^+} h(w)N(\mathrm{d}w) - X_t(f)\right\}\right]$$
$$= \exp\left\{ - \int_{E^\circ} \mathbf{Q}_x\left(1 - \exp\{-h(w) - w_r(V_{t-r}f)\}\right)\mu(\mathrm{d}x)\right\}$$
$$= \mathbf{P}\left[\exp\left\{ - \int_{W_0^+} h(w)N(\mathrm{d}w) - X_r(V_{t-r}f)\right\}\right].$$

Then $\{(X_t, \mathscr{G}_t) : t \geq 0\}$ is a Markov process with transition semigroup $(Q_t)_{t\geq 0}$.

$\square$

Example 8.1 Suppose that $z \mapsto \phi(z)$ is a branching mechanism given by (3.1) satisfying $\phi'(z) \to \infty$ as $z \to \infty$. By Theorem 3.10, the corresponding CB-process has cumulant semigroup admitting the representation (3.15). Let D_0 be the set of positive càdlàg paths $\{w(t) : t \geq 0\}$ satisfying $w(t) = w(0) = 0$ for $t \geq \inf\{s > 0 : w(s) = 0\}$. By Theorem 8.22, the entrance law $(l_t)_{t>0}$ defined by (3.15) corresponds an excursion law $\mathbf{Q}_l(\mathrm{d}w)$ that can be identified as a σ-finite measure on D_0.

Example 8.2 Let E be a complete separable metric space. Suppose that ξ is a càdlàg Borel right process in E satisfying Condition 4.7. We shall construct a variation of the ξ-Brownian snake. Let $R_{a,b}((u,y), \mathrm{d}(w,z))$ be defined as in Section 4.2. For $g \in C([0,\infty), \mathbb{R}_+)$ and $(w_0, z_0) \in S$ satisfying $g(0) = z_0$ let $\mathbf{Q}^g_{(w_0,z_0)}$ denote the law on $S^{[0,\infty)}$ of the time-inhomogeneous Markov process started at (w_0, z_0) whose transition kernel from time $r \geq 0$ to time $t \geq r$ is

$$R_{m(r,t),g(t)}((u,y), \mathrm{d}(w,z)), \quad (u,y) \in S, (w,z) \in S,$$

where $m(r,t) = \inf_{r \leq s \leq t} g(s)$. Let $\mathbf{n}(\mathrm{d}g)$ denote Itô's excursion law, which is the Kuznetsov measure corresponding to the entrance law (A.29) for the absorbing-barrier Brownian motion. We think of $\mathbf{n}(\mathrm{d}g)$ as a σ-finite measure carried by the set C_0 of positive continuous paths $\{g(t) : t \geq 0\}$ such that $g(0) = g(t) = 0$ for every $t \geq \inf\{s > 0 : g(s) = 0\}$. For $x \in E$ let $\mathbf{N}_x$ be the σ-finite measure on $C_0 \times S^{[0,\infty)}$ defined by

$$\mathbf{N}_x(\mathrm{d}g, \mathrm{d}(\eta, \zeta)) = \mathbf{n}(\mathrm{d}g)\mathbf{Q}^g_{(w_0,0)}(\mathrm{d}(\eta, \zeta)), \tag{8.57}$$

where $w_0(t) = x$ for all $t \geq 0$. Roughly speaking, under $\mathbf{N}_x$ the process $\{(\eta_s, \zeta_s) : s \geq 0\}$ behaves as the ξ-Brownian snake. The only difference is that $\{\zeta_s : s \geq 0\}$ is distributed according to Itô's excursion law. Let $\tau_0(\zeta) = \inf\{s > 0 : \zeta_s = 0\}$ and let $\{2l_s(y, \zeta) : s \geq 0, y \geq 0\}$ be the local time of $\{\zeta_s : s \geq 0\}$. For $t \geq 0$ we define the measure $\nu_t(\mathrm{d}y) = \nu_t(g, \eta, \zeta, \mathrm{d}y)$ on E by

$$\nu_t(f) = \int_0^{\tau_0(\zeta)} f(\eta_s(\zeta_s))\mathrm{d}l_s(t, \zeta), \quad f \in C(E)^+. \tag{8.58}$$

Then $\{\nu_t : t \geq 0\}$ is continuous in $M(E)$. It was proved in Le Gall (1999, p.63) that

$$\nu_t(f) = \lim_{\varepsilon \to 0} \frac{1}{\varepsilon} \int_0^{\tau_0(\zeta)} 1_{[t,t+\varepsilon]}(\zeta_s)f(\eta_s(\zeta_s))\mathrm{d}s$$

in $\mathbf{N}_x$-measure. Consequently, for any $h \in C(\mathbb{R}_+)^+$ with bounded support,

$$\int_0^\infty h(t)\nu_t(f)\mathrm{d}t = \lim_{k \to \infty} \sum_{i=0}^\infty h(i/k) \int_0^{\tau_0(\zeta)} 1_{[i/k,(i+1)/k]}(\zeta_s)f(\eta_s(\zeta_s))\mathrm{d}s$$

$$= \int_0^{\tau_0(\zeta)} h(\zeta_s)f(\eta_s(\zeta_s))\mathrm{d}s.$$

From Proposition 3 of Le Gall (1999, p.59) it follows that

$$\mathbf{N}_x\left[1 - \exp\left\{-\int_0^\infty h(t)\nu_t(f)\mathrm{d}t\right\}\right] = u_0(x),$$

where $(r, x) \mapsto u_r(x)$ solves the integral equation

$$u_r(x) + \int_r^\infty \mathbf{P}_{r,x}\left[u_s(\xi_s)^2\right]\mathrm{d}s = \int_r^\infty \mathbf{P}_{r,x}\left[h(s)f(\xi_s)\right]\mathrm{d}s.$$

Thus $\{\nu_t : t > 0\}$ under $\mathbf{N}_x$ is distributed on W_0^+ according to the excursion law $\mathbf{Q}_x$ defined by (5.26) and (8.53) with binary local branching mechanism $\phi(z) = z^2$. This gives a representation of $\mathbf{Q}_x$ using $\mathbf{N}_x$. In view of (8.58), the result of Theorem 8.22 is obviously true for the binary local branching. It was pointed out in Le Gall (1999, p.56) that $\mathbf{N}_x$ can be understood as an excursion law of the ξ-Brownian snake from the state $(w_0, 0)$.

Example 8.3 Let us continue the discussion in the last example. Suppose that $\mu \in M(E)$ and $N(\mathrm{d}x, \mathrm{d}g, \mathrm{d}(\eta, \zeta))$ is a Poisson random measure on $E \times C_0 \times S^{[0,\infty)}$ with intensity $\mu(\mathrm{d}x)\mathbf{N}_x(\mathrm{d}g, \mathrm{d}(\eta, \zeta))$. Let $\nu_t = \nu_t(g, \eta, \zeta)$ be given by (8.58) and define the measure-valued process $\{X_t : t \geq 0\}$ by $X_0 = \mu$ and

$$X_t = \int_E \int_{C_0} \int_{S^{[0,\infty)}} \nu_t N(\mathrm{d}x, \mathrm{d}g, \mathrm{d}(\eta, \zeta)), \quad t > 0. \tag{8.59}$$

Then $\{X_t : t \geq 0\}$ is a (ξ, ϕ)-superprocess with local branching mechanism $\phi(z) = z^2$; see Le Gall (1999, pp.61–62). This follows from Theorem 8.24 and the fact that $\{\nu_t : t > 0\}$ is distributed under $\mathbf{N}_x$ according to the excursion law $\mathbf{Q}_x$ of the (ξ, ϕ)-superprocess. Let $\{(x_i, g_i, (\eta^i, \zeta^i)) : i = 1, 2, \ldots\}$ be the countable support of $N(\mathrm{d}x, \mathrm{d}g, \mathrm{d}(\eta, \zeta))$. Let

$$T = \sup\{t \geq 0 : X_t > 0\} = \inf\{t \geq 0 : X_t = 0\}$$

denote the extinction time of $\{X_t : t \geq 0\}$. Then there is a unique label $k \geq 1$ so that

$$T = \sup_{0 \leq s \leq \tau_0(\zeta^k)} \zeta^k(s) = \sup_{i \geq 1} \sup_{0 \leq s \leq \tau_0(\zeta^i)} \zeta^i(s).$$

Let S be the unique value in $[0, \tau_0(\zeta^k)]$ so that $\zeta^k(S) = T$. It is easy to see that

$$\lim_{t \uparrow T} X_t(1)^{-1} X_t = \delta_{\eta_S^k(T)}.$$

This gives a description of the superprocess near its extinction time.

8.5 Super-Absorbing-Barrier Brownian Motions

We first consider a bounded domain E in the Euclidean space $\mathbb{R}^d$ with twice continuously differentiable boundary ∂E. Let ξ be an absorbing-barrier Brownian motion in E. Let $(P_t)_{t \geq 0}$ denote the transition semigroup of ξ. It is well-known that $P_t(x, \mathrm{d}y)$ has a density $p_t(x, y)$ for $t > 0$, which is the fundamental solution of the heat equation on E with Dirichlet boundary condition. Moreover, $p_t(x, y) = p_t(y, x)$ is continuously differentiable in x and y to the boundary ∂E; see, e.g., Friedman (1964, p.83). We shall use ∂_z to denote the operator of inward normal differentiation at $z \in \partial E$. Clearly,

$$h(x) = \int_0^1 P_s 1(x)\mathrm{d}s, \quad x \in E \tag{8.60}$$

defines a bounded strictly positive excessive function for $(P_t)_{t \geq 0}$ and $h(x) \to 0$ as $x \to z \in \partial E$. Let $M_h(E)$ denote the set of σ-finite measures μ on E such that $\mu(h) < \infty$.

Lemma 8.25 *The function h is continuously differentiable to the boundary and $z \mapsto \partial_z h$ is bounded above and bounded away from zero on ∂E.*

Proof. We only give the proof for $d \geq 2$ since the result for $d = 1$ is well-known. We shall use superscripts to indicate the dependence of the objects on the domain E. The arguments consist of three steps.

Step 1. By Friedman (1964, p.83) and dominated convergence it is easy to see that h is continuously differentiable to the boundary and

$$\partial_z h^E = \int_0^1 \partial_z P_s 1 ds = \int_0^1 ds \int_E \partial_z p_s^E(\cdot, x) dx, \quad z \in \partial E. \quad (8.61)$$

Let $(\mathscr{F}_t, \xi(t), \mathbf{P}_x)$ be a standard Brownian motion on $\mathbb{R}^d$ and let τ_{E^c} denote its hitting time of E^c. For $x \in E$ we have $\mathbf{P}_x$-a.s. $\xi(\tau_{E^c}) \in \partial E$ and

$$\partial_z p_s^E(\cdot, x) ds\sigma(dz) = \partial_z p_s^E(x, \cdot) ds\sigma(dz) = 2\mathbf{P}_x\{\tau_{E^c} \in ds, \xi(\tau_{E^c}) \in dz\},$$

where $\sigma(dz)$ is the volume element on ∂E; see, e.g., Hsu (1986, p.110). Since E is bounded, integrating both sides of the equality above and using (8.61) we see that

$$0 < \int_{\partial E} \partial_z h^E \sigma(dz) = 2 \int_E \mathbf{P}_x\{\tau_{E^c} \leq 1\} dx < \infty. \quad (8.62)$$

Step 2. For $w \in \mathbb{R}^d$ and $r > 0$ let $B_w(r) = \{x \in \mathbb{R}^d : |x - w| < r\}$. By (8.61), (8.62) and the spatial homogeneity and symmetry of the Brownian motion, there is a constant $0 < c(r) < \infty$ such that $\partial_z h^{B_w(r)} = c(r)$ for every $z \in \partial B_w(r)$. Consequently, the theorem holds for $E = B_w(r)$. In the general case, the smoothness of the boundary ∂E implies the existence of a constant $r > 0$ so that for each $z \in \partial E$ there exists $w \in E$ so that $|w - z| = r$ and $B_w(r) \subset E$. Then using the absorbing-barrier Brownian motion we have

$$\begin{aligned}
p_s^E(x, y) dy &= \mathbf{P}_x\{\tau_{E^c} > s, \xi(s) \in dy\} \\
&\geq \mathbf{P}_x\{\tau_{B_w(r)^c} > s, \xi(s) \in dy\} \\
&= p_s^{B_w(r)}(x, y) dy
\end{aligned}$$

for every $s > 0$ and $x, y \in B_w(r)$. It follows that $\partial_z p_s^E(\cdot, y) \geq \partial_z p_s^{B_w(r)}(\cdot, y)$ for every $s > 0$ and $y \in B_w(r)$. In view of (8.61) we have $\partial_z h^E \geq \partial_z h^{B_w(r)} = c(r)$. Then $\partial_z h^E$ is bounded away from zero.

Step 3. For $w \in \mathbb{R}^d$ and $0 < r < \rho < \infty$ let $B_w(r, \rho) = \{x \in \mathbb{R}^d : r < |x - w| < \rho\}$. By the spatial homogeneity and symmetry of the Brownian motion, there are constants $0 < \alpha(r, \rho), \beta(r, \rho) < \infty$ such that

$$\partial_z h^{B_w(r,\rho)} = \begin{cases} \alpha(r, \rho) & \text{if } |z - w| = r, \\ \beta(r, \rho) & \text{if } |z - w| = \rho. \end{cases}$$

We can find constants $0 < r < \rho < \infty$ so that for each $z \in \partial E$ there exists $w \in E^c$ satisfying $|w - z| = r$ and $B_w(r, \rho) \supset E$. By a comparison argument as in Step 2 we get $\partial_z h^E \leq \partial_z h^{B_w(r,\rho)} = \alpha(r, \rho)$. $\qquad \square$

Theorem 8.26 *For any $\mu \in M_h(E)$ and $\eta \in M(\partial E)$, we can define an entrance law $\kappa \in \mathscr{K}(P)$ by*

$$\kappa_t(f) = \mu(P_t f) + \int_{\partial E} \partial_z P_t f \, \eta(\mathrm{d}z), \quad t > 0, f \in B(E). \tag{8.63}$$

Moreover, every $\kappa \in \mathscr{K}(P)$ has the representation (8.63) for some $\mu \in M_h(E)$ and $\eta \in M(\partial E)$.

Proof. If $\kappa = (\kappa_t)_{t>0}$ is given by (8.63), then clearly $\kappa \in \mathscr{K}(P)$. For the converse, suppose that $\kappa \in \mathscr{K}(P)$. Let $(T_t)_{t \geq 0}$ be the h-transform of Doob defined from $(P_t)_{t \geq 0}$ and the function (8.60) and let $\gamma = (\gamma_t)_{t>0}$ be the bounded entrance law for $(T_t)_{t \geq 0}$ defined by $\gamma_t(f) = \kappa_t(hf)$ for $t > 0$ and $f \in C(E)$. By the smoothness of the transition density $p_t(x, y)$, we can extend $(T_t)_{t \geq 0}$ to a transition semigroup $(\bar{T}_t)_{t \geq 0}$ on $\bar{E} := E \cup \partial E$ by letting $\bar{T}_0 \bar{f} \equiv \bar{f}$ and

$$\bar{T}_t \bar{f}(x) = \begin{cases} h(x)^{-1} P_t(hf)(x) & \text{if } x \in E, \\ (\partial_x h)^{-1} \partial_x P_t(hf) & \text{if } x \in \partial E \end{cases} \tag{8.64}$$

for $t > 0$, where $\bar{f} \in C(\bar{E})$ and $f = \bar{f}|_E$. Here $\bar{T}_t$ maps $C(\bar{E})$ to itself. Moreover, since $\bar{E}$ is compact, it is easy to show that $t \mapsto \bar{T}_t \bar{f}$ is strongly continuous. By Theorem 1.14, the family $(\gamma_t)_{t>0}$ is relatively compact if we regard them as measures on $\bar{E}$. Choosing a sequence $r_n \to 0$ such that γ_{r_n} converges weakly to some $\gamma_0 \in M(\bar{E})$ as $n \to \infty$ we get

$$\gamma_t(f) = \lim_{n \to \infty} \gamma_{r_n}(\bar{T}_{t-r_n} f) = \lim_{n \to \infty} \gamma_{r_n}(\bar{T}_t f) = \gamma_0(\bar{T}_t f),$$

and hence

$$\kappa_t(f) = \int_E h(x)^{-1} P_t f(x) \gamma_0(\mathrm{d}x) + \int_{\partial E} (\partial_x h)^{-1} \partial_x P_t f \gamma_0(\mathrm{d}x).$$

Then (8.63) follows with $\mu(\mathrm{d}x) = h(x)^{-1} \gamma_0(\mathrm{d}x)$ and $\eta(\mathrm{d}x) = (\partial_x h)^{-1} \gamma_0(\mathrm{d}x)$. The extension to $f \in B(E)$ is immediate. $\qquad \square$

Let ϕ be a branching mechanism on E given by (2.26) or (2.27) and let $(Q_t)_{t \geq 0}$ be the transition semigroup of the super-absorbing-barrier Brownian motion defined by (2.32) and (2.33). For the entrance law $\kappa \in \mathscr{K}(P)$ given by (8.63) we have

$$S_t(\kappa, f) = \mu(V_t f) + \int_{\partial E} \partial_z V_t f \, \eta(\mathrm{d}z), \quad t > 0, f \in B(E)^+.$$

Then Theorems 8.11 and 8.28 imply the following:

Theorem 8.27 *For any $\mu \in M_h(E)$ and $\eta \in M(\partial E)$, we can define an entrance law $K \in \mathscr{K}_m^1(Q)$ by*

$$\int_{M(E)} e^{-\nu(f)} K_t(\mathrm{d}\nu) = \exp\{-\mu(V_t f) - \eta(\partial . V_t f)\}. \tag{8.65}$$

Moreover, every $K \in \mathcal{K}_m^1(Q)$ has the representation (8.65) for $\mu \in M_h(E)$ and $\eta \in M(\partial E)$.

The above theorem gives a complete characterization for minimal probability entrance laws for the super-absorbing-barrier Brownian motion. Using Theorems 8.20 and 8.28 we can also give a characterization for its infinitely divisible probability entrance laws. Those results can be extended to some unbounded domains. For simplicity we only discuss briefly the extensions to the positive half line $E_0 := (0, \infty)$. In the remainder of this section, let $(P_t)_{t \geq 0}$ be the transition semigroup of the absorbing-barrier Brownian motion in E_0 given by (A.27) and (A.28). In this case, the function $x \mapsto h(x)$ defined by (8.60) is a smooth function on E_0 with $h(0+) = 0$.

Theorem 8.28 *For any $\mu \in M_h(E_0)$ and any $\alpha \geq 0$, we can define an entrance law $\kappa \in \mathcal{K}(P)$ by*

$$\kappa_t(f) = \mu(P_t f) + \alpha \partial_0 P_t f, \qquad t > 0, f \in B(E_0). \tag{8.66}$$

Moreover, every $\kappa \in \mathcal{K}(P)$ has the representation (8.66) with $\mu \in M_h(E_0)$ and $\alpha \geq 0$.

Proof. The proof is very similar to that of Theorem 8.26, so we only describe the difference. It suffices to prove each $\kappa \in \mathcal{K}(P)$ has the representation (8.66). Let $(T_t)_{t \geq 0}$ and its bounded entrance law $(\gamma_t)_{t > 0}$ be defined as in the proof of Theorem 8.26. By the smoothness of $p_t(x, y)$, we now extend $(T_t)_{t \geq 0}$ to a strongly continuous transition semigroup $(\bar{T}_t)_{t \geq 0}$ on $\bar{E}_0 := [0, \infty]$ by letting $\bar{T}_0 \bar{f} \equiv \bar{f}$ and

$$\bar{T}_t \bar{f}(x) = \begin{cases} h(x)^{-1} P_t(hf)(x) & \text{if } 0 < x < \infty, \\ (\partial_0 h)^{-1} \partial_0 P_t(hf) & \text{if } x = 0, \\ \bar{f}(\infty) & \text{if } x = \infty \end{cases} \tag{8.67}$$

for $t > 0$, where $\bar{f} \in C(\bar{E}_0)$ and $f = \bar{f}|_{E_0}$. Then $\gamma_t(f) = \gamma_0(\bar{T}_t \bar{f})$ for some $\gamma_0 \in M(\bar{E}_0)$. Since ∞ is a trap for $(\bar{T}_t)_{t \geq 0}$, we must have $\gamma_0(\{\infty\}) = 0$ and then (8.66) follows. $\square$

Let $(Q_t)_{t \geq 0}$ be the transition semigroup of the super-absorbing-barrier Brownian motion over E_0 with branching mechanism given by (2.26) or (2.27). By easy applications of Theorems 8.11 and 8.28 we get the following:

Theorem 8.29 *For any $\mu \in M_h(E_0)$ and $\alpha \geq 0$, we can define $K \in \mathcal{K}_m^1(Q)$ by*

$$\int_{M(E_0)} e^{-\nu(f)} K_t(d\nu) = \exp\{-\mu(V_t f) - \alpha \partial_0 V_t f\}. \tag{8.68}$$

Moreover, every $K \in \mathcal{K}_m^1(Q)$ has the representation (8.68) with $\mu \in M_h(E_0)$ and $\alpha \geq 0$.

Example 8.4 Let W_0^+ be the space of right continuous paths from $(0, \infty)$ to $M(E_0)$ having zero as a trap. Let $(\mathscr{A}^0, \mathscr{A}_t^0)$ be the natural σ-algebras on W_0^+ generated by the coordinate process. By Theorem 8.29, for each $u > 0$,

$$\int_{M(E_0)} e^{-\nu(f)} K_t^u(d\nu) = \exp\{-u\partial_0 V_s f\}, \quad t > 0, f \in B(E_0)^+, \quad (8.69)$$

defines a probability entrance law $(K_t^u)_{t>0}$ for the super-absorbing-barrier Brownian motion. Let $\mathbf{Q}^u$ be the corresponding Kuznetsov measure on W_0^+. Recall that the branching mechanism ϕ is given by (2.26). Suppose that

$$\sup_{x \in E_0} h(x)^{-1}\left\{\eta(x, h) + \int_{M(E_0)^\circ} \left[\nu(h) \wedge \nu(h)^2 + \nu_x(h)\right] H(x, d\nu)\right\} < \infty. \quad (8.70)$$

Then for any $\varepsilon > 0$ we have

$$w_t((0, \varepsilon]) \to \infty \text{ and } w_t([\varepsilon, \infty)) \to 0 \text{ as } t \to 0 \quad (8.71)$$

for $\mathbf{Q}^u$-a.e. $w \in W_0^+$. To see this let $(\bar{T}_t)_{t \geq 0}$ be defined as in the proof of Theorem 8.28 and use the same notation for its restriction to $\mathbb{R}_+$. Under condition (8.70) we can define a branching mechanism $\bar{\psi}$ on $\mathbb{R}_+$ by $\bar{\psi}(x, \bar{f}) = h(x)^{-1}\phi(x, hf)$ for $x > 0$ and $\bar{\psi}(0, \bar{f}) = 0$, where $\bar{f} \in B(\mathbb{R}_+)^+$ and $f = \bar{f}|_{E_0}$. Let $(\bar{U}_t)_{t \geq 0}$ be the cumulant semigroup defined from $(\bar{T}_t)_{t \geq 0}$ and $\bar{\psi}$. Then for $t > 0$ we have

$$\bar{U}_t\bar{f}(x) = \begin{cases} h(x)^{-1}V_t(hf)(x) & \text{if } x > 0, \\ (\partial_0 h)^{-1}\partial_0 V_t(hf) & \text{if } x = 0. \end{cases} \quad (8.72)$$

Theorem 5.12 implies that $(\bar{U}_t)_{t \geq 0}$ determines a Borel right semigroup on $M(\mathbb{R}_+)$. We then define the measure-valued process $\{\bar{X}_t : t > 0\}$ by $\bar{X}_t(\{0\}) = 0$ and $\bar{X}_t(dx) = h(x)w_t(dx)$ for $x > 0$. It is not hard to see that $\{\bar{X}_t : t > 0\}$ is a superprocess in $M(\mathbb{R}_+)$ with cumulant semigroup $(\bar{U}_t)_{t \geq 0}$. Using (8.67) one may check $\partial_0 V_s(hf) = \partial_0 h\bar{U}_t\bar{f}(0)$. From (8.69) we have

$$\mathbf{Q}^u \exp\{-\bar{X}_t(\bar{f})\} = \exp\{-u\partial_0 h\bar{U}_t\bar{f}(0)\}, \quad t > 0, \bar{f} \in B(\mathbb{R}_+)^+.$$

This implies $\mathbf{Q}^u\{\bar{X}_{0+} = u\partial_0 h\delta_0\} = 1$. Since $h(0+) = 0$ and $h(x) > 0$ for $x > 0$, we have (8.71) for $\mathbf{Q}^u$-a.e. $w \in W_0^+$.

8.6 Notes and Comments

The structures of entrance laws for Dawson–Watanabe superprocesses were investigated in Dynkin (1989b), Fitzsimmons (1988) and Li (1995/6, 1996, 1998b). Those for branching particle systems were studied in Li (1998a). Evans (1992) gave a characterization for the entrance laws of a conditioned superprocess.

Theorem 8.2 first appeared in Li (1995/6). A slightly different form of Theorem 8.6 was given in Dawson (1993, p.195). The proof of Theorem 8.11 follows Dynkin (1989b) and Li (1996, 1998b). Fitzsimmons (1988) proved Theorem 8.13 for local branching mechanisms. Theorem 8.20 was proved in Li (1996) for local branching mechanisms and in Li (1998b) for decomposable branching mechanisms.

The existence of excursion laws of Dawson–Watanabe superprocesses was first observed by El Karoui and Roelly (1991). The presentation of Theorem 8.22 given here is in the spirit of Li (2002). A special form of this result was proved in Li and Shiga (1995) using a theorem of Perkins (1992), which asserts that a conditioned Dawson–Watanabe superprocess is a generalized Fleming–Viot superprocess. The proof of Lemma 8.25 was taken from Li (1998a). This results can also be obtained from the general estimates of heat kernels given in Zhang (2002). The proofs of Theorems 8.26 and 8.28 follow those of Li and Shiga (1995). It was showed in van der Hofstad (2006) that the canonical measure of the super-Brownian motion is a natural candidate for the scaling limits of various random systems; see also Hara and Slade (2000) and van der Hofstad and Slade (2003).

The representation (8.59) of binary branching superprocesses was established in Le Gall (1991, 1999). This representation relies on the fact that the genealogical structure of Feller's branching diffusion can be coded by the Brownian excursions. A similar coding for the structures of CB-processes with general branching mechanisms was given in Le Gall and Le Jan (1998a, 1998b) using spectrally positive Lévy processes. Bertoin et al. (1997) provided a snake-like construction of superprocesses based on subordination. A direct construction of superprocesses with general branching mechanisms was given by Duquesne and Le Gall (2002) using the so-called Lévy snakes.

The snake representation has become a powerful tool in the study of the superprocesses and the associated nonlinear partial differential equations; see Duquesne and Le Gall (2002), Le Gall (1999, 2005) and the references therein. This technique also played an important role in the work of Le Gall (2007) on scaling limits of large planar maps; see also Le Gall and Paulin (2008). A Schilder type theorem on large deviations of the super-Brownian was established recently in Xiang (2010) using the snake representation; see also Fleischmann et al. (1996) and Schied (1996). The behavior of superprocesses near extinction was studied by Liu and Mueller (1989) and Tribe (1992); see also Le Gall (1999, p.72). The tool of Lévy snakes was used in Abraham and Delmas (2008) in the construction of fragmentation processes.

Chapter 9
Structures of Independent Immigration

In this chapter we study independent immigration structures associated with MB-processes. We first give a formulation of the structures using skew convolution semigroups. Those semigroups are in one-to-one correspondence with infinitely divisible probability entrance laws for the MB-processes. We shall see that an immigration superprocess has a Borel right realization if the corresponding skew convolution semigroup is determined by a closable infinitely divisible probability entrance law. The trajectories of the immigration processes are constructed using stochastic integrals with respect to Poisson random measures determined by entrance rules. We discuss briefly martingale problems for the immigration superprocesses, from which some stochastic equations of one-dimensional CBI-processes are derived.

9.1 Skew Convolution Semigroups

Suppose that E is a Lusin topological space and $(Q_t)_{t \geq 0}$ is the transition semigroup of an MB-process given by (2.3) and (2.5). Let $(N_t)_{t \geq 0}$ be a family of probability measures on $M(E)$. We call $(N_t)_{t \geq 0}$ a *skew convolution semigroup* (SC-semigroup) associated with $(Q_t)_{t \geq 0}$ provided

$$N_{r+t} = (N_r Q_t) * N_t, \qquad r, t \geq 0. \tag{9.1}$$

The above equation is of interest because of the following:

Theorem 9.1 *The relation* (9.1) *is satisfied if and only if*

$$Q_t^N(\mu, \cdot) := Q_t(\mu, \cdot) * N_t, \qquad t \geq 0, \ \mu \in M(E) \tag{9.2}$$

defines a Markov semigroup $(Q_t^N)_{t \geq 0}$ *on* $M(E)$.

Proof. It is easy to see that (9.2) really defines a probability kernel on $M(E)$. If $(N_t)_{t \geq 0}$ satisfies (9.1), for any $r, t \geq 0$ and $f \in B(E)^+$ we have

$$\int_{M(E)} e^{-\nu(f)} Q_{r+t}^N(\mu, d\nu)$$

$$= \int_{M(E)} e^{-\nu_1(f)} Q_{r+t}(\mu, d\nu_1) \int_{M(E)} e^{-\nu_2(f)} N_{r+t}(d\nu_2)$$

$$= \int_{M(E)} Q_r(\mu, d\gamma) \int_{M(E)} e^{-\nu_1(f)} Q_t(\gamma, d\nu_1)$$

$$\cdot \int_{M(E)} e^{-\nu_2(f)} (N_r Q_t)(d\nu_2) \int_{M(E)} e^{-\nu_3(f)} N_t(d\nu_3)$$

$$= \int_{M(E)} Q_r^N(\mu, d\gamma) \int_{M(E)} e^{-\nu_1(f)} Q_t(\gamma, d\nu_1) \int_{M(E)} e^{-\nu_3(f)} N_t(d\nu_3)$$

$$= \int_{M(E)} Q_r^N(\mu, d\gamma) \int_{M(E)} e^{-\nu(f)} Q_t^N(\gamma, d\nu),$$

where we used Theorem 2.1 for the third equality. Then $(Q_t^N)_{t\geq 0}$ satisfies the Chapman–Kolmogorov equation. For the converse, suppose the kernels $(Q_t^N)_{t\geq 0}$ constitute a transition semigroup. Since $Q_t(0, \cdot) = \delta_0$, we have $Q_t^N(0, \cdot) = N_t$. Consequently, for any $r, t \geq 0$ and $f \in B(E)^+$,

$$\int_{M(E)} e^{-\nu(f)} N_{r+t}(d\nu)$$

$$= \int_{M(E)} N_r(d\mu) \int_{M(E)} e^{-\nu(f)} Q_t^N(\mu, d\nu)$$

$$= \int_{M(E)} N_r(d\mu) \int_{M(E)} e^{-\nu_1(f)} Q_t(\mu, d\nu_1) \int_{M(E)} e^{-\nu_2(f)} N_t(d\nu_2)$$

$$= \int_{M(E)} e^{-\nu_1(f)} N_r Q_t(d\nu_1) \int_{M(E)} e^{-\nu_2(f)} N_t(d\nu_2).$$

Thus $(N_t)_{t\geq 0}$ satisfies (9.1). $\qquad\qquad\qquad\qquad\qquad\qquad\qquad\qquad\qquad\qquad \Box$

Suppose that T is an interval on the real line and $(\mathscr{G}_t)_{t\in T}$ is a filtration. If $\{(Y_t, \mathscr{G}_t) : t \in T\}$ is a Markov process in $M(E)$ with transition semigroup $(Q_t^N)_{t\geq 0}$ given by (9.2), we call it an *immigration process* associated with $(Q_t)_{t\geq 0}$ or the corresponding MB-process. In the special case where $(Q_t)_{t\geq 0}$ is the transition semigroup of a Dawson–Watanabe superprocess, we also call Y an *immigration superprocess*. The intuitive meaning of the model is clear in view of (9.1) and (9.2). From (9.2) we see that the population at any time $t \geq 0$ is made up of two parts; the native part generated by the mass $\mu \in M(E)$ has distribution $Q_t(\mu, \cdot)$ and the immigration in the time interval $(0, t]$ gives the distribution N_t. In a similar way, the equation (9.1) decomposes the population immigrating to E during the time interval $(0, r + t]$ into two parts; the immigration in the interval $(r, r + t]$ gives the distribution N_t while the immigration in the interval $(0, r]$ generates the distribution N_r at time r and gives the distribution $N_r Q_t$ at time $r + t$. It is not hard to understand that (9.2) gives a general formulation of the immigration independent of the state of the population.

Theorem 9.2 *Suppose that* $\{(X_t, \mathscr{F}_t) : t \in T\}$ *and* $\{(Y_t, \mathscr{G}_t) : t \in T\}$ *are two independent immigration processes associated with* $(Q_t)_{t \geq 0}$ *corresponding to the SC-semigroups* $(M_t)_{t \geq 0}$ *and* $(N_t)_{t \geq 0}$, *respectively. Let* $Z_t = X_t + Y_t$ *and* $\mathscr{H}_t = \sigma(\mathscr{F}_t \cup \mathscr{G}_t)$. *Then* $\{(Z_t, \mathscr{H}_t) : t \in T\}$ *is an immigration process corresponding to the SC-semigroup* $(L_t)_{t \geq 0}$ *defined by* $L_t = M_t * N_t$.

Proof. By Theorem 2.1 it is simple to show that $(L_t)_{t \geq 0}$ is really an SC-semigroup associated with $(Q_t)_{t \geq 0}$. Let $(Q_t^L)_{t \geq 0}$ be defined by (2.2) with $(N_t)_{t \geq 0}$ replaced by $(L_t)_{t \geq 0}$. By a modification of the proof of Theorem 2.3 one shows $\{(Z_t, \mathscr{H}_t) : t \in T\}$ is a Markov process with transition semigroup $(Q_t^L)_{t \geq 0}$. $\qquad\square$

Corollary 9.3 *Let* $\{(X_t, \mathscr{F}_t) : t \in T\}$ *be an MB-process with transition semigroup* $(Q_t)_{t \geq 0}$ *and let* $\{(Y_t, \mathscr{G}_t) : t \in T\}$ *be an immigration process with transition semigroup* $(Q_t^N)_{t \geq 0}$. *Suppose the two processes are independent of each other. Let* $Z_t = X_t + Y_t$ *and* $\mathscr{H}_t = \sigma(\mathscr{F}_t \cup \mathscr{G}_t)$. *Then* $\{(Z_t, \mathscr{H}_t) : t \in T\}$ *is an immigration process with transition semigroup* $(Q_t^N)_{t \geq 0}$.

Theorem 9.4 *Suppose that* $t \mapsto V_t f(x)$ *is locally bounded and right continuous pointwise for every* $f \in C(E)^+$. *If* $(K_s)_{s > 0}$ *is an infinitely divisible probability entrance law for* $(Q_t)_{t \geq 0}$ *satisfying*

$$- \int_0^t \log L_{K_s}(1) \mathrm{d}s < \infty, \qquad t \geq 0, \tag{9.3}$$

then an SC-semigroup $(N_t)_{t \geq 0}$ *associated with* $(Q_t)_{t \geq 0}$ *is defined by*

$$\log L_{N_t}(f) = \int_0^t \big[\log L_{K_s}(f) \big] \mathrm{d}s, \qquad t \geq 0, f \in B(E)^+. \tag{9.4}$$

Conversely, for every SC-semigroup $(N_t)_{t \geq 0}$ *associated with* $(Q_t)_{t \geq 0}$ *there is an infinitely divisible probability entrance law* $(K_s)_{s > 0}$ *for* $(Q_t)_{t \geq 0}$ *satisfying (9.3) so that the Laplace functionals of* $(N_t)_{t \geq 0}$ *are given by (9.4).*

Proof. If $(K_s)_{s > 0}$ is an infinitely divisible probability entrance law for $(Q_t)_{t \geq 0}$ satisfying (9.3), then (9.4) clearly defines a family of infinitely divisible probability measures $(N_t)_{t \geq 0}$ on $M(E)$. By the entrance law property of $(K_s)_{s > 0}$ it is easy to show that (9.1) holds, so $(N_t)_{t \geq 0}$ is an SC-semigroup associated with $(Q_t)_{t \geq 0}$. To prove the converse, let $(N_t)_{t \geq 0}$ be an SC-semigroup associated with $(Q_t)_{t \geq 0}$ and let

$$J_t(f) = - \log \int_{M(E)} \mathrm{e}^{-\nu(f)} N_t(\mathrm{d}\nu), \qquad t \geq 0, f \in B(E)^+. \tag{9.5}$$

From (9.1) we have

$$J_{r+t}(f) = J_t(f) + J_r(V_t f), \qquad r, t \geq 0, f \in B(E)^+. \tag{9.6}$$

Then $t \mapsto J_t(f)$ is increasing for every $f \in B(E)^+$. Observe also $J_t(f_1) \leq J_t(f_2)$ for $f_1 \leq f_2 \in B(E)^+$. For $f \in C(E)^+$ we have $V_t f(x) \to f(x)$ pointwise as

$t \to 0$. Then letting $t \to 0$ and $r \to 0$ in (9.6) and using (9.5) and dominated convergence we get

$$\lim_{t\to 0} J_t(f) = \lim_{t\to 0} J_t(f) + \lim_{r\to 0} J_r(f).$$

Thus $\lim_{t\to 0} J_t(f) = 0$ first for $f \in C(E)^+$ and then for all $f \in B(E)^+$. Let $q = q(l, f) > 0$ be such that $\|V_t f\| \le q$ for all $0 \le t \le l$. For $0 \le c_1 < d_1 < \cdots < c_n < d_n < \cdots \le l$, set $\sigma_n = \sum_{i=1}^n (d_i - c_i)$. We claim that

$$\sum_{i=1}^n [J_{d_i}(f) - J_{c_i}(f)] \le J_{\sigma_n}(q).$$

For $n = 1$ this follows from (9.6). If the above inequality is true when n is replaced by $n - 1$, using (9.6) we have

$$\sum_{i=1}^n [J_{d_i}(f) - J_{c_i}(f)] = \sum_{i=1}^{n-1} [J_{d_i}(f) - J_{c_i}(f)] + J_{d_n-c_n}(V_{c_n} f)$$
$$\le J_{\sigma_{n-1}}(q) + J_{d_n-c_n}(V_{\sigma_{n-1}} q) = J_{\sigma_n}(q).$$

Thus $t \mapsto J_t(f)$ is absolutely continuous. That is, we have $J_t(f) = \int_0^t I_s(f)ds$ for a positive Borel function $s \mapsto I_s(f)$ on $[0, \infty)$. Let $S_2(E, r)$ be defined as in Section 1.2. Then there is a set $A \subset [0, \infty)$ with full Lebesgue measure so that for all $s \in A$ and $f \in S_2(E, r)$,

$$I_s(f) = \lim_{r\to 0} \frac{1}{r} \Big[J_{s+r}(f) - J_s(f) \Big] = \lim_{r\to 0} \frac{1}{r} J_r(V_s f)$$
$$= \lim_{r\to 0} \frac{1}{r} \Big(1 - \exp\{-J_r(V_s f)\} \Big)$$
$$= \lim_{r\to 0} \frac{1}{r} \int_{M(E)} N_r(d\mu) \int_{M(E)^\circ} (1 - e^{-\nu(f)}) Q_s(\mu, d\nu).$$

In view of (9.5) we have $\lim_{n\to\infty} J_t(1/n) = 0$ decreasingly. Thus, by taking a smaller set $A \subset [0, \infty)$ with full Lebesgue measure, we may assume $\lim_{n\to\infty} I_s(1/n) = 0$ for all $s \in A$. By Proposition 1.31 there is $U_s \in \mathscr{I}(E)$ such that $I_s(f) = U_s(f)$ for all $s \in A$ and $f \in S_2(E, r)$. Consequently,

$$J_t(f) = \int_0^t U_s(f)ds, \qquad t \ge 0,$$

first for $f \in S_2(E, r)$ and then for all $f \in B(E)^+$. Now the equation (9.6) yields

$$\int_0^r U_{s+t}(f)ds = \int_0^r U_s(V_t f)ds, \qquad r, t \ge 0, f \in B(E)^+.$$

Then for any $t \ge 0$ and $f \in B(E)^+$ there is a subset $A_t(f)$ of $[0, \infty)$ with full Lebesgue measures so that $U_{s+t}(f) = U_s(V_t f)$ for all $s \in A_t(f)$. By Fubini's theorem, there are subsets $B(f)$ and $B_s(f)$ of $[0, \infty)$ with full Lebesgue measures such that

$$U_{s+t}(f) = U_s(V_t f), \qquad s \in B(f), t \in B_s(f).$$

Since U_{s+t} and $U_s \circ V_t$ are determined by their restrictions to the countable set $S_2(E, r)$, for $B \subset [0, \infty)$ and $B_s \subset [s, \infty)$ with full Lebesgue measures we have

$$U_t = U_s \circ V_{t-s}, \qquad s \in B, t \in B_s.$$

Choose a sequence $\{s_n\} \subset B$ with $s_n \to 0$. For any $t > s_n > s_{n+1} > 0$, we may take $s \in B_{s_n} \cap B_{s_{n+1}} \cap [0, t]$ to see

$$\begin{aligned}
U_{s_n} \circ V_{t-s_n} &= U_{s_n} \circ V_{s-s_n} \circ V_{t-s} = U_s \circ V_{t-s} \\
&= U_{s_{n+1}} \circ V_{s-s_{n+1}} \circ V_{t-s} = U_{s_{n+1}} \circ V_{t-s_{n+1}}.
\end{aligned}$$

Then $W_t := U_{s_n} \circ V_{t-s_n} \in \mathscr{I}(E)$ is independent of $n \geq 1$ and $W_t = U_t$ for almost all $t > 0$. Thus we have

$$J_t(f) = \int_0^t W_s(f) \mathrm{d}s, \qquad t \geq 0, f \in B(E)^+.$$

Moreover, it is easy to see that $W_{r+t} = W_r \circ V_t$ for all $r, t > 0$. Let K_s be the infinitely divisible probability measure on $M(E)$ such that $- \log L_{K_s} = W_s$. Then $(K_s)_{s>0}$ is an entrance law for $(Q_t)_{t \geq 0}$ and (9.4) holds. $\qquad \square$

It is clear that (9.4) establishes a one-to-one correspondence between the SC-semigroup $(N_t)_{t \geq 0}$ and the infinitely divisible probability entrance law $(K_s)_{s>0}$ satisfying (9.3). If $(K_s)_{s>0}$ can be extended to a closed entrance law $(K_s)_{s \geq 0}$, we say the SC-semigroup and the corresponding immigration process are *regular*. For the regular SC-semigroup we have

$$\int_{M(E)} \mathrm{e}^{-\nu(f)} N_t(\mathrm{d}\nu) = \exp\left\{ -\int_0^t I(V_s f) \mathrm{d}s \right\}, \qquad f \in B(E)^+, \qquad (9.7)$$

where $I = - \log L_{K_0}$. The functional $I \in \mathscr{I}(E)$ is called the *immigration mechanism* of the regular immigration process. From (9.4) we obtain

$$\int_{M(E)} \nu(1) N_t(\mathrm{d}\nu) = \int_0^t \mathrm{d}s \int_{M(E)^\circ} \nu(1) K_s(\mathrm{d}\nu), \qquad t \geq 0. \qquad (9.8)$$

Then a combination of Theorem 9.4 and the results of Section 8.3 gives characterizations of SC-semigroups associated with the (ξ, ϕ)-superprocess under the first-moment assumption.

9.2 The Canonical Entrance Rule

Suppose that E is a Lusin topological space and $(Q_t)_{t\geq 0}$ is the transition semigroup of an MB-process given by (2.3) and (2.5). Suppose that $t \mapsto V_t f(x)$ is locally bounded and right continuous pointwise for every $f \in C(E)^+$. Recall that $(Q_t^\circ)_{t\geq 0}$ denotes the restriction of $(Q_t)_{t\geq 0}$ to $M(E)^\circ$.

Theorem 9.5 *If $(N_t)_{t\geq 0}$ is an SC-semigroup associated with $(Q_t)_{t\geq 0}$, each proba-bility measure N_t is infinitely divisible. Moreover, we have the canonical represen-tation $N_t = I(\gamma_t, G_t)$, where $t \mapsto \gamma_t$ is an increasing path from $[0, \infty)$ to $M(E)$ and $(G_t)_{t\geq 0}$ is an entrance rule for $(Q_t^\circ)_{t\geq 0}$.*

Proof. By Theorem 9.4 there is an infinitely divisible probability entrance law $(K_s)_{s>0}$ such that (9.4) holds. By Theorem 8.2 we may assume $(K_s)_{s>0}$ is given by (8.2) and (8.3). Then we have $N_t = I(\gamma_t, G_t)$ with

$$\gamma_t = \int_0^t \eta_s ds \text{ and } G_t = \int_0^t H_s ds. \tag{9.9}$$

In particular, the map $t \mapsto \gamma_t(dx)$ is increasing. For $t > r \geq 0$ the relation (9.1) implies

$$\gamma_t = \gamma_{t-r} + \int_E \gamma_r(dx)\lambda_{t-r}(x, \cdot) \tag{9.10}$$

and

$$G_t = G_{t-r} + G_r Q_{t-r}^\circ + \int_E \gamma_r(dx)L_{t-r}(x, \cdot). \tag{9.11}$$

Thus $G_r Q_{t-r}^\circ \leq G_t$. By the second equality in (9.9) we have

$$G_r Q_{t-r}^\circ = \int_0^r H_s Q_{t-r}^\circ ds = G_t Q_{t-r}^\circ - \int_r^t H_s Q_{t-r}^\circ ds.$$

It follows that $G_r Q_{t-r}^\circ \to G_t$ as $r \to t-$. Then $(G_t)_{t\geq 0}$ is an entrance rule for $(Q_t^\circ)_{t\geq 0}$. $\qquad\square$

In the situation of the above theorem, we call $(G_t)_{t\geq 0}$ the *canonical entrance rule* of the SC-semigroup $(N_t)_{t\geq 0}$. The following theorem gives a general representation of the canonical entrance rule.

Theorem 9.6 *The canonical entrance rule $(G_t)_{t\geq 0}$ of an SC-semigroup associated with $(Q_t)_{t\geq 0}$ has the representation*

$$G_t = \int_0^t G_{t-s}^s \zeta(ds), \qquad t \geq 0, \tag{9.12}$$

where $\zeta(\mathrm{d}s)$ is a diffuse Radon measure on $[0,\infty)$ and $\{(G_t^s)_{t>0} : s \geq 0\}$ is a family of entrance laws for $(Q_t^\circ)_{t\geq 0}$.

Proof. By Theorem A.39, there is a Radon measure $\zeta(\mathrm{d}s)$ on $[0,\infty)$ and a family of entrance laws $\{(G_t^s)_{t>0} : s \geq 0\}$ for $(Q_t^\circ)_{t\geq 0}$ such that

$$G_t = \int_{[0,t)} G_{t-s}^s \, \zeta(\mathrm{d}s), \qquad t \geq 0. \tag{9.13}$$

For $f \in C(E)^+$ we can use the second expression in (9.9) to see

$$t \mapsto \int_{M(E)^\circ} (1 - \mathrm{e}^{-\nu(f)}) G_t(\mathrm{d}\nu) = \int_0^t \mathrm{d}s \int_{M(E)^\circ} (1 - \mathrm{e}^{-\nu(f)}) H_s(\mathrm{d}\nu)$$

is continuous on $[0,\infty)$. Then for any $r \geq 0$ we have

$$\begin{aligned}
0 &= \lim_{t\to 0+} \int_{M(E)^\circ} (1 - \mathrm{e}^{-\nu(f)}) G_{r+t}(\mathrm{d}\nu) - \int_{M(E)^\circ} (1 - \mathrm{e}^{-\nu(f)}) G_r(\mathrm{d}\nu) \\
&= \lim_{t\to 0+} \int_{[0,r+t)} \zeta(\mathrm{d}s) \int_{M(E)^\circ} (1 - \mathrm{e}^{-\nu(f)}) G_{r+t-s}^s(\mathrm{d}\nu) \\
&\qquad - \int_{[0,r)} \zeta(\mathrm{d}s) \int_{M(E)^\circ} (1 - \mathrm{e}^{-\nu(f)}) G_{r-s}^s(\mathrm{d}\nu) \\
&= \lim_{t\to 0+} \int_{[r,r+t)} \zeta(\mathrm{d}s) \int_{M(E)^\circ} (1 - \mathrm{e}^{-\nu(f)}) G_{r+t-s}^s(\mathrm{d}\nu) \\
&\qquad + \lim_{t\to 0+} \int_{[0,r)} \zeta(\mathrm{d}s) \int_{M(E)^\circ} (1 - \mathrm{e}^{-\nu(V_t f)}) G_{r-s}^s(\mathrm{d}\mu) \\
&\qquad - \int_{[0,r)} \zeta(\mathrm{d}s) \int_{M(E)^\circ} (1 - \mathrm{e}^{-\nu(f)}) G_{r-s}^s(\mathrm{d}\nu) \\
&= \lim_{t\to 0+} \int_{[r,r+t)} \zeta(\mathrm{d}s) \int_{M(E)^\circ} (1 - \mathrm{e}^{-\nu(f)}) G_{r+t-s}^s(\mathrm{d}\nu) \\
&\geq \limsup_{t\to 0+} \zeta(\{r\}) \int_{M(E)^\circ} (1 - \mathrm{e}^{-\nu(f)}) G_t^r(\mathrm{d}\nu).
\end{aligned}$$

Now suppose that $\zeta(\{r\}) > 0$ for some $r \geq 0$. From the above it follows that

$$\limsup_{t\to 0+} \int_{M(E)^\circ} (1 - \mathrm{e}^{-\nu(f)}) G_t^r(\mathrm{d}\nu) = 0. \tag{9.14}$$

For $t \geq 0$ let $q_t = \sup_{0\leq s\leq t} \|V_s f\|$. Then for $0 < s < t$ we have

$$\begin{aligned}
\int_{M(E)^\circ} (1 - \mathrm{e}^{-\nu(f)}) G_t^r(\mathrm{d}\nu) &= \int_{M(E)^\circ} (1 - \mathrm{e}^{-\nu(V_{t-s}f)}) G_s^r(\mathrm{d}\nu) \\
&\leq \int_{M(E)^\circ} (1 - \mathrm{e}^{-\nu(q_t)}) G_s^r(\mathrm{d}\nu).
\end{aligned}$$

By (9.14) the right-hand side tends to zero as $s \to 0$, so $(G_t^r)_{t>0}$ is trivial. Therefore we can remove all the atoms of $\zeta(ds)$ and obtain (9.12) from (9.13). $\qquad \square$

Corollary 9.7 *For every SC-semigroup* $(N_t)_{t\geq 0}$ *associated with* $(Q_t)_{t\geq 0}$, *we have the decomposition*

$$- \log L_{N_t}(f) = \gamma_t(f) + \int_0^t \zeta(ds) \int_{M(E)^\circ} (1 - e^{-\nu(f)}) G_{t-s}^s(d\nu),$$

where $t \mapsto \gamma_t$ *is an increasing path from* $[0, \infty)$ *to* $M(E)$, $\zeta(ds)$ *is a diffuse Radon measure on* $[0, \infty)$ *and* $\{(G_t^s)_{t>0} : s \geq 0\}$ *is a family of entrance laws for* $(Q_t^\circ)_{t\geq 0}$.

Given two probability measures F_1 and F_2 on $M(E)$, we write $F_1 \preceq F_2$ if $F_1 * G = F_2$ for another probability measure G on $M(E)$. Clearly, the measure G is unique if it exists. Let $\mathscr{E}^*(Q)$ denote the set of probabilities F on $M(E)$ satisfying $FQ_t \preceq F$ for all $t \geq 0$.

Theorem 9.8 *For each* $F \in \mathscr{E}^*(Q)$ *there is a unique SC-semigroup* $(N_t)_{t\geq 0}$ *associated with* $(Q_t)_{t\geq 0}$ *such that* $FQ_t * N_t = F$ *for all* $t \geq 0$.

Proof. Since $F \in \mathscr{E}^*(Q)$, for each $t \geq 0$ there is a unique probability measure N_t on $M(E)$ satisfying $F = (FQ_t) * N_t$. By Theorem 2.1, for $r, t \geq 0$ we have

$$(FQ_{r+t}) * N_{r+t} = F = (FQ_t) * N_t = \{[(FQ_r) * N_r]Q_t\} * N_t$$
$$= (FQ_{r+t}) * (N_r Q_t) * N_t.$$

Then (9.1) holds, that is, $(N_t)_{t\geq 0}$ is an SC-semigroup associated with $(Q_t)_{t\geq 0}$. $\quad \square$

The infinitely divisible probabilities in $\mathscr{E}^*(Q)$ are closely related with excessive measures for $(Q_t^\circ)_{t\geq 0}$. Let $\mathscr{E}(Q^\circ)$ denote the class of all excessive measures H for $(Q_t^\circ)_{t\geq 0}$ satisfying

$$\int_{M(E)^\circ} [1 \wedge \nu(1)] H(d\nu) < \infty. \tag{9.15}$$

Proposition 9.9 *Let* $F = I(\eta, H)$ *be an infinitely divisible probability measure on* $M(E)$. *Then* $F \in \mathscr{E}^*(Q)$ *if and only if* (η, H) *satisfy*

$$\int_E \eta(dx)\lambda_t(x, \cdot) \leq \eta \quad and \quad \int_E \eta(dx)L_t(x, \cdot) + HQ_t^\circ \leq H. \tag{9.16}$$

In particular, if $H \in \mathscr{E}(Q^\circ)$, *then* $F = I(0, H) \in \mathscr{E}^*(Q)$.

Proof. By Proposition 2.6 we have $FQ_t = I(\eta_t, H_t)$, where

$$\eta_t = \int_E \eta(dx)\lambda_t(x, \cdot) \quad and \quad H_t = \int_E \eta(dx)L_t(x, \cdot) + HQ_t^\circ.$$

Then $FQ_t \preceq F$ holds if and only if (9.16) is satisfied. The second assertion is immediate. $\qquad \square$

We write $F \in \mathscr{E}_i^*(Q)$ if $F \in \mathscr{E}^*(Q)$ is a stationary distribution for $(Q_t)_{t \geq 0}$, and write $F \in \mathscr{E}_p^*(Q)$ if $F \in \mathscr{E}^*(Q)$ and $\lim_{t \to \infty} FQ_t = \delta_0$. Clearly, we have $\delta_0 \in \mathscr{E}_i^*(Q)$, but there can be other non-trivial stationary distributions.

Theorem 9.10 *Let $F \in \mathscr{E}^*(Q)$ and let $(N_t)_{t \geq 0}$ be the SC-semigroup defined in Theorem 9.8. Then $F = F_i * F_p$, where $F_i = \lim_{t \to \infty} FQ_t \in \mathscr{E}_i^*(Q)$ and $F_p = \lim_{t \to \infty} N_t \in \mathscr{E}_p^*(Q)$.*

Proof. By the definition of $\mathscr{E}^*(Q)$ we have $FQ_{r+t} \preceq FQ_t$ for $r, t \geq 0$. Thus for every $f \in B(E)^+$ the limits

$$L_{F_i}(f) = \uparrow \lim_{t \to \infty} L_{FQ_t}(f) \quad \text{and} \quad L_{F_p}(f) = \downarrow \lim_{t \to \infty} L_{N_t}(f)$$

exist and they are the Laplace functionals of two probability measures F_i and F_p on $M(E)$. Clearly, $F_i \in \mathscr{E}_i^*(Q)$ and $F = F_i * F_p$. On the other hand,

$$F_i * F_p = F = (FQ_t) * N_t = F_i * (F_pQ_t) * N_t,$$

so $F_p = (F_pQ_t) * N_t$. Therefore $F_p \in \mathscr{E}^*(Q)$ and $\lim_{t \to \infty} F_pQ_t = \delta_0$. $\qquad\square$

It is simple to see that the measure $F_p \in \mathscr{E}_p^*(Q)$ in Theorem 9.10 is a stationary distribution of the transition semigroup $(Q_t^N)_{t \geq 0}$ defined from $(Q_t)_{t \geq 0}$ and $(N_t)_{t \geq 0}$. Then the above theorem shows that any $F \in \mathscr{E}^*(Q)$ can be decomposed as the convolution of a stationary distribution of $(Q_t)_{t \geq 0}$ with a stationary distribution of an immigration process associated with $(Q_t)_{t \geq 0}$.

9.3 Regular Immigration Superprocesses

In this section we state some basic properties of regular immigration superprocesses. Their proofs are not provided since they are similar to those for the superprocesses without immigration given before. Suppose that ξ is a Borel right process in a Lusin topological space E with transition semigroup $(P_t)_{t \geq 0}$. Let ϕ be a branching mechanism given by (2.26) or (2.27) and let $(V_t)_{t \geq 0}$ be the cumulant semigroup of the (ξ, ϕ)-superprocess defined by (2.33). Suppose that $I \in \mathscr{I}(E)$ is an immigration mechanism given by

$$I(f) = \eta(f) + \int_{M(E)^\circ} \left(1 - e^{-\nu(f)}\right) H(d\nu), \quad f \in B(E)^+, \qquad (9.17)$$

where $\eta \in M(E)$ and $\nu(1)H(d\nu)$ is a finite measure on $M(E)^\circ$. Let $(Q_t^N)_{t \geq 0}$ be the transition semigroup of the regular immigration superprocess defined by

$$\int_{M(E)} e^{-\nu(f)} Q_t^N(\mu, d\nu) = \exp\left\{ -\mu(V_t f) - \int_0^t I(V_s f) ds \right\}. \qquad (9.18)$$

This corresponds to the regular SC-semigroup given by (9.7). Recall that $c_0 = \|b^-\| + \|\gamma(\cdot, 1)\|$.

Proposition 9.11 *Let $(Q_t^N)_{t\geq 0}$ be defined by (9.18). Then for $t \geq 0$, $\mu \in M(E)$ and $f \in B(E)$ we have*

$$\int_{M(E)} \nu(f)Q_t^N(\mu, d\nu) = \mu(\pi_t f) + \int_0^t \Gamma(\pi_s f)ds, \qquad (9.19)$$

where $t \mapsto \pi_t f$ is defined by (2.35) and

$$\Gamma(f) = \eta(f) + \int_{M(E)^\circ} \nu(f)H(d\nu). \qquad (9.20)$$

Corollary 9.12 *Suppose that $(Y_t, \mathscr{G}_t, \mathbf{P})$ is a realization of the immigration superprocess with transition semigroup $(Q_t^N)_{t\geq 0}$ such that $\mathbf{P}[Y_0(1)] < \infty$. Let $\alpha \geq 0$ and let $f \in B(E)^+$ be an α-super-mean-valued function for $(P_t)_{t\geq 0}$ satisfying $\varepsilon := \inf_{x\in E} f(x) > 0$. Then for any $\beta \geq \alpha + c_0\varepsilon^{-1}\|f\|$ the process*

$$Z_t(f) := e^{-2\beta t}Y_t(f) + (2\beta)^{-1}e^{-2\beta t}\Gamma(f), \qquad t \geq 0$$

is a $(\mathscr{G}_t)$-supermartingale.

Let $\mathbf{F}$ be the set of functions $f \in B(E)$ that are finely continuous relative to ξ. Fix $\beta > 0$ and let $(A, \mathscr{D}(A))$ be the weak generator of $(P_t)_{t\geq 0}$ defined by $\mathscr{D}(A) = U^\beta\mathbf{F}$ and $Af = \beta f - g$ for $f = U^\beta g \in \mathscr{D}(A)$.

Theorem 9.13 *Suppose that $(Y_t, \mathscr{G}_t, \mathbf{P})$ is a progressive realization of the immigration superprocess with transition semigroup $(Q_t^N)_{t\geq 0}$ such that $\mathbf{P}[Y_0(1)] < \infty$. Then for any $f \in \mathscr{D}(A)$, the process*

$$M_t(f) := Y_t(f) - Y_0(f) - \int_0^t \big[Y_s(Af + \gamma f - bf) + \Gamma(f)\big]ds$$

is a $(\mathscr{G}_t)$-martingale.

Proposition 9.14 *Suppose that the kernel $H(x, d\nu)$ in (2.26) and the measure $H(d\nu)$ in (9.17) satisfy*

$$\sup_{x\in E} \int_{M(E)^\circ} \nu(f)^2 H(x, d\nu) + \int_{M(E)^\circ} \nu(f)^2 H(d\nu) < \infty. \qquad (9.21)$$

Then for $t \geq 0$, $\mu \in M(E)$ and $f \in B(E)$ we have

$$\int_{M(E)} \nu(f)^2 Q_t^N(\mu, d\nu) = \left(\mu(\pi_t f) + \int_0^t \Gamma(\pi_s f)ds\right)^2$$
$$+ \int_0^t ds \int_E q(x, \pi_s f)\mu\pi_{t-s}(dx)$$

$$+ \int_0^t du \int_0^u ds \int_E q(x, \pi_s f) \Gamma \pi_{u-s}(dx)$$

$$+ \int_0^t ds \int_{M(E)^\circ} \nu(\pi_s f)^2 H(d\nu), \qquad (9.22)$$

where $q(x, f)$ is defined by (2.59).

Theorem 9.15 *The immigration superprocess with transition semigroup $(Q_t^N)_{t\geq 0}$ has a right realization in $M(E)$. If ξ is a Hunt process, then the immigration super-process has a Hunt realization in $M(E)$.*

Let $\xi = (\Omega, \mathscr{F}, \mathscr{F}_{r,t}, \xi_t, \mathbf{P}_{r,x})$ and $Y = (W, \mathscr{G}, \mathscr{G}_{r,t}, Y_t, \mathbf{Q}_{r,\mu}^N)$ be respectively right realizations of the underlying spatial motion and the immigration superprocess from an arbitrary initial time $r \geq 0$.

Theorem 9.16 *Suppose that $t \geq 0$ and $\lambda \in M([0,t])$. Let $(s, x) \mapsto f_s(x)$ be a bounded positive Borel function on $[0,t] \times E$. Then we have*

$$\mathbf{Q}_{r,\mu}^N \exp \left\{ - \int_{[r,t]} Y_s(f_s) \lambda(ds) \right\} = \exp \left\{ - \mu(u_r) - \int_r^t I(u_s) ds \right\}$$

for every $0 \leq r \leq t$, where $(r, x) \mapsto u_r(x)$ is the unique bounded positive solution on $[0,t] \times E$ of (5.26).

We can also extend the immigration superprocesses to the state space of tempered measures. Let $\alpha \geq 0$ and let $h \in p\mathscr{B}(E)$ be a strictly positive α-excessive function for ξ. Recall that $M_h(E)$ is the space of measures μ on E satisfying $\mu(h) < \infty$ and $B_h(E)$ is the set of Borel functions f on E satisfying $|f| \leq \text{const} \cdot h$.

Theorem 9.17 *Let $(V_t)_{t\geq 0}$ be the cumulant semigroup on $B_h(E)^+$ defined in Theorem 6.3. Suppose that $\eta \in M_h(E)$ and $\nu(h)H(d\nu)$ is a finite measure on $M_h(E)^\circ := M_h(E) \setminus \{0\}$ and write*

$$I(f) = \eta(f) + \int_{M_h(E)^\circ} \left(1 - e^{-\nu(f)}\right) H(d\nu), \quad f \in B_h(E)^+. \qquad (9.23)$$

Then a Borel right transition semigroup $(Q_t^N)_{t\geq 0}$ on $M_h(E)$ is defined by

$$\int_{M_h(E)} e^{-\nu(f)} Q_t^N(\mu, d\nu) = \exp \left\{ - \mu(V_t f) - \int_0^t I(V_s f) ds \right\}. \qquad (9.24)$$

If, in addition, the semigroup $(\tilde{P}_t)_{t\geq 0}$ given by (6.10) has a Hunt realization, then $(Q_t^N)_{t\geq 0}$ has a Hunt realization.

We now discuss briefly martingale problems for the immigration superprocesses, which generalize those of superprocesses without immigration. In the remainder of this section, we assume E is a locally compact separable metric space and ξ is a Hunt process in E with transition semigroup $(P_t)_{t\geq 0}$. Suppose that $(P_t)_{t\geq 0}$ preserves $C_0(E)$ and $t \mapsto P_t f$ is continuous in the supremum norm for every $f \in C_0(E)$. Let

A denote the strong generator of $(P_t)_{t\geq 0}$ with domain $D_0(A) \subset C_0(E)$. Let ϕ be a branching mechanism given by (2.26) or (2.27) satisfying Conditions 7.1 and 7.2. By Theorem 9.15 the immigration superprocess with transition semigroup $(Q_t^N)_{t\geq 0}$ given by (9.18) has a càdlàg realization. Let L_0 be the generator defined by (7.12). For $F \in \mathcal{D}_0$ define

$$LF(\mu) = L_0F(\mu) + \int_E F'(\mu; x)\eta(\mathrm{d}x)$$
$$+ \int_{M(E)^\circ} [F(\mu + \nu) - F(\mu)]H(\mathrm{d}\nu). \qquad (9.25)$$

Suppose that $(W, \mathcal{G}, \mathcal{G}_t, \mathbf{P})$ is a filtered probability space satisfying the usual hypotheses and $\{Y_t : t \geq 0\}$ is a càdlàg process in $M(E)$ that is adapted to $(\mathcal{G}_t)_{t\geq 0}$ and satisfies $\mathbf{P}[Y_0(1)] < \infty$. For this process we consider the following properties:

(1) For every $T \geq 0$ and $f \in C_0(E)^+$,

$$\exp\left\{ -Y_t(V_{T-t}f) - \int_0^{T-t} I(V_sf)\mathrm{d}s \right\}, \qquad 0 \leq t \leq T$$

is a martingale.

(2) For every $f \in D_0(A)^+$,

$$H_t(f) := \exp\left\{ -Y_t(f) + \int_0^t [Y_s(Af - \phi(f)) + I(f)]\mathrm{d}s \right\}, \qquad t \geq 0$$

is a local martingale.

(3) (a) The process $\{Y_t : t \geq 0\}$ has no negative jumps. Let $N(\mathrm{d}s, \mathrm{d}\nu)$ be the optional random measure on $[0, \infty) \times M(E)^\circ$ defined by

$$N(\mathrm{d}s, \mathrm{d}\nu) = \sum_{s>0} 1_{\{\Delta Y_s \neq 0\}} \delta_{(s, \Delta Y_s)}(\mathrm{d}s, \mathrm{d}\nu),$$

where $\Delta Y_s = Y_s - Y_{s-}$, and let $\hat{N}(\mathrm{d}s, \mathrm{d}\nu)$ denote the predictable compensator of $N(\mathrm{d}s, \mathrm{d}\nu)$. Then $\hat{N}(\mathrm{d}s, \mathrm{d}\nu) = \mathrm{d}sK(Y_{s-}, \mathrm{d}\nu) + \mathrm{d}sH(\mathrm{d}\nu)$ with

$$K(\mu, \mathrm{d}\nu) = \int_E \mu(\mathrm{d}x)H(x, \mathrm{d}\nu).$$

(b) Let Γ be defined by (9.20) and let $\tilde{N}(\mathrm{d}s, \mathrm{d}\nu) = N(\mathrm{d}s, \mathrm{d}\nu) - \hat{N}(\mathrm{d}s, \mathrm{d}\nu)$. For any $f \in D_0(A)$,

$$Y_t(f) = Y_0(f) + M_t^c(f) + M_t^d(f) + \int_0^t [Y_s(Af + \gamma f - bf) + \Gamma(f)]\mathrm{d}s,$$

where $\{M_t^c(f) : t \geq 0\}$ is a continuous local martingale with quadratic variation $2Y_t(cf^2)\mathrm{d}t$ and

$$M_t^d(f) = \int_0^t \int_{M(E)^\circ} \nu(f)\tilde{N}(ds, d\nu), \quad t \geq 0,$$

is a purely discontinuous local martingale.

(4) For every $F \in \mathscr{D}_0$ we have

$$F(Y_t) = F(Y_0) + \int_0^t LF(Y_s)ds + \text{local mart.}$$

(5) For every $G \in C^2(\mathbb{R})$ and $f \in D_0(A)$,

$$\begin{aligned}
G(Y_t(f)) = G(Y_0(f)) &+ \int_0^t \Big\{ G'(Y_s(f))Y_s(Af + \gamma f - bf) \\
&+ \int_E Y_s(dx) \int_{M(E)^\circ} \Big[G(Y_s(f) + \nu(f)) - G(Y_s(f)) \\
&\quad - \nu(f)G'(Y_s(f)) \Big] H(x, d\nu) + G''(Y_s(f))Y_s(cf^2) \\
&+ G'(Y_s(f))\eta(f) + \int_{M(E)^\circ} \Big[G(Y_s(f) + \nu(f)) \\
&\quad - G(Y_s(f)) \Big] H(d\nu) \Big\} ds + \text{local mart.}
\end{aligned}$$

Theorem 9.18 *The above properties* (1), (2), (3), (4) *and* (5) *are equivalent to each other. Those properties hold if and only if* $\{(Y_t, \mathscr{G}_t) : t \geq 0\}$ *is an immigration superprocess with transition semigroup* $(Q_t^N)_{t \geq 0}$ *given by* (9.18).

Corollary 9.19 *Let* $\{(Y_t, \mathscr{G}_t) : t \geq 0\}$ *be a càdlàg realization of the immigration superprocess satisfying* $\mathbf{P}[Y_0(1)] < \infty$. *Then for every* $T \geq 0$ *and* $f \in D_0(A)$ *there is a constant* $C(T, f) \geq 0$ *such that*

$$\begin{aligned}
\mathbf{P}\Big[\sup_{0 \leq t \leq T} |Y_t(f)| \Big] \leq C(T, f) \Big\{ &\mathbf{P}[Y_0(1)] + \Gamma(1) \\
&+ \sqrt{\mathbf{P}[Y_0(1)]} + \sqrt{\Gamma(1)} \Big\}.
\end{aligned}$$

Corollary 9.20 *Suppose that* (9.21) *holds and* $\{(Y_t, \mathscr{G}_t) : t \geq 0\}$ *is a càdlàg realization of the immigration superprocess such that* $\mathbf{P}[Y_0(1)] < \infty$. *Then for every* $f \in D_0(A)$,

$$M_t(f) = Y_t(f) - Y_0(f) - t\Gamma(f) - \int_0^t Y_s(Af + \gamma f - bf)ds \qquad (9.26)$$

is a square integrable $(\mathscr{G}_t)$-*martingale with increasing process*

$$\langle M(f) \rangle_t = \int_0^t \Big[\int_E q(x, f)Y_s(dx) + \int_{M(E)^\circ} \nu(f)^2 H(d\nu) \Big] ds,$$

where $q(x, f)$ *is defined by* (2.59).

Corollary 9.21 *Suppose that $H(x, M(E)^\circ) = H(M(E)^\circ) = 0$ for all $x \in E$. Then any càdlàg realization $\{(Y_t, \mathscr{G}_t) : t \geq 0\}$ of the immigration superprocess is a.s. continuous. Conversely, if $\{(Y_t, \mathscr{G}_t) : t \geq 0\}$ is a continuous process in $M(E)$ and if for every $f \in D_0(A)$,*

$$M_t(f) = Y_t(f) - Y_0(f) - t\eta(f) - \int_0^t Y_s(Af + \gamma f - bf)\mathrm{d}s$$

is a $(\mathscr{G}_t)$-local martingale with increasing process

$$\langle M(f)\rangle_t = 2\int_0^t \mathrm{d}s \int_E c(x)f(x)^2 Y_s(\mathrm{d}x),$$

then $\{(Y_t, \mathscr{G}_t) : t \geq 0\}$ is a realization of the immigration superprocess.

Theorem 9.22 *Suppose that (9.21) holds and $\{(Y_t, \mathscr{G}_t) : t \geq 0\}$ is a càdlàg realization of the immigration superprocess such that $\mathbf{P}[Y_0(1)] < \infty$. Then the $(\mathscr{G}_t)$-martingales defined in (9.26) induce a unique worthy $(\mathscr{G}_t)$-martingale measure $\{M_t(B) : t \geq 0; B \in \mathscr{B}(E)\}$ satisfying*

$$M_t(f) = \int_0^t \int_E f(x)M(\mathrm{d}s, \mathrm{d}x), \qquad t \geq 0, f \in D_0(A) \qquad (9.27)$$

and having covariance measure defined by

$$\begin{aligned}
\eta(\mathrm{d}s, \mathrm{d}x, \mathrm{d}y) = {} & \mathrm{d}s \int_E 2c(z)\delta_z(\mathrm{d}x)\delta_z(\mathrm{d}y)Y_s(\mathrm{d}z) \\
& + \mathrm{d}s \int_E Y_s(\mathrm{d}z) \int_{M(E)^\circ} \nu(\mathrm{d}x)\nu(\mathrm{d}y)H(z, \mathrm{d}\nu) \\
& + \mathrm{d}s \int_{M(E)^\circ} \nu(\mathrm{d}x)\nu(\mathrm{d}y)H(\mathrm{d}\nu).
\end{aligned}$$

Theorem 9.23 *Suppose that (9.21) holds and $\{(Y_t, \mathscr{G}_t) : t \geq 0\}$ is a càdlàg realization of the immigration superprocess such that $\mathbf{P}[Y_0(1)] < \infty$. Let $\{M_t(B) : t \geq 0; B \in \mathscr{B}(E)\}$ be the worthy $(\mathscr{G}_t)$-martingale measure defined by (9.27). Then for any $t \geq 0$ and $f \in B(E)$ we have a.s.*

$$Y_t(f) = Y_0(\pi_t f) + \int_0^t \Gamma(\pi_{t-s}f)\mathrm{d}s + \int_0^t \int_E \pi_{t-s}f(x)M(\mathrm{d}s, \mathrm{d}x),$$

where $t \mapsto \pi_t f$ is defined by (2.35).

If $P_t 1 \in C(E)$ for every $t \geq 0$ and there exists $A1 \in C(E)$ such that (7.26) holds by the uniform convergence, we can extend the operator A to the linear span $D(A)$ of $D_0(A)$ and the constant functions. In this case, the results of Theorem 9.18 and its corollaries remain true with $D_0(A)$ replaced by $D(A)$.

We can also give martingale problem formulations of the immigration superprocess on the tempered space. Suppose that $h \in D_0(A)$ is strictly positive and there is a constant $\alpha > 0$ such that $Ah \leq \alpha h$. Recall that $C_h(E)$ is the set of continuous functions f on E satisfying $|f| \leq$ const $\cdot h$ and $D_h(A) = \{f \in D_0(A) \cap C_h(E) : Af \in C_h(E)\}$. Let ϕ be a branching mechanism given as in Section 6.1 with $\rho = h$ and let $(Q_t^N)_{t \geq 0}$ be the transition semigroup on $M_h(E)$ defined by (9.23) and (9.24). Suppose that $f \mapsto h^{-1}\phi(\cdot, hf) - \alpha f$ satisfies the conditions for the branching mechanism specified at the beginning of Section 7.1. Then we have:

Theorem 9.24 *Let $(W, \mathscr{G}, \mathscr{G}_t, \mathbf{P})$ be a filtered probability space satisfying the usual hypotheses and let $\{Y_t : t \geq 0\}$ be a càdlàg process in $M_h(E)$ that is adapted to $(\mathscr{G}_t)_{t \geq 0}$ and satisfies $\mathbf{P}[Y_0(h)] < \infty$. Then Theorem 9.18 still holds when $M(E)$, $C_0(E)$ and $D_0(A)$ are replaced by $M_h(E)$, $C_h(E)$ and $D_h(A)$, respectively.*

Theorem 9.25 *Let $\eta \in M_h(E)$ and let $(Q_t^\eta)_{t \geq 0}$ denote the transition semigroup defined by (9.24) with $I(f) = \eta(f)$ for $f \in B_h(E)^+$. Then for every $\mu \in M_h(E)$ the immigration superprocess in $M_h(E)$ with transition semigroup $(Q_t^\eta)_{t \geq 0}$ has a càdlàg realization $\{Y_t : t \geq 0\}$ with initial value $Y_0 = \mu$.*

9.4 Constructions of the Trajectories

Suppose that E is a Lusin topological space. Let $(V_t)_{t \geq 0}$ and $(Q_t)_{t \geq 0}$ denote respectively the cumulant and transition semigroups of a general Borel right MB-process in $M(E)$. Suppose that $t \mapsto V_t f(x)$ is locally bounded and right continuous pointwise for every $f \in C(E)^+$. Let $(N_t)_{t \geq 0}$ be an SC-semigroup associated with $(Q_t)_{t \geq 0}$ defined by (9.4) from the infinitely divisible probability entrance law $(K_s)_{s > 0}$ and let $I_s = -\log L_{K_s}$ for $s > 0$. Then the transition semigroup $(Q_t^N)_{t \geq 0}$ of the corresponding immigration superprocess is defined by

$$\int_{M(E)} e^{-\nu(f)} Q_t^N(\mu, d\nu) = \exp\left\{ -\mu(V_t f) - \int_0^t I_s(f) ds \right\} \qquad (9.28)$$

for $f \in B(E)^+$. By Corollary 9.7 we can write

$$\int_0^t I_s(f) ds = \gamma_t(f) + \int_0^t \zeta(ds) \int_{M(E)^\circ} (1 - e^{-\nu(f)}) G_{t-s}^s(d\nu), \qquad (9.29)$$

where $t \mapsto \gamma_t$ is an increasing path from $[0, \infty)$ to $M(E)$, $\zeta(ds)$ is a diffuse Radon measure on $[0, \infty)$ and $\{(G_t^s)_{t > 0} : s \geq 0\}$ is a family of entrance laws for $(Q_t^\circ)_{t \geq 0}$.

Recall that $(Q_t^\circ)_{t \geq 0}$ denotes the restriction of $(Q_t)_{t \geq 0}$ to $M(E)^\circ$. Let W_0^+ be the space of right continuous paths from $(0, \infty)$ to $M(E)$ that have zero as a trap. Let $(\mathscr{A}^0, \mathscr{A}_t^0)$ be the natural σ-algebras on W_0^+ generated by the coordinate process $\{w_t : t > 0\}$. By Theorem A.40 to each $(G_t^s)_{t > 0}$ there corresponds a σ-finite measure $\mathbf{Q}^s(dw)$ on $(W_0^+, \mathscr{A}^0)$ such that

$$\mathbf{Q}^s\{w_{t_1} \in d\nu_1, w_{t_2} \in d\nu_2, \ldots, w_{t_n} \in d\nu_n\}$$

$$= G_{t_1}^s(d\nu_1)Q_{t_2-t_1}^\circ(\nu_1, d\nu_2)\cdots Q_{t_n-t_{n-1}}^\circ(\nu_{n-1}, d\nu_n) \qquad (9.30)$$

for $\{t_1 < \cdots < t_n\} \subset (0, \infty)$ and $\{\nu_1, \ldots, \nu_n\} \subset M(E)^\circ$. Let $N(ds, dw)$ be a Poisson random measure on $(0, \infty) \times W_0^+$ with intensity $\zeta(ds)\mathbf{Q}^s(dw)$. We assume this random measure is defined on some probability space $(\Omega, \mathscr{G}, \mathbf{P})$ and define the measure-valued process

$$Y_t = \gamma_t + \int_{(0,t)} \int_{W_0^+} w_{t-s} N(ds, dw), \qquad t \geq 0. \qquad (9.31)$$

Theorem 9.26 *For $t \geq 0$ let $\mathscr{G}_t$ be the σ-algebra generated by the collection of random variables $\{N((0, u] \times A) : A \in \mathscr{A}_{t-u}^0, 0 \leq u < t\}$. Then $\{(Y_t, \mathscr{G}_t) : t \geq 0\}$ is an immigration process with transition semigroup $(Q_t^N)_{t\geq0}$ and one-dimensional distributions $(N_t)_{t\geq0}$.*

Proof. It is easy to show that $\{Y_t : t \geq 0\}$ has one-dimensional distributions $(N_t)_{t\geq0}$. Let Y_t^G denote the second term on the right-hand side of (9.31). Let $t \geq r > u \geq 0$ and let h be a bounded positive function on W_0^+ measurable relative to $\mathscr{A}_{r-u}^0$. For $f \in B(E)^+$ we can see as in the proof of Theorem 8.24 that

$$\mathbf{P}\left[\exp\left\{-\int_0^u \int_{W_0^+} h(w)N(ds, dw) - Y_t^G(f)\right\}\right]$$

$$= \mathbf{P}\left[\exp\left\{-\int_0^t \int_{W_0^+} [h(w)1_{\{s\leq u\}} + w_{t-s}(f)]N(ds, dw)\right\}\right]$$

$$= \exp\left\{-\int_0^t \mathbf{Q}^s\left(1 - \exp\left\{-h(w)1_{\{s\leq u\}} - w_{t-s}(f)\right\}\right)\zeta(ds)\right\}$$

$$= \exp\left\{-\int_0^u \mathbf{Q}^s\left(1 - \exp\left\{-h(w) - w_{r-s}(V_{t-r}f)\right\}\right)\zeta(ds)\right\}$$

$$\cdot \exp\left\{-\int_u^r \mathbf{Q}^s\left(1 - \exp\left\{-w_{r-s}(V_{t-r}f)\right\}\right)\zeta(ds)\right\}$$

$$\cdot \exp\left\{-\int_r^t \mathbf{Q}^s\left(1 - \exp\left\{-w_{t-s}(f)\right\}\right)\zeta(ds)\right\}$$

$$= \mathbf{P}\left[\exp\left\{-\int_0^u \int_{W_0^+} h(w)N(ds, dw) - Y_r^G(V_{t-r}f)\right\}\right]$$

$$\cdot \exp\left\{-\int_r^t \zeta(ds)\int_{M(E)^\circ}(1 - e^{-\nu(f)})G_{t-s}^s(d\nu)\right\}, \qquad (9.32)$$

where we have used the Markov property (9.30) for the third equality. By Theorem 8.2 we may assume $(K_t)_{t>0}$ has the representation (8.2). Then the second equation in (8.3) implies

$$\int_0^r H_{s+t-r}ds = \int_0^r H_s Q_{t-r}^\circ ds + \int_0^r ds \int_E \eta_s(dx)L_{t-r}(x, \cdot).$$

From (9.9), (9.13) and the above equation it follows that

$$\int_r^t G_{t-s}^s \zeta(\mathrm{d}s) = \int_0^t G_{t-s}^s \zeta(\mathrm{d}s) - \int_0^r G_{r-s}^s Q_{t-r}^\circ \zeta(\mathrm{d}s)$$

$$= \int_0^t H_s \mathrm{d}s - \int_0^r H_s Q_{t-r}^\circ \mathrm{d}s$$

$$= \int_0^t H_s \mathrm{d}s - \int_0^r H_{s+t-r} \mathrm{d}s + \int_0^r \mathrm{d}s \int_E \eta_s(\mathrm{d}x) L_{t-r}(x, \cdot)$$

$$= \int_0^{t-r} H_s \mathrm{d}s + \int_0^r \mathrm{d}s \int_E \eta_s(\mathrm{d}x) L_{t-r}(x, \cdot)$$

$$= G_{t-r} + \int_E \gamma_r(\mathrm{d}x) L_{t-r}(x, \cdot). \tag{9.33}$$

By (9.10) and (9.33), for $f \in B(E)^+$ we have

$$\gamma_t(f) + \int_r^t \zeta(\mathrm{d}s) \int_{M(E)^\circ} \left(1 - \mathrm{e}^{-\nu(f)}\right) G_{t-s}^s(\mathrm{d}\nu)$$

$$= \gamma_{t-r}(f) + \int_E \gamma_r(\mathrm{d}x) \lambda_{t-r}(x, f) + \int_{M(E)^\circ} \left(1 - \mathrm{e}^{-\nu(f)}\right) G_{t-r}(\mathrm{d}\nu)$$

$$+ \int_E \gamma_r(\mathrm{d}x) \int_{M(E)^\circ} \left(1 - \mathrm{e}^{-\nu(f)}\right) L_{t-r}(x, \mathrm{d}\nu)$$

$$= \gamma_{t-r}(f) + \gamma_r(V_{t-r}f) + \int_{M(E)^\circ} \left(1 - \mathrm{e}^{-\nu(f)}\right) G_{t-r}(\mathrm{d}\nu)$$

$$= \gamma_r(V_{t-r}f) + \int_0^{t-r} I_s(f)\mathrm{d}s, \tag{9.34}$$

where the last equality holds by (9.29). Since $\{\gamma_t : t \geq 0\}$ is deterministic, from (9.32) and (9.34) we have

$$\mathbf{P}\left[\exp\left\{-\int_0^u \int_{W_0^+} h(w)N(\mathrm{d}s, \mathrm{d}w) - Y_t(f)\right\}\right]$$

$$= \mathbf{P}\left[\exp\left\{-\int_0^u \int_{W_0^+} h(w)N(\mathrm{d}s, \mathrm{d}w) - Y_r(V_{t-r}f)\right\}\right.$$

$$\left. \cdot \exp\left\{-\int_0^{t-r} I_s(f)\mathrm{d}s\right\}\right],$$

That shows $\{(Y_t, \mathscr{G}_t) : t \geq 0\}$ is a Markov process in $M(E)$ with transition semigroup $(Q_t^N)_{t \geq 0}$. $\qquad\square$

The above theorem gives an explicit construction of the trajectories of the immigration process. From (9.31) we see that, except the deterministic part $\{\gamma_t : t \geq 0\}$, both the entry times and the evolutions of the immigrants are decided by the Poisson random measure $N(\mathrm{d}s, \mathrm{d}w)$. By the proof of Theorem 9.26, the second term on the right-hand side of (9.31) is a Markov process with inhomogeneous transition semigroup $(Q_{r,t} : t \geq r \geq 0)$ defined by

$$\int_{M(E)} e^{-\nu(f)} Q_{r,t}(\mu, d\nu) = \exp\left\{ -\mu(V_{t-r}f) - \int_r^t I_{s,t}(f)\zeta(ds) \right\}, \quad (9.35)$$

where

$$I_{s,t}(f) = \int_{M(E)^\circ} (1 - e^{-\nu(f)}) G_{t-s}^s(d\nu).$$

By Theorem 9.26 and Corollary 9.3 we have the following:

Corollary 9.27 *Let $(Y_t)_{t\geq 0}$ be defined by (9.31) and let $(\mathscr{G}_t)_{t\geq 0}$ be defined as in Theorem 9.26. Suppose that $\{(X_t, \mathscr{F}_t) : t \geq 0\}$ is an MB-process independent of $\{N(ds, dw)\}$ with transition semigroup $(Q_t)_{t\geq 0}$. Let $Z_t = X_t + Y_t$ and $\mathscr{H}_t = \sigma(\mathscr{F}_t \cup \mathscr{G}_t)$. Then $\{(Z_t, \mathscr{H}_t) : t \geq 0\}$ is an immigration process with transition semigroup $(Q_t^N)_{t\geq 0}$ defined by (9.28) and (9.29).*

Let $\eta \in M(E)$ and let $(Q_t^\eta)_{t\geq 0}$ denote the transition semigroup defined by (9.18) with $I(f) = \eta(f)$ for $f \in B(E)^+$. Recall that for each $x \in E^\circ$ there is an excursion law $\mathbf{Q}_x$ on W_0^+, which is the Kuznetsov measure corresponding to the entrance law $\{L_t(x, \cdot) : t > 0\}$ defined by (2.9). Suppose that $N^\eta(ds, dw)$ is a Poisson random measure on $(0, \infty) \times W_0^+$ with intensity

$$ds \int_{E^\circ} \eta(dx) \mathbf{Q}_x(dw).$$

We define the process

$$Y_t = \int_{(0,t)} \int_{W_0^+} w_{t-s} N^\eta(ds, dw), \quad t \geq 0. \quad (9.36)$$

Theorem 9.28 *Suppose that $\eta(dx)$ is carried by E°. For $t \geq 0$ let $\mathscr{G}_t$ be the σ-algebra generated by the collection of random variables $\{N^\eta((0, u] \times A) : A \in \mathscr{A}_{t-u}^0, 0 \leq u < t\}$. Then $\{(Y_t, \mathscr{G}_t) : t \geq 0\}$ is an immigration superprocess with transition semigroup $(Q_t^\eta)_{t\geq 0}$.*

Proof. This follows essentially from the calculations in the proof of Theorem 9.26; see also the proof of Theorem 8.24. $\qquad\square$

We next assume $(Q_t)_{t\geq 0}$ is the transition semigroup of a (ξ, ϕ)-superprocess, where ξ is a Borel right process in E and ϕ is given by (2.26) or (2.27). Suppose that $X = (W, \mathscr{F}, \mathscr{F}_t, X_t, \mathbf{Q}_\mu)$ is a right realization of the (ξ, ϕ)-superprocess. Let $\nu(1)H(d\nu)$ be a finite measure on $M(E)$ and let $(Q_t^H)_{t\geq 0}$ denote the transition semigroup defined by (9.18) with

$$I(f) = \int_{M(E)^\circ} (1 - e^{-\nu(f)}) H(d\nu), \quad f \in B(E)^+. \quad (9.37)$$

Clearly, we can define a σ-finite measure $\mathbf{Q}_H$ on W by

$$\mathbf{Q}_H(\mathrm{d}w) = \int_{M(E)^\circ} \mathbf{Q}_\mu(\mathrm{d}w) H(\mathrm{d}\mu).$$

Suppose that $N^H(\mathrm{d}s, \mathrm{d}w)$ is a Poisson random measure on $(0, \infty) \times W$ with intensity $\mathrm{d}s\mathbf{Q}_H(\mathrm{d}w)$ and define

$$Y_t = \int_{(0,t]} \int_W X_{t-s}(w) N^H(\mathrm{d}s, \mathrm{d}w), \qquad t \geq 0. \tag{9.38}$$

Theorem 9.29 *For $t \geq 0$ let $\mathscr{G}_t$ be the σ-algebra generated by the family of random variables $\{N^H((0, u] \times A) : A \in \mathscr{F}_{t-u}, 0 \leq u \leq t\}$ and let $(\bar{\mathscr{G}}_t)$ be the augmentation of $(\mathscr{G}_t)$. Then $\{(Y_t, \bar{\mathscr{G}}_{t+}) : t \geq 0\}$ is an a.s. right continuous realization of the immigration superprocess in $M(E)$ with transition semigroup $(Q_t^H)_{t\geq0}$.*

Proof. Step 1. By calculations similar to those in (9.32) it is easy to show that $(Y_t, \mathscr{G}_t)$ is an immigration superprocess with transition semigroup $(Q_t^H)_{t\geq0}$. For $k \geq 1$ let $W_k = \{w \in W : \langle X_0(w), 1\rangle \geq 1/k\}$ and define $\{Y_k(t) : t \geq 0\}$ by the right-hand side of (9.38) with W replaced by W_k. Let $\mathscr{C}_t^k$ be the σ-algebra generated by

$$\{N^H((0, u] \times A) : A \subset W_k, A \in \mathscr{F}_{t-u}, 0 \leq u \leq t\}.$$

Then $\{(Y_k(t), \mathscr{C}_t^k) : t \geq 0\}$ is an immigration superprocess with transition semigroup $(Q_k(t))_{t\geq0}$ given by (9.18) and (9.37) with $M(E)^\circ$ replaced by $M_k := \{\mu \in M(E)^\circ : \mu(1) \geq 1/k\}$. Let $\mathscr{D}_t^k$ be the σ-algebra generated by

$$\{N^H((0, u] \times A) : A \subset W_k^c, A \in \mathscr{F}_{t-u}, 0 \leq u \leq t\}.$$

By the property of the Poisson random measure, the σ-algebras $\mathscr{C}_t^k$ and $\mathscr{D}_t^k$ are independent and $\mathscr{G}_t = \sigma(\mathscr{C}_t^k \cup \mathscr{D}_t^k)$. Thus $\{(Y_k(t), \mathscr{G}_t) : t \geq 0\}$ is also an immigration superprocess with transition semigroup $(Q_k(t))_{t\geq0}$. It is easy to see that $\mathbf{Q}_H(W_k) = H(M_k) < \infty$. Then $N^H((0, t] \times W_k) < \infty$ a.s. for every $t \geq 0$, and hence $\{Y_k(t) : t \geq 0\}$ is a.s. right continuous in $M(E)$. Since $Y_k(t) \to Y_t$ increasingly, we conclude that $t \mapsto Y_t(f)$ is a.s. right lower semi-continuous for every $f \in C(E)^+$.

Step 2. For any constants $q \geq 0$ and $\beta \geq c_0$ we define the positive a.s. right continuous process

$$M_k(t) := e^{-2\beta t}\langle Y_k(t), q\rangle + (2\beta)^{-1}e^{-2\beta t}\int_{M_k(E)} \langle \nu, q\rangle H(\mathrm{d}\nu).$$

By Corollary 9.12 one sees that $\{M_k(t) : t \geq 0\}$ is a $(\mathscr{C}_t^k)$-supermartingale. It follows that $\{M_k(t) : t \geq 0\}$ is also a $(\mathscr{G}_t)$-supermartingale. Then $\{M_k(t) : t \geq 0\}$ is a $(\bar{\mathscr{G}}_{t+})$-supermartingale by Dellacherie and Meyer (1982, p.69). Note that $M_k(t)$ increases as $k \to \infty$ to

$$M(t) := e^{-2\beta t} \langle Y_t, q \rangle + (2\beta)^{-1} e^{-2\beta t} \int_{M(E)^\circ} \langle \nu, q \rangle H(d\nu).$$

Thus $\{M(t) : t \geq 0\}$ is an a.s. right continuous $(\mathscr{G}_{t+})$-supermartingale; see Del-lacherie and Meyer (1982, p.79). In particular, we conclude that $t \mapsto \langle Y_t, q \rangle$ is a.s. right continuous for every $q \geq 0$. For any $f \in C(E)^+$, choose a constant $q \geq \|f\|$. The arguments above imply that $t \mapsto \langle Y_t, q \rangle$ is a.s. right continuous and both $t \mapsto \langle Y_t, f \rangle$ and $t \mapsto \langle Y_t, q - f \rangle$ are a.s. right lower semi-continuous. Those clearly yield the a.s. right continuity of $t \mapsto \langle Y_t, f \rangle$. Then $t \mapsto Y_t$ is a.s. right continuous in $M(E)$.

Step 3. Let $t \geq 0$ and $f \in B(E)^+$. By Theorem A.16 and the strong Markov property of X one sees that $s \mapsto \exp\{-\langle X_s(w), V_{t-s}f \rangle\}$ is right continuous on $[0, t]$ for $\mathbf{Q}_H$-a.e. $w \in W$. Then

$$s \mapsto 1 - \exp\left\{ - \langle Y_k(s), V_{t-s}f \rangle - \int_0^{t-s} dr \int_{M_k} (1 - e^{-\langle \nu, V_r f \rangle}) H(d\nu) \right\}$$

is an a.s. right continuous $(\mathscr{G}_s)$-martingale on $[0, t]$. As in the second step it is easy to show

$$s \mapsto 1 - \exp\left\{ - \langle Y_s, V_{t-s}f \rangle - \int_0^{t-s} dr \int_{M(E)^\circ} (1 - e^{-\langle \nu, V_r f \rangle}) H(d\nu) \right\} \quad (9.39)$$

is an a.s. right continuous $(\bar{\mathscr{G}}_{s+})$-supermartingale on $[0, t]$. By the Markov property of $\{(Y_t, \mathscr{G}_t) : t \geq 0\}$ the right-hand side of (9.39) has expectation $\mathbf{P}[1 - \exp\{-\langle Y_t, f \rangle\}]$, which is independent of $s \in [0, t]$. Then the process is actually a $(\bar{\mathscr{G}}_{s+})$-martingale on $[0, t]$. That gives the desired Markov property of $\{(Y_t, \bar{\mathscr{G}}_{t+}) : t \geq 0\}$. $\qquad\square$

Example 9.1 Let us consider the case where ξ is the absorbing-barrier Brownian motion in $E_0 := (0, \infty)$ with transition semigroup defined by (A.27) and (A.28). Let $\eta \in M(E_0)$ and let $\alpha \geq 0$ be a constant. We can use the notation of Section 8.5 to define the transition semigroup $(Q_t^{\eta, \alpha})_{t \geq 0}$ of an immigration superprocess in $M(E_0)$ by

$$\int_{M(E_0)} e^{-\nu(f)} Q_t^{\eta, \alpha}(\mu, d\nu)$$
$$= \exp\left\{ - \mu(V_t f) - \int_0^t [\eta(V_s f) + \alpha \partial_0 V_s f] ds \right\}.$$

Example 9.2 In the situation of the above example, let $uF(du)$ be a non-trivial finite measure on $(0, \infty)$. We define another transition semigroup $(Q_t^F)_{t \geq 0}$ on $M(E_0)$ by

$$\int_{M(E_0)} e^{-\nu(f)} Q_t^F(\mu, d\nu) = \exp\left\{ - \mu(V_t f) - \int_0^t I_s(f) ds \right\},$$

where

$$I_s(f) = \int_0^\infty (1 - e^{-u\partial_0 V_s f}) F(du).$$

The corresponding immigration superprocess can be constructed in the following way. From (8.69) one can see there is an entrance law $H \in \mathscr{K}(Q^\circ)$ so that

$$\int_{M(E_0)^\circ} (1 - e^{-\nu(f)}) H_s(d\nu) = I_s(f), \quad s > 0, f \in B(E_0)^+.$$

Let $\mathbf{Q}_F(dw)$ be the corresponding Kuznetsov measure on W_0^+, the space of right continuous paths from $(0, \infty)$ to $M(E_0)$ with zero as a trap. Suppose $N_F(ds, dw)$ is a Poisson random measure on $(0, \infty) \times W_0^+$ with intensity $ds\mathbf{Q}_F(dw)$. By Theorem 9.26,

$$Y_t := \int_{(0,t)} \int_{W_0^+} w_{t-s} N_F(ds, dw), \quad t \geq 0 \tag{9.40}$$

is an immigration superprocess in $M(E_0)$ with transition semigroup $(Q_t^F)_{t \geq 0}$. Let $\{(s_i, w_i) : i = 1, 2, \ldots\}$ be an enumeration of the atoms of $N_F(ds, dw)$. It is easy to see that

$$\mathbf{Q}_F(dw) = \int_0^\infty \mathbf{Q}^u(dw) F(du), \quad w \in W_0^+,$$

where $\mathbf{Q}^u(dw)$ is as in Example 8.4. Let h be defined by (8.60) and assume (8.70) holds. Then by (8.71) and (9.40), for every $\varepsilon > 0$ we have a.s.

$$\lim_{t \to s_i} Y_t((0, \varepsilon]) \geq \lim_{t \to s_i} w_{i,t-s_i}((0, \varepsilon]) = \infty. \tag{9.41}$$

If $F(du)$ is an infinite measure on $(0, \infty)$, then $\{s_i : i = 1, 2, \ldots\} \cap (r, t)$ is a.s. infinite for every $t > r \geq 0$.

Example 9.3 Let h be defined by (8.60) and assume (8.70) holds. The immigration superprocess constructed in the last example is certainly not right continuous. Indeed, the transition semigroup $(Q_t^F)_{t \geq 0}$ has no right continuous realization. Otherwise, suppose that $\{Z_t : t \geq 0\}$ is such a realization with $Z_0 = 0$. Given $\mu \in M(E_0)$ we define $\mu^h \in M(E_0)$ by $\mu^h(dx) = h(x)\mu(dx)$ for $x \in E_0$. In an obvious way, we also regard μ^h as a measure in $M(\mathbb{R}_+)$. Then $\{Z_t^h : t \geq 0\}$ is an a.s. right continuous immigration superprocess in $M(\mathbb{R}_+)$ with semigroup $(\bar{Q}_t^F)_{t \geq 0}$ given by

$$\int_{M(\mathbb{R}_+)} e^{-\nu(\bar{f})} \bar{Q}_t^F(\mu, d\nu) = \exp\left\{ -\mu(\bar{U}_t\bar{f}) - \int_0^t \bar{I}_s(\bar{f})ds \right\},$$

where $(\bar{U}_t)_{t \geq 0}$ is defined by (8.72) and

$$\bar{I}_s(\bar{f}) = \int_0^\infty (1 - e^{-u\partial_0 h\bar{U}_s\bar{f}(0)})F(du), \quad \bar{f} \in B(\mathbb{R}_+)^+.$$

It is easily seen that

$$\mathbf{P}\{Z_t^h(\{0\}) = 0 \text{ for all } t \geq 0\} = 1. \tag{9.42}$$

For any $w \in W_0^+$ define

$$w_0^h = \begin{cases} \lim_{t\to 0} w_t^h & \text{if the limit exists in } M(\mathbb{R}_+), \\ 0 & \text{if the limit above does not exist.} \end{cases}$$

By the discussions in Example 8.4 one can see $\{w_t^h : t \geq 0\}$ under $\mathbf{Q}_F$ is a super-process in $M(\mathbb{R}_+)$ with cumulant semigroup $(\bar{U}_t)_{t\geq 0}$ and $w_0^h(\{0\}) > 0$ for $\mathbf{Q}_F$-a.e. $w \in W_0^+$. Then by Theorem 9.29,

$$\bar{Y}_t = \int_{(0,t]} \int_{W_0^+} w_{t-s}^h N_F(ds, dw), \qquad t \geq 0, \tag{9.43}$$

defines another a.s. right continuous realization of $(\bar{Q}_t^F)_{t\geq 0}$ with $\bar{Y}_0 = 0$. Let $S_0 = \inf\{t \geq 0 : \bar{Y}_t(\{0\}) > 0\}$. In view of (9.43) we have

$$\mathbf{P}\{S_0 \leq a\} = 1 - e^{-aF(E_0)}, \qquad u > 0. \tag{9.44}$$

However, an application of Theorem 9.15 shows that $(\bar{Q}_t^F)_{t\geq 0}$ is a Borel right semi-group on $M(\mathbb{R}_+)$, so (9.42) and (9.44) are in contradiction.

9.5 One-Dimensional Stochastic Equations

Suppose that (ϕ, ψ) are given respectively by (3.1) and (3.26) with $un(du)$ being a finite measure on $(0, \infty)$. Let $(Q_t^\gamma)_{t\geq 0}$ be the transition semigroup defined by (3.3) and (3.29). By Theorem 9.18 we have the following characterization of the CBI-process with transition semigroup $(Q_t^\gamma)_{t\geq 0}$.

Theorem 9.30 *Suppose that $\{y(t) : t \geq 0\}$ is a positive càdlàg process such that $\mathbf{P}[y(0)] < \infty$. Then $\{y(t) : t \geq 0\}$ is a CBI-process with transition semigroup $(Q_t^\gamma)_{t\geq 0}$ if and only if for every $f \in C^2(\mathbb{R}_+)$ we have*

$$f(y(t)) = f(y(0)) + \int_0^t Lf(y(s))ds + \text{local mart.},$$

where

$$Lf(x) = cxf''(x) + x\int_0^\infty [f(x+z) - f(x) - zf'(x)]m(dz)$$

$$+ (\beta - bx)f'(x) + \int_0^\infty [f(x+z) - f(x)]n(\mathrm{d}z). \qquad (9.45)$$

From the above theorem we can derive some stochastic equations for the CBI-processes. Let $\{B(t)\}$ be a standard Brownian motion and let $\{N_0(\mathrm{d}s, \mathrm{d}z, \mathrm{d}u)\}$ and $\{N_1(\mathrm{d}s, \mathrm{d}z)\}$ be Poisson random measures on $(0, \infty)^3$ and $(0, \infty)^2$ with intensities $\mathrm{d}s m(\mathrm{d}z)\mathrm{d}u$ and $\mathrm{d}s n(\mathrm{d}z)$, respectively. Suppose that $\{B(t)\}$, $\{N_0(\mathrm{d}s, \mathrm{d}z, \mathrm{d}u)\}$ and $\{N_1(\mathrm{d}s, \mathrm{d}z)\}$ are defined on a complete probability space and are independent of each other. Let us consider the stochastic integral equation

$$y(t) = y(0) + \int_0^t \sqrt{2cy(s)}\mathrm{d}B(s) + \int_0^t \int_0^\infty \int_0^{y(s-)} z\tilde{N}_0(\mathrm{d}s, \mathrm{d}z, \mathrm{d}u)$$
$$+ \int_0^t (\beta - by(s))\mathrm{d}s + \int_0^t \int_0^\infty zN_1(\mathrm{d}s, \mathrm{d}z), \qquad (9.46)$$

where $\tilde{N}_0(\mathrm{d}s, \mathrm{d}z, \mathrm{d}u) = N_0(\mathrm{d}s, \mathrm{d}z, \mathrm{d}u) - \mathrm{d}s m(\mathrm{d}z)\mathrm{d}u$. We understand the third term on the right-hand side as an integral over the set $\{(s, z, u) : 0 < s \le t, 0 < z < \infty, 0 < u \le y(s-)\}$ and give similar interpretations for other integrals with respect to Poisson random measures in this section.

Theorem 9.31 *There is a unique positive weak solution to (9.46) and the solution is a CBI-process with transition semigroup* $(Q_t^\gamma)_{t\ge0}$.

Proof. Suppose that $\{y(t)\}$ is a càdlàg realization of the CBI-process with transition semigroup given by (3.3) and (3.29). By Theorem 9.18 the process has no negative jumps and the random measure

$$N(\mathrm{d}s, \mathrm{d}z) := \sum_{s>0} 1_{\{y(s)\ne y(s-)\}}\delta_{(s,y(s)-y(s-))}(\mathrm{d}s, \mathrm{d}z)$$

has predictable compensator

$$\hat{N}(\mathrm{d}s, \mathrm{d}z) = y(s-)\mathrm{d}s m(\mathrm{d}z) + \mathrm{d}s n(\mathrm{d}z)$$

and

$$y(t) = y(0) + t\left[\beta + \int_0^\infty un(\mathrm{d}u)\right] - \int_0^t by(s)\mathrm{d}s$$
$$+ M^c(t) + \int_0^t \int_0^\infty z\tilde{N}(\mathrm{d}s, \mathrm{d}z), \qquad (9.47)$$

where $\tilde{N}(\mathrm{d}s, \mathrm{d}z) = N(\mathrm{d}s, \mathrm{d}z) - \hat{N}(\mathrm{d}s, \mathrm{d}z)$ and $t \mapsto M^c(t)$ is a continuous local martingale with quadratic variation $2cy(t)\mathrm{d}t$. By representation theorems for semimartingales, we have equation (9.46) on an extension of the original probability space; see, e.g., Ikeda and Watanabe (1989, p.90 and p.93). That proves the existence of a weak solution to (9.46). Conversely, if $\{y(t)\}$ is a positive solution to (9.46), one can use Itô's formula to see the process is a solution of the martingale problem associated with the generator L defined by (9.45). By Theorem 9.30 we see $\{y(t)\}$ is a CBI-process with transition semigroup $(Q_t^\gamma)_{t\geq0}$. That implies the weak uniqueness of the solution to (9.46). □

Theorem 9.32 *Suppose that* $m(dz) = \sigma z^{-1-\alpha}dz$ *for constants* $\sigma \geq 0$ *and* $1 < \alpha < 2$. *Then the CBI-process with transition semigroup* $(Q_t^\gamma)_{t\geq0}$ *is the unique positive weak solution of*

$$dy(t) = \sqrt{2cy(t)}dB(t) + \sqrt[\alpha]{\sigma y(t-)}dz_0(t) - by(t)dt + dz_1(t), \quad (9.48)$$

where $\{B(t)\}$ *is a standard Brownian motion,* $\{z_0(t)\}$ *is a one-sided* α-stable *process with Lévy measure* $z^{-1-\alpha}dz$, $\{z_1(t)\}$ *is an increasing Lévy process determined by* (β, n), *and* $\{B(t)\}$, $\{z_0(t)\}$ *and* $\{z_1(t)\}$ *are independent of each other.*

Proof. We assume $\sigma > 0$, for otherwise the proof is easier. Let us consider the CBI-process $\{y(t)\}$ given by (9.46) with $\{N_0(ds, dz, du)\}$ being a Poisson random measure on $(0, \infty)^3$ with intensity $\sigma z^{-1-\alpha}dsdzdu$. We define the random measure $\{N(ds, dz)\}$ on $(0, \infty)^2$ by

$$N((0,t] \times B) = \int_0^t \int_0^\infty \int_0^{y(s-)} 1_{\{y(s-)>0\}} 1_B\left(\frac{z}{\sqrt[\alpha]{\sigma y(s-)}}\right) N_0(ds, dz, du)$$
$$+ \int_0^t \int_0^\infty \int_0^{1/\sigma} 1_{\{y(s-)=0\}} 1_B(z) N_0(ds, dz, du).$$

It is easy to compute that $\{N(ds, dz)\}$ has predictable compensator

$$\hat{N}((0,t] \times B) = \int_0^t \int_0^\infty 1_{\{y(s-)>0\}} 1_B\left(\frac{z}{\sqrt[\alpha]{\sigma y(s-)}}\right) \frac{\sigma y(s-)dsdz}{z^{1+\alpha}}$$
$$+ \int_0^t \int_0^\infty 1_{\{y(s-)=0\}} 1_B(z) \frac{dsdz}{z^{1+\alpha}}$$
$$= \int_0^t \int_0^\infty 1_B(z) \frac{dsdz}{z^{1+\alpha}}.$$

Thus $\{N(ds, dz)\}$ is a Poisson random measure with intensity $z^{-1-\alpha}dsdz$; see, e.g., Ikeda and Watanabe (1989, p.93). Now define the Lévy processes

$$z_0(t) = \int_0^t \int_0^\infty z\tilde{N}(ds, dz) \quad \text{and} \quad z_1(t) = \beta t + \int_0^t \int_0^\infty zN_1(ds, dz),$$

where $\tilde{N}(ds, dz) = N(ds, dz) - \hat{N}(ds, dz)$. It is easy to see that

$$\int_0^t \sqrt[\alpha]{\sigma y(s-)}dz_0(s) = \int_0^t \int_0^\infty \sqrt[\alpha]{\sigma y(s-)}\, z\tilde{N}(ds, dz)$$
$$= \int_0^t \int_0^\infty \int_0^{y(s-)} z\tilde{N}_0(ds, dz, du).$$

Then we get (9.48) from (9.46). Conversely, if $\{y(t)\}$ is a solution of (9.48), one can use Itô's formula to see that $\{y(t)\}$ solves the martingale problem associated with the generator L defined by (9.45) with $m(\mathrm{d}z) = \sigma z^{-1-\alpha}\mathrm{d}z$. Then $\{y(t)\}$ is a CBI-process with transition semigroup $(Q_t^\gamma)_{t\geq 0}$ and the solution of (9.48) is unique in law.   □

9.6 Notes and Comments

The concept of SC-semigroups was introduced in Li (1995/6), where Theorem 9.4 was proved. Most of the results in Section 9.2 can be found in Li (2002). Theorem 9.26 was also taken from Li (2002). Theorem 9.29 was essentially proved in Li (1996). Example 9.2 can also be found in Li (1996). Theorem 9.30 was given in Kawazu and Watanabe (1971) under more general assumptions. By the results of Dawson and Li (2006) and Fu and Li (2010), both (9.46) and (9.48) have unique strong solutions; see also Bertoin and Le Gall (2006) for a result on weak solutions of a special case of (9.46). We refer to Bertoin (1996) and Sato (1999) for the general theory of Lévy processes. By a modification of the arguments in Section 6.3, one can obtain a special immigration superprocess from a two-type superprocess without immigration.

The structure of immigration was studied in Li and Shiga (1995) in the setting of measure-valued diffusions. In particular, the transition semigroup $(Q_t^{\eta,\alpha})_{t\geq 0}$ given in Example 9.1 was considered in Li and Shiga (1995) for binary local branching with constant branching rate. There it was proved the corresponding immigration superprocess $\{Y_t : t \geq 0\}$ has a continuous density field $\{Y(t,x) : t > 0, x > 0\}$ satisfying the following stochastic partial differential equation:

$$\frac{\partial}{\partial t}Y(t,x) = \sqrt{Y(t,x)}\dot{W}(t,x) + \frac{1}{2}\Delta Y(t,x) + \dot{\eta}(x) - \alpha\dot{\delta}_0,$$

where $\{W(t,x) : t \geq 0, x > 0\}$ is a time–space white noise based on the Lebesgue measure and the dot denotes derivative in the distribution sense. The random measures $\{Y_t : t \geq 0\}$ have bounded supports if and only if both $\mathrm{supp}(Y_0)$ and $\mathrm{supp}(\eta)$ are bounded. In this case, let $R_t = \cup_{0\leq s\leq t}\mathrm{supp}(Y_s)$ and $\hat{R}_t = \sup\{x > 0 : x \in R_t\}$. It was proved in Li and Shiga (1995) that the distribution of $t^{-1/3}\hat{R}_t$ converges as $t \to \infty$ to the Fréchet distribution $F(z) = \mathrm{e}^{-\gamma z^{-3}}$ with

$$\gamma = \frac{1}{18}\left(\frac{\Gamma(1/3)\Gamma(1/6)}{\Gamma(1/2)}\right)^3\left(\alpha + \int_0^\infty x\eta(\mathrm{d}x)\right).$$

A central limit theorem for the super-Brownian motion with immigration was also given in Li and Shiga (1995). The corresponding large and moderate deviations were studied in Zhang (2004a, 2004b). A number of functional central limit theorems for related processes were proved in Zhang (2005a, 2008), which gave rise to distribution-valued Gaussian processes. Li (1998a) studied immigration structures associated with branching particle systems. A large-deviation principle for a Brownian branching immigration particle system was established in Zhang (2005b). The spectral properties for immigration superprocesses were investigated in Stannat (2003). The equilibrium behavior of a population with immigration on a hierarchical group was studied in Dawson et al. (2004a).

Chapter 10
State-Dependent Immigration Structures

In this chapter we investigate the structures of state-dependent immigration. We first construct some immigration superprocesses by summing up excursions according to Poisson random measures. Based on those constructions, we prove the existence and uniqueness of solutions to some stochastic integral equations, which define the interactive immigration structures. We shall deal with processes with càdlàg paths. Throughout this chapter, we assume E is a locally compact separable metric space and ξ is a Hunt process in E with Feller transition semigroup $(P_t)_{t\geq0}$. Let A denote the strong generator of $(P_t)_{t\geq0}$ with domain $D_0(A) \subset C_0(E)$. Let $D(A)$ be the linear span of $D_0(A) \cup \{1\}$ and extend A to $D(A)$ such that $A1 = 0$. Let ϕ be a branching mechanism given by (2.26) or (2.27). Suppose that $\nu(1)^2 H(x, d\nu)$ is a bounded kernel from E to $M(E)^\circ$ and that Conditions 7.1, 7.2 and 8.5 are satisfied. We use $\| \cdot \|$ to denote both the supremum norm of functions and the total variation norm of signed measures.

10.1 Deterministic Immigration Rates

By Theorem 8.6 the cumulant semigroup of the (ξ, ϕ)-superprocess admits the representation (2.9) for all $x \in E$. Then Corollary 8.4 implies that for any $x \in E$ the family of measures $\{L_t(x, \cdot) : t > 0\}$ on $M(E)^\circ$ constitute an entrance law for the restricted semigroup $(Q_t^\circ)_{t\geq0}$. Let D_0 be the set of paths $w \in D([0,\infty), M(E))$ that have zero as a trap. We equip this space with the natural σ-algebras $(\mathscr{A}^0, \mathscr{A}_t^0)$ generated by the coordinate process $\{w_t : t \geq 0\}$. By Theorem 8.22 the excursion law $\mathbf{Q}_{L(x)}$ determined by $\{L_t(x, \cdot) : t > 0\}$ is actually carried by the excursions $w \in D_0$ satisfying $w_0 = 0$ and $\lim_{t\to0} w_t(1)^{-1}w_t = \delta_x$ by weak convergence.

Suppose that F is a Lusin topological space. Let $\lambda(dy)$ be a σ-finite Borel measure on F and let $\kappa(y, dx)$ be a bounded kernel from F to E. In view of (9.17) and (9.18), given any λ-integrable function $\rho \in \mathscr{B}(F)^+$, we can define the transition semigroup $(Q_t^\rho)_{t\geq0}$ of an immigration superprocess by

Z. Li, *Measure-Valued Branching Markov Processes*,
Probability and Its Applications, DOI 10.1007/978-3-642-15004-3_10,
© Springer-Verlag Berlin Heidelberg 2011

$$\int_{M(E)} e^{-\langle \nu, f \rangle} Q_t^\rho(\mu, d\nu) = \exp\left\{ -\langle \mu, V_t f \rangle - \int_0^t \langle \lambda, \rho\kappa V_s f \rangle ds \right\}. \quad (10.1)$$

Let $\{(X_t, \mathscr{F}_t) : t \geq 0\}$ be a càdlàg realization of the (ξ, ϕ)-superprocess with deterministic initial value $X_0 = \mu \in M(E)$ and let $\{N(ds, dy, du, dw)\}$ be a Poisson random measure on $(0, \infty) \times F \times (0, \infty) \times D_0$ with intensity $ds\lambda(dy)du\mathbf{Q}_\kappa(y, dw)$, where

$$\mathbf{Q}_\kappa(y, dw) = \int_E \kappa(y, dx)\mathbf{Q}_{L(x)}(dw), \quad y \in F, w \in D_0.$$

We assume $\{(X_t, \mathscr{F}_t) : t \geq 0\}$ and $\{N(ds, dy, du, dw)\}$ are defined on a complete probability space $(\Omega, \mathscr{G}, \mathbf{P})$ and are independent of each other. For $t \geq 0$ let $\mathscr{G}_t$ be the σ-algebra generated by $\mathscr{F}_t$ and the family of random variables

$$\{N((0, s] \times U \times A) : U \in \mathscr{B}(F \times (0, \infty)), A \in \mathscr{A}_{t-s}^0, 0 \leq s \leq t\}.$$

Let $(\bar{\mathscr{G}}_t)_{t \geq 0}$ be the augmentation of the filtration $(\mathscr{G}_t)_{t \geq 0}$. Given λ-integrable functions $p \leq q \in \mathscr{B}(F)^+$, we define

$$Y_t = X_t + \int_0^t \int_F \int_{p(y)}^{q(y)} \int_{D_0} w_{t-s} N(ds, dy, du, dw), \quad t \geq 0. \quad (10.2)$$

We understand the second term on the right-hand side as an integral over the set $\{(s, y, u, w) : 0 < s \leq t, y \in F, p(y) < u \leq q(y), w \in D_0\}$ and give similar interpretations for all Poisson integrals in the sequel.

Theorem 10.1 *The process $\{Y_t : t \geq 0\}$ defined by (10.2) has a càdlàg modification and $\{(Y_t, \bar{\mathscr{G}}_{t+}) : t \geq 0\}$ is an immigration superprocess with transition semigroup $(Q_t^\rho)_{t \geq 0}$ defined by (10.1) with $\rho = q - p$.*

Proof. We first note that the construction (10.2) can be simplified. In fact, we have

$$Y_t = X_t + \int_0^t \int_{D_0} w_{t-s} N_\rho(ds, dw), \quad t \geq 0,$$

where $\{N_\rho(ds, dw)\}$ is the random measure on $(0, \infty) \times D_0$ defined by

$$N_\rho((0, t] \times A) = \int_0^t \int_F \int_{p(y)}^{q(y)} \int_A N(ds, dy, du, dw).$$

It is easy to see that $\{N_\rho(ds, dw)\}$ is a Poisson random measure with intensity $ds\mathbf{Q}^\rho(dw)$, where

$$\mathbf{Q}^\rho(dw) = \int_F \rho(y)\lambda(dy) \int_E \kappa(y, dx)\mathbf{Q}_{L(x)}(dw).$$

By Theorem 9.26 one sees that $\{(Y_t, \mathscr{G}_t) : t \geq 0\}$ is an immigration superprocess with transition semigroup $(Q_t^\rho)_{t \geq 0}$. By Theorem 9.15 this process has a càdlàg realization. Since E is locally compact and separable, it can be metrized as a complete metric space. Following Parthasarathy (1967, pp.46–47) one can show that $M(E)$ can also be metrized as a complete metric space, so Proposition A.7 implies $\{Y_t : t \geq 0\}$ has a càdlàg modification. By the strong continuity of $(V_t)_{t \geq 0}$ on $C_0(E)^+$ one can see that $\{(Y_t, \mathscr{G}_{t+}) : t \geq 0\}$ is an immigration superprocess with transition semigroup $(Q_t^\rho)_{t \geq 0}$. $\qquad\square$

The process constructed in (10.2) deals with the case where immigration occurs continuously at rate $\rho(y)$ with respect to the measure $\lambda(dy)$. The kernel $\kappa(y, dx)$ transfers the immigration affect from F to the underlying space E. We can also consider the case where immigration only occurs in a fixed interval. Fix $r \geq 0$ and for $t \geq r$ let

$$Z_t = X_t + \int_0^r \int_F \int_{p(y)}^{q(y)} \int_{D_0} w_{t-s} N(ds, dy, du, dw). \qquad (10.3)$$

By a reorganization of the arguments in the proofs of Theorems 9.26 and 10.1 it is easy to show the following:

Theorem 10.2 *The process $\{Z_t : t \geq r\}$ defined by (10.3) has a càdlàg modification and $\{(Z_t, \mathscr{G}_{t+}) : t \geq r\}$ is a (ξ, ϕ)-superprocess.*

10.2 Stochastic Immigration Rates

Let $\{X_t : t \geq 0\}$ and $\{N(ds, dy, du, dw)\}$ be given as in the first section. Let $\mathscr{L}_\lambda^1(F)$ denote the set of two-parameter processes $h = \{h_t(y) : t \geq 0, y \in F\}$ defined on $(\Omega, \mathscr{G}, \mathbf{P})$ that are predictable relative to the augmented filtration $(\mathscr{G}_{t+})$ and satisfy

$$\|h\|_{\lambda,n} := \int_0^n \mathbf{P}[\langle \lambda, |h_s|\rangle] ds < \infty, \qquad n \geq 1. \qquad (10.4)$$

We identify h_1 and $h_2 \in \mathscr{L}_\lambda^1(F)$ if $\|h_1 - h_2\|_{\lambda,n} = 0$ for every $n \geq 1$ and define the metric d_1 on $\mathscr{L}_\lambda^1(F)$ by

$$d_1(h_1, h_2) = \sum_{n=1}^\infty \frac{1}{2^n} (1 \wedge \|h_1 - h_2\|_{\lambda,n}). \qquad (10.5)$$

We say $h = \{h_s(y) : s \geq 0, y \in F\}$ is *left continuous* if for every $(\omega, y) \in \Omega \times F$ the mapping $s \mapsto h_s(\omega, y)$ is left continuous on $(0, \infty)$. Recall that a two-parameter process $\{q_s(y) : s \geq 0, y \in F\}$ is called a step process if it is of the form

$$q_s(y) = g_0(y)1_{\{r_0\}}(s) + \sum_{i=0}^{\infty} g_i(y)1_{(r_i, r_{i+1}]}(s), \qquad (10.6)$$

where $\{0 = r_0 < r_1 < r_2 < \cdots\}$ is a sequence increasing to infinity and each $(\omega, y) \mapsto g_i(\omega, y)$ is a $\bar{\mathscr{G}}_{r_i+} \times \mathscr{B}(F)$-measurable function. Let $\mathscr{L}_\lambda^0(F)$ denote the set of step processes $q \in \mathscr{L}_\lambda^1(F)$.

Proposition 10.3 *The metric space $(\mathscr{L}_\lambda^1(F), d_1)$ is complete and $\mathscr{L}_\lambda^0(F)$ is a dense subset of $\mathscr{L}_\lambda^1(F)$.*

We shall not give the proof of the above proposition since it is a simple modification of that of Proposition 7.22. In the sequel, we only deal with left continuous processes in $\mathscr{L}_\lambda^1(F)$. Given left continuous processes $p \le q \in \mathscr{L}_\lambda^1(F)^+$ define

$$Y_t = X_t + \int_0^t \int_F \int_{p_s(y)}^{q_s(y)} \int_{D_0} w_{t-s} N(ds, dy, du, dw), \quad t \ge 0. \qquad (10.7)$$

Setting $\rho_s(y) = q_s(y) - p_s(y)$ we have:

Theorem 10.4 *The process $\{Y_t : t \ge 0\}$ defined by (10.7) has a càdlàg modification in $M(E)$ with the following property: For every $G \in C^2(\mathbb{R})$ and $f \in D(A)$ we have*

$$\begin{aligned}
G(\langle Y_t, f \rangle) = G(\langle Y_0, f \rangle) + \int_0^t \Big\{ & G'(\langle Y_s, f \rangle)\langle Y_s, Af + \gamma f - bf \rangle \\
& + \int_E Y_s(dx) \int_{M(E)^\circ} \Big[G(\langle Y_s, f \rangle + \langle \nu, f \rangle) - G(\langle Y_s, f \rangle) \\
& - \langle \nu, f \rangle G'(\langle Y_s, f \rangle) \Big] H(x, d\nu) + G''(\langle Y_s, f \rangle)\langle Y_s, cf^2 \rangle \\
& + G'(\langle Y_s, f \rangle)\langle \lambda, \rho_s \kappa f \rangle \Big\} ds + (\bar{\mathscr{G}}_{t+})\text{-local mart.} \qquad (10.8)
\end{aligned}$$

Corollary 10.5 *Let $\{Y_t : t \ge 0\}$ be defined by (10.7). Then $t \mapsto \mathbf{P}[\langle Y_t, 1 \rangle]$ is locally bounded and*

$$\mathbf{P}[\langle Y_t, f \rangle] = \langle \mu, \pi_t f \rangle + \int_0^t \mathbf{P}[\langle \lambda, \rho_s \kappa \pi_{t-s} f \rangle] ds, \quad t \ge 0, f \in B(E),$$

where $t \mapsto \pi_t f(x)$ is defined by (2.35).

Corollary 10.6 *Let $\{Y_t : t \ge 0\}$ be defined by (10.7). Then for every $f \in D(A)$,*

$$M_t(f) = \langle Y_t, f \rangle - \langle Y_0, f \rangle - \int_0^t \langle Y_s, Af + \gamma f - bf \rangle ds - \int_0^t \langle \lambda, \rho_s \kappa f \rangle ds$$

is a square-integrable $(\bar{\mathscr{G}}_{t+})$-martingale with increasing process

$$\langle M(f) \rangle_t = \int_0^t ds \int_E q(x, f) Y_s(dx),$$

where $q(x, f)$ is defined by (2.59).

Corollary 10.7 *There is a unique worthy $(\bar{\mathscr{G}}_{t+})$-martingale measure $\{M_t(B) : t \geq 0; B \in \mathscr{B}(E)\}$ satisfying*

$$M_t(f) = \int_0^t \int_E f(x)M(ds, dx), \quad t \geq 0, f \in D(A) \tag{10.9}$$

and having covariance measure defined by

$$\eta(ds, dx, dy) = ds \int_E 2c(z)\delta_z(dx)\delta_z(dy)X_s(dz)$$
$$+ ds \int_E X_s(dz) \int_{M(E)^\circ} \nu(dx)\nu(dy)H(z, d\nu).$$

Corollary 10.8 *Let $\{M_t(B) : t \geq 0; B \in \mathscr{B}(E)\}$ be the worthy $(\bar{\mathscr{G}}_{t+})$-martingale measure defined by (10.9). Then for any $t \geq 0$ and $f \in B(E)$ we have a.s.*

$$\langle Y_t, f \rangle = \langle Y_0, \pi_t f \rangle + \int_0^t \langle \lambda, \rho_s \kappa \pi_{t-s} f \rangle ds + \int_0^t \int_E \pi_{t-s} f(x) M(ds, dx),$$

where $t \mapsto \pi_t f(x)$ is defined by (2.35).

In view of (10.8), we can interpret $\{Y_t : t \geq 0\}$ as an immigration superprocess with *stochastic immigration rate* given by $\{\rho_s(y) : s \geq 0, y \in F\}$. To prove Theorem 10.4 and its corollaries we need some preparations.

Lemma 10.9 *The results of Theorem 10.4 and its corollaries hold for $p \leq q \in \mathscr{L}^0_\lambda(F)^+$.*

Proof. Clearly, we can represent $p_s(y)$ and $q_s(y)$ in the form of (10.6) using the same sequence $\{0 = r_0 < r_1 < r_2 < \cdots\}$. Then under $\mathbf{P}\{\cdot|\bar{\mathscr{G}}_{r_i+}\}$ we can think of $y \mapsto p_{r_{i+1}}(y)$ and $y \mapsto q_{r_{i+1}}(y)$ as deterministic functions. Moreover, under the same conditional law the restriction of $N(ds, dy, du, dw)$ to $\{s > r_i\}$ is still a Poisson random measure with intensity $ds\lambda(dy)du\mathbf{Q}_\kappa(y, dw)$. By Theorem 10.1 we conclude that $\{Y_t : 0 \leq t \leq r_1\}$ has a càdlàg modification and $\{(Y_t, \bar{\mathscr{G}}_{t+}) : 0 \leq t \leq r_1\}$ is an immigration superprocess under $\mathbf{P}\{\cdot|\bar{\mathscr{G}}_{0+}\}$ with deterministic immigration rate $y \mapsto \rho_{r_1}(y)$. Next we define

$$Y_0(t) = X_t + \int_0^{r_1} \int_F \int_{p_{r_1}(y)}^{q_{r_1}(y)} \int_{D_0} w_{t-s}N(ds, dy, du, dw), \quad t \geq r_1.$$

By Theorem 10.2 one sees that $\{Y_0(t) : t \geq r_1\}$ has a càdlàg modification and $\{(Y_0(t), \bar{\mathscr{G}}_{t+}) : t \geq r_1\}$ under $\mathbf{P}\{\cdot|\bar{\mathscr{G}}_{r_1+}\}$ is a (ξ, ϕ)-superprocess. By Theorem 10.1,

$$Y_t = Y_0(t) + \int_{r_1}^t \int_F \int_{p_{r_2}(y)}^{q_{r_2}(y)} \int_{D_0} w_{t-s}N(ds, dy, du, dw), \quad r_1 \leq t \leq r_2$$

has a càdlàg modification and under $\mathbf{P}\{\cdot|\mathscr{G}_{r_1+}\}$ it is an immigration superprocess relative to $\{\mathscr{G}_{t+} : r_1 \leq t \leq r_2\}$ with deterministic immigration rate $y \mapsto \rho_{r_2}(y)$. Using the above arguments successively one can see that $\{(Y_t, \mathscr{G}_{t+}) : r_i \leq t \leq r_{i+1}\}$ under $\mathbf{P}\{\cdot|\mathscr{G}_{r_i+}\}$ is an immigration superprocess with deterministic immigration rate $y \mapsto \rho_{r_{i+1}}(y)$. Consequently, Theorem 10.4 and its corollaries follow from the results in Section 9.3. $\hfill\square$

Lemma 10.10 *Suppose that $\{\rho_k\} \subset \mathscr{L}_\lambda^0(F)^+$ and $\lim_{k\to\infty} \rho_k = \rho$ in $\mathscr{L}_\lambda^1(F)$. For $k \geq 1$ we define the càdlàg process $\{Z_k(t) : t \geq 0\}$ in $M(E)$ by*

$$Z_k(t) = X_t + \int_0^t \int_F \int_0^{\rho_k(s,y)} \int_{D_0} w_{t-s} N(ds, dy, du, dw). \qquad (10.10)$$

Then there is a subsequence $\{k_n\} \subset \{k\}$ and a càdlàg process $\{Z(t) : t \geq 0\}$ in $M(E)$ so that

$$\lim_{n\to\infty} \sup_{0 \leq s \leq t} \|Z_{k_n}(s) - Z(s)\| = 0 \qquad (10.11)$$

holds a.s. for every $t \geq 0$. Moreover, the characterizations given in Theorem 10.4 and its corollaries hold for the process $\{Z(t) : t \geq 0\}$.

Proof. For fixed $k, j \geq 1$ set $p(s,y) = \rho_j(s,y) \wedge \rho_k(s,y)$ and $q(s,y) = \rho_j(s,y) \vee \rho_k(s,y)$. Then we have

$$Z_k(t) - Z_j(t) = \int_0^t \int_F \int_{p(s,y)}^{q(s,y)} \int_{D_0} w_{t-s} 1_{\{\rho_j(s,y) < \rho_k(s,y)\}} N(ds, dy, du, dw)$$
$$- \int_0^t \int_F \int_{p(s,y)}^{q(s,y)} \int_{D_0} w_{t-s} 1_{\{\rho_j(s,y) > \rho_k(s,y)\}} N(ds, dy, du, dw).$$

It follows that

$$\|Z_k(t) - Z_j(t)\| \leq \langle Z_{j,k}(t), 1 \rangle, \qquad t \geq 0, \qquad (10.12)$$

where

$$Z_{j,k}(t) = \int_0^t \int_F \int_{p(s,y)}^{q(s,y)} \int_{D_0} w_{t-s} N(ds, dy, du, dw).$$

By applying Corollary 10.6 to the above process and using Doob's martingale inequality we get

$$\mathbf{P}\Big[\sup_{0 \leq s \leq t} \langle Z_{j,k}(s), 1 \rangle\Big] \leq (\|\gamma\| + \|b\|) \int_0^t \mathbf{P}[\langle Z_{j,k}(s), 1 \rangle] ds$$
$$+ 2\sqrt{\|q(\cdot, 1)\|} \Big(\int_0^t \mathbf{P}[\langle Z_{j,k}(s), 1 \rangle] ds\Big)^{\frac{1}{2}}$$
$$+ \|\kappa\| \int_0^t \mathbf{P}[\langle \lambda, |\rho_k(s, \cdot) - \rho_j(s, \cdot)| \rangle] ds,$$

where $\|\kappa\| = \|\kappa(\cdot, 1)\|$. By Corollary 10.5 there are locally bounded functions $t \mapsto C_1(t)$ and $t \mapsto C_2(t)$ so that

$$\mathbf{P}\left[\sup_{0 \leq s \leq t} \langle Z_{j,k}(s), 1 \rangle\right] \leq C_1(t) \int_0^t \mathbf{P}[\langle \lambda, |\rho_k(s, \cdot) - \rho_j(s, \cdot)| \rangle] \mathrm{d}s$$

$$+ C_2(t) \left(\int_0^t \mathbf{P}[\langle \lambda, |\rho_k(s, \cdot) - \rho_j(s, \cdot)| \rangle] \mathrm{d}s \right)^{\frac{1}{2}}.$$

It follows that

$$\lim_{j,k \to \infty} \mathbf{P}\left[\sup_{0 \leq s \leq t} \langle Z_{j,k}(s), 1 \rangle\right] = 0, \tag{10.13}$$

and hence there is a subsequence $\{k_n\} \subset \{k\}$ so that

$$\mathbf{P}\left[\sup_{0 \leq s \leq n} \langle Z_{k_n, k_{n+1}}(s), 1 \rangle\right] \leq 1/2^n, \qquad n \geq 1.$$

Consequently, for every $t \geq 0$,

$$\mathbf{P}\left[\sum_{n=1}^{\infty} \sup_{0 \leq s \leq t} \langle Z_{k_n, k_{n+1}}(s), 1 \rangle\right] = \sum_{n=1}^{\infty} \mathbf{P}\left[\sup_{0 \leq s \leq t} \langle Z_{k_n, k_{n+1}}(s), 1 \rangle\right] < \infty.$$

Thus we have a.s.

$$\sum_{n=1}^{\infty} \sup_{0 \leq s \leq t} \|Z_{k_n}(s) - Z_{k_{n+1}}(s)\| \leq \sum_{n=1}^{\infty} \sup_{0 \leq s \leq t} \langle Z_{k_n, k_{n+1}}(s), 1 \rangle < \infty.$$

That yields the existence of a càdlàg process $\{Z(t) : t \geq 0\}$ in $M(E)$ so that (10.11) holds a.s. for every $t \geq 0$. By Lemma 9.9 the results of Theorem 10.4 and its corollaries hold for each $\{Z_k(t) : t \geq 0\}$. Then (10.11) implies they also hold for $\{Z(t) : t \geq 0\}$. See also Jacod and Shiryaev (2003, p.385) to get the result of Corollary 10.6. $\qquad \square$

Lemma 10.11 *Let $\{Z_k(t) : t \geq 0\}$ and $\{Z(t) : t \geq 0\}$ be given as Lemma 10.10. Then for every $t \geq 0$ and $f \in B(E)$ we have*

$$\lim_{k \to \infty} \mathbf{P}\left[\sup_{0 \leq s \leq t} |\langle Z_k(s), f \rangle - \langle Z(s), f \rangle|\right] = 0. \tag{10.14}$$

Proof. Let $\{Z_{j,k}(t) : t \geq 0\}$ be defined as in the proof of Lemma 10.10. From (10.12) it follows that

$$\mathbf{P}\left[\sup_{0 \leq s \leq t} |\langle Z_j(s), f \rangle - \langle Z_k(s), f \rangle|\right] \leq \|f\| \mathbf{P}\left[\sup_{0 \leq s \leq t} |\langle Z_{j,k}(s), 1 \rangle|\right].$$

By (10.13) the right-hand side tend to zero as $j, k \to \infty$. Then (10.14) follows from (10.11) by an elementary argument. $\qquad \square$

Lemma 10.12 *Let $\rho \in \mathscr{L}^1_\lambda(F)^+$ be left continuous and assume there is a sequence $\{\rho_k\} \subset \mathscr{L}^0_\lambda(F)^+$ so that $\lim_{k\to\infty} d_1(\rho_k, \rho) = 0$ and*

$$\rho_s(\omega, y) = \lim_{k\to\infty} \rho_k(\omega, s, y), \quad \omega \in \Omega, s \geq 0, y \in F.$$

Let $\{Z_k(t) : t \geq 0\}$ be the càdlàg process in $M(E)$ defined by (10.10). Then the process

$$Z(t) = X_t + \int_0^t \int_F \int_0^{\rho_s(y)} \int_{D_0} w_{t-s} N(ds, dy, du, dw) \qquad (10.15)$$

has a càdlàg version and there is a subsequence $\{k_n\} \subset \{k\}$ so that (10.11) holds a.s. for every $t \geq 0$.

Proof. Fix a strictly positive deterministic function $h \in B(F)^+$ so that $\langle \lambda, h \rangle < \infty$. For $\alpha \geq 0$ and $k \geq 1$ let $\rho_k^\alpha(s, y) = \rho_k(s, y) + \alpha h(y)$ and let $\{Z_k^\alpha(t) : t \geq 0\}$ be defined by (10.10) with $\rho_k(s, y)$ replaced by $\rho_k^\alpha(s, y)$. By Lemma 10.10 there is a subsequence $\{k_n\} \subset \{k\}$ and a càdlàg process $\{Z^\alpha(t) : t \geq 0\}$ in $M(E)$ so that

$$\lim_{n\to\infty} \sup_{0\leq s\leq t} \|Z_{k_n}^\alpha(s) - Z^\alpha(s)\| = 0 \qquad (10.16)$$

holds a.s. for every $t \geq 0$. For any $\alpha > 0$ it is easy to see that

$$Z^0(t) \leq Z(t) \leq X^\alpha(t) \leq Z^\alpha(t),$$

where

$$X^\alpha(t) = X_t + \int_0^t \int_F \int_{(0,\rho_s^\alpha(y))} \int_{D_0} w_{t-s} N(ds, dy, du, dw)$$

with $\rho_s^\alpha(y) = \rho_s(y) + \alpha h(y)$. For any $t \geq 0$ and $f \in B(E)$ we clearly have

$$\mathbf{P}[\langle Z^\alpha(t), f \rangle] = \langle \mu, \pi_t f \rangle + \int_0^t \mathbf{P}[\langle \lambda, \rho_s^\alpha \kappa \pi_{t-s} f \rangle] ds.$$

It follows that $\mathbf{P}[\langle Z^\alpha(t), 1 \rangle] \to \mathbf{P}[\langle Z^0(t), 1 \rangle]$ as $\alpha \to 0$. Then we have a.s. $Z^0(t) = Z(t)$ for every $t \geq 0$. In other words, the càdlàg process $\{Z^0(t) : t \geq 0\}$ is a modification of $\{Z(t) : t \geq 0\}$. $\qquad \square$

Proof (of Theorem 10.4 and its corollaries). By the left continuity of $s \mapsto q_s(y)$ it is easy to define $\{q_k\} \subset \mathscr{L}^0_\lambda(F)^+$ so that $\lim_{k\to\infty} d_1(q, q_k) = 0$ and

$$q(\omega, s, y) = \lim_{k\to\infty} q_k(\omega, s, y), \quad \omega \in \Omega, s \geq 0, y \in F.$$

Similarly, there is a sequence $\{p_k\} \subset \mathscr{L}^0_\lambda(F)^+$ so that $\lim_{k\to\infty} d_1(p, p_k) = 0$ and $p(\omega, s, y) = \lim_{k\to\infty} p_k(\omega, s, y)$ for all $(\omega, s, y) \in \Omega \times [0, \infty) \times F$. Let $\{Y_k(t) : t \geq 0\}$ be defined by (10.7) with $p_s(y) = p_k(s, y)$ and $q_s(y) = q_k(s, y)$. Clearly, we can write $\{Y_k(t) : t \geq 0\}$ as the difference of two processes in the form of (10.10). A similar decomposition can be given for $\{Y_t : t \geq 0\}$. By Lemma 10.12 one sees $\{Y_t : t \geq 0\}$ has a càdlàg version and there is a subsequence $\{k_n\} \subset \{k\}$ so that (10.11) holds a.s. for every $t \geq 0$. Then the results of Theorem 10.4 and its corollaries follow immediately. $\square$

10.3 Interactive Immigration Rates

Let $\{X_t : t \geq 0\}$ and $\{N(ds, dy, du, dw)\}$ be given as in the first section. For any $R \geq 0$ let $M_R(E) = \{\nu \in M(E) : \nu(1) \leq R\}$. Suppose that $(\nu, y) \mapsto q(\nu, y)$ is a positive Borel function on $M(E) \times F$ such that $\nu \mapsto q(\nu, y)$ is continuous for every $y \in F$. In addition, we assume there is a constant $K \geq 0$ so that

$$\langle \lambda, q(\nu, \cdot) \rangle \leq K(1 + \langle \nu, 1 \rangle), \qquad \nu \in M(E), \tag{10.17}$$

and for each $R > 0$ there is a constant $K_R \geq 0$ so that

$$\langle \lambda, |q(\nu_2, \cdot) - q(\nu_1, \cdot)| \rangle \leq K_R \|\nu_2 - \nu_1\| \tag{10.18}$$

for $\nu_1, \nu_2 \in M_R(E)$. Let us consider the stochastic integral equation

$$Y_t = X_t + \int_0^t \int_F \int_0^{q(Y_{s-}, y)} \int_{D_0} w_{t-s} N(ds, dy, du, dw), \quad t \geq 0. \tag{10.19}$$

By a solution of (10.19) we mean a càdlàg process $\{Y_t : t \geq 0\}$ in $M(E)$ that is adapted to the filtration $(\mathscr{G}_{t+})$ and satisfies the equation with probability one.

Theorem 10.13 *There is a unique solution $\{Y_t : t \geq 0\}$ of (10.19). Moreover, $\{(Y_t, \mathscr{G}_{t+}) : t \geq 0\}$ is a strong Markov process and for $G \in C^2(\mathbb{R})$ and $f \in D(A)$ we have*

$$G(\langle Y_t, f \rangle) = G(\langle Y_0, f \rangle) + \int_0^t \Big\{ G'(\langle Y_s, f \rangle) \langle Y_s, Af + \gamma f - bf \rangle$$
$$+ \int_E Y_s(dx) \int_{M(E)^\circ} \Big[G(\langle Y_s, f \rangle + \langle \nu, f \rangle) - G(\langle Y_s, f \rangle)$$
$$- \langle \nu, f \rangle G'(\langle Y_s, f \rangle) \Big] H(x, d\nu) + G''(\langle Y_s, f \rangle) \langle Y_s, cf^2 \rangle$$
$$+ G'(\langle Y_s, f \rangle) \langle \lambda, q(Y_s, \cdot) \kappa f \rangle \Big\} ds + (\mathscr{G}_{t+})\text{-local mart.} \tag{10.20}$$

In view of (10.19) and (10.20), we may interpret $\{Y_t : t \geq 0\}$ as an immigration superprocess with *interactive immigration rate* given by the two-parameter process $(s, y) \mapsto q(Y_{s-}, y)$. The uniqueness of solution of the martingale problem (10.20) still remains open. Then the theorem above shows the advantage of the stochastic equation (10.19) in constructing the interactive immigration superprocess as a Markov process. Recall that $\|\pi_t f\| \leq e^{c_0 t} \|f\|$ for all $t \geq 0$ and $f \in B(E)$.

Lemma 10.14 *Let $R \geq 0$ be a constant and let q_1 and q_2 be positive Borel functions on $M(E) \times F$ satisfying (10.17) and (10.18) and assume*

$$q_1(\nu, y) = q_2(\nu, y) = q(\nu, y), \quad y \in F, \nu \in M_R(E).$$

Suppose that $\{Y_1(t) : t \geq 0\}$ and $\{Y_2(t) : t \geq 0\}$ are solutions of (10.19) with q replaced by q_1 and q_2, respectively. Let

$$\tau = \inf\{t \geq 0 : \langle Y_1(t), 1 \rangle \geq R \text{ or } \langle Y_2(t), 1 \rangle \geq R\}.$$

Then $\{Y_1(t \wedge \tau) : t \geq 0\}$ and $\{Y_2(t \wedge \tau) : t \geq 0\}$ are indistinguishable.

Proof. For $s \geq 0$ and $y \in F$ let

$$q_*(s, y) = q_1(Y_1(s), y) \wedge q_2(Y_2(s), y)$$

and

$$q^*(s, y) = q_1(Y_1(s), y) \vee q_2(Y_2(s), y).$$

Then $\|Y_1(t \wedge \tau) - Y_2(t \wedge \tau)\| \leq \langle Z(t), 1 \rangle$, where

$$\begin{aligned}
Z(t) &= \int_0^{t \wedge \tau} \int_F \int_{q_*(s-,y)}^{q^*(s-,y)} \int_{D_0} w_{t-s} N(ds, dy, du, dw) \\
&= \int_0^t \int_F \int_{1_{\{s \leq \tau\}} q_*(s-,y)}^{1_{\{s \leq \tau\}} q^*(s-,y)} \int_{D_0} w_{t-s} N(ds, dy, du, dw).
\end{aligned}$$

We next use Corollary 10.5 to obtain

$$\begin{aligned}
\mathbf{P}[\langle Z(t), 1 \rangle] &= \mathbf{P}\left[\int_0^t \langle \lambda, 1_{\{s \leq \tau\}} [q^*(s-, \cdot) - q_*(s-, \cdot)] \kappa \pi_{t-s} 1 \rangle ds \right] \\
&\leq \|\kappa\| e^{c_0 t} \mathbf{P}\left[\int_0^t \langle \lambda, 1_{\{s \leq \tau\}} |q(Y_1(s), \cdot) - q(Y_2(s), \cdot)| \rangle ds \right] \\
&\leq K_R \|\kappa\| e^{c_0 t} \mathbf{P}\left[\int_0^t 1_{\{s \leq \tau\}} \|Y_1(s) - Y_2(s)\| ds \right] \\
&\leq K_R \|\kappa\| e^{c_0 t} \int_0^t \mathbf{P}[\langle Z(s), 1 \rangle] ds,
\end{aligned}$$

where $\|\kappa\| = \|\kappa(\cdot, 1)\|$. By Gronwall's inequality one sees $\mathbf{P}[\langle Z(t), 1 \rangle] = 0$ for all $t \geq 0$. Then $\{Y_1(t \wedge \tau) : t \geq 0\}$ and $\{Y_2(t \wedge \tau) : t \geq 0\}$ are indistinguishable. $\square$

Lemma 10.15 *There is at most one solution of (10.19).*

Proof. Suppose $\{Y_1(t) : t \geq 0\}$ and $\{Y_2(t) : t \geq 0\}$ are two solutions of (10.19). Let $\tau_n = \inf\{t \geq 0 : \langle Y_1(t), 1 \rangle \geq n \text{ or } \langle Y_2(t), 1 \rangle \geq n\}$. By Lemma 10.14, the processes $\{Y_1(t \wedge \tau_n) : t \geq 0\}$ and $\{Y_2(t \wedge \tau_n) : t \geq 0\}$ are indistinguishable for every $n \geq 1$. Then $\{Y_1(t) : t \geq 0\}$ and $\{Y_2(t) : t \geq 0\}$ are indistinguishable. □

Lemma 10.16 *The results of Theorem 10.13 hold if there is a universal constant* $K \geq 0$ *so that*

$$\langle \lambda, |q(\nu_2, \cdot) - q(\nu_1, \cdot)| \rangle \leq K \|\nu_2 - \nu_1\|, \quad \nu_1, \nu_2 \in M(E). \quad (10.21)$$

Proof. By Lemma 10.15 the uniqueness of solution holds for (10.19). We shall prove the existence of the solution using an approximating argument. Let $Y_0(t) = X_t$ and define inductively

$$Y_k(t) = X_t + \int_0^t \int_F \int_0^{q(Y_{k-1}(s-),y)} \int_{D_0} w_{t-s} N(\mathrm{d}s, \mathrm{d}y, \mathrm{d}u, \mathrm{d}w). \quad (10.22)$$

In the remainder of this proof we write $q_k(s, y) = q(Y_{k-1}(s-), y)$ for simplicity. Let $Y_{0,1}(t) = Y_1(t) - Y_0(t)$ and for $k > j \geq 1$ let

$$Y_{j,k}(t) = \int_0^t \int_F \int_{q_j(s,y) \wedge q_k(s,y)}^{q_j(s,y) \vee q_k(s,y)} \int_{D_0} w_{t-s} N(\mathrm{d}s, \mathrm{d}y, \mathrm{d}u, \mathrm{d}w). \quad (10.23)$$

Then $\|Y_k(t) - Y_j(t)\| \leq \langle Y_{j,k}(t), 1 \rangle$. By Corollary 10.5 it is easy to see that

$$\mathbf{P}[\langle Y_{k,k+1}(t), 1 \rangle] = \mathbf{P}\left[\int_0^t \langle \lambda, |q_{k+1}(s, \cdot) - q_k(s, \cdot)| \kappa \pi_{t-s} 1 \rangle \mathrm{d}s\right]$$

$$\leq \|\kappa\| e^{c_0 t} \mathbf{P}\left[\int_0^t \langle \lambda, |q_{k+1}(s, \cdot) - q_k(s, \cdot)| \rangle \mathrm{d}s\right].$$

Then we use (10.21) to see

$$\langle \lambda, |q_{k+1}(s, \cdot) - q_k(s, \cdot)| \rangle \leq K \|Y_k(s-) - Y_{k-1}(s-)\|$$
$$\leq K \langle Y_{k-1,k}(s-), 1 \rangle. \quad (10.24)$$

It follows that

$$\mathbf{P}[\langle Y_{k,k+1}(t), 1 \rangle] \leq K \|\kappa\| e^{c_0 t} \mathbf{P}\left[\int_0^t \langle Y_{k-1,k}(s-), 1 \rangle \mathrm{d}s\right]$$

$$\leq K \|\kappa\| e^{c_0 t} \int_0^t \mathbf{P}[\langle Y_{k-1,k}(s), 1 \rangle] \mathrm{d}s.$$

Using Corollary 10.5 and (10.17) we get

$$\mathbf{P}[\langle Y_{0,1}(t), 1 \rangle] = \mathbf{P}\left[\int_0^t \langle \lambda, q(X_{s-}, \cdot) \kappa \pi_{t-s} 1 \rangle \mathrm{d}s\right]$$

$$\leq K \|\kappa\| e^{c_0 t} \int_0^t \mathbf{P}[1 + \langle X_s, 1 \rangle] \mathrm{d}s.$$

By a standard argument, one shows

$$\sum_{k=1}^{\infty} \sup_{0 \le s \le t} \mathbf{P}[\langle Y_{k-1,k}(s), 1\rangle] < \infty.$$

In view of (10.24) we have

$$\sum_{k=1}^{\infty} \sup_{0 \le s \le t} \mathbf{P}[\langle \lambda, |q_{k+1}(s, \cdot) - q_k(s, \cdot)|\rangle] < \infty.$$

It follows that

$$\lim_{j,k \to \infty} \int_0^t \mathbf{P}[\langle \lambda, |q_k(s, \cdot) - q_j(s, \cdot)|\rangle] ds = 0.$$

Then there exists $\rho \in \mathscr{L}_\lambda^1(F)^+$ so that $d_1(q_k, \rho) \to 0$ as $k \to \infty$. By the arguments in the proof of Lemma 10.10 there is a subsequence $\{k_n\} \subset \{k\}$ and a càdlàg process $\{Y(t) : t \ge 0\}$ so that

$$\lim_{n \to \infty} \sup_{0 \le s \le t} \|Y_{k_n}(s) - Y(s)\| = 0$$

holds a.s. for every $t \ge 0$. Now the continuity of $\nu \mapsto q(\nu, y)$ implies

$$\lim_{n \to \infty} q_{k_n+1}(s, y) = \lim_{n \to \infty} q(Y_{k_n}(s-), y) = q(Y(s-), y)$$

for all $s \ge 0$ and $y \in F$ on an event with full probability. Then we may identify $(s, y) \mapsto \rho(s, y)$ and $(s, y) \mapsto q(Y(s-), y)$ as elements of $\mathscr{L}_\lambda^1(F)$. The arguments in the proof of Lemma 10.11 yield

$$\lim_{k \to \infty} \mathbf{P}\left[\sup_{0 \le s \le t} |\langle Y(s), f\rangle - \langle Y_k(s), f\rangle| \right] = 0$$

for every $t \ge 0$ and $f \in B(E)$. By letting $k \to \infty$ along $\{k_n + 1\}$ in (10.22) and arguing as in the proof of Lemma 10.12 we see $\{Y(t) : t \ge 0\}$ is a solution of (10.19). The martingale problem characterization (10.20) follows by Theorem 10.4. $\qquad \square$

We now use a localization method to establish Theorem 10.13 under the general conditions (10.17) and (10.18). For any integer $n \ge 1$ it is easy to define a continuously differentiable function $z \mapsto a_n(z)$ on $[0, \infty)$ such that

$$a_n(z) = \begin{cases} 1 & \text{for } 0 \le z \le n-1, \\ n/z & \text{for } z \ge n+1. \end{cases}$$

In addition, we can also assume $0 \le a_n(z) \le 1 \wedge (n/z)$ and $-1/z \le a_n'(z) \le 0$ for all $z > 0$.

Lemma 10.17 *Suppose that* $(\nu, y) \mapsto q(\nu, y)$ *is a positive Borel function on* $M(E) \times F$ *satisfying* (10.18). *Then* $q_n(\nu, y) := q(a_n(\langle \nu, 1 \rangle)\nu, y)$ *defines a positive Borel function on* $M(E) \times F$ *satisfying*

$$\langle \lambda, |q_n(\nu_2, \cdot) - q_n(\nu_1, \cdot)| \rangle \leq 2K_n \|\nu_2 - \nu_1\|, \quad \nu_1, \nu_2 \in M(E). \quad (10.25)$$

Proof. For $\nu_1, \nu_2 \in M(E)$ satisfying $\langle \nu_1, 1 \rangle \leq \langle \nu_2, 1 \rangle$ we can find a constant $s \geq 0$ between $\langle \nu_1, 1 \rangle$ and $\langle \nu_2, 1 \rangle$ so that

$$\langle \nu_1, 1 \rangle |a_n(\langle \nu_2, 1 \rangle) - a_n(\langle \nu_1, 1 \rangle)| \leq \langle \nu_1, 1 \rangle |a_n'(s)||\langle \nu_2, 1 \rangle - \langle \nu_1, 1 \rangle|$$
$$\leq \|\nu_2 - \nu_1\|.$$

Let $\gamma = \nu_1 + \nu_2$ and let g_{ν_1} and g_{ν_2} denote respectively the densities of ν_1 and ν_2 with respect to γ. It follows that

$$
\begin{aligned}
\text{l.h.s. of } (10.25) &\leq K_n \|a_n(\langle \nu_2, 1 \rangle)\nu_2 - a_n(\langle \nu_1, 1 \rangle)\nu_1\| \\
&= K_n \langle \gamma, |a_n(\langle \nu_2, 1 \rangle)g_{\nu_2} - a_n(\langle \nu_1, 1 \rangle)g_{\nu_1}| \rangle \\
&\leq K_n a_n(\langle \nu_2, 1 \rangle)\langle \gamma, |g_{\nu_2} - g_{\nu_1}| \rangle \\
&\quad + K_n |a_n(\langle \nu_2, 1 \rangle) - a_n(\langle \nu_1, 1 \rangle)|\langle \gamma, g_{\nu_1} \rangle \\
&\leq K_n \|\nu_2 - \nu_1\| + K_n |a_n(\langle \nu_2, 1 \rangle) - a_n(\langle \nu_1, 1 \rangle)|\langle \nu_1, 1 \rangle \\
&\leq 2K_n \|\nu_2 - \nu_1\|.
\end{aligned}
$$

That proves the desired inequality.   □

Proof (of Theorem 10.13). By Lemma 10.15 the uniqueness of solution holds for (10.19). For each integer $n \geq 1$ let q_n be defined as in Lemma 10.17. By Lemma 10.16 there is an unique solution $\{Y_n(t) : t \geq 0\}$ to (10.19) if q is replaced by q_n. For each $n \geq 1$ define the stopping time $\sigma_n = \inf\{t \geq 0 : \langle Y_n(t), 1 \rangle \geq n - 1\}$. Then Lemma 10.14 implies that $\{Y_m(t \wedge \sigma_m) : t \geq 0\}$ and $\{Y_n(t \wedge \sigma_m) : t \geq 0\}$ are indistinguishable for any $n \geq m \geq 1$. It is easy to show $\{\sigma_n\}$ is an increasing sequence. Since $(P_t)_{t \geq 0}$ is a Feller semigroup, the result of Corollary 7.14 is also true for $f = 1$. Consequently,

$$t \mapsto C(t) := \mathbf{P}[\langle X_{t \wedge \sigma_n}, 1 \rangle]$$

is a locally bounded function on $[0, \infty)$. From (10.17) it follows that

$$
\begin{aligned}
&\mathbf{P}[1 + \langle Y_n(t \wedge \sigma_n), 1 \rangle] \\
&= \mathbf{P}[1 + \langle X_{t \wedge \sigma_n}, 1 \rangle] + \mathbf{P}\left[\int_0^{t \wedge \sigma_n} \langle \lambda, q(Y_n(s-), \cdot)\kappa \pi_{t-s} 1 \rangle ds \right] \\
&\leq 1 + C(t) + K\|\kappa\|e^{c_0 t} \int_0^t \mathbf{P}[1 + \langle Y_n(s \wedge \sigma_n), 1 \rangle] ds,
\end{aligned}
$$

By Gronwall's inequality, there is a locally bounded function $t \mapsto B(t)$ on $[0, \infty)$ independent of $n \geq 1$ such that

$$\mathbf{P}[1 + \langle Y_n(t \wedge \sigma_n), 1\rangle] \leq B(t), \qquad t \geq 0. \tag{10.26}$$

The right continuity of $t \mapsto Y_n(t)$ implies $\langle Y_n(\sigma_n), 1\rangle \geq n-1$. By (10.26) we have $n\mathbf{P}\{\sigma_n \leq t\} \leq B(t)$ and so $\mathbf{P}\{\sigma_n \leq t\} \to 0$ as $n \to \infty$. That implies $\sigma_n \to \infty$ increasingly. Thus there is a càdlàg process $\{Y_t : t \geq 0\}$ such that a.s. $Y_n(t) = Y_t$ for every $0 \leq t \leq \sigma_n$. By (10.26) and Fatou's lemma we get $\mathbf{P}[1 + \langle Y_t, 1\rangle] \leq B(t)$ for every $t \geq 0$. It is then easy to show $(s, y) \mapsto q(Y_{s-}, y)$ belongs to $\mathscr{L}_\lambda^1(F)$. Clearly the equation (10.19) holds. The martingale problem characterization (10.20) follows from Theorem 10.4. The uniqueness of the solution of (10.19) implies the strong Markov property of $\{(Y_t, \bar{\mathscr{G}}_{t+}) : t \geq 0\}$. □

Example 10.1 Suppose that $p(x, \mathrm{d}y)$ is a bounded kernel from E to F absolutely continuous with respect to $\lambda(\mathrm{d}y)$ with density $p(x, y)$ so that $\sup_{x \in E} p(x, y) < \infty$ and $x \mapsto p(x, y)$ is continuous for every $y \in F$. Suppose that $r \in C(M(E) \times F)^+$ is a function satisfying

$$|r(\nu_2, y) - r(\nu_1, y)| \leq K\|\nu_2 - \nu_1\|, \quad \nu_1, \nu_2 \in M(E), y \in F$$

for a constant $K \geq 0$. For $\nu \in M(E)$ and $z \in F$ let

$$q(\nu, y) = r(\nu, y)\langle \nu, p(\cdot, y)\rangle = r(\nu, y)\int_E p(x, y)\nu(\mathrm{d}x).$$

Then $\nu \mapsto q(\nu, y)$ is continuous on $M(E)$ and

$$\langle \lambda, q(\nu, \cdot)\rangle = \int_F r(\nu, y)\lambda(\mathrm{d}y)\int_E p(x, y)\nu(\mathrm{d}x) \leq \|r\|\|p\|\langle \nu, 1\rangle,$$

where $\|p\| = \sup_{x \in E} p(x, 1)$. For $\nu_1, \nu_2 \in M(E)$ we have

$$\begin{aligned}
\langle \lambda, |q(\nu_2, \cdot) - q(\nu_1, \cdot)|\rangle &= \int_F |\langle r(\nu_2, y)\nu_2 - r(\nu_1, y)\nu_1, p(\cdot, y)\rangle|\lambda(\mathrm{d}y) \\
&\leq \int_F |r(\nu_2, y) - r(\nu_1, y)|\langle \nu_2, p(\cdot, y)\rangle \lambda(\mathrm{d}y) \\
&\quad + \int_F r(\nu_1, y)|\langle \nu_2 - \nu_1, p(\cdot, y)\rangle|\lambda(\mathrm{d}y) \\
&\leq K\|\nu_2 - \nu_1\|\int_F \langle \nu_2, p(\cdot, y)\rangle \lambda(\mathrm{d}y) \\
&\quad + \|r\|\int_F |\langle |\nu_2 - \nu_1|, p(\cdot, y)\rangle|\lambda(\mathrm{d}y) \\
&\leq K\|\nu_2 - \nu_1\|\|p\|\langle \nu_2, 1\rangle + \|r\|\|p\|\|\nu_2 - \nu_1\|.
\end{aligned}$$

Then $(\nu, y) \mapsto q(\nu, y)$ satisfies conditions (10.17) and (10.18). Let $\{Y_t : t \geq 0\}$ be the unique càdlàg solution of (10.19). By Theorem 10.13 it is easy to see

$$G(\langle Y_t, f\rangle) = G(\langle Y_0, f\rangle) + \int_0^t \Big\{ G'(\langle Y_s, f\rangle)\langle Y_s, Af + \gamma f - bf\rangle$$

$$+ \int_E Y_s(\mathrm{d}x) \int_{M(E)^\circ} \Big[G(\langle Y_s, f \rangle + \langle \nu, f \rangle) - G(\langle Y_s, f \rangle)$$

$$- \langle \nu, f \rangle G'(\langle Y_s, f \rangle) \Big] H(x, \mathrm{d}\nu) + G''(\langle Y_s, f \rangle) \langle Y_s, cf^2 \rangle$$

$$+ G'(\langle Y_s, f \rangle) \langle Y_s, \kappa_0(Y_s, \cdot, f) \rangle \Big\} \mathrm{d}s + (\mathscr{G}_{t+})\text{-local mart.}$$

for $G \in C^2(\mathbb{R})$ and $f \in D(A)$, where

$$\kappa_0(\nu, x, f) = \int_F r(\nu, y) p(x, y) \kappa f(y) \lambda(\mathrm{d}y). \tag{10.27}$$

We can interpret $\{Y_t : t \geq 0\}$ as a superprocess with an extra interactive non-local branching mechanism given by (10.27).

10.4 General Interactive Immigration

In this section, we give some generalizations of the immigration models considered in the previous sections. Suppose that F_0 and F_1 are Lusin topological spaces. Let $\lambda_0(\mathrm{d}y)$ and $\lambda_1(\mathrm{d}y)$ be σ-finite Borel measures on F_0 and F_1, respectively. Let $\kappa(y, \mathrm{d}x)$ be a bounded kernel from F_0 to E and let $K(y, \mathrm{d}\nu)$ be a kernel from F_1 to $M(E)^\circ$ satisfying

$$\sup_{y \in F_1} \int_{M(E)^\circ} \langle \nu, 1 \rangle K(y, \mathrm{d}\nu) < \infty. \tag{10.28}$$

Suppose that $(D_0, \mathscr{A}^0, \mathscr{A}_t^0, \mathbf{Q}_\nu)$ is the canonical càdlàg realization of the (ξ, ϕ)-superprocess. Let $\{(X_t, \mathscr{F}_t) : t \geq 0\}$ be a càdlàg (ξ, ϕ)-superprocess with deterministic initial state $X_0 = \mu \in M(E)$. For $i = 0, 1$ let $\{N_i(\mathrm{d}s, \mathrm{d}y, \mathrm{d}u, \mathrm{d}w)\}$ be a Poisson random measure on $(0, \infty) \times F_i \times (0, \infty) \times D_0$ with intensity $\mathrm{d}s\lambda_i(\mathrm{d}y)\mathrm{d}u\mathbf{Q}_i(y, \mathrm{d}w)$, where

$$\mathbf{Q}_0(y, \mathrm{d}w) = \int_E \kappa(y, \mathrm{d}x) \mathbf{Q}_{L(x)}(\mathrm{d}w), \quad y \in F_0, w \in D_0,$$

and

$$\mathbf{Q}_1(y, \mathrm{d}w) = \int_{M(E)^\circ} K(y, \mathrm{d}\nu) \mathbf{Q}_\nu(\mathrm{d}w), \quad y \in F_1, w \in D_0.$$

We assume that the process $\{(X_t, \mathscr{F}_t) : t \geq 0\}$ and the Poisson random measures $\{N_0(\mathrm{d}s, \mathrm{d}y, \mathrm{d}u, \mathrm{d}w)\}$ and $\{N_1(\mathrm{d}s, \mathrm{d}y, \mathrm{d}u, \mathrm{d}w)\}$ are defined on a complete probability space $(\Omega, \mathscr{G}, \mathbf{P})$ and are independent of each other. For $t \geq 0$ let $\mathscr{G}_t$ be the σ-algebra generated by $\mathscr{F}_t$ and the collection of random variables

$$\{N_i((0,s] \times U \times A) : U \in \mathscr{B}(F_i \times (0,\infty)), A \in \mathscr{A}_{t-s}^0, 0 \le s \le t; i = 0, 1\}.$$

Let $(\bar{\mathscr{G}}_t)$ denote the augmentation of $(\mathscr{G}_t)$. Let $\mathscr{L}_{\lambda_0}^1(F_0)$ be the set of processes $\{h_t(y) : t \ge 0, y \in F_0\}$ that are predictable relative to the filtration $(\bar{\mathscr{G}}_{t+})$ and satisfy (10.4) with $\lambda = \lambda_0$. Let $\mathscr{L}_{\lambda_1}^1(F_1)$ be defined similarly. Given left continuous processes $\rho_0 \in \mathscr{L}_{\lambda_0}^1(F_0)^+$ and $\rho_1 \in \mathscr{L}_{\lambda_1}^1(F_1)^+$, for $t \ge 0$ we define

$$Y_t = X_t + \int_0^t \int_{F_0} \int_0^{\rho_0(s,y)} \int_{D_0} w_{t-s} N_0(ds, dy, du, dw)$$
$$+ \int_0^t \int_{F_1} \int_0^{\rho_1(s,y)} \int_{D_0} w_{t-s} N_1(ds, dy, du, dw). \tag{10.29}$$

Theorem 10.18 *The process $\{Y_t : t \ge 0\}$ defined by (10.29) has a càdlàg modification in $M(E)$ with the following property: For every $G \in C^2(\mathbb{R})$ and $f \in D(A)$ we have*

$$G(\langle Y_t, f \rangle) = G(\langle Y_0, f \rangle) + \int_0^t \Big\{ G'(\langle Y_s, f \rangle) \langle Y_s, Af + \gamma f - bf \rangle$$
$$+ \int_E Y_s(dx) \int_{M(E)^\circ} \Big[G(\langle Y_s, f \rangle + \langle \nu, f \rangle)$$
$$- G(\langle Y_s, f \rangle) - \langle \nu, f \rangle G'(\langle Y_s, f \rangle) \Big] H(x, d\nu)$$
$$+ G''(\langle Y_s, f \rangle) \langle Y_s, cf^2 \rangle + G'(\langle Y_s, f \rangle) \langle \lambda_0, \rho_0(s, \cdot) \kappa f \rangle$$
$$+ \int_{F_1} \rho_1(s, y) \lambda_1(dy) \int_{M(E)^\circ} \Big[G(\langle Y_s, f \rangle + \langle \nu, f \rangle)$$
$$- G(\langle Y_s, f \rangle) \Big] K(y, d\nu) \Big\} ds + (\bar{\mathscr{G}}_{t+})\text{-local mart.} \tag{10.30}$$

Proof. By making an extension of $(\Omega, \mathscr{G}, \mathbf{P})$ we can assume $N_1(ds, dy, du, dw)$ is the projection of a Poisson random measure $\bar{N}_1(ds, dy, du, d\nu, dw)$ on $(0, \infty) \times F_1 \times (0, \infty) \times M(E)^\circ \times D_0$ with intensity $ds\lambda_1(dy)duK(y, d\nu)\mathbf{Q}_\nu(dw)$. We can also assume $\{\bar{N}_1(ds, dy, du, d\nu, dw)\}$ is independent of $\{(X_t, \mathscr{F}_t) : t \ge 0\}$ and $\{N_0(ds, dy, du, dw)\}$. For every $k \ge 1$ define

$$Y_k(t) = X_t + \int_0^t \int_{F_0} \int_0^{\rho_0(s,y)} \int_{D_0} w_{t-s} N_0(ds, dy, du, dw)$$
$$+ \int_0^t \int_{F_1} \int_0^{\rho_1(s,y)} \int_{M_k^\circ} \int_{D_0} w_{t-s} \bar{N}_1(ds, dy, du, d\nu, dw),$$

where $M_k^\circ = \{\nu \in M(E)^\circ : \langle \nu, 1 \rangle \le k\}$. Following the arguments in Section 10.2, one can show that the results of the theorem hold for $\{Y_k(t) : t \ge 0\}$ with $M(E)^\circ$ replaced by M_k°. (This cutoff is necessary in defining the covariance measure of the induced martingale measure.) In particular, we have

$$G(\langle Y_k(t), f \rangle) = G(\langle Y_k(0), f \rangle) + \int_0^t \Big\{ G'(\langle Y_k(s), f \rangle) \langle Y_k(s), Af + \gamma f - bf \rangle$$

$$
\begin{aligned}
&+ \int_E Y_k(s, dx) \int_{M(E)^\circ} \Big[G(\langle Y_k(s), f \rangle + \langle \nu, f \rangle) \\
&- G(\langle Y_k(s), f \rangle) - \langle \nu, f \rangle G'(\langle Y_k(s), f \rangle) \Big] H(x, d\nu) \\
&+ G''(\langle Y_k(s), f \rangle) \langle Y_k(s), cf^2 \rangle + G'(\langle Y_k(s), f \rangle) \langle \lambda_0, \rho_0(s, \cdot) \kappa f \rangle \\
&+ \int_{F_1} \rho_1(s, y) \lambda_1(dy) \int_{M_k^\circ} \Big[G(\langle Y_k(s), f \rangle + \langle \nu, f \rangle) \\
&- G(\langle Y_k(s), f \rangle) \Big] K(y, d\nu) \Big\} ds + (\mathscr{G}_{t+})\text{-local mart.} \quad (10.31)
\end{aligned}
$$

For any $k \geq 1$ let $T_k = \inf\{t \geq 0 : \bar{N}_1(A_k(t)) \geq 1\}$, where

$$
A_k(t) = \{(s, y, u, \nu, w) : 0 < s \leq t, \langle \nu, 1 \rangle > k, 0 < u \leq \rho_1(s, y), w \in D_0\}.
$$

It is easily seen that $Y_t = Y_k(t)$ for $0 \leq t < T_k$. Observe also that

$$
\begin{aligned}
\mathbf{P}[\bar{N}_1(A_k(t))] &= \mathbf{P}\left[\int_0^t ds \int_{F_1} \rho_1(s, y) \lambda_1(dy) \int_{\{\langle \nu, 1 \rangle > k\}} K(y, d\nu) \right] \\
&\leq \frac{1}{k} \int_0^t ds \int_{F_1} \mathbf{P}[\rho_1(s, y)] \lambda_1(dy) \int_{M(E)^\circ} \langle \nu, 1 \rangle K(y, d\nu).
\end{aligned}
$$

The right-hand side tends to zero as $k \to \infty$. Then $\bar{N}_1(A_k(t)) \to 0$ decreasingly and hence $T_k \to \infty$ increasingly as $k \to \infty$. From (10.31) we get (10.30). $\quad\square$

Corollary 10.19 *Let $\{Y_t : t \geq 0\}$ be defined by (10.29). Then $t \mapsto \mathbf{P}[\langle Y_t, 1 \rangle]$ is locally bounded and*

$$
\begin{aligned}
\mathbf{P}[\langle Y_t, f \rangle] = \langle \mu, \pi_t f \rangle &+ \int_0^t ds \int_{F_0} \mathbf{P}[\rho_0(s, y)] \kappa \pi_{t-s} f(y) \lambda_0(dy) \\
&+ \int_0^t ds \int_{F_1} \mathbf{P}[\rho_1(s, y)] \lambda_1(dy) \int_{M(E)^\circ} \langle \nu, \pi_{t-s} f \rangle K(y, d\nu)
\end{aligned}
$$

for any $t \geq 0$ and $f \in B(E)$.

Corollary 10.20 *Let $\{Y_t : t \geq 0\}$ be defined by (10.29). If, in addition to (10.28), we assume*

$$
\sup_{y \in F_1} \int_{M(E)^\circ} \langle \nu, 1 \rangle^2 K(y, d\nu) < \infty, \quad (10.32)
$$

then for every $f \in D(A)$,

$$
\begin{aligned}
M_t(f) = \langle Y_t, f \rangle - \langle Y_0, f \rangle &- \int_0^t \langle Y_s, Af + \gamma f - bf \rangle ds \\
&- \int_0^t \langle \lambda_0, \rho_0(s, \cdot) \kappa f \rangle ds \\
&- \int_0^t ds \int_{F_1} \rho_1(s, y) \lambda_1(dy) \int_{M(E)^\circ} \langle \nu, f \rangle K(y, d\nu)
\end{aligned}
$$

is a square-integrable $(\bar{\mathscr{G}}_{t+})$-martingale with increasing process

$$\langle M(f)\rangle_t = \int_0^t \left[\int_E q(x,f)Y_s(dx) + \int_{F_1} \rho_1(s,y)\lambda_1(dy) \int_{M(E)^\circ} \langle \nu,f\rangle^2 K(y,d\nu) \right] ds,$$

where $q(x,f)$ is defined by (2.59).

Let $(\nu,y) \mapsto q_0(\nu,y)$ be a positive Borel function on $M(E) \times F_0$ satisfying
(10.17) and (10.18) with $\lambda = \lambda_0$. Let $(\nu,y) \mapsto q_1(\nu,y)$ be a positive Borel function
on $M(E) \times F_1$ satisfying conditions (10.17) and (10.18) with $\lambda = \lambda_1$. In addition,
assume that $\nu \mapsto q_0(\nu,y)$ is continuous for every $y \in F_0$ and $\nu \mapsto q_1(\nu,y)$ is
continuous for every $y \in F_1$. We consider the equation

$$Y_t = X_t + \int_0^t \int_{F_0} \int_0^{q_0(Y_{s-},y)} \int_{D_0} w_{t-s}N_0(ds,dy,du,dw)$$
$$+ \int_0^t \int_{F_1} \int_0^{q_1(Y_{s-},y)} \int_{D_0} w_{t-s}N_1(ds,dy,du,dw). \tag{10.33}$$

Theorem 10.21 *There is a unique càdlàg solution $\{Y_t : t \geq 0\}$ of (10.33). Moreover, $\{(Y_t,\bar{\mathscr{G}}_{t+}) : t \geq 0\}$ is a strong Markov process and for $G \in C^2(\mathbb{R})$ and
$f \in D(A)$ we have*

$$G(\langle Y_t,f\rangle) = G(\langle Y_0,f\rangle) + \int_0^t \Big\{ G'(\langle Y_s,f\rangle)\langle Y_s, Af + \gamma f - bf\rangle$$
$$+ \int_E Y_s(dx) \int_{M(E)^\circ} \Big[G(\langle Y_s,f\rangle + \langle \nu,f\rangle)$$
$$- G(\langle Y_s,f\rangle) - \langle \nu,f\rangle G'(\langle Y_s,f\rangle) \Big] H(x,d\nu)$$
$$+ G''(\langle Y_s,f\rangle)\langle Y_s, cf^2\rangle + G'(\langle Y_s,f\rangle)\langle \lambda_0, q_0(Y_s,\cdot)\kappa f\rangle$$
$$+ \int_{F_1} q_1(Y_s,y)\lambda_1(dy) \int_{M(E)^\circ} \Big[G(\langle Y_s,f\rangle + \langle \nu,f\rangle)$$
$$- G(\langle Y_s,f\rangle) \Big] K(y,d\nu) \Big\} ds + (\bar{\mathscr{G}}_{t+})\text{-local mart.} \tag{10.34}$$

Proof. Let $\{\bar{N}_1(ds,dy,du,d\nu,dw)\}$ be as in the proof of Theorem 10.18. For every
$k \geq 1$ one can use the arguments in Section 10.3 to show there is a unique solution
$\{Y_k(t) : t \geq 0\}$ of

$$Y_t = X_t + \int_0^t \int_{F_0} \int_0^{q_0(Y_{s-},y)} \int_{D_0} w_{t-s}N_0(ds,dy,du,dw)$$
$$+ \int_0^t \int_{F_1} \int_0^{q_1(Y_{s-},y)} \int_{M_k^\circ} \int_{D_0} w_{t-s}\bar{N}_1(ds,dy,du,d\nu,dw),$$

where $M_k^\circ = \{\nu \in M(E)^\circ : \langle \nu,1\rangle \leq k\}$. Let $\tau_k = \inf\{t \geq 0 : \langle Y_k(t),1\rangle \geq k\}$.
By Gronwall's inequality one shows

$$\mathbf{P}[\langle Y_k(t \wedge \tau_k), 1\rangle] \le B(t), \quad t \ge 0, \tag{10.35}$$

where $t \mapsto B(t)$ is a locally bounded function independent of $k \ge 1$. By a standard argument one shows $\tau_k \to \infty$ increasingly as $k \to \infty$. Clearly, for any $n \ge k \ge 1$ the two processes $\{Y_k(t) : t \ge 0\}$ and $\{Y_n(t) : t \ge 0\}$ coincide in the time interval $[0, \tau_k)$. Then we can define a process $\{Y_t : t \ge 0\}$ which agrees with $\{Y_k(t) : t \ge 0\}$ on $[0, \tau_k)$. By (10.35) and Fatou's lemma we have

$$\mathbf{P}[\langle Y_t, 1\rangle] \le B(t), \quad t \ge 0.$$

It follows that $(s, y) \mapsto q_0(Y_{s-}, y)$ belongs to $\mathscr{L}_{\lambda_0}^1(F_0)^+$ and $(s, y) \mapsto q_1(Y_{s-}, y)$ belongs to $\mathscr{L}_{\lambda_1}^1(F_1)^+$. Then one can see $\{Y_t : t \ge 0\}$ solves (10.33). The uniqueness of the solution follows as that for (10.19). Then $\{(Y_t, \bar{\mathscr{G}}_{t+}) : t \ge 0\}$ is a strong Markov process. The characterization (10.34) follows from Theorem 10.18. We omit the details. □

Example 10.2 Let $p_i(x, \mathrm{d}y) = p_i(x, y)\lambda_i(\mathrm{d}y)$ be a bounded kernel from E to F_i for $i = 0$ and 1. We assume $\sup_{x \in E} p_i(x, y) < \infty$ and $x \mapsto p_i(x, y)$ is continuous for every $y \in F_i$. Let

$$q_i(\nu, y) = \langle \nu, p_i(\cdot, y)\rangle = \int_E p_i(x, y)\nu(\mathrm{d}x), \quad \nu \in M(E), y \in F_i.$$

By the calculations in Example 10.1 one sees that $\nu \mapsto q_i(\nu, y)$ satisfies conditions (10.17) and (10.18) with $\lambda = \lambda_i$. Let $\{Y_t : t \ge 0\}$ be the unique càdlàg solution of (10.33). Then for $G \in C^2(\mathbb{R})$ and $f \in D(A)$ we have

$$\begin{aligned}
G(\langle Y_t, f\rangle) = {} & G(\langle Y_0, f\rangle) + \int_0^t \bigg\{ G'(\langle Y_s, f\rangle)\langle Y_s, Af + \gamma f - bf\rangle \\
& + \int_E Y_s(\mathrm{d}x) \int_{M(E)^\circ} \Big[G(\langle Y_s, f\rangle + \langle \nu, f\rangle) \\
& - G(\langle Y_s, f\rangle) - \langle \nu, f\rangle G'(\langle Y_s, f\rangle) \Big] H(x, \mathrm{d}\nu) \\
& + G''(\langle Y_s, f\rangle)\langle Y_s, cf^2\rangle + G'(\langle Y_s, f\rangle)\langle Y_s, \kappa_0 f\rangle \\
& + \int_E Y_s(\mathrm{d}x) \int_{M(E)^\circ} \Big[G(\langle Y_s, f\rangle + \langle \nu, f\rangle) \\
& - G(\langle Y_s, f\rangle) \Big] K_1(x, \mathrm{d}\nu) \bigg\} \mathrm{d}s + (\bar{\mathscr{G}}_{t+})\text{-local mart.,}
\end{aligned}$$

where

$$\kappa_0(x, \mathrm{d}z) = \int_{F_0} p_0(x, \mathrm{d}y)\kappa(y, \mathrm{d}z), \quad x, z \in E,$$

and

$$K_1(x, \mathrm{d}\nu) = \int_{F_1} p_1(x, \mathrm{d}y) K(y, \mathrm{d}\nu), \qquad x \in E, \nu \in M(E)^\circ.$$

By Theorem 7.13 it is easy to see that $\{(Y_t, \bar{\mathscr{G}}_{t+}) : t \geq 0\}$ is a superprocess with branching mechanism

$$(x, f) \mapsto \phi(x, f) - \kappa_0(x, f) - \int_{M(E)^\circ} \left(1 - \mathrm{e}^{-\nu(f)}\right) K_1(x, \mathrm{d}\nu).$$

This provides a method of changing the branching mechanism of the superprocess using immigration. One can also introduce an interactive structure as in (10.27).

Example 10.3 Let $(\nu, y) \mapsto q_i(\nu, y)$ be defined as in the last example and let $z \mapsto g_i(z)$ be a Lipschitz function on $[0, \infty)$ satisfying $g_i(z) \leq$ const $\cdot z$. Then $(\nu, y) \mapsto g_i(q_i(\nu, y))$ satisfies conditions (10.17) and (10.18) with $\lambda = \lambda_i$.

10.5 Notes and Comments

The approach of stochastic integral equations in constructing interactive immigration superprocess was suggested by Shiga (1990), who studied a stochastic equation involving Poisson processes on the space of one-dimensional excursions. However, Shiga (1990) only considered the trivial spatial motion. A generalization of his result to non-trivial spatial motions was given in Fu and Li (2004) for binary local branching. The results in this chapter generalize those in Fu and Li (2004) and Shiga (1990). Abraham and Delmas (2009) gave a method of changing the branching mechanism of a CB-process using immigration, which is different from that in Example 10.2.

A class of *immigration superprocess with dependent spatial motion* were constructed in Li et al. (2005). Using the techniques developed in Kurtz and Xiong (1999), they gave a characterization of the conditional cumulant semigroup of the superprocess in terms of a stochastic partial differential equation driven by a time–space white noise. The approach was stimulated by Xiong (2004), who established a similar characterization for the model of Skoulakis and Adler (2001).

Let $\sigma \in C^2(\mathbb{R})^+$ and $h \in C^1(\mathbb{R})$ and assume both h and h' are square-integrable. Let $\lambda \in M(\mathbb{R})$ and let ρ be defined by (7.58). Suppose that $(\nu, x) \mapsto q(\nu, x)$ is a positive Borel function on $M(\mathbb{R}) \times \mathbb{R}$ such that there is a constant $K > 0$ such that

$$\langle \lambda, q(\nu, \cdot) \rangle \leq K(1 + \langle \nu, 1 \rangle), \quad \nu \in M(\mathbb{R}),$$

and for each $R > 0$ there is a constant $K_R > 0$ such that

$$\langle \lambda, |q(\nu_2, \cdot) - q(\nu_1, \cdot)| \rangle \leq K_R \|\nu_2 - \nu_1\|$$

for $\nu_1, \nu_2 \in M(\mathbb{R})$ satisfying $\langle \nu_1, 1 \rangle \leq R$ and $\langle \nu_2, 1 \rangle \leq R$. By considering a stochastic equation similar to (10.19) involving one-dimensional excursions carried by a stochastic flow, Dawson and Li (2003) constructed an *interactive immigration superprocess with dependent spatial motion*, which is a diffusion process $\{X_t : t \geq 0\}$ in $M(\mathbb{R})$ satisfying the martingale problem: For each $f \in C^2(\mathbb{R})$,

$$M_t(f) = \langle X_t, f \rangle - \langle X_0, f \rangle - \frac{1}{2}\rho(0) \int_0^t \langle X_s, f'' \rangle ds$$
$$- \int_0^t ds \int_{\mathbb{R}} q(Y_s, x) f(x) \lambda(dx)$$

is a continuous martingale with quadratic variation process

$$\langle M(f) \rangle_t = \int_0^t \langle X_s, \sigma f^2 \rangle ds + \int_0^t ds \int_{\mathbb{R}} \langle X_s, h(z - \cdot) f' \rangle^2 dz.$$

Rescaling limits of the superprocess with dependent spatial motion were investigated in Dawson et al. (2004c) and Li et al. (2004), which led to a *superprocess with coalescing spatial motion*. Zhou (2007) gave a characterization of the conditional Laplace functional of the latter. A lattice branching-coalescing particle system was studied in Athreya and Swart (2005, 2009).

Chapter 11
Generalized Ornstein–Uhlenbeck Processes

Generalized Ornstein–Uhlenbeck processes constitute a large class of explicit examples of Markov processes in infinite-dimensional spaces with rich mathematical structures. Those processes may have non-trivial invariant measures, which make them become better candidates for infinite-dimensional reference processes than Lévy processes. In this chapter, we first give a formulation of generalized Ornstein–Uhlenbeck processes in Hilbert spaces using the generalized Mehler semigroups introduced by Bogachev and Röckner (1995) and Bogachev et al. (1996). Then we give a systematic exploration of the structures of the generalized Mehler semigroups. Since such a semigroup can be defined by a linear semigroup and a skew convolution semigroup, we mainly discuss the latter. We shall see that a skew convolution semigroup is always formed with infinitely divisible probability measures. The key result is a characterization for the skew convolution semigroups in terms of infinitely divisible probability entrance laws. For centered skew convolution semigroups with finite second-moments, those entrance laws can be closed by probability measures on an enlarged Hilbert space. We also give some constructions for the generalized Ornstein–Uhlenbeck processes determined by closed entrance laws and study the corresponding Langevin type equations.

11.1 Generalized Mehler Semigroups

Suppose that H is a real separable Hilbert space with inner product $\langle \cdot, \cdot \rangle$. We say a bounded linear operator S on this space is *symmetric* if $\langle Sa, b \rangle = \langle a, Sb \rangle$ for all $a, b \in H$, and say it is *positive definite* if $\langle Sa, a \rangle \geq 0$ for all $a \in H$. A bounded linear operator S on H is called a *trace class operator* if there is an orthonormal basis $\{e_1, e_2, \ldots\}$ of H such that

$$\mathrm{Tr}(S) := \sum_i \langle Se_i, e_i \rangle < \infty.$$

Z. Li, *Measure-Valued Branching Markov Processes*,
Probability and Its Applications, DOI 10.1007/978-3-642-15004-3_11,
© Springer-Verlag Berlin Heidelberg 2011

The sum $\mathrm{Tr}(S)$ is called the *trace* of S, which is independent of the choice of the orthonormal basis $\{e_1, e_2, \ldots\}$. In most cases, we consider an infinite-dimensional space H. Given a probability measure ν on H, let $\hat{\nu}$ denote its *characteristic functional* defined by

$$\hat{\nu}(a) = \int_H e^{i\langle x, a \rangle} \nu(\mathrm{d}x), \qquad a \in H. \tag{11.1}$$

It is well-known that $\hat{\nu}$ determines ν uniquely. The infinite divisibility of the probability measure can be defined as in Section 1.4. If ν is an infinitely divisible probability measure on H, then $\hat{\nu}(a) \neq 0$ for all $a \in H$ and there is a unique continuous function $\log \hat{\nu}$ on H such that $\log \hat{\nu}(0) = 0$ and $\hat{\nu}(a) = \exp\{\log \hat{\nu}(a)\}$ for $a \in H$; see, e.g., Linde (1986, p.20) and Parthasarathy (1967, p.171). Let

$$K(x, a) = e^{i\langle x, a \rangle} - 1 - i\langle x, a \rangle 1_{\{\|x\| \leq 1\}}, \qquad x, a \in H.$$

Proposition 11.1 *A probability measure ν on H is infinitely divisible if and only if $\psi := -\log \hat{\nu}$ is uniquely represented by*

$$\psi(a) = -i\langle b, a \rangle + \frac{1}{2}\langle Ra, a \rangle - \int_{H^\circ} K(x, a) M(\mathrm{d}x), \tag{11.2}$$

where $b \in H$, R is a symmetric positive definite trace class operator on H, and M is a σ-finite (Lévy) measure on $H^\circ := H \setminus \{0\}$ satisfying

$$\int_{H^\circ} (1 \wedge \|x\|^2) M(\mathrm{d}x) < \infty. \tag{11.3}$$

Proof. See, e.g., Linde (1986, p.84) and Parthasarathy (1967, p.181). □

We write $\nu = I(b, R, M)$ if ν is an infinitely divisible probability measure with $\psi = -\log \hat{\nu}$ given by (11.2). Under the stronger moment condition

$$\int_{H^\circ} (\|x\| \wedge \|x\|^2) M(\mathrm{d}x) < \infty, \tag{11.4}$$

we can define

$$\beta = \int_H x \nu(\mathrm{d}x) = b + \int_{H^\circ} x 1_{\{\|x\| > 1\}} M(\mathrm{d}x)$$

and rewrite (11.2) into

$$\psi(a) = -i\langle \beta, a \rangle + \frac{1}{2}\langle Ra, a \rangle - \int_{H^\circ} K_1(x, a) M(\mathrm{d}x), \tag{11.5}$$

where

$$K_1(x, a) := e^{i\langle x, a \rangle} - 1 - i\langle x, a \rangle, \qquad x, a \in H.$$

In this case, we write $\nu = I_1(\beta, R, M)$.

Let $(T_t)_{t\geq 0}$ be a strongly continuous semigroup of bounded linear operators on H. Then the semigroup $(T_t^*)_{t\geq 0}$ of dual operators of $(T_t)_{t\geq 0}$ is also strongly continuous on H; see, e.g., Pazy (1983, p.41). A family of probability measures $(\gamma_t)_{t\geq 0}$ on H is called a *skew convolution semigroup* (SC-semigroup) associated with $(T_t)_{t\geq 0}$ if the following equation is satisfied:

$$\gamma_{r+t} = (\gamma_r \circ T_t^{-1}) * \gamma_t, \qquad r, t \geq 0. \tag{11.6}$$

In this case, we can define a transition semigroup $(Q_t^\gamma)_{t\geq 0}$ on H by

$$Q_t^\gamma(x, \cdot) = \delta_{T_t x}(\cdot) * \gamma_t(\cdot), \tag{11.7}$$

which is called a *generalized Mehler semigroup* associated with $(T_t)_{t\geq 0}$. A Markov process in H is called a *generalized Ornstein–Uhlenbeck process* (generalized OU-process) if it has transition semigroup $(Q_t^\gamma)_{t\geq 0}$. Observe that (11.6) is equivalent to

$$\hat{\gamma}_{r+t}(a) = \hat{\gamma}_r(T_t^* a)\hat{\gamma}_t(a), \qquad r, t \geq 0, a \in H. \tag{11.8}$$

Proposition 11.2 *If $(\gamma_t)_{t\geq 0}$ is an SC-semigroup associated with $(T_t)_{t\geq 0}$, then each probability measure γ_t is infinitely divisible.*

Proof. This proof is a simplification of that of Schmuland and Sun (2001). By (11.6) we have $\gamma_0 = \gamma_0 * \gamma_0$ and so $\gamma_0 = \delta_0$, which is certainly infinitely divisible. From (11.8) it follows that

$$|\hat{\gamma}_{r+t}(a)| = |\hat{\gamma}_r(T_t^* a)||\hat{\gamma}_t(a)|, \qquad r, t \geq 0, a \in H. \tag{11.9}$$

Then $t \mapsto |\hat{\gamma}_t(a)|$ is decreasing and hence the limit $|\hat{\gamma}_{0+}(a)| := \lim_{t\to 0}|\hat{\gamma}_t(a)|$ exists. Observe also that $\lim_{t\to 0}\hat{\gamma}_r(T_t^* a) = \hat{\gamma}_r(a)$. Then we may let $t \to 0$ and $r \to 0$ in (11.9) to see that $|\hat{\gamma}_{0+}(a)| = |\hat{\gamma}_{0+}(a)|^2$. It thus follows that $|\hat{\gamma}_{0+}(a)| = 1$ or 0. For any $t > 0$ and any integer $k \geq 1$ we have

$$(\gamma_{t-t/k} \circ T_{t/k}^{-1}) * \gamma_{t/k} = \gamma_t.$$

By Parthasarathy (1967, p.72), there is a sequence $\{x_k\} \subset H$ so that $\gamma_{t/k} * \delta_{x_k} \to \nu$ weakly as $k \to \infty$, where ν is a probability measure on H. Consequently,

$$|\hat{\nu}(a)| = \lim_{k\to\infty}|\hat{\gamma}_{t/k}(a)| = |\hat{\gamma}_{0+}(a)| = 1 \text{ or } 0, \qquad a \in H.$$

Since $\alpha \mapsto |\hat{\nu}(\alpha a)|$ is continuous and $|\hat{\nu}(0)| = 1$, we must have $|\hat{\nu}(a)| = 1$. Then the symmetrization of ν is the unit mass concentrated at zero. It follows that $\nu = \delta_x$ for some $x \in H$; see, e.g., Linde (1986, p.23). Setting $y_k = x_k - x$ we have $\gamma_{t/k} * \delta_{y_k} \to \delta_0$ as $k \to \infty$. Since $t \mapsto \|T_t\|$ is a locally bounded function, the probability measures

$$\nu_{k,j} := (\gamma_{t/k} * \delta_{y_k}) \circ T^{-1}_{t-jt/k}, \quad 1 \le j \le k, k \ge 1,$$

form a uniform infinitesimal triangular array. By applying (11.6) inductively we see that

$$\gamma_t = \prod_{j=1}^{k} *(\gamma_{t/k} \circ T^{-1}_{t-jt/k}) = \left(\prod_{j=1}^{k} *\nu_{k,j} \right) * \delta_{z_k},$$

where $z_k = -\sum_{j=1}^{k} T_{t-jt/k} y_k$. It then follows that each γ_t is infinitely divisible; see de Acosta et al. (1978) and Parthasarathy (1967, p.199).   □

By Propositions 11.1 and 11.2, we may write $\gamma_t = I(b_t, R_t, M_t)$ for $t \ge 0$. It is not hard to show that (11.8) is satisfied if and only if we have

$$R_{r+t} = T_t R_r T_t^* + R_t, \quad M_{r+t} = (M_r \circ T_t^{-1})|_{H^\circ} + M_t, \qquad (11.10)$$

and

$$b_{r+t} = b_t + T_t b_r + \int_{H^\circ} \left(1_{\{\|T_t x\| \le 1\}} - 1_{\{\|x\| \le 1\}} \right) T_t x M_r(dx) \qquad (11.11)$$

for all $r, t \ge 0$. If the stronger moment condition (11.4) is satisfied for each M_t, we write $\gamma_t = I_1(\beta_t, R_t, M_t)$, where

$$\beta_t = \int_H x \nu_t(dx) = b_t + \int_{H^\circ} x 1_{\{\|x\|>1\}} M_t(dx).$$

In this case, the property (11.8) holds if and only if we have (11.10) and

$$\beta_{r+t} = \beta_t + T_t \beta_r, \qquad r, t \ge 0. \qquad (11.12)$$

Suppose that ν_0 is a probability measure on H and let $\psi_0 := -\log \hat{\nu}_0$. Since the dual semigroup $(T_t^*)_{t \ge 0}$ is also strongly continuous, for each $t \ge 0$ we can define an infinitely divisible probability measure γ_t on H by

$$\hat{\gamma}_t(a) = \exp \left\{ -\int_0^t \psi_0(T_s^* a) ds \right\}, \quad a \in H. \qquad (11.13)$$

It is easy to show that $(\gamma_t)_{t \ge 0}$ is an SC-semigroup associated with $(T_t)_{t \ge 0}$, which is called a *regular SC-semigroup*. The corresponding generalized Mehler semigroup is given by

$$\int_H e^{i\langle y, a \rangle} Q_t^\gamma(x, dy) = \exp \left\{ i\langle x, T_t^* a \rangle - \int_0^t \psi_0(T_s^* a) ds \right\}. \qquad (11.14)$$

Example 11.1 Let $b > 0$ and $c > 0$ be constants and let $\{B(t) : t \ge 0\}$ be a standard Brownian motion. A *classical OU-process* is the solution of the *Langevin equation*

$$dX(t) = c\,dB(t) - bX(t)dt, \qquad t \geq 0.$$

Given the initial value $X(0) = x$, we have

$$X(t) = e^{-bt}x + c\int_0^t e^{-b(t-s)}dB(s), \quad t \geq 0.$$

The second term on the right-hand side has Gaussian distribution $\gamma_t := N(0, \sigma_t^2)$ with

$$\sigma_t^2 = c^2\int_0^t e^{-2b(t-s)}ds = \frac{c^2}{2b}(1 - e^{-2bt}).$$

Then $\{X(t) : t \geq 0\}$ has transition semigroup $(Q_t^\gamma)_{t\geq 0}$ given by

$$Q_t^\gamma f(x) = \int_{\mathbb{R}} f(e^{-bt}x + y)\gamma_t(dy), \qquad t \geq 0, x \in \mathbb{R}.$$

Setting $\mu = N(0, c^2/2b)$ we obtain the *classical Mehler formula*

$$Q_t^\gamma f(x) = \int_{\mathbb{R}} f\big(e^{-bt}x + \sqrt{1 - e^{-2bt}}y\big)\mu(dy).$$

11.2 Gaussian Type Semigroups

We say an SC-semigroup $(\gamma_t)_{t\geq 0}$ is of the *Gaussian type* if each γ_t is a centered Gaussian probability measure. By the discussions of the first section, a Gaussian type SC-semigroup $(\gamma_t)_{t\geq 0}$ associated with $(T_t)_{t\geq 0}$ is given by

$$\log \hat{\gamma}_t(a) = -\frac{1}{2}\langle R_t a, a\rangle, \qquad t \geq 0, a \in H, \tag{11.15}$$

where R_t is a symmetric positive definite trace class operator on H satisfying the first equation in (11.10).

Theorem 11.3 *A family of centered Gaussian probability measures* $(\gamma_t)_{t\geq 0}$ *on* H *given by* (11.15) *form an SC-semigroup associated with* $(T_t)_{t\geq 0}$ *if and only if* $(R_t)_{t\geq 0}$ *is given by*

$$\langle R_t a, a\rangle = \int_0^t \langle U_s a, a\rangle ds, \qquad t \geq 0, a \in H, \tag{11.16}$$

where $(U_s)_{s>0}$ *is a family of symmetric positive definite trace class operators on* H *satisfying* $U_{s+t} = T_t U_s T_t^*$ *for all* $s, t > 0$ *and*

$$\int_0^t \operatorname{Tr} U_s \mathrm{d}s < \infty, \qquad t \geq 0.$$

Lemma 11.4 *If the family $(\gamma_t)_{t\geq 0}$ given by (11.15) is an SC-semigroup associated with $(T_t)_{t\geq 0}$, then the function $t \mapsto \langle R_t a, b \rangle$ is absolutely continuous in $t \geq 0$ for all $a, b \in H$.*

Proof. Since $(T_t)_{t\geq 0}$ is strongly continuous, there are constants $B \geq 1$ and $c \geq 0$ such that $\|T_t\| \leq Be^{ct}$. From (11.6) we have

$$\int_H \|x\|^2 \gamma_{r+t}(\mathrm{d}x) = \int_H \|T_t x\|^2 \gamma_r(\mathrm{d}x) + \int_H \|x\|^2 \gamma_t(\mathrm{d}x). \qquad (11.17)$$

It follows that

$$g(t) := \int_H \|x\|^2 \gamma_t(\mathrm{d}x), \qquad t \geq 0 \qquad (11.18)$$

is an increasing function. We claim

$$\sum_{j=1}^n [g(t_j) - g(r_j)] \leq B^2 e^{2cl} g(\sigma_n) \qquad (11.19)$$

for $0 < r_1 < t_1 < \cdots < r_n < t_n \leq l$, where $\sigma_n = \sum_{j=1}^n (t_j - r_j)$. When $n = 1$, this follows from (11.17). Now let us assume (11.19) holds for $n - 1$. By applying (11.17) twice we have

$$\begin{aligned}
\sum_{j=1}^n [g(t_j) - g(r_j)] &\leq B^2 e^{2cl} \int_H \|x\|^2 \gamma_{\sigma_{n-1}}(\mathrm{d}x) + [g(t_n) - g(r_n)] \\
&= B^2 e^{2cl} \int_H \|x\|^2 \gamma_{\sigma_{n-1}}(\mathrm{d}x) + \int_H \|T_{r_n} x\|^2 \gamma_{t_n - r_n}(\mathrm{d}x) \\
&\leq B^2 e^{2cl} \int_H \|x\|^2 \gamma_{\sigma_{n-1}}(\mathrm{d}x) \\
&\quad + B^2 e^{2cl} \int_H \|T_{\sigma_{n-1}} x\|^2 \gamma_{t_n - r_n}(\mathrm{d}x) \\
&= B^2 e^{2cl} \int_H \|x\|^2 \gamma_{\sigma_n}(\mathrm{d}x),
\end{aligned}$$

which gives (11.19). Letting $r \to 0$ and $t \to 0$ in (11.17) and using the fact that g is an increasing function one sees that $g(t) \to 0$ as $t \to 0$. Then (11.19) implies that g is absolutely continuous in $t \geq 0$. By the first equality in (11.10) we see $t \mapsto \langle R_t a, a \rangle$ is increasing for any $a \in H$. For $t \geq r \geq 0$ we use (11.10) again to see

$$\langle R_t a, a \rangle - \langle R_r a, a \rangle = \langle R_{t-r} T_r^* a, T_r^* a \rangle = \int_H \langle x, T_r^* a \rangle^2 \gamma_{t-r}(\mathrm{d}x)$$

$$\leq \|a\|^2 \int_H \|T_r x\|^2 \gamma_{t-r}(\mathrm{d}x) = \|a\|^2 [g(t) - g(r)].$$

Then $\langle R_t a, a \rangle$ is absolutely continuous in $t \geq 0$. A polarization argument shows $\langle R_t a, b \rangle$ is absolutely continuous in $t \geq 0$ for all $a, b \in H$. $\qquad\square$

Lemma 11.5 *If the family* $(\gamma_t)_{t \geq 0}$ *given by* (11.15) *is an SC-semigroup associated with* $(T_t)_{t \geq 0}$, *then there is a family of symmetric positive definite trace class operators* $(U_s)_{s > 0}$ *on* H *such that* (11.16) *holds.*

Proof. Let $\{e_n : n = 1, 2, \dots\}$ be an orthonormal basis of the space H. By Lemma 11.4, there are locally integrable Borel functions $A_{m,n}$ on $[0, \infty)$ such that

$$\langle R_t e_m, e_n \rangle = \int_0^t A_{m,n}(s)\mathrm{d}s, \qquad t \geq 0, \ m, n \geq 1. \tag{11.20}$$

The symmetry of R_t implies

$$\int_0^t A_{m,n}(s)\mathrm{d}s = \int_0^t A_{n,m}(s)\mathrm{d}s. \tag{11.21}$$

Since R_t is positive definite, for any $a \in \mathrm{span}\{e_1, e_2, \dots\}$ we have

$$\langle R_t a, a \rangle = \int_0^t \sum_{m,n=1}^{\infty} A_{m,n}(s)\langle a, e_m \rangle\langle a, e_n \rangle \mathrm{d}s \geq 0. \tag{11.22}$$

(The sum only contains finitely many nontrivial terms!) In addition, since R_t is a trace class operator, we get

$$\int_0^t \left(\sum_{n=1}^{\infty} A_{n,n}(s) \right) \mathrm{d}s = \sum_{n=1}^{\infty} \langle R_t e_n, e_n \rangle = \mathrm{Tr}(R_t) < \infty. \tag{11.23}$$

Let F be the subset of $[0, \infty)$ consisting of all $s \geq 0$ such that $A_{m,n}(s) = A_{n,m}(s)$ for $m, n \geq 1$ and

$$\sum_{n=1}^{\infty} A_{n,n}(s) < \infty \quad \text{and} \quad \sum_{m,n=1}^{\infty} A_{m,n}(s)\langle a, e_m \rangle\langle a, e_n \rangle \geq 0 \tag{11.24}$$

for $a \in \mathrm{span}\{e_1, e_2, \dots\}$ with rational coefficients. As observed in the proof of Lemma 11.4, the function $t \mapsto \langle R_t a, a \rangle$ is increasing. From (11.21), (11.22) and (11.23) we conclude that F has full Lebesgue measure. For any $s \in F$,

$$U_s a = \sum_{m,n=1}^{\infty} A_{m,n}(s)\langle a, e_m \rangle e_n \tag{11.25}$$

defines a symmetric positive definite linear operator on $\mathrm{span}\{e_1, e_2, \dots\}$. Taking $b = x e_m + y e_n$ with rational x and y, we get

$$\langle U_s b, b \rangle = (x, \ y) \begin{pmatrix} A_{m,m}(s) & A_{m,n}(s) \\ A_{n,m}(s) & A_{n,n}(s) \end{pmatrix} \begin{pmatrix} x \\ y \end{pmatrix} \geq 0,$$

so that the 2×2 matrix above is positive definite. Therefore, its determinant is positive, that is,

$$A_{m,n}(s)^2 \leq A_{m,m}(s) A_{n,n}(s). \tag{11.26}$$

This combined with Schwarz's inequality gives

$$\begin{aligned}
\|U_s a\|^2 &= \sum_{n=1}^{\infty} \left(\sum_{m=1}^{\infty} A_{m,n}(s) \langle a, e_m \rangle \right)^2 \\
&\leq \sum_{n=1}^{\infty} \sum_{m=1}^{\infty} A_{m,n}(s)^2 \sum_{m=1}^{\infty} \langle a, e_m \rangle^2 \\
&\leq \left(\sum_{n=1}^{\infty} A_{n,n}(s) \right)^2 \|a\|^2
\end{aligned}$$

for $a \in \mathrm{span}\{e_1, e_2, \dots\}$. Then U_s is a bounded operator and can be extended to the entire space H. In fact, U_s is a trace class operator since

$$\mathrm{Tr}(U_s) = \sum_{n=1}^{\infty} \langle U_s e_n, e_n \rangle = \sum_{n=1}^{\infty} A_{n,n}(s) < \infty.$$

For $s \notin F$ we let $U_s = 0$. By (11.22) and (11.25), for $a \in \mathrm{span}\{e_1, e_2, \dots\}$ we have

$$\langle R_t a, a \rangle = \int_0^t \langle U_s a, a \rangle \mathrm{d}s. \tag{11.27}$$

Since $s \mapsto \mathrm{Tr}(U_s)$ is locally integrable, by dominated convergence we see that (11.27) holds for all $a \in H$. □

Proof (of Theorem 11.3). If $(\gamma_t)_{t \geq 0}$ is a family of probability measures given by (11.15) and (11.16), it is clearly an SC-semigroup associated with $(T_t)_{t \geq 0}$. For the converse, suppose the family $(\gamma_t)_{t \geq 0}$ given by (11.15) form an SC-semigroup associated with $(T_t)_{t \geq 0}$. Let $(U_s)_{s > 0}$ be provided by Lemma 11.5. Note that (11.16) and the first equation of (11.10) imply

$$\int_0^r \langle U_{s+t} a, a \rangle \mathrm{d}s = \int_0^r \langle U_s T_t^* a, T_t^* a \rangle \mathrm{d}s, \qquad r, t \geq 0, a \in H.$$

Since H is separable, by Fubini's theorem, there are subsets B and B_s of $[0, \infty)$ with full Lebesgue measure such that

$$U_{s+t} = T_t U_s T_t^*, \qquad s \in B, t \in B_s.$$

As in the proof of Theorem 8.11, we can choose a decreasing sequence $s_n \in B$ with $s_n \to 0$ and redefine $(U_t)_{t>0}$ by

$$U_t := T_{t-s_n} U_{s_n} T^*_{t-s_n}, \qquad t > s_n.$$

With this modification, the family of operators $(U_t)_{t>0}$ satisfy $U_{r+t} = T_t U_r T^*_t$ for all $r, t > 0$ while (11.16) remains unchanged. □

11.3 Non-Gaussian Type Semigroups

Suppose that $(\gamma_t)_{t \geq 0}$ a family of infinitely divisible probability measures on H such that $\gamma_t = I(b_t, R_t, M_t)$. We say the linear part $(b_t)_{t \geq 0}$ is *absolutely continuous* if there exists an H-valued path $(c_s)_{s>0}$ such that

$$\langle b_t, a \rangle = \int_0^t \langle c_s, a \rangle ds, \qquad t \geq 0, a \in H. \tag{11.28}$$

Proposition 11.6 *If $(\gamma_t)_{t \geq 0}$ is an SC-semigroup given by $\gamma_t = I(b_t, R_t, M_t)$, we can write*

$$\int_{H^\circ} K(x, a) M_t(dx) = \int_0^t ds \int_{H^\circ} K(x, a) L_s(dx), \ t \geq 0, a \in H, \tag{11.29}$$

where $L_s(dx)$ is a σ-finite kernel from $(0, \infty)$ to H° satisfying $L_{r+t} = (L_r \circ T_t^{-1})|_{H^\circ}$ for $r, t > 0$ and

$$\int_0^t ds \int_H (1 \wedge \|x\|^2) L_s(dx) < \infty, \quad t \geq 0. \tag{11.30}$$

Proof. Let $c \geq 1$ and $b \geq 0$ be as in the proof of Lemma 11.4. From the second equation in (11.10) we see that $t \mapsto M_t$ is increasing. Let

$$h(t) := \int_{H^\circ} (1 \wedge \|x\|^2) M_t(dx), \quad t \geq 0.$$

By (11.10) for $r, t \geq 0$ we have

$$h(r+t) - h(r) = \int_{H^\circ} (1 \wedge \|T_r x\|^2) M_t(dx),$$

which is bounded above by $c^2 e^{2br} h(t)$. As in the proof of Lemma 11.4, one sees that $h(t)$ is absolutely continuous in $t \geq 0$. Observe that $t \mapsto \nu_t(dx) := (1 \wedge \|x\|^2) M_t(dx)$ defines an increasing family of finite measures, so $t \mapsto \nu_t(B)$ determines a locally bounded Borel measure $\nu(ds, B)$ on $[0, \infty)$ for each $B \in \mathscr{B}(H^\circ)$. A monotone class argument shows that $\nu(A, \cdot)$ is a Borel measure on H° for each

$A \in \mathcal{B}([0, \infty))$, so that $\nu(\cdot, \cdot)$ is a bimeasure. By Ethier and Kurtz (1986, p.502), there is a probability kernel $J_s(\mathrm{d}x)$ from $[0, \infty)$ to H° such that

$$\nu(A, B) = \int_A J_s(B)\nu(\mathrm{d}s, H^\circ) = \int_A J_s(B)\mathrm{d}h(s) = \int_A J_s(B)h'(s)\mathrm{d}s,$$

where $h'(s)$ is a Radon–Nikodym derivative of $\mathrm{d}h(s)$ relative to the Lebesgue measure. Defining the σ-finite kernel $L_s(\mathrm{d}x) := (1 \wedge \|x\|^2)^{-1}h'(s)J_s(\mathrm{d}x)$ we obtain (11.29). By the second equation of (11.10) one can modify the definition of $(L_t)_{t>0}$ so that $L_{r+t} = (L_r \circ T_t^{-1})|_{H^\circ}$ is satisfied for all $r, t > 0$. $\qquad\square$

We say a family of σ-finite measures $(\nu_s)_{s>0}$ on H is an *entrance law* for the semigroup $(T_t)_{t\geq 0}$ if it satisfies $\nu_{r+t} = \nu_r \circ T_t^{-1}$ for all $r, t > 0$. In fact, that means $(\nu_s)_{s>0}$ is an entrance law for the deterministic Markov process $\{T_t x : t \geq 0\}$ according to the standard definition.

Theorem 11.7 *Let $(\gamma_t)_{t\geq 0}$ be a family of infinitely divisible probability measures on H with absolutely continuous linear part. Then $(\gamma_t)_{t\geq 0}$ is an SC-semigroup associated with $(T_t)_{t\geq 0}$ if and only if there is an infinitely divisible probability entrance law $(\nu_s)_{s>0}$ for $(T_t)_{t\geq 0}$ such that*

$$\hat{\gamma}_t(a) = \exp\left\{ \int_0^t \log \hat{\nu}_s(a)\mathrm{d}s \right\}, \quad t \geq 0, a \in H. \tag{11.31}$$

Proof. If $(\gamma_t)_{t\geq 0}$ is given by (11.31), it is clearly an SC-semigroup associated with $(T_t)_{t\geq 0}$. Conversely, suppose that $(\gamma_t)_{t\geq 0}$ is an SC-semigroup associated with $(T_t)_{t\geq 0}$ and write $\gamma_t = I(b_t, R_t, M_t)$ for $t \geq 0$. It is easy to see that

$$\log \hat{\gamma}_t^g(a) = -\frac{1}{2}\langle R_t a, a\rangle, \quad t \geq 0, a \in H, \tag{11.32}$$

defines a Gaussian type SC-semigroup $(\gamma_t^g)_{t\geq 0}$. Let $(U_s)_{s>0}$ and $(L_s)_{s>0}$ be provided by Theorem 11.3 and Proposition 11.6, respectively. Suppose that $\langle b_t, a\rangle = \int_0^t \langle c_s, a\rangle \mathrm{d}s$. By (11.11), we can modify the definition of $(c_s)_{s>0}$ so that

$$c_{r+t} = T_t c_r + \int_{H^\circ} \left(1_{\{\|T_t x\| \leq 1\}} - 1_{\{\|x\| \leq 1\}}\right)T_t x L_r(\mathrm{d}x), \quad r, t > 0.$$

Then we have (11.31) with $\nu_s = I(c_s, U_s, L_s)$ for $s > 0$. $\qquad\square$

Example 11.2 Let $t \mapsto b_t$ be a real-valued discontinuous function satisfying $b_{r+t} = b_r + b_t$ for all $r, t \geq 0$; see, e.g., Sato (1999, p.37). It is simple to check that $(\delta_{b_t})_{t\geq 0}$ is a classical convolution semigroup, which cannot be represented in the form (11.31). This example shows that some condition on the linear part $t \mapsto b_t$ has to be imposed to get the representation (11.31) of the SC-semigroup.

Example 11.3 Let μ be the uniform distribution on $[0, 2\pi)$ and consider the Hilbert space $L^2([0, 2\pi), \mu)$ equipped with the inner product $\langle \cdot, \cdot\rangle$ defined by

$$\langle f, h \rangle = \frac{1}{2\pi} \int_0^{2\pi} f(x)h(x)\mathrm{d}x.$$

For $t \geq 0$ and $f \in L^2([0, 2\pi), \mu)$ let $T_t f(x) = f(x + t) \pmod{2\pi}$. By using approximation by continuous functions it is not hard to show that

$$\lim_{t \to 0} \int_0^{2\pi} |f(x + t) - f(x)|^2 \mathrm{d}x = 0.$$

Then $(T_t)_{t \geq 0}$ is a strongly continuous semigroup on $L^2([0, 2\pi), \mu)$. Set $b_t = f - T_t f$. It is easy to show that $(\delta_{b_t})_{t \geq 0}$ is an SC-semigroup associated with $(T_t)_{t \geq 0}$. For any $f \in L^2([0, 2\pi), \mu)$ we have the Fourier expansion

$$f(x) = \sum_{n=-\infty}^{\infty} \hat{f}(n)e^{inx}, \qquad x \in [0, 2\pi), \tag{11.33}$$

where

$$\hat{f}(n) = \frac{1}{2\pi} \int_0^{2\pi} f(x)e^{-inx}\mathrm{d}x, \qquad n = 0, \pm 1, \pm 2, \ldots; \tag{11.34}$$

see, e.g., Conway (1990, p.21). Clearly, the n-th Fourier coefficient of $T_t f$ is $\hat{f}(n)e^{int}$. Since both f and $T_t f$ are real functions, from (11.33) and (11.34) we obtain

$$\langle f, b_t \rangle = \|f\|^2 - \frac{1}{2\pi} \int_0^{2\pi} f(x)T_t f(x)\mathrm{d}x$$
$$= \|f\|^2 - \sum_{n=-\infty}^{\infty} \hat{f}(n)\hat{f}(-n)e^{-int}$$
$$= \|f\|^2 - |\hat{f}(0)|^2 - 2\sum_{n=1}^{\infty} |\hat{f}(n)|^2 \cos(nt).$$

Now let us take the particular function $f \in L^2([0, 2\pi), \mu)$ given by (11.33) with

$$\hat{f}(n) = \begin{cases} 2^{-k/2} & \text{if } |n| = 2^k \text{ and } k \geq 1, \\ 0 & \text{otherwise.} \end{cases}$$

Then we have

$$f(x) = 2\sum_{k=1}^{\infty} \frac{1}{2^{k/2}} \cos(2^k x), \qquad x \in [0, 2\pi).$$

It follows that

$$\|f\|^2 = \frac{2}{\pi} \sum_{k=1}^{\infty} \frac{1}{2^k} \int_0^{2\pi} \cos^2(2^k x)\mathrm{d}x = 2$$

and

$$\langle f, b_t \rangle = 2 - 2 \sum_{k=1}^{\infty} 2^{-k} \cos(2^k t),$$

which is Weierstrass's nowhere differentiable continuous function; see, e.g., Hewitt and Stromberg (1965, p.258). Therefore $(\delta_{b_t})_{t \geq 0}$ cannot be represented in the form (11.31).

11.4 Extensions of Centered Semigroups

In this section, we give some characterizations for centered SC-semigroups with finite second-moments. In particular, we shall see any SC-semigroup of this type can be extended to a regular SC-semigroup on an enlarged Hilbert space. Since $(T_t)_{t \geq 0}$ is strongly continuous, there are constants $B \geq 0$ and $c_0 \geq 0$ such that $\|T_t\| \leq Be^{c_0 t}$ for every $t \geq 0$. Let $(U_\alpha)_{\alpha > c_0}$ denote the resolvent of $(T_t)_{t \geq 0}$ and let A denote its generator with domain $\mathscr{D}(A) = U_\alpha H \subset H$.

A path $\tilde{x} = \{\tilde{x}(s) : s > 0\}$ taking values in H is called an *entrance path* for the semigroup $(T_t)_{t \geq 0}$ if it satisfies $\tilde{x}(r + t) = T_t \tilde{x}(r)$ for all $r, t > 0$. Let E denote the set of all entrance paths for $(T_t)_{t \geq 0}$. We say $\tilde{x} \in E$ is *closable* if there is an element $\tilde{x}(0) \in H$ such that $\tilde{x}(s) = T_s \tilde{x}(0)$ for all $s > 0$, and say it is *locally square integrable* if

$$\int_0^t \|\tilde{x}(s)\|^2 \mathrm{d}s < \infty, \qquad t \geq 0. \tag{11.35}$$

Lemma 11.8 *Let $\tilde{x} \in E$. Then* (11.35) *holds if and only if*

$$\int_0^\infty e^{-2bs} \|\tilde{x}(s)\|^2 \mathrm{d}s < \infty, \qquad b > c_0. \tag{11.36}$$

Proof. It is easy to see that (11.36) implies (11.35). For the converse, assume (11.35) holds. Then we have

$$\int_0^\infty e^{-2bs} \|\tilde{x}(s)\|^2 \mathrm{d}s = \sum_{k=0}^{\infty} e^{-2kb} \int_0^1 e^{-2bs} \|T_k \tilde{x}(s)\|^2 \mathrm{d}s$$

$$\leq \sum_{k=0}^{\infty} B^2 e^{-2k(b-c_0)} \int_0^1 e^{-2bs} \|\tilde{x}(s)\|^2 \mathrm{d}s.$$

The right-hand side is finite for every $b > c_0$. $\square$

Let $\tilde{H}$ denote the set of all locally square integrable entrance paths for $(T_t)_{t \geq 0}$. We call $\tilde{H}$ the *entrance space* for $(T_t)_{t \geq 0}$. For any fixed $b > c_0$ we can define an inner product on $\tilde{H}$ by

$$\langle \tilde{x}, \tilde{y} \rangle_\sim := \int_0^\infty e^{-2bs} \langle \tilde{x}(s), \tilde{y}(s) \rangle ds, \quad \tilde{x}, \tilde{y} \in \tilde{H}. \tag{11.37}$$

Let $\| \cdot \|_\sim$ be the norm induced by this inner product.

Lemma 11.9 *For every $t > 0$ the projection $\pi_t : \tilde{x} \mapsto \tilde{x}(t)$ from $\tilde{H}$ to H is a bounded linear operator.*

Proof. The linearity of π_t is obvious. For any $\tilde{x} \in \tilde{H}$ we have

$$\|\pi_t \tilde{x}\|^2 = t^{-1} \int_0^t \|\tilde{x}(t)\|^2 ds = t^{-1} \int_0^t \|T_{t-s}\tilde{x}(s)\|^2 ds$$
$$\leq B^2 t^{-1} e^{2bt} \int_0^t e^{-2bs} \|\tilde{x}(s)\|^2 ds \leq B^2 t^{-1} e^{2bt} \|\tilde{x}\|_\sim^2.$$

Then π_t is a bounded operator. $\qquad\square$

Lemma 11.10 *The norm $\| \cdot \|_\sim$ is complete and $(\tilde{H}, \langle \cdot, \cdot \rangle_\sim)$ is a Hilbert space.*

Proof. Suppose $\{\tilde{x}_n\} \subset \tilde{H}$ is a Cauchy sequence under the norm $\| \cdot \|_\sim$, that is,

$$\|\tilde{x}_n - \tilde{x}_m\|_\sim^2 = \int_0^\infty e^{-2bs} \|\tilde{x}_n(s) - \tilde{x}_m(s)\|^2 ds \to 0$$

as $m, n \to \infty$. By Lemma 11.9, for each $t > 0$ the limit $\tilde{x}(t) = \lim_{n\to\infty} \tilde{x}_n(t)$ exists in H. Since T_s is a bounded linear operator on H, for $s > 0$ we have

$$T_s \tilde{x}(t) = \lim_{n\to\infty} T_s \tilde{x}_n(t) = \lim_{n\to\infty} \tilde{x}_n(t + s) = \tilde{x}(t + s).$$

Then $\tilde{x} = \{\tilde{x}(t) : t > 0\}$ is an entrance path for $(T_t)_{t\geq0}$. For $\varepsilon > 0$ choose $N \geq 1$ such that

$$\int_0^\infty e^{-2bs} \|\tilde{x}_n(s) - \tilde{x}_m(s)\|^2 ds < \varepsilon, \quad m, n \geq N.$$

By Fatou's lemma we get

$$\int_0^\infty e^{-2bs} \|\tilde{x}_n(s) - \tilde{x}(s)\|^2 ds \leq \varepsilon, \quad n \geq N.$$

It follows that

$$\int_0^\infty e^{-2bs} \|\tilde{x}(s)\|^2 ds \leq \int_0^\infty e^{-2bs} \|\tilde{x}(s) - \tilde{x}_n(s)\|^2 ds$$
$$+ \int_0^\infty e^{-2bs} \|\tilde{x}_n(s)\|^2 ds < \infty.$$

Then $\tilde{x} \in \tilde{H}$ and $\lim_{n\to\infty} \|\tilde{x}_n - \tilde{x}\|_\sim^2 = 0$. $\qquad\square$

Lemma 11.11 *The linear operator $J : x \mapsto \{T_s x : s > 0\}$ from H to $\tilde{H}$ is a continuous dense embedding and $(\tilde{H}, \|\cdot\|_\sim)$ is separable.*

Proof. Since $x = \lim_{t \to 0} T_t x$, the map J is injective. For any $x \in H$,

$$\|Jx\|_\sim^2 = \int_0^\infty e^{-2bs}\|T_s x\|^2 ds \leq B^2\|x\|^2 \int_0^\infty e^{-2(b-c_0)s} ds.$$

Thus J is a continuous embedding. For an arbitrary $\tilde{x} \in \tilde{H}$ we have

$$\begin{aligned}
\|J\tilde{x}(t) - \tilde{x}\|_\sim^2 &= \int_0^\infty e^{-2bs}\|T_t\tilde{x}(s) - \tilde{x}(s)\|^2 ds \\
&= \int_0^r e^{-2bs}\|T_t\tilde{x}(s) - \tilde{x}(s)\|^2 ds \\
&\quad + \int_r^\infty e^{-2bs}\|T_{s-r}[T_t\tilde{x}(r) - \tilde{x}(r)]\|^2 ds \\
&\leq 2(B^2 e^{2c_0 t} + 1)\int_0^r e^{-2bs}\|\tilde{x}(s)\|^2 ds \\
&\quad + B^2 e^{-2c_0 r}\|T_t\tilde{x}(r) - \tilde{x}(r)\|^2 \int_r^\infty e^{-2(b-c_0)s} ds.
\end{aligned}$$

Observe that the first integral on the right-hand side goes to zero as $r \to 0$ and for fixed $r > 0$ the second term goes to zero as $t \to 0$. Then we have $\|J\tilde{x}(t) - \tilde{x}\|_\sim \to 0$ as $t \to 0$, and hence JH is dense in $\tilde{H}$. Since H is separable, so is $\tilde{H}$. □

Lemma 11.12 *The Borel σ-algebra $\mathscr{B}(\tilde{H})$ on $(\tilde{H}, \|\cdot\|_\sim)$ is also generated by the projections $\{\pi_t : t > 0\}$ from $\tilde{H}$ to H.*

Proof. By Lemma 11.9, each π_t is continuous. Then $\sigma(\{\pi_t : t > 0\}) \subset \mathscr{B}(\tilde{H})$. On the other hand, we have

$$\|\tilde{x} - \tilde{z}\|_\sim^2 = \lim_{n \to \infty} \frac{1}{n} \sum_{i=1}^{n^2} e^{-2bi/n}\|\tilde{x}(i/n) - \tilde{z}(i/n)\|^2, \quad \tilde{x}, \tilde{z} \in \tilde{H}.$$

Then for any fixed $\tilde{z} \in \tilde{H}$, the function $\tilde{x} \mapsto \|\tilde{x} - \tilde{z}\|_\sim$ on $\tilde{H}$ is measurable with respect to $\sigma(\{\pi_t : t > 0\})$. Consequently, every open ball $B(\tilde{z}, \varepsilon) := \{\tilde{x} \in \tilde{H} : \|\tilde{x} - \tilde{z}\|_\sim < \varepsilon\}$ belongs to $\sigma(\{\pi_t : t > 0\})$. Since $\tilde{H}$ is separable, all open sets in $\tilde{H}$ are contained in $\sigma(\{\pi_t : t > 0\})$ and hence $\mathscr{B}(\tilde{H}) \subset \sigma(\{\pi_t : t > 0\})$. □

Theorem 11.13 *A family $(\gamma_t)_{t\geq 0}$ of centered probability measures on H satisfying the second-moment condition*

$$\int_{H^\circ} \|x\|^2 \gamma_t(dx) < \infty, \quad t \geq 0 \tag{11.38}$$

is an SC-semigroup associated with $(T_t)_{t\geq 0}$ if and only if its characteristic functionals are given by (11.31) with $(\nu_s)_{s>0}$ being a centered infinitely divisible probability entrance law for $(T_t)_{t\geq 0}$ satisfying

$$\int_0^t ds \int_{H^\circ} \|x\|^2 \nu_s(dx) < \infty, \quad t \geq 0. \qquad (11.39)$$

Proof. It is well-known that the second-moment of a centered infinitely divisible probability measure only involves the Gaussian covariance operator and the Lévy measure. If the centered infinitely divisible probability measures $(\gamma_t)_{t\geq 0}$ and $(\nu_s)_{s>0}$ are related by (11.31), the Gaussian covariance operators and Lévy measures of $(\gamma_t)_{t\geq 0}$ can be represented as integrals of those of $(\nu_s)_{s>0}$. This observation yields

$$\int_{H^\circ} \langle x, a \rangle^2 \gamma_t(dx) = \int_0^t ds \int_{H^\circ} \langle x, a \rangle^2 \nu_s(dx), \quad t \geq 0, a \in H.$$

Let $\{e_n : n = 1, 2, \dots\}$ be an orthonormal basis of H. Applying the above equation to each e_n and taking the summation we see

$$\int_{H^\circ} \|x\|^2 \gamma_t(dx) = \int_0^t ds \int_{H^\circ} \|x\|^2 \nu_s(dx), \quad t \geq 0. \qquad (11.40)$$

In particular, conditions (11.38) and (11.39) are equivalent for the infinitely divisible probability measures $(\gamma_t)_{t\geq 0}$ and $(\nu_s)_{s>0}$ related by (11.31). Then the desired result follows by Theorem 11.7. $\qquad \square$

Theorem 11.14 *A family $(\gamma_t)_{t\geq 0}$ of centered probability measures on H satisfying (11.38) is an SC-semigroup associated with $(T_t)_{t\geq 0}$ if and only if its characteristic functionals are given by*

$$\hat{\gamma}_t(a) = \exp\left\{ \int_0^t \left[\log \int_{\tilde{H}} e^{i\langle \tilde{x}(s), a \rangle} \lambda_0(d\tilde{x}) \right] ds \right\}, \quad a \in H, \qquad (11.41)$$

where λ_0 is a centered infinitely divisible probability measure on $\tilde{H}$ satisfying

$$\int_{\tilde{H}} \|\tilde{x}\|_{\sim}^2 \lambda_0(d\tilde{x}) < \infty. \qquad (11.42)$$

Proof. Suppose that $(\gamma_t)_{t\geq 0}$ is a family of centered probability measures on H defined by (11.41) and (11.42). Let ν_s be the image of λ_0 induced by the projection π_s from $\tilde{H}$ to H. Then $(\nu_s)_{s>0}$ is a centered infinitely divisible probability entrance law for $(T_t)_{t\geq 0}$ satisfying (11.39) and the relation (11.31) holds. By Theorem 11.13, $(\gamma_t)_{t\geq 0}$ is a centered SC-semigroup satisfying (11.38). Conversely, suppose that $(\gamma_t)_{t\geq 0}$ is a centered SC-semigroup associated with $(T_t)_{t\geq 0}$ satisfying (11.38). Let $(\nu_s)_{s>0}$ be the entrance law given by Theorem 11.13. Then $(\nu_s)_{s>0}$ is a probability entrance law for the continuous Markov process $\{T_t x : t \geq 0\}$ with deterministic motion. Let E_0 be the set of continuous paths $\{w(t) : t > 0\}$ from $(0, \infty)$ to H. We endow E_0 with the σ-algebra $\mathscr{E}_0$ generated by the coordinate process. Then there is a unique probability measure λ_0 on $(E_0, \mathscr{E}_0)$ under which $\{w(t) : t > 0\}$ is a Markov process with the same transition semigroup as the process $\{T_t x : t \geq 0\}$

and ν_s is the image of λ_0 under $w \mapsto w(s)$. It follows that

$$\hat{\gamma}_t(a) = \exp\left\{ \int_0^t \left[\log \int_{E_0} e^{i\langle w(s), a\rangle} \lambda_0(dw) \right] ds \right\}, \quad a \in H. \quad (11.43)$$

Because of the special deterministic motion mechanism of the process $\{T_t x : t \geq 0\}$ we may assume that λ_0 is supported by the space E of the entrance paths. Let $\mathscr{E}_0(E)$ and $\mathscr{E}_0(\tilde{H})$ denote respectively the traces of $\mathscr{E}_0$ on E and $\tilde{H}$. Since $w \mapsto \|w(s)\|^2$ is clearly a non-negative $\mathscr{E}_0(E)$-measurable function on E,

$$w \mapsto \|w\|_{\sim}^2 := \int_0^\infty e^{-2bs} \|w(s)\|^2 ds$$

is an $\mathscr{E}_0(E)$-measurable function on E taking values in $[0, \infty]$. Since $(\nu_s)_{s>0}$ satisfies (11.39), we have

$$\begin{aligned}
\int_E \|w\|_{\sim}^2 \lambda_0(dw) &= \int_E \lambda_0(dw) \int_0^\infty e^{-2bs} \|w(s)\|^2 ds \\
&= \int_0^\infty ds \int_H e^{-2bs} \|x\|^2 \nu_s(dx) \\
&= \sum_{n=0}^\infty \int_0^1 ds \int_H e^{-2b(n+s)} \|T_n x\|^2 \nu_s(dx) \\
&\leq B^2 \sum_{n=0}^\infty e^{-2(b-c_0)n} \int_0^1 ds \int_H e^{-2bs} \|x\|^2 \nu_s(dx) < \infty.
\end{aligned}$$

Then λ_0 is actually supported by $\tilde{H}$ and (11.42) holds. By Lemma 11.12 we have $\mathscr{B}(\tilde{H}) = \mathscr{E}_0(\tilde{H})$, so we can regard λ_0 as a probability measure on $(\tilde{H}, \mathscr{B}(\tilde{H}))$. Now we get (11.41) from (11.43). Because each ν_s is a centered infinitely divisible probability measure, so is λ_0. □

Theorem 11.15 *All centered SC-semigroups associated with $(T_t)_{t\geq 0}$ satisfying (11.38) are regular if and only if all of its locally square integrable entrance paths are closable.*

Proof. Suppose that all entrance paths $\tilde{x} \in \tilde{H}$ are closable and $(\gamma_t)_{t\geq 0}$ is an SC-semigroup given by (11.41). To each $\tilde{x} \in \tilde{H}$ there corresponds some $\tilde{x}(0) \in H$ such that $\tilde{x}(s) = T_s \tilde{x}(0)$ for all $s > 0$. This element is apparently determined by $\tilde{x}$ uniquely. Letting ν_0 be the image of λ_0 under the map $\tilde{x} \mapsto \tilde{x}(0)$, we get (11.31). Conversely, if $\tilde{x} = \{\tilde{x}(s) : s > 0\} \in \tilde{H}$ is not closable, then

$$\hat{\gamma}_t(a) = \exp\left\{ -\frac{1}{2} \int_0^t \langle \tilde{x}(s), a\rangle^2 ds \right\}, \quad t \geq 0, a \in H,$$

defines an irregular SC-semigroup for $(T_t)_{t\geq 0}$. □

We now discuss how to extend a centered SC-semigroup on H to a regular one on the entrance space $\tilde{H}$. Given the semigroup $(T_t)_{t\geq 0}$, we can define a semigroup of linear operators $(\tilde{T}_t)_{t\geq0}$ on $\tilde{H}$ by $\tilde{T}_0\tilde{x} = \tilde{x}$ and $\tilde{T}_t\tilde{x} = J\tilde{x}(t)$ for $t > 0$ and $\tilde{x} \in \tilde{H}$. It follows that

$$(\tilde{T}_t\tilde{x})(s) = \tilde{x}(t+s) = T_t(\tilde{x}(s)), \qquad s, t > 0. \tag{11.44}$$

In view of (11.37) we have

$$\|\tilde{T}_t\tilde{x}\|_{\sim}^2 = \int_0^\infty e^{-2bs}\|\tilde{x}(t+s)\|^2 ds \leq \|T_t\|^2 \int_0^\infty e^{-2bs}\|\tilde{x}(s)\|^2 ds.$$

Then $\|\tilde{T}_t\|_{\sim} \leq \|T_t\|$ for every $t \geq 0$. Let $(\tilde{U}_\alpha)_{\alpha>c_0}$ denote the resolvent of $(\tilde{T}_t)_{t\geq0}$ and let $\tilde{A}$ denote its generator with domain $\mathscr{D}(\tilde{A}) = \tilde{U}_\alpha\tilde{H} \subset \tilde{H}$.

Lemma 11.16 *Let J be defined as in Lemma 11.11. Then $JT_tx = \tilde{T}_tJx$ for all $t \geq 0$ and $x \in H$ and $(\tilde{T}_t)_{t\geq0}$ is a strongly continuous semigroup of linear operators on $\tilde{H}$.*

Proof. For $t \geq 0$ and $x \in H$ we have

$$JT_tx = \{T_sT_tx : s > 0\} = \{T_tT_sx : s > 0\} = \tilde{T}_tJx,$$

giving the first assertion. By the proof of Lemma 11.11 we have

$$\lim_{t\to0}\|\tilde{T}_t\tilde{x} - \tilde{x}\|_{\sim} = \lim_{t\to0}\|J\tilde{x}(t) - \tilde{x}\|_{\sim} = 0.$$

Then $(\tilde{T}_t)_{t\geq0}$ is strongly continuous. □

Lemma 11.17 *We have $\tilde{U}_\alpha\tilde{x} = \{U_\alpha\tilde{x}(s) : s > 0\}$ and $\tilde{A}\tilde{U}_\alpha\tilde{x} = \{AU_\alpha\tilde{x}(s) : s > 0\}$ for all $\tilde{x} \in \tilde{H}$.*

Proof. The first assertion follows as we observe that, for $\alpha > c_0$,

$$\tilde{U}_\alpha\tilde{x}(s) = \int_0^\infty e^{-\alpha s}\tilde{T}_t\tilde{x}(s)dt = \int_0^\infty e^{-\alpha s}T_t\tilde{x}(s)dt = U_\alpha\tilde{x}(s).$$

The second follows from the equality $\tilde{A}\tilde{U}_\alpha\tilde{x} = \alpha\tilde{U}_\alpha\tilde{x} - \tilde{x}$. □

Theorem 11.18 *All entrance paths for $(\tilde{T}_t)_{t\geq0}$ are closable.*

Proof. Suppose that $\bar{x} = \{\tilde{x}_u : u > 0\}$ is an entrance path for $(\tilde{T}_t)_{t\geq0}$, where each $\tilde{x}_u = \{\tilde{x}_u(s) : s > 0\} \in \tilde{H}$ is an entrance path for $(T_t)_{t\geq0}$. In view of (11.44) we have

$$\{\tilde{x}_{t+u}(s) : s > 0\} = \tilde{x}_{t+u} = \tilde{T}_t\tilde{x}_u = \{\tilde{x}_u(t+s) : s > 0\}. \tag{11.45}$$

Set $\tilde{x}_0 = \{\tilde{x}_{s/2}(s/2) : s > 0\}$. By (11.45) we have

$$T_t(\tilde{x}_{s/2}(s/2)) = \tilde{x}_{s/2}(t+s/2) = \tilde{x}_{(s+t)/2}((s+t)/2).$$

Then $\tilde{x}_0$ is an entrance path for $(T_t)_{t \geq 0}$. Moreover,

$$(\tilde{T}_u \tilde{x}_0)(s) = T_u(\tilde{x}_{s/2}(s/2)) = \tilde{x}_{s/2}(u + s/2) = \tilde{x}_u(s),$$

and hence $\tilde{T}_u \tilde{x}_0 = \tilde{x}_u$. Thus $\bar{x} = \{\tilde{x}_u : u > 0\}$ is closed by $\tilde{x}_0$.   □

Theorem 11.19 *Let* $(\gamma_t)_{t \geq 0}$ *be a centered SC-semigroup given by* (11.41) *and* (11.42). *Let* $\tilde{\gamma}_t = \gamma_t \circ J^{-1}$ *for* $t \geq 0$. *Then* $(\tilde{\gamma}_t)_{t \geq 0}$ *is a regular centered SC-semigroup associated with* $(\tilde{T}_t)_{t \geq 0}$ *and*

$$\int_{\tilde{H}} e^{i \langle \tilde{x}, \tilde{a} \rangle \sim} \tilde{\gamma}_t(d\tilde{x}) = \exp \left\{ \int_0^t \left[\log \int_{\tilde{H}} e^{i \langle \tilde{x}, \tilde{T}_s^* \tilde{a} \rangle \sim} \lambda_0(d\tilde{x}) \right] ds \right\} \quad (11.46)$$

for every $t \geq 0$ *and* $\tilde{a} \in \tilde{H}$.

Proof. It is not hard to show $(\tilde{\gamma}_t)_{t \geq 0}$ is an SC-semigroup associated with $(\tilde{T}_t)_{t \geq 0}$. Since $(T_t)_{t \geq 0}$ is a strongly continuous semigroup, for any $\tilde{a} = \{\tilde{a}(s) : s > 0\} \in \tilde{H}$ we can use dominated convergence and (11.41) to see

$$\int_H \exp\left\{ i \int_0^\infty e^{-2bs} \langle T_s x, \tilde{a}(s) \rangle ds \right\} \gamma_t(dx)$$

$$= \lim_{n \to \infty} \int_H \exp\left\{ i \sum_{k=1}^\infty n^{-1} e^{-2bk/n} \langle T_{k/n} x, \tilde{a}(k/n) \rangle \right\} \gamma_t(dx)$$

$$= \lim_{n \to \infty} \int_H e^{i \langle x, a_n \rangle} \gamma_t(dx)$$

$$= \lim_{n \to \infty} \exp\left\{ \int_0^t \left[\log \int_{\tilde{H}} e^{i \langle \tilde{x}(s), a_n \rangle} \lambda_0(d\tilde{x}) \right] ds \right\},$$

where

$$a_n = \sum_{k=1}^\infty n^{-1} e^{-2bk/n} T_{k/n}^* \tilde{a}(k/n).$$

By the strong continuity of $(T_t^*)_{t \geq 0}$ we have

$$\lim_{n \to \infty} \langle \tilde{x}(s), a_n \rangle = \int_0^\infty e^{-2br} \langle \tilde{x}(s), T_r^* \tilde{a}(r) \rangle dr$$

$$= \int_0^\infty e^{-2br} \langle \tilde{x}(r), T_s^* \tilde{a}(r) \rangle dr.$$

Then another application of the dominated convergence gives

$$\int_H \exp\left\{ i \int_0^\infty e^{-2bs} \langle T_s x, \tilde{a}(s) \rangle ds \right\} \gamma_t(dx)$$

$$= \exp\left\{ \int_0^t \left[\log \int_{\tilde{H}} \exp\left\{ i \int_0^\infty e^{-2br} \langle \tilde{x}(r), T_s^* \tilde{a}(r) \rangle dr \right\} \lambda_0(d\tilde{x}) \right] ds \right\}$$

$$= \exp\left\{ \int_0^t \left[\log \int_{\tilde{H}} e^{i\langle \tilde{x}, \tilde{T}_s^* \tilde{a} \rangle} {\sim} \lambda_0(\mathrm{d}\tilde{x}) \right] \mathrm{d}s \right\}.$$

That proves (11.46). □

Theorem 11.20 *Let* $(\tilde{\gamma}_t)_{t \geq 0}$ *be a centered SC-semigroup associated with* $(\tilde{T}_t)_{t \geq 0}$ *satisfying*

$$\int_{\tilde{H}^\circ} \|\tilde{x}\|^2 \tilde{\gamma}_t(\mathrm{d}\tilde{x}) < \infty, \quad t \geq 0. \tag{11.47}$$

Then there is a centered SC-semigroup $(\gamma_t)_{t \geq 0}$ *associated with* $(T_t)_{t \geq 0}$ *satisfying* (11.38) *and* $\tilde{\gamma}_t = \gamma_t \circ J^{-1}$ *for each* $t \geq 0$.

Proof. By Theorems 11.15 and 11.18, any centered SC-semigroup associated with $(\tilde{T}_t)_{t \geq 0}$ is regular, so $(\tilde{\gamma}_t)_{t \geq 0}$ has the expression (11.46) for an infinitely divisible probability λ_0 on $\tilde{H}$. Then we get $(\gamma_t)_{t \geq 0}$ by Theorem 11.14, which clearly satisfies the requirements. □

By Theorems 11.19 and 11.20, centered SC-semigroups associated with $(T_t)_{t \geq 0}$ are in one-to-one correspondence with centered regular SC-semigroups associated with $(\tilde{T}_t)_{t \geq 0}$. Therefore we may reduce some analysis of irregular SC-semigroups to those of regular ones. The consideration of centered SC-semigroups is not a serious restriction. In fact, if $(\gamma_t)_{t \geq 0}$ is an arbitrary SC-semigroup satisfying condition (11.38), we can define

$$b_t = \int_{H^\circ} x \gamma_t(\mathrm{d}x)$$

and $\gamma_t^c = \delta_{-b_t} * \gamma_t$ for $t \geq 0$. It is easy to check that both $(\delta_{b_t})_{t \geq 0}$ and $(\gamma_t^c)_{t \geq 0}$ are SC-semigroups associated with $(T_t)_{t \geq 0}$. Therefore $(\gamma_t)_{t \geq 0}$ can always be decomposed as the convolution of a degenerate SC-semigroup and a centered one.

11.5 Construction of the Processes

In this section, we prove that the generalized OU-processes corresponding to a regular SC-semigroup has a càdlàg realization in a suitable extension of the space. Let $(T_t)_{t \geq 0}$ be a strongly continuous semigroup on H with generator $(A, \mathscr{D}(A))$. Then $\mathscr{D}(A)$ with the inner product norm $\| \cdot \|_A$ defined by

$$\|x\|_A^2 = \|x\|^2 + \|Ax\|^2, \quad x \in \mathscr{D}(A) \tag{11.48}$$

is a Hilbert space and $\mathscr{D}(A) \subset H$ is a continuous embedding.

Proposition 11.21 *There is a Hilbert space* $(\bar{H}, \| \cdot \|_-)$ *and a strongly continuous semigroup* $(\bar{T}_t)_{t \geq 0}$ *on* $(\bar{H}, \| \cdot \|_-)$ *with generator* $(\bar{A}, \mathscr{D}(\bar{A}))$ *such that:*

(1) $H \subset \bar{H}$ with dense continuous embedding;
(2) each T_t is the restriction of $\bar{T}_t$ to H;
(3) $H \subset \mathscr{D}(\bar{A})$ with continuous embedding.

Proof. Recall that there are constants $B \geq 0$ and $c_0 \geq 0$ so that $\|T_t\| \leq Be^{c_0 t}$ for every $t \geq 0$. Let $(U^\alpha)_{\alpha > c_0}$ denote the resolvent of $(T_t)_{t \geq 0}$. Fix $b > c_0$ and define an inner product on H by

$$\langle x, y \rangle_- = \langle U^b x, U^b y \rangle, \quad x, y \in H. \tag{11.49}$$

Let $\| \cdot \|_-$ be the corresponding norm and let $\bar{H}$ be the completion of H with respect to this norm. From (11.49) we get

$$\|x\|_- \leq \|U^b\| \|x\|, \quad x \in H, \tag{11.50}$$

so the identity mapping I from $(H, \| \cdot \|)$ to $(\bar{H}, \| \cdot \|_-)$ is a continuous dense embedding. Consequently, the linear semigroup $(T_t)_{t \geq 0}$ can be uniquely extended to a strongly continuous semigroup $(\bar{T}_t)_{t \geq 0}$ on $\bar{H}$, which satisfies $\|\bar{T}_t\| \leq Be^{bt}$ for every $t \geq 0$. In addition, we have $\mathscr{D}(\bar{A}) \subset \bar{H}$ with continuous embedding. For any $x \in \mathscr{D}(A) \subset \mathscr{D}(\bar{A})$ we have $x = U^b y$ for some $y \in H$. By (11.49) we have

$$\|\bar{A}x\|_- = \|AU^b U^b y\| = \|(bU^b - 1)U^b y\| \leq (b\|U^b\| + 1)\|x\|. \tag{11.51}$$

Since $\mathscr{D}(A)$ is a dense subset of $(H, \| \cdot \|)$, for $x \in H$ we can find $\{x_n\} \subset \mathscr{D}(A)$ so that $\lim_{n \to \infty} x_n = x$ in H and hence in $\bar{H}$. Then $\{x_n\}$ is a Cauchy sequence in both H and $\bar{H}$. From (11.48), (11.50) and (11.51) we see $\{x_n\}$ is also a Cauchy sequence in $\mathscr{D}(\bar{A})$. Since $\mathscr{D}(\bar{A})$ is complete and $\mathscr{D}(\bar{A}) \subset \bar{H}$ is a continuous embedding, we also have $\lim_{n \to \infty} x_n = x$ in $\mathscr{D}(\bar{A})$. That proves $H \subset \mathscr{D}(\bar{A})$. Since $\mathscr{D}(A)$ is dense in H, from (11.51) we have $\|\bar{A}x\|_- \leq (b\|U^b\| + 1)\|x\|$ for all $x \in H$. Then $H \subset \mathscr{D}(\bar{A})$ is a continuous embedding. $\qquad\square$

Now let ν_0 be an infinitely divisible probability measure on H and let $\psi_0(a) = -\log \hat{\nu}_0(a)$ for $a \in H$. A convolution semigroup $(\mu_t)_{t \geq 0}$ on H is given by

$$\hat{\mu}_t(a) = \exp\{-t\psi_0(a)\}, \quad a \in H. \tag{11.52}$$

Let $(P_t)_{t \geq 0}$ be the transition semigroup on H defined by

$$P_t f(x) = \int_H f(x + y)\mu_t(\mathrm{d}y), \quad x \in H, f \in C(H). \tag{11.53}$$

A càdlàg Markov process in H with transition semigroup $(P_t)_{t \geq 0}$ is called a *Lévy process*. In view of (11.53), a Lévy process is translation invariant and has independent increments. The existence of such a process is given by the following:

Proposition 11.22 *There is a Lévy process in H with transition semigroup $(P_t)_{t \geq 0}$ defined by (11.53).*

Proof. By (11.52) and Parthasarathy (1967, p.189) it is easy to see that $\lim_{t \to 0} \mu_t = \delta_0$ by the weak convergence. In particular, we have

$$\lim_{t \to 0} \sup_{x \in H} P_t(x, B(x, \varepsilon)^c) = \lim_{t \to 0} \mu_t(B(0, \varepsilon)^c) = 0, \qquad (11.54)$$

where $B(x, \varepsilon)^c$ denotes the complement of the open ball centered at $x \in H$ with radius $\varepsilon > 0$. Then the result follows by the general theory of stochastic processes; see, e.g., Wentzell (1981, p.170). $\qquad \square$

Suppose that $(\bar{H}, \| \cdot \|_-)$ is an extension of $(H, \| \cdot \|)$ with the three properties in Proposition 11.21. Let $(\gamma_t)_{t \geq 0}$ be the regular SC-semigroup associated with $(T_t)_{t \geq 0}$ defined by (11.13). We can certainly regard ν_0 and γ_t as infinitely divisible probability measures on the enlarged space $\bar{H}$. Then $(\gamma_t)_{t \geq 0}$ is also an SC-semigroup associated with $(\bar{T}_t)_{t \geq 0}$. Let

$$\bar{\psi}_0(\bar{a}) = -\log \int_{\bar{H}} e^{i \langle \bar{x}, \bar{a} \rangle} - \nu_0(d\bar{x}), \qquad \bar{a} \in \bar{H}^* \subset H. \qquad (11.55)$$

From (11.13) it is not hard to show that

$$\int_{\bar{H}} e^{i \langle \bar{x}, \bar{a} \rangle} - \gamma_t(d\bar{x}) = \exp \left\{ - \int_0^t \bar{\psi}_0(\bar{T}_s^* \bar{a}) ds \right\}, \qquad \bar{a} \in \bar{H}^*, \qquad (11.56)$$

where $(\bar{T}_t^*)_{t \geq 0}$ denotes the dual semigroup of $(\bar{T}_t)_{t \geq 0}$. Let $(\bar{Q}_t^\gamma)_{t \geq 0}$ be the generalized Mehler semigroup defined by (11.14) from $(\bar{T}_t)_{t \geq 0}$ and $(\gamma_t)_{t \geq 0}$.

By Proposition 11.22, the transition semigroup $(P_t)_{t \geq 0}$ defined by (11.53) has a càdlàg realization $\{Y_t : t \geq 0\}$ in $H \subset \mathscr{D}(\bar{A})$ with $Y_0 = 0$. Since $s \mapsto \bar{A} Y_s$ is right continuous, for any $\bar{x} \in \bar{H}$,

$$\bar{Z}_t = \bar{T}_t \bar{x} + Y_t + \int_0^t \bar{T}_{t-s} \bar{A} Y_s ds \qquad (11.57)$$

defines a càdlàg process $\{\bar{Z}_t : t \geq 0\}$ in $\bar{H}$.

Lemma 11.23 *For any $t \geq 0$ the random variable $\bar{Z}_t$ defined by (11.57) has distribution $\bar{Q}_t^\gamma(\bar{x}, \cdot)$ on $\bar{H}$.*

Proof. By dominated convergence and the right continuity of $s \mapsto \bar{A} Y_s$ we get

$$\bar{Z}_n(t) := \bar{T}_t \bar{x} + Y_t + \sum_{k=1}^n \int_{(k-1)t/n}^{kt/n} \bar{T}_{t-s} \bar{A} Y_{kt/n} ds \to \bar{Z}_t \qquad (11.58)$$

in $\bar{H}$ as $n \to \infty$. Observe that

$$\bar{Z}_n(t) = \bar{T}_t \bar{x} + Y_t + \sum_{k=1}^n (\bar{T}_{(n-k+1)t/n} - \bar{T}_{(n-k)t/n}) Y_{kt/n}$$

$$= \bar{T}_t \bar{x} + \sum_{k=1}^{n} \bar{T}_{(n-k+1)t/n}(Y_{kt/n} - Y_{(k-1)t/n}),$$

and hence

$$\mathbf{E} \exp \left\{ i \langle \bar{Z}_n(t), \bar{a} \rangle_- \right\} = \exp \left\{ i \langle \bar{x}, \bar{T}_t^* \bar{a} \rangle - \frac{t}{n} \sum_{k=1}^{n} \psi_0 \left(\bar{T}_{(n-k+1)t/n}^* \bar{a} \right) \right\}.$$

Since $(\bar{T}_t)_{t \geq 0}$ is strongly continuous, so is $(\bar{T}_t^*)_{t \geq 0}$. Then $s \mapsto \psi_0(\bar{T}_s^* \bar{a})$ is continuous on $[0, \infty)$ for each $\bar{a} \in \bar{H}$. By letting $n \to \infty$ in the equality above we obtain

$$\mathbf{E} \exp \left\{ i \langle \bar{Z}_t, \bar{a} \rangle_- \right\} = \exp \left\{ i \langle \bar{x}, \bar{T}_t^* \bar{a} \rangle - \int_0^t \psi_0 \left(\bar{T}_{t-s}^* \bar{a} \right) \mathrm{d}s \right\}.$$

That gives the desired result.   □

Theorem 11.24 *The process* $\{ \bar{Z}_t : t \geq 0 \}$ *defined by* (11.57) *is a càdlàg strong Markov process in* $\bar{H}$ *with transition semigroup* $(\bar{Q}_t^\gamma)_{t \geq 0}$.

Proof. In view of (11.57), the process $\{ \bar{Z}_t : t \geq 0 \}$ is adapted to the filtration $(\mathscr{F}_t)_{t \geq 0}$ generated by $\{ Y_t : t \geq 0 \}$. For $r, t \geq 0$ we have

$$\bar{Z}_{r+t} - \bar{T}_t \bar{Z}_r = Y_{r+t} - \bar{T}_t Y_r + \int_r^{r+t} \bar{T}_{r+t-s} \bar{A} Y_s \mathrm{d}s$$

$$= (Y_{r+t} - Y_r) + \int_r^{r+t} \bar{T}_{r+t-s} \bar{A}(Y_s - Y_r) \mathrm{d}s.$$

Since $\{ Y_{r+t} - Y_r : t \geq 0 \}$ given $\mathscr{F}_r$ is a process with independent increments and has the same law as $\{ Y_t : t \geq 0 \}$, an application of Lemma 11.23 shows that

$$\mathbf{E} \left[\exp \left\{ i \langle \bar{Z}_{r+t}, \bar{a} \rangle_- \right\} \middle| \mathscr{F}_r \right] = \exp \left\{ i \langle \bar{Z}_r, \bar{T}_t^* \bar{a} \rangle_- - \int_0^t \psi_0(\bar{T}_s^* \bar{a}) \mathrm{d}s \right\}.$$

Therefore $\{ \bar{Z}_t : t \geq 0 \}$ is a Markov process with transition semigroup $(\bar{Q}_t^\gamma)_{t \geq 0}$. Since $(\bar{Q}_t^\gamma)_{t \geq 0}$ preserves $C(H)$, the strong Markov property follows by a standard argument.   □

From the theorem above we easily obtain a construction of the generalized OU-process corresponding to the generalized Mehler semigroup $(Q_t^\gamma)_{t \geq 0}$. Indeed, for any $x \in H$ we have $\bar{T}_t x = T_t x \in H$ and hence $\bar{Z}_t \in H$ a.s. for every $t \geq 0$. Then $\{ \bar{Z}_t : t \geq 0 \}$ is also a Markov process with transition semigroup $(Q_t^\gamma)_{t \geq 0}$. However, this process usually does not have a càdlàg version in H. In other words, to get the sample path regularity, we need to observe the process in the enlarged state space $\bar{H}$ with a weaker topology. A similar phenomenon has been observed in Example 9.2.

Suppose that E is a real separable Hilbert space containing H as a subspace and A is the generator of a semigroup of bounded linear operators $(T_t)_{t \geq 0}$ on E with domain $\mathscr{D}(A) \supset H$. Let $\{Y_t : t \geq 0\}$ be a Lévy process in H. We say a stochastic process $\{X_t : t \geq 0\}$ in H solves a *Langevin type equation* provided

$$X_t = X_0 + \int_0^t AX_s ds + Y_t, \quad t \geq 0. \tag{11.59}$$

As usual, we may write the equation into the differential form

$$dX_t = AX_t dt + dY_t, \quad t \geq 0. \tag{11.60}$$

Theorem 11.25 *Let $\{Y_t : t \geq 0\}$ be a Lévy process in H with transition semigroup $(P_t)_{t \geq 0}$ defined by (11.53). Then the generalized OU-process $\{\bar{Z}_t : t \geq 0\}$ defined by (11.57) satisfies the stochastic equation*

$$\bar{Z}_t = \bar{x} + Y_t + \bar{A}\left(\int_0^t \bar{Z}_s ds\right), \quad t \geq 0. \tag{11.61}$$

Proof. From (11.57) we have

$$\int_0^t \bar{Z}_s ds = \int_0^t \bar{T}_s \bar{x} ds + \int_0^t Y_s ds + \int_0^t du \int_0^u \bar{T}_{u-s} \bar{A} Y_s ds$$
$$= \int_0^t \bar{T}_s \bar{x} ds + \int_0^t Y_s ds + \int_0^t (\bar{T}_{t-s} Y_s - Y_s) ds$$
$$= \int_0^t \bar{T}_s \bar{x} ds + \int_0^t \bar{T}_{t-s} Y_s ds.$$

From (11.48) we see that $\bar{A} : \mathscr{D}(\bar{A}) \to \bar{H}$ is a bounded operator, so the above equation implies

$$\bar{A}\left(\int_0^t \bar{Z}_s ds\right) = \bar{A}\left(\int_0^t \bar{T}_s \bar{x} ds\right) + \lim_{n \to \infty} \sum_{k=1}^n \bar{A} \int_{(k-1)t/n}^{kt/n} \bar{T}_{t-s} Y_{kt/n} ds$$
$$= \bar{T}_t \bar{x} - \bar{x} + \lim_{n \to \infty} \sum_{k=1}^n \int_{(k-1)t/n}^{kt/n} \bar{A} \bar{T}_{t-s} Y_{kt/n} ds$$
$$= \bar{T}_t \bar{x} - \bar{x} + \int_0^t \bar{T}_{t-s} \bar{A} Y_s ds$$
$$= \bar{Z}_t - \bar{x} - Y_t.$$

That proves (11.61). $\qquad\square$

One naturally wishes to exchange the order of the integral and the operation of the generator in (11.61). To do so, we need a further extension of the domain of the generator. Let $(\tilde{H}, \|\cdot\|_{\sim})$ be an extension of $(\bar{H}, \|\cdot\|_{-})$ with the properties in Proposition 11.21. Let $(\tilde{T}_t)_{t \geq 0}$ and $\tilde{A}$ be the corresponding extensions of $(\bar{T}_t)_{t \geq 0}$ and $\bar{A}$, respectively.

Theorem 11.26 *The generalized OU-process* $\{\bar{Z}_t : t \geq 0\}$ *defined by* (11.57) *is càdlàg in* $\mathscr{D}(\tilde{A})$ *and satisfies the Langevin type equation*

$$\bar{Z}_t = \bar{x} + Y_t + \int_0^t \tilde{A}\bar{Z}_s \mathrm{d}s, \quad t \geq 0. \tag{11.62}$$

Proof. By Proposition 11.21, we have $H \subset \mathscr{D}(\tilde{A})$ with continuous embedding. Since $\bar{A}$ is a bounded operator from $\mathscr{D}(\bar{A})$ to $\bar{H}$, the process $s \mapsto \bar{A}Y_s$ is càdlàg in $\bar{H}$. Similarly we find $s \mapsto \tilde{A}\bar{A}Y_s$ is càdlàg in $\tilde{H}$. By Theorem 11.24 the process $t \mapsto \bar{Z}_t$ is càdlàg in $\bar{H}$, so it is càdlàg in $\mathscr{D}(\tilde{A})$ and

$$\tilde{A}\bar{Z}_t = \tilde{A}\bar{T}_t\bar{x} + \bar{A}Y_t + \int_0^t \tilde{T}_{t-s}\tilde{A}\bar{A}Y_s \mathrm{d}s.$$

Moreover, we have

$$\bar{A}\Big(\int_0^t \bar{Z}_s \mathrm{d}s \Big) = \tilde{A}\Big(\int_0^t \bar{Z}_s \mathrm{d}s \Big) = \int_0^t \tilde{A}\bar{Z}_s \mathrm{d}s.$$

Then (11.62) follows from (11.61). $\qquad\qquad\qquad\qquad\qquad\qquad\qquad\square$

Theorem 11.27 *If* $\{\bar{Z}_t : t \geq 0\}$ *is a càdlàg process in* $\mathscr{D}(\tilde{A})$ *satisfying* (11.62), *then it is given by* (11.57). *Consequently, the pathwise uniqueness holds for the equation* (11.62).

Proof. From (11.62) we have

$$\int_0^t \tilde{T}_{t-s}\bar{Z}_s \mathrm{d}s = \int_0^t \bar{T}_{t-s}\bar{x}\mathrm{d}s + \int_0^t \bar{T}_{t-s}Y_s \mathrm{d}s + \int_0^t \mathrm{d}u \int_0^u \tilde{T}_{t-u}\tilde{A}\bar{Z}_s \mathrm{d}s$$

$$= \int_0^t \bar{T}_{t-s}\bar{x}\mathrm{d}s + \int_0^t \bar{T}_{t-s}Y_s \mathrm{d}s + \int_0^t (\tilde{T}_{t-s}\bar{Z}_s - \bar{Z}_s)\mathrm{d}s,$$

and hence

$$\int_0^t \bar{Z}_s \mathrm{d}s = \int_0^t \bar{T}_{t-s}\bar{x}\mathrm{d}s + \int_0^t \bar{T}_{t-s}Y_s \mathrm{d}s.$$

It follows that

$$\int_0^t \tilde{A}\bar{Z}_s \mathrm{d}s = \bar{A}\Big(\int_0^t \bar{T}_{t-s}\bar{x}\mathrm{d}s \Big) + \int_0^t \bar{T}_{t-s}\bar{A}Y_s \mathrm{d}s$$

$$= \bar{T}_t\bar{x} - \bar{x} + \int_0^t \bar{T}_{t-s}\bar{A}Y_s \mathrm{d}s.$$

By using (11.62) again we obtain (11.57). $\qquad\qquad\qquad\qquad\qquad\qquad\square$

11.6 Notes and Comments

The concept of the generalized Mehler semigroup was introduced by Bogachev and Röckner (1995) and Bogachev et al. (1996) as a generalization of the classical Mehler formula; see, e.g., Malliavin (1997, p.25). The subject has become a very interesting field of research.

Proposition 11.2 was first proved by Schmuland and Sun (2001). The current form of Theorem 11.3 is due to Dawson et al. (2004b), which extends an earlier result of Bogachev et al. (1996) in the setting of cylindrical measures. By a theorem of Keller-Ressel et al. (2010), every stochastically continuous Ornstein–Uhlenbeck process in a finite-dimensional space is regular. Theorem 11.7 was first proved in Dawson et al. (2004b). The main reference of Sections 11.4 and 11.5 is Dawson and Li (2004). See also Fuhrman and Röckner (2000) for the construction of the process.

A set of generalized Ornstein–Uhlenbeck processes were defined using Langevin type equations in Chojnowska-Michalik (1987), where the following mild form of (11.59) was considered:

$$X_t = T_t X_0 + \int_0^t T_{t-s} \mathrm{d}Y_s, \quad t \geq 0. \tag{11.63}$$

If the Lévy process $\{Y_t : t \geq 0\}$ has transition semigroup given by (11.52) and (11.53), then $\{X_t : t \geq 0\}$ has transition semigroup $(Q_t^\gamma)_{t \geq 0}$ given by (11.14); see, e.g., Applebaum (2007).

Let us consider the regular SC-semigroup defined by (11.13) with $a \mapsto \psi_0(a)$ given by the right-hand side of (11.2). It was proved in Fuhrman and Röckner (2000) that the corresponding generalized Mehler semigroup $(Q_t^\gamma)_{t \geq 0}$ is weakly continuous on the space of uniformly continuous bounded functions. The notion of weak continuity was introduced in Cerrai (1994), where it was shown that the strong continuity fails even in the Gaussian case. The generator of $(Q_t^\gamma)_{t \geq 0}$ was defined in Fuhrman and Röckner (2000) by the resolvent. Lescot and Röckner (2002) characterized the generator as a pseudo-differential operator. The existence and uniqueness of invariant measures for generalized OU-processes were studied in Chojnowska-Michalik (1987) and Fuhrman and Röckner (2000).

The mixed topology on $C(H)$ is by definition the finest locally convex topology that agrees on bounded sets with the uniform convergence on compact sets in H. The semigroup $(Q_t^\gamma)_{t \geq 0}$ is strongly continuous on $C(H)$ with this topology. Applebaum (2007) gave an explicit representation of the generator of $(Q_t^\gamma)_{t \geq 0}$ as a semigroup on $C(H)$, which is closable and has a convenient invariant core of cylinder functions. The mixed topology was already used to study Gaussian type Mehler semigroups in Goldys and Kocan (2001) and Goldys and van Neerven (2003).

The mild form (11.63) of the Langevin type equation makes sense even when $\{Y_t : t \geq 0\}$ is a Lévy process in some larger space $E \supset H$. Priola and Zabczyk (2010) considered the case where $\{Y_t : t \geq 0\}$ is a cylindrical stable process. Suppose that $A : \mathscr{D}(A) \to H$ is a self-adjoint operator and $\{e_1, e_2, \ldots\}$ is an orthonormal basis of H such that $A e_n = \gamma_n e_n$ for every $n \geq 1$ with $\gamma_n > 0$ and

$\gamma_n \to \infty$ as $n \to \infty$. Then each e_n is an eigenvector of A. Let $\{Y_t : t \geq 0\}$ be a cylindrical stable process given by

$$Y_t = \sum_{n=1}^{\infty} \beta_n y_n(t) e_n,$$

where $\{y_n(t) : t \geq 0\}$, $n = 1, 2, \ldots$ are i.i.d. one-dimensional α-stable processes with $0 < \alpha < 2$ and $\beta_1, \beta_2, \ldots$ are strictly positive constants. It was proved in Priola and Zabczyk (2010) that for any $X_0 = x \in H$ the generalized OU-process $\{X_t : t \geq 0\}$ defined by (11.63) takes values in H if and only if

$$\sum_{n=1}^{\infty} \frac{\beta_n^{\alpha}}{\gamma_n} < \infty,$$

and in this case $\{X_t : t \geq 0\}$ is a stochastically continuous Markov process.

Suppose that $\{Y_t : t \geq 0\}$ is a Lévy process in H and $x \mapsto b(x)$ is an operator on H. A generalization of the Langevin type equation (11.59) is the following:

$$dX_t = AX_t dt + b(X_t) dt + dY_t, \quad t \geq 0. \tag{11.64}$$

This equation was studied in Lescot and Röckner (2004) under certain regularity conditions. Their approach was to construct the transition semigroup of the solution by applying the perturbation theory to the generalized Mehler semigroup in the space $L^2(H, \mu)$, where μ is the invariant measure for the solution of the equation with $b = 0$. Priola and Zabczyk (2010) studied the Markov property, irreducibility and strong Feller property of the solution to (11.64) for a cylindrical stable noise.

We refer the reader to Röckner and Wang (2003) and Wang (2005) for some powerful Harnack and functional inequalities of generalized Mehler semigroups.

Chapter 12
Small-Branching Fluctuation Limits

A typical class of generalized OU-processes arise as small-branching fluctuation limits of subcritical immigration superprocesses around their equilibrium means. In this chapter, we first establish such a fluctuation limit theorem in the space of Schwartz distributions. A stronger result is then proved which shows that the convergence actually holds in a suitable weighted Sobolev space. To avoid complicated regularity assumptions, we only consider the case where the spatial motion is a Brownian motion with killing.

12.1 The Brownian Immigration Superprocess

We first introduce the Brownian immigration superprocess to be considered. Let $b > 0$ be a constant and let ξ be a killed Brownian motion in $\mathbb{R}^d$ with generator $A := \Delta/2 - b$ and transition semigroup $(P_t^b)_{t \geq 0}$. Then ξ has finite potential operator U given by

$$Uf(x) = \int_0^\infty P_t^b f(x) dt = \int_0^\infty e^{-bt} P_t f(x) dt, \quad f \in B(\mathbb{R}^d),$$

where $(P_t)_{t \geq 0}$ is the transition semigroup of the standard Brownian motion in $\mathbb{R}^d$. Let ϕ be a critical local branching mechanism on $\mathbb{R}^d$ given by

$$\phi(x, z) = c(x)z^2 + \int_0^\infty (e^{-zu} - 1 + zu)m(x, du), \tag{12.1}$$

where $c \in C(\mathbb{R}^d)^+$ and $u^2 m(x, du)$ is a bounded kernel from $\mathbb{R}^d$ to $(0, \infty)$. We assume $x \mapsto u^2 m(x, du)$ is continuous by weak convergence on $(0, \infty)$. The cumulant semigroup $(V_t)_{t \geq 0}$ of the (ξ, ϕ)-superprocess is defined by

$$V_t f(x) = P_t^b f(x) - \int_0^t P_{t-s}^b \phi(V_s f)(x) ds, \quad t \geq 0, x \in \mathbb{R}^d. \tag{12.2}$$

Z. Li, *Measure-Valued Branching Markov Processes*,
Probability and Its Applications, DOI 10.1007/978-3-642-15004-3_12,
© Springer-Verlag Berlin Heidelberg 2011

Clearly, the actual branching mechanism of the (ξ, ϕ)-superprocess is strictly sub-critical because of the killing rate $b > 0$ in the underlying spatial motion.

We fix a constant $p > d$ and let $h_p(x) = (1 + |x|^2)^{-p/2}$ for $x \in \mathbb{R}^d$, where $|\cdot|$ denotes the Euclidean norm. It is easy to find a constant $\alpha > 0$ so that h_p is α-excessive relative to $(P_t^b)_{t \geq 0}$. Let $C_p(\mathbb{R}^d)$ denote the set of continuous functions $f \in C_0(\mathbb{R}^d)$ satisfying $|f| \leq \text{const} \cdot h_p$. Let $M_p(\mathbb{R}^d)$ be the space of σ-finite measures μ on $\mathbb{R}^d$ satisfying $\langle \mu, h_p \rangle < \infty$. We endow $M_p(\mathbb{R}^d)$ with the topology defined by the convention:

$$\mu_n \to \mu \text{ in } M_p(\mathbb{R}^d) \text{ if and only if } \langle \mu_n, f \rangle \to \langle \mu, f \rangle \text{ for all } f \in C_p(\mathbb{R}^d).$$

In this chapter, we denote the Lebesgue measure on $\mathbb{R}^d$ by λ, which clearly belongs to $M_p(\mathbb{R}^d)$. Given $\eta \in M_p(\mathbb{R}^d)$ we define the transition semigroup $(Q_t^\eta)_{t \geq 0}$ on $M_p(\mathbb{R}^d)$ by

$$\int_{M_p(\mathbb{R}^d)} e^{-\langle \nu, f \rangle} Q_t^\eta(\mu, d\nu) = \exp\left\{ -\langle \mu, V_t f \rangle - \int_0^t \langle \eta, V_s f \rangle ds \right\}. \quad (12.3)$$

By Theorem 9.25 there is a càdlàg realization $Y = (W, \mathscr{G}, \mathscr{G}_t, Y_t, \mathbf{Q}_\mu^\eta)$ of the immi-gration superprocess in $M_p(\mathbb{R}^d)$ with transition semigroup $(Q_t^\eta)_{t \geq 0}$. The results of Propositions 9.11 and 9.14 extend immediately to the present case. Then for $t \geq 0$ and $f \in C_p(\mathbb{R}^d)$ we have

$$\mathbf{Q}_\mu^\eta[\langle Y_t, f \rangle] = \langle \mu, P_t^b f \rangle + \int_0^t \langle \eta, P_s^b f \rangle ds, \quad (12.4)$$

and

$$\begin{aligned}
\mathbf{Q}_\mu^\eta[\langle Y_t, f \rangle^2] &= \left(\langle \mu, P_t^b f \rangle + \int_0^t \langle \eta, P_s^b f \rangle ds \right)^2 \\
&\quad + \int_0^t \langle \mu, P_{t-s}^b [\phi''(\cdot, 0)(P_s^b f)^2] \rangle ds \\
&\quad + \int_0^t du \int_0^u \langle \eta, P_{u-s}^b [\phi''(\cdot, 0)(P_s^b f)^2] \rangle ds, \quad (12.5)
\end{aligned}$$

where $\phi''(x, 0)$ is defined by (2.64).

Proposition 12.1 *Suppose that $\eta(dx)$ is absolutely continuous with respect to the Lebesgue measure $\lambda(dx)$ with bounded density $x \mapsto \eta'(x)$. Then $(Q_t^\eta)_{t \geq 0}$ has a stationary distribution F^η defined by*

$$\int_{M_p(\mathbb{R}^d)} e^{-\langle \nu, f \rangle} F^\eta(d\nu) = \exp\left\{ -\int_0^\infty \langle \eta, V_s f \rangle ds \right\}, \quad (12.6)$$

where $f \in C_p(\mathbb{R}^d)^+$. Moreover, we have $Q_t^\eta(0, \cdot) \to F^\eta$ by weak convergence as $t \to \infty$.

Proof. Clearly, the mapping $\nu(\mathrm{d}x) \mapsto h_p(x)\nu(\mathrm{d}x)$ induces a homeomorphism between $M_p(\mathbb{R}^d)$ and $M(\mathbb{R}^d)$. Let $G_t^\eta(\mathrm{d}\nu)$ denote the image of $Q_t^\eta(0, \mathrm{d}\nu)$ under the above mapping. For any $f \in B(\mathbb{R}^d)^+$ we have

$$V_t(fh_p)(x) \leq P_t^b(fh_p)(x) \leq \|f\| P_t^b h_p(x), \quad t \geq 0, x \in \mathbb{R}^d,$$

where $\|\cdot\|$ denotes the supremum norm. Since $\lambda(\mathrm{d}x)$ is an invariant measure for the Brownian motion, we have

$$\int_0^\infty \langle \eta, V_s(fh_p)\rangle \mathrm{d}s \leq \|f\| \int_0^\infty \langle \eta, P_s^b h_p \rangle \mathrm{d}s$$
$$\leq \|f\eta'\| \int_0^\infty e^{-bs} \langle \lambda, h_p \rangle \mathrm{d}s < \infty.$$

It follows that

$$\lim_{t\to\infty} \int_{M(\mathbb{R}^d)} e^{-\langle \nu, f\rangle} G_t^\eta(\mathrm{d}\nu) = \lim_{t\to\infty} \int_{M_p(\mathbb{R}^d)} e^{-\langle \nu, fh_p\rangle} Q_t^\eta(0, \mathrm{d}\nu)$$
$$= \exp\left\{ -\int_0^\infty \langle \eta, V_s(fh_p)\rangle \mathrm{d}s \right\}, \qquad (12.7)$$

and the right-hand side is continuous in $f \in B(\mathbb{R}^d)^+$ with respect to bounded pointwise convergence. By Theorem 1.20, it is the Laplace functional of a probability measure G^η on $M(\mathbb{R}^d)$ and $G_t^\eta \to G^\eta$ weakly as $t \to \infty$. Then (12.6) defines a probability measure F^η on $M_p(\mathbb{R}^d)$ and $Q_t^\eta(0, \cdot) \to F^\eta$ weakly as $t \to \infty$. It is easily seen that F^η is a stationary distribution of $(Q_t^\eta)_{t\geq 0}$. $\qquad\square$

Under the condition of Proposition 12.1, the measure potential $\zeta := \eta U$ is the mean of F^η. In fact, from (12.6) we have

$$\int_{M_p(\mathbb{R}^d)} \langle \nu, f\rangle F^\eta(\mathrm{d}\nu) = \int_0^\infty \langle \eta, P_s^b f\rangle \mathrm{d}s = \langle \zeta, f\rangle.$$

Observe also that

$$\langle \zeta, f\rangle = \langle \zeta, P_t^b f\rangle + \int_0^t \langle \eta, P_s^b f\rangle \mathrm{d}s, \qquad t \geq 0. \qquad (12.8)$$

12.2 Stochastic Processes in Nuclear Spaces

Suppose that E is an infinite-dimensional real linear space and $\|\cdot\|_0 \leq \|\cdot\|_1 \leq \|\cdot\|_2 \leq \cdots$ is a sequence of Hilbertian norms on E. Let E_n be the completion of E relative to $\|\cdot\|_n$ and let $\langle\cdot,\cdot\rangle_n$ denote the inner product in E_n. Then we have $E_0 \supset E_1 \supset E_2 \supset \cdots$. The sequence of norms $\|\cdot\|_0 \leq \|\cdot\|_1 \leq \|\cdot\|_2 \leq \cdots$ induces a topology on the set $E_\infty := \bigcap_{n=0}^\infty E_n$, which is compatible with the metric ρ defined by

$$\rho(x,y) = \sum_{k=0}^{\infty} \frac{\|y-x\|_k}{2^k(1+\|y-x\|_k)}, \qquad x,y \in E_\infty. \qquad (12.9)$$

Proposition 12.2 *The metric space* (E,ρ) *is complete if and only if* $E = E_\infty$.

Proof. Suppose that (E,ρ) is complete. If $x \in E_\infty$, for each $n \geq 0$ there is $x_n \in E$ such that $\|x_n - x\|_n < 1/2^{n+1}$ implying $\|x_n - x\|_k < 1/2^{n+1}$ for $0 \leq k \leq n$. It follows that

$$\rho(x_n, x) \leq \sum_{k=0}^{n} \frac{1}{2^k}\|x_n - x\|_k + \sum_{k=n+1}^{\infty} \frac{1}{2^k} < \frac{2}{2^{n+1}} + \frac{1}{2^n} = \frac{1}{2^{n-1}}.$$

Then we have $x \in E$, proving $E = E_\infty$. For the converse, suppose that $E = E_\infty$. If $\{x_k\}$ is a Cauchy sequence in (E,ρ), it is a Cauchy sequence relative to each norm $\|\cdot\|_n$. Then there is $y_n \in E_n$ so that $\|x_k - y_n\|_n \to 0$ as $k \to \infty$. By relations $\|\cdot\|_0 \leq \|\cdot\|_1 \leq \|\cdot\|_2 \leq \cdots$, we must have $y_n = y_0$ for every $n \geq 0$. Then $\|x_k - y_0\|_n \to 0$ as $k \to \infty$ for every $n \geq 0$. From (12.9) it follows that $\rho(x_k, y_0) \to 0$ as $k \to \infty$. Thus (E,ρ) is complete. $\qquad \square$

Proposition 12.3 *Let* f *be a linear map of* E *into a normed linear space* $(F, \|\cdot\|)$. *Then* f *is continuous relative to the metric defined by* (12.9) *if and only if it is continuous relative to one of the norms* $\|\cdot\|_n$.

Proof. If f is continuous relative to one of the norms $\|\cdot\|_n$, it is clearly continuous relative to the metric ρ defined by (12.9). Conversely, suppose that f is a continuous linear map of (E,ρ) into $(F, \|\cdot\|)$. Then there is a neighborhood G of zero such that $\|f(y)\| < 1$ for all $y \in G$. Consequently, there exists $n \geq 0$ and $\delta > 0$ such that $\{x \in E : \|x\|_n < \delta\} \subset G$. Therefore $\|x\|_n < \delta$ implies $\|f(x)\| < 1$, and so $\|x\|_n < \delta\varepsilon$ implies $\|f(x)\| < \varepsilon$ for every $\varepsilon > 0$. That gives the continuity of f relative to $\|\cdot\|_n$. $\qquad \square$

By Proposition 12.3 the space E has dual $E' := \bigcup_{n=0}^{\infty} E_{-n}$, where E_{-n} denotes the dual space of E_n. Let $\langle\cdot,\cdot\rangle$ denote the duality between E and E'. A subset B of E is said to be *bounded* if it is bounded in each norm $\|\cdot\|_n$, that is, $\sup_{x \in B} \|x\|_n < \infty$ for each $n \geq 0$. For any bounded set $B \subset E$ define the semi-norm p_B on E' by

$$p_B(f) = \sup\{|f(x)| : x \in B\}, \qquad f \in E'. \qquad (12.10)$$

We endow E' with the topology generated by the collection of semi-norms $\{p_B : B \subset E \text{ is bounded}\}$, which is called the *strong topology*.

For every $n \geq 0$ let $\{e_1^n, e_2^n, \ldots\}$ be an orthonormal basis of E_n and let $\|\cdot\|_{-n}$ be the norm of E_{-n} defined by

$$\|f\|_{-n}^2 = \sum_{k=1}^{\infty} \langle f, e_k^n\rangle^2, \qquad f \in E_{-n}.$$

We identify E_{-0} with E_0, but not E_{-n} with E_n for $n \geq 1$. We call E or (E, ρ) a *countably Hilbert nuclear space* or simply a *nuclear space* if the following conditions are satisfied:

(1) E is separable with respect to $\| \cdot \|_n$ for every $n \geq 0$;
(2) for every $m \geq 0$ there exists $n > m$ and an orthonormal basis $\{e_1^n, e_2^n, \ldots\}$ of E_n so that

$$\sum_{k=1}^{\infty} \|e_k^n\|_m^2 < \infty; \tag{12.11}$$

(3) the metric space (E, ρ) is complete.

It is well-known that the above property (2) is equivalent to the embedding operator $\pi_{n,m}$ of E_n into E_m being *Hilbert–Schmidt*; see, e.g., Kallianpur and Xiong (1995, p.18). For any orthonormal basis $\{f_1^m, f_2^m, \ldots\}$ of E_m we have

$$\begin{aligned}
\sum_{k=1}^{\infty} \|e_k^n\|_m^2 &= \sum_{k=1}^{\infty} \sum_{i=1}^{\infty} \langle \pi_{n,m} e_k^n, f_i^m \rangle_m^2 \\
&= \sum_{i=1}^{\infty} \sum_{k=1}^{\infty} \langle e_k^n, \pi_{-m,-n} f_i^m \rangle_n^2 \\
&= \sum_{i=1}^{\infty} \|f_i^m\|_{-n}^2.
\end{aligned}$$

Then the value on the left-hand side of (12.11) does not depend on the choice of the orthonormal basis $\{e_1^n, e_2^n, \ldots\}$. If E is a nuclear space, we have

$$E' = \bigcup_{n=0}^{\infty} E_{-n} \supset \cdots \supset E_{-2} \supset E_{-1} \supset E_0 \supset E_1 \supset E_2 \supset \cdots \supset \bigcap_{n=0}^{\infty} E_n = E.$$

The following two theorems were established in Mitoma (1983); see also Walsh (1986, pp.361–365).

Theorem 12.4 *Let E be a nuclear space with strong dual E'. Let $\{(Y_k(t))_{t \geq 0} : k \geq 1\}$ be a sequence of processes with sample paths in the space $D([0, \infty), E')$. If for each $x \in E$ the sequence of real processes $\{(\langle Y_k(t), x \rangle)_{t \geq 0} : k \geq 1\}$ is tight in $D([0, \infty), \mathbb{R})$, then $\{(Y_k(t))_{t \geq 0} : k \geq 1\}$ is tight in $D([0, \infty), E')$.*

Theorem 12.5 *Let $\{(Y_k(t))_{t \geq 0} : k \geq 1\}$ be a sequence of processes satisfying the conditions of Theorem 12.4. Suppose that $n > m \geq 0$ and $E_n \subset E_m$ is a Hilbert–Schmidt embedding. If for every $t \geq 0$, $\rho > 0$ and $\varepsilon > 0$ there exists $\delta > 0$ such that $x \in E$ and $\|x\|_m \leq \delta$ imply*

$$\sup_{k \geq 1} \mathbf{P} \left\{ \sup_{0 \leq s \leq t} |\langle Y_k(s), x \rangle| \geq \rho \right\} \leq \varepsilon,$$

then each process $(Y_k(t))_{t\geq 0}$ has sample paths a.s. in $D([0,\infty), E_{-n})$ and the sequence $\{(Y_k(t))_{t\geq 0} : k \geq 1\}$ is tight in $D([0,\infty), E_{-n})$.

We next consider a typical example of the nuclear space. Let $\mathbb{N} = \{0, 1, 2, \ldots\}$. Let $C^\infty(\mathbb{R}^d)$ be the set of bounded infinitely differentiable functions on $\mathbb{R}^d$ with bounded derivatives. Let $\mathscr{S}(\mathbb{R}^d) \subset C^\infty(\mathbb{R}^d)$ denote the *Schwartz space* of rapidly decreasing functions. That is, a function $f \in \mathscr{S}(\mathbb{R}^d)$ is infinitely differentiable and for every $k \geq 0$ and every $\alpha = (\alpha_1, \ldots, \alpha_d) \in \mathbb{N}^d$ we have

$$\lim_{|x|\to\infty} |x|^k |\partial^\alpha f(x)| = 0, \tag{12.12}$$

where $|\cdot|$ denotes the Euclidean norm and

$$\partial^\alpha f(x) = \frac{\partial^{\alpha_1 + \cdots + \alpha_d}}{\partial x_1^{\alpha_1} \cdots \partial x_d^{\alpha_d}} f(x_1, \ldots, x_d).$$

We first define an increasing sequence of norms $\{p_0, p_1, p_2, \ldots\}$ on $\mathscr{S}(\mathbb{R}^d)$ by

$$p_n(f) = \sum_{0 \leq \bar{\alpha} \leq n} \sup_{x \in \mathbb{R}^d} (1 + |x|^2)^{n/2} |\partial^\alpha f(x)|, \tag{12.13}$$

where $\bar{\alpha} = \alpha_1 + \cdots + \alpha_d$. The norms $\{p_0, p_1, p_2, \ldots\}$ are not Hilbertian. We also define the Hilbertian norms $\{q_0, q_1, q_2, \ldots\}$ on $\mathscr{S}(\mathbb{R}^d)$ by

$$q_n(f)^2 = \sum_{0 \leq \bar{\alpha} \leq n} \int_{\mathbb{R}^d} (1 + |x|^2)^n |\partial^\alpha f(x)|^2 \mathrm{d}x. \tag{12.14}$$

Proposition 12.6 *For every $n \geq 0$ there is a constant $b(n) > 0$ so that*

$$q_n(f) \leq b(n)p_{n+d}(f) \quad and \quad p_n(f) \leq b(n)q_{n+d}(f), \quad f \in \mathscr{S}(\mathbb{R}^d).$$

Proof. For any $n \geq 0$ and $\alpha \in \mathbb{N}^d$ satisfying $0 \leq \bar{\alpha} \leq n$ we have

$$(1 + |x|^2)^n |\partial^\alpha f(x)|^2 \leq \sup_{y \in \mathbb{R}^d} \left[(1 + |y|^2)^{n+d} |\partial^\alpha f(y)|^2 \right] \frac{1}{(1 + |x|^2)^d}$$

$$\leq p_{n+d}(f)^2 \frac{1}{(1 + |x|^2)^d}.$$

It follows that $q_n(f) \leq b_1(n)p_{n+d}(f)$ for a constant $b_1(n) > 0$. On the other hand, for $d = 1$ and $0 \leq k \leq n$ we have

$$(1 + x^2)^{n/2} |f^{(k)}(x)| = \left| \int_{-\infty}^x \left[(1 + y^2)^{n/2} f^{(k)}(y) \right]' \mathrm{d}y \right|$$

$$\leq \int_{\mathbb{R}} |(1 + y^2)^{n/2} f^{(k+1)}(y)| \mathrm{d}y$$

$$+ \frac{n}{2} \int_{\mathbb{R}} |(1 + y^2)^{n/2} f^{(k)}(y)| \mathrm{d}y$$

$$\leq \left(\int_{\mathbb{R}} \frac{\mathrm{d}y}{1+y^2} \right)^{\frac{1}{2}} \left[\left(\int_{\mathbb{R}} (1+y^2)^{n+1} f^{(k+1)}(y)^2 \mathrm{d}y \right)^{\frac{1}{2}} \right.$$

$$\left. + \frac{n}{2} \left(\int_{\mathbb{R}} (1+y^2)^{n+1} f^{(k)}(y)^2 \mathrm{d}y \right)^{\frac{1}{2}} \right]$$

$$\leq \sqrt{\pi} \left(1 + \frac{n}{2} \right) q_{n+1}(f).$$

Then there is a constant $b_2(n) > 0$ so that $p_n(f) \leq b_2(n) q_{n+1}(f)$. The inequality for higher dimensions follows similarly. $\qquad\square$

By Proposition 12.6 the sequences of norms $\{p_n\}$ and $\{q_n\}$ induce the same topology on $\mathscr{S}(\mathbb{R}^d)$. To show this is a nuclear space let us introduce another sequence of Hilbertian norms. The *Hermite polynomials* on $\mathbb{R}$ are given by

$$g_k(x) = (-1)^k e^{x^2} \frac{\mathrm{d}^k}{\mathrm{d}x^k} e^{-x^2}, \quad k = 0, 1, 2, \dots.$$

Based on those we define the *Hermite functions*

$$h_k(x) = \frac{1}{\sqrt[4]{\pi}\sqrt{2^k k!}} e^{-x^2/2} g_k(x), \quad k = 0, 1, 2, \dots.$$

For $x \in \mathbb{R}^d$ and $\alpha \in \mathbb{N}^d$ let $h_\alpha(x) = h_{\alpha_1}(x_1)\cdots h_{\alpha_d}(x_d)$. Then $h_\alpha \in \mathscr{S}(\mathbb{R}^d)$ and $\{h_\alpha : \alpha \in \mathbb{N}^d\}$ is an orthonormal basis of $L^2(\mathbb{R}^d)$. Let $\langle \cdot, \cdot \rangle$ denote the inner product of $L^2(\mathbb{R}^d)$. For $f \in \mathscr{S}(\mathbb{R}^d)$ we write

$$f(x) = \sum_{\alpha \in \mathbb{N}^d} \langle f, h_\alpha \rangle h_\alpha(x), \quad x \in \mathbb{R}^d \tag{12.15}$$

and define

$$\|f\|_n^2 = \sum_{\alpha \in \mathbb{N}^d} (2\bar{\alpha} + d)^{2n} \langle f, h_\alpha \rangle^2 \tag{12.16}$$

for $n = 0, \pm 1, \pm 2, \dots$. For any $g, f \in \mathscr{S}(\mathbb{R}^d)$ we have

$$\langle g, f \rangle = \sum_{\alpha \in \mathbb{N}^d} \langle g, h_\alpha \rangle \langle f, h_\alpha \rangle,$$

and $\langle g, f \rangle \leq \|g\|_{-n} \|f\|_n$ by Schwarz's inequality. Let $H_n(\mathbb{R}^d)$ be the completion of $\mathscr{S}(\mathbb{R}^d)$ with respect to $\|\cdot\|_n$, which we refer to as a *weighted Sobolev space*. By approximation we can extend $\langle \cdot, \cdot \rangle$ to a bilinear form between $H_{-n}(\mathbb{R}^d)$ and $H_n(\mathbb{R}^d)$. Let $\langle \cdot, \cdot \rangle_n$ denote the inner product of $H_n(\mathbb{R}^d)$. For $g, f \in H_n(\mathbb{R}^d)$ we have

$$\langle g, f \rangle_n = \sum_{\alpha \in \mathbb{N}^d} (2\bar{\alpha} + d)^{2n} \langle g, h_\alpha \rangle \langle f, h_\alpha \rangle = \langle \pi_n g, f \rangle,$$

where

$$\pi_n g = \sum_{\alpha \in \mathbb{N}^d} (2\bar{\alpha} + d)^{2n} \langle g, h_\alpha \rangle h_\alpha \in H_{-n}(\mathbb{R}^d).$$

Then $H_{-n}(\mathbb{R}^d)$ and $H_n(\mathbb{R}^d)$ are dual spaces with the duality $\langle \cdot, \cdot \rangle$.

Proposition 12.7 *For every $n \geq 0$ there is a constant $c(n) > 0$ so that*

$$q_n(f) \leq c(n)\|f\|_n \quad \text{and} \quad \|f\|_n \leq c(n)q_{2n}(f), \qquad f \in \mathscr{S}(\mathbb{R}^d).$$

Proof. We only give the proof for the case $d = 1$. The proof in the general case is based on similar ideas with more complicated calculations. It is easy to show that

$$h'_k(x) = \sqrt{\frac{k}{2}} h_{k-1}(x) - \sqrt{\frac{k+1}{2}} h_{k+1}(x) \tag{12.17}$$

and

$$x h_k(x) = \sqrt{\frac{k}{2}} h_{k-1}(x) + \sqrt{\frac{k+1}{2}} h_{k+1}(x) \tag{12.18}$$

with $h_{-1}(x) = 0$ by convention. For $f \in \mathscr{S}(\mathbb{R})$ we have

$$f(x) = \sum_{k=0}^{\infty} \langle f, h_k \rangle h_k(x), \quad x \in \mathbb{R}. \tag{12.19}$$

For $0 \leq k, l \leq n$ one can use (12.17) and (12.18) to see

$$x^l f^{(k)}(x) = \sum_{i=0}^{\infty} \sum_{j=(i-2n)^+}^{i+2n} a_n(i, j, k, l)(i + 2n)^n \langle f, h_j \rangle h_i(x),$$

where $(i - 2n)^+ = 0 \vee (i - 2n)$ and $\{a_n(i, j, k, l) : i, j, k, l \geq 0\}$ is a countable set bounded by some $b_0(n) > 0$. Then there are constants $b_i(n) > 0$ so that

$$\int_{\mathbb{R}} (1 + x^2)^n f^{(k)}(x)^2 dx \leq b_1(n) \sum_{i=0}^{\infty} \sum_{j=(i-2n)^+}^{i+2n} (i + 2n)^{2n} \langle f, h_j \rangle^2$$

$$= b_1(n) \sum_{j=0}^{\infty} \sum_{i=(j-2n)^+}^{j+2n} (i + 2n)^{2n} \langle f, h_j \rangle^2$$

$$\leq b_2(n) \sum_{j=0}^{\infty} (j + 4n)^{2n} \langle f, h_j \rangle^2$$

$$\leq b_3(n) \sum_{j=0}^{\infty} (2j + 1)^{2n} \langle f, h_j \rangle^2.$$

That gives the first inequality. Using (12.17) and (12.18) one can show

$$x^2 h_k(x) - h_k''(x) = (2k+1)h_k(x). \tag{12.20}$$

For $f \in \mathscr{S}(\mathbb{R})$ given by (12.19) we have

$$x^2 f(x) - f''(x) = \sum_{k=0}^{\infty} (2k+1)\langle f, h_k \rangle h_k(x).$$

Let us denote the above function by $f_1(x)$ and define $f_n(x) = x^2 f_{n-1}(x) - f_{n-1}''(x)$ for $n \geq 2$ inductively. It is simple to see that

$$f_n(x) = \sum_{k=0}^{\infty} (2k+1)^n \langle f, h_k \rangle h_k(x).$$

Then there is a constant $c(n) > 0$ so that

$$\|f\|_n = \left(\int_{\mathbb{R}} f_n(x)^2 \mathrm{d}x \right)^{\frac{1}{2}} \leq c(n) q_{2n}(f).$$

That gives the second inequality. □

Proposition 12.8 *For $n > m + d/2$ the embedding $H_n(\mathbb{R}^d) \subset H_m(\mathbb{R}^d)$ is Hilbert–Schmidt.*

Proof. It is easy to see that $\{(2\bar{\alpha} + d)^{-n} h_\alpha : \alpha \in \mathbb{N}^d\}$ is an orthonormal basis of $\mathscr{S}(\mathbb{R}^d)$ with respect to the norm $\| \cdot \|_n$ and

$$\sum_{\alpha \in \mathbb{N}^d} \|(2\bar{\alpha} + d)^{-n} h_\alpha\|_m^2 = \sum_{\alpha \in \mathbb{N}^d} (2\bar{\alpha} + d)^{2(m-n)} < \infty$$

for any $n > m + d/2$. □

Let $\mathscr{S}(\mathbb{R}^d)$ be endowed with the metric ρ defined by (12.9) and let $\mathscr{S}'(\mathbb{R}^d)$ denote its dual endowed with the strong topology. The elements of $\mathscr{S}'(\mathbb{R}^d)$ are called *Schwartz distributions*.

Theorem 12.9 *Both $\mathscr{S}(\mathbb{R}^d)$ and $\mathscr{S}'(\mathbb{R}^d)$ are nuclear spaces.*

Proof. By Propositions 12.6 and 12.7 the two families of norms $\{p_0, p_1, p_2, \ldots\}$ and $\{\| \cdot \|_0, \| \cdot \|_1, \| \cdot \|_2, \ldots\}$ are equivalent. Then it is easily seen that $\mathscr{S}(\mathbb{R}^d)$ is complete under the metric ρ defined by (12.9). Let $\mathscr{G}$ be the collection of functions $f \in \mathscr{S}(\mathbb{R}^d)$ having the decomposition

$$f(x) = \sum_{\bar{\alpha} \leq n} r_\alpha h_\alpha(x), \quad x \in \mathbb{R}^d$$

for all possible finite sets of rational coefficients $\{r_\alpha : \bar{\alpha} \leq n\}$. Clearly, $\mathscr{G}$ is dense in H_n for every $n \geq 0$. In other words, each H_n is separable. By Proposition 12.8 the embedding $H_n(\mathbb{R}^d) \subset H_m(\mathbb{R}^d)$ is Hilbert–Schmidt for $n > m+d/2$, so $\mathscr{S}(\mathbb{R}^d)$ is a nuclear space. Since $\mathscr{S}(\mathbb{R}^d)$ is clearly a Fréchet space, its strong dual $\mathscr{S}'(\mathbb{R}^d)$ is also a nuclear space; see, e.g., Treves (1967, p.523).   □

12.3 Fluctuation Limits in the Schwartz Space

Let (ξ, ϕ, η) be the parameters given as in the first section and assume $\eta(dx)$ is absolutely continuous with respect to the Lebesgue measure $\lambda(dx)$ with a bounded density. For any integer $k \geq 1$ let $\phi_k(x, z) = \phi(x, z/k)$ and suppose that $\{Y_k(t) : t \geq 0\}$ is a càdlàg immigration superprocess in $M_p(\mathbb{R}^d)$ with parameters (ξ, ϕ_k, η). Then each $\{Y_k(t) : t \geq 0\}$ has equilibrium mean $\zeta := \eta U$. We are interested in the asymptotic fluctuating behavior of the immigration processes around this mean. For simplicity, we assume $Y_k(0) = \zeta$, so (12.4) and (12.8) imply

$$\mathbf{E}\langle Y_k(t), f \rangle = \langle \zeta, f \rangle, \quad t \geq 0, f \in C_p(\mathbb{R}^d).$$

Then we define the centered $\mathscr{S}'(\mathbb{R}^d)$-valued process $\{Z_k(t) : t \geq 0\}$ by

$$Z_k(t) := k[Y_k(t) - \zeta], \qquad t \geq 0. \tag{12.21}$$

Since $t \mapsto \langle Z_k(t), f \rangle$ is càdlàg for every $f \in \mathscr{S}(\mathbb{R}^d)$, the process $t \mapsto Z_k(t)$ is càdlàg in the strong topology of $\mathscr{S}'(\mathbb{R}^d)$; see, e.g., Treves (1967, p.358). Recall that $C^2(\mathbb{R})$ denotes the set of bounded continuous real functions on $\mathbb{R}$ with bounded continuous derivatives up to the second order.

Lemma 12.10 *For any $G \in C^2(\mathbb{R})$ and $f \in \mathscr{S}(\mathbb{R}^d)$ we have*

$$G(\langle Z_k(t), f \rangle) = \int_0^t G'(\langle Z_k(s), f \rangle)\langle Z_k(s), Af \rangle ds$$
$$+ \int_0^t G''(\langle Z_k(s), f \rangle)\langle Y_k(s), cf^2 \rangle ds$$
$$+ \int_0^t ds \int_{\mathbb{R}^d} l(x, Z_k(s))Y_k(s, dx) + mart.,$$

where

$$l(x, \mu) = \int_0^\infty \Big[G(\langle \mu, f \rangle + uf(x)) - G(\langle \mu, f \rangle)$$
$$- G'(\langle \mu, f \rangle)uf(x) \Big] m(x, du).$$

Proof. Let $F_k(\nu) = G(\langle \nu, kf \rangle - \langle \zeta, kf \rangle)$ for $\nu \in M_p(\mathbb{R}^d)$. Then $G(\langle Z_k(t), f \rangle) = F_k(Y_k(t))$. By Theorem 9.24,

$$G(\langle Z_k(t), f\rangle) = \int_0^t G'(\langle Y_k(s), kf\rangle - \langle \zeta, kf\rangle)\langle Y_k(s), kAf\rangle \mathrm{d}s$$
$$+ \int_0^t G'(\langle Y_k(s), kf\rangle - \langle \zeta, kf\rangle)\langle \eta, kf\rangle \mathrm{d}s$$
$$+ \int_0^t G''(\langle Y_k(s), kf\rangle - \langle \zeta, kf\rangle)\langle Y_k(s), cf^2\rangle \mathrm{d}s$$
$$+ \int_0^t \mathrm{d}s \int_{\mathbb{R}^d} l(x, Z_k(s))Y_k(s, \mathrm{d}x) + \text{local mart.}$$

Using (12.5) one can see the local martingale above is actually a square-integrable martingale. Observe that

$$\langle \eta, f\rangle = -\langle \eta, UAf\rangle = -\langle \zeta, Af\rangle. \tag{12.22}$$

Then we obtain the desired equality. $\qquad\square$

Lemma 12.11 *Let $\phi''(x, 0)$ be given by (2.64). Then for any $t \geq 0$ and $f \in C_p(\mathbb{R}^d)$ we have*

$$\mathbf{E}[\langle Z_k(t), f\rangle^2] = \int_0^t \langle \zeta, \phi''(\cdot, 0)(P_s^b f)^2\rangle \mathrm{d}s, \tag{12.23}$$

Proof. In view of (12.4) and (12.5) we have

$$\mathbf{E}\big[\langle Z_k(t), f\rangle^2\big] = \mathbf{E}\bigg[\bigg(\langle Y_k(t), kf\rangle - \langle \zeta, kP_t^b f\rangle - \int_0^t \langle \eta, kP_s^b f\rangle \mathrm{d}s\bigg)^2\bigg]$$
$$= \int_0^t \langle \zeta, P_{t-s}^b[\phi''(\cdot, 0)(P_s^b f)^2]\rangle \mathrm{d}s$$
$$+ \int_0^t \mathrm{d}s \int_0^{t-s} \langle \eta, P_u^b[\phi''(\cdot, 0)(P_s^b f)^2]\rangle \mathrm{d}u$$
$$= \int_0^t \langle \zeta, \phi''(\cdot, 0)(P_s^b f)^2\rangle \mathrm{d}s,$$

where for the last equality we also used (12.8) to the function $\phi''(\cdot, 0)(P_s^b f)^2$. $\qquad\square$

Lemma 12.12 *Let ζ' be a bounded density of $\zeta(\mathrm{d}x)$ with respect to the Lebesgue measure. Then for any $t \geq 0$ and $f \in \mathscr{S}(\mathbb{R}^d)$ we have*

$$\sup_{k\geq 1} \mathbf{E}\Big[\sup_{0\leq s\leq t} \langle Z_k(s), f\rangle^2\Big] \leq t\|\zeta'\|\|\phi''(\cdot, 0)\|\big[8\langle \lambda, f^2\rangle + t^2\langle \lambda, (Af)^2\rangle\big].$$

Proof. By Theorem 9.24 we have

$$\langle Y_k(t), f\rangle = \langle \zeta, f\rangle + M_k(t, f) + \int_0^t \big[\langle Y_k(s), Af\rangle + \langle \eta, f\rangle\big]\mathrm{d}s, \tag{12.24}$$

where $\{M_k(t, f) : t \geq 0\}$ is a càdlàg martingale with increasing process

$$\langle M_k(f)\rangle_t = \frac{1}{k^2}\int_0^t \langle Y_k(s), \phi''(\cdot,0)f^2\rangle \mathrm{d}s.$$

From (12.22) and (12.24) we get

$$\langle Z_k(t), f\rangle = k M_k(t, f) + \int_0^t \langle Z_k(s), Af\rangle \mathrm{d}s.$$

Then by (12.5) and (12.23),

$$
\begin{aligned}
\mathbf{E}&\left[\sup_{0\le s\le t} \langle Z_k(s), f\rangle^2\right]\\
&\le 2k^2\mathbf{E}\left[\sup_{0\le s\le t}|M_k(s, f)|^2\right] + 2\mathbf{E}\left[\left(\int_0^t |\langle Z_k(s), Af\rangle|\mathrm{d}s\right)^2\right]\\
&\le 8\int_0^t \mathbf{E}[\langle Y_k(s), \phi''(\cdot,0)f^2\rangle]\mathrm{d}s + 2t\int_0^t \mathbf{E}[\langle Z_k(s), Af\rangle^2]\mathrm{d}s\\
&\le 8t\|\phi''(\cdot,0)\|\langle\zeta, f^2\rangle + 2t\|\phi''(\cdot,0)\|\int_0^t \mathrm{d}s\int_0^s \langle\zeta, (P_u^b Af)^2\rangle \mathrm{d}u\\
&\le 8t\|\zeta'\|\|\phi''(\cdot,0)\|\langle\lambda, f^2\rangle + 2t\|\zeta'\|\|\phi''(\cdot,0)\|\int_0^t \mathrm{d}s\int_0^s \langle\lambda, (P_u^b Af)^2\rangle \mathrm{d}u\\
&\le t\|\zeta'\|\|\phi''(\cdot,0)\|\left[8\langle\lambda, f^2\rangle + t^2\langle\lambda, (Af)^2\rangle\right].
\end{aligned}
$$

That gives the desired estimate.   □

Lemma 12.13 *The sequence $\{(Z_k(t))_{t\ge 0} : k \ge 1\}$ is tight in $D([0,\infty), \mathscr{S}'(\mathbb{R}^d))$.*

Proof. By Theorem 12.4 we only need to prove the sequence $\{\langle Z_k(t), f\rangle : t \ge 0; k \ge 1\}$ is tight in $D([0,\infty), \mathbb{R})$ for every $f \in \mathscr{S}(\mathbb{R}^d)$. By Lemma 12.12 and Chebyshev's inequality we have

$$\sup_{k\ge 1} \mathbf{P}\left[\sup_{0\le s\le t}|\langle Z_k(s), f\rangle| \ge \alpha\right] \to 0$$

as $\alpha \to \infty$. Then $\{\langle Z_k(t), f\rangle : t \ge 0\}$ satisfies the compact containment condition of Ethier and Kurtz (1986, p.142). Let $G \in C^\infty(\mathbb{R})$ and let $l(x, \mu)$ be defined as in Lemma 12.10. By Taylor's expansion we have

$$|l(x, \mu)| \le \|G''\| \int_0^\infty u^2 m(x, \mathrm{d}u)|f(x)|^2.$$

Then it is easy to show

$$
\begin{aligned}
\sup_{k\ge 1} \mathbf{E}\bigg[\int_0^t \bigg| G'(\langle Z_k(s), f\rangle)\langle Z_k(s), Af\rangle &+ G''(\langle Z_k(s), f\rangle)\langle Y_k(s), cf^2\rangle\\
&+ \int_{\mathbb{R}^d} l(x, Z_k(s))Y_k(s, \mathrm{d}x)\bigg|^2 \mathrm{d}s\bigg] < \infty.
\end{aligned}
$$

By Lemma 12.10 and Ethier and Kurtz (1986, p.145) we infer that $\{G(\langle Z_k(t), f\rangle) : t \geq 0; k \geq 1\}$ is tight. The tightness of $\{\langle Z_k(t), f\rangle : t \geq 0; k \geq 1\}$ then follows by Ethier and Kurtz (1986, p.142). $\qquad\square$

Lemma 12.14 *Let* $\{Z_0(t) : t \geq 0\}$ *be any limit point of* $\{Z_k(t) : t \geq 0; k \geq 1\}$ *in the sense of distribution on* $D([0, \infty), \mathscr{S}'(\mathbb{R}^d))$. *Then for* $G \in C^\infty(\mathbb{R})$ *and* $f \in \mathscr{S}(\mathbb{R}^d)$ *we have*

$$G(\langle Z_0(t), f\rangle) = \int_0^t \Big[G'(\langle Z_0(s), f\rangle)\langle Z_0(s), Af\rangle + G''(\langle Z_0(s), f\rangle)\langle \zeta, cf^2\rangle$$
$$+ \int_{\mathbb{R}^d} l(x, Z_0(s))\zeta(\mathrm{d}x)\Big] \mathrm{d}s + mart.$$

Proof. By passing to a subsequence and using the Skorokhod representation, we may assume $\{Z_k(t) : t \geq 0\}$ and $\{Z_0(t) : t \geq 0\}$ are defined on the same probability space and $\{Z_k(t) : t \geq 0\}$ converges a.s. to $\{Z_0(t) : t \geq 0\}$ in the topology of $D([0, \infty), \mathscr{S}'(\mathbb{R}^d))$. Then $\{\langle Z_k(t), f\rangle : t \geq 0\}$ converges a.s. to $\{\langle Z_0(t), f\rangle : t \geq 0\}$ in the topology of $D([0, \infty), \mathbb{R})$. Consequently, we have a.s. $\langle Z_k(t), f\rangle \to \langle Z_0(t), f\rangle$ for a.e. $t \geq 0$; see, e.g., Ethier and Kurtz (1986, p.118). Note also that $\{Y_k(t) : t \geq 0\}$ converges a.s. to the deterministic constant process $\{Y_0(t) = \zeta : t \geq 0\}$ in the topology of $D([0, \infty), \mathscr{S}'(\mathbb{R}^d))$. From Lemma 12.10 we have

$$G(\langle Z_k(t), f\rangle) = \int_0^t G'(\langle Z_k(s), f\rangle)\langle Z_k(s), Af\rangle \mathrm{d}s$$
$$+ \int_0^t G''(\langle Z_k(s), f\rangle)\langle Y_k(s), cf^2\rangle \mathrm{d}s$$
$$+ \int_0^t \langle Y_k(s), l(\cdot, Z_0(s)) + l_k(s, \cdot)\rangle \mathrm{d}s + mart., \quad (12.25)$$

where $l_k(s, x) = l(x, Z_k(s)) - l(x, Z_0(s))$. By applying the mean-value theorem to the function

$$z \mapsto H(x, u, z) := G(z + uf(x)) - G(z) - G'(z)uf(x)$$

we get

$$l_k(s, x) = \langle Z_k(s) - Z_0(s), f\rangle \int_0^\infty H_z'(x, u, \theta_s)m(x, \mathrm{d}u),$$

where

$$\langle Z_k(s), f\rangle \wedge \langle Z_0(s), f\rangle \leq \theta_s \leq \langle Z_k(s), f\rangle \vee \langle Z_0(s), f\rangle.$$

By Taylor's expansion,

$$|H_z'(x, u, \theta_s)| = |G'(\theta_s + uf(x)) - G'(\theta_s) - G''(\theta_s)uf(x)|$$

$$\leq \frac{1}{2}\|G^{(3)}\|u^2 f(x)^2.$$

It follows that

$$|l_k(s,x)| \leq \frac{1}{2}\|G^{(3)}\|\|f(x)^2|\langle Z_k(s) - Z_0(s), f\rangle| \int_0^\infty u^2 m(x,\mathrm{d}u).$$

Then we have

$$\langle |l_k(s,\cdot)|, Y_k(s)\rangle \leq C|\langle Z_k(s) - Z_0(s), f\rangle|\langle Y_k(s), f^2\rangle, \qquad (12.26)$$

where

$$C = \frac{1}{2}\|G^{(3)}\| \sup_{x\in\mathbb{R}^d} \int_0^\infty u^2 m(x,\mathrm{d}u).$$

For $n \geq 1$ let

$$\tau_n = \inf\left\{t \geq 0 : \sup_{k\geq 1} \int_0^t \langle Z_k(s) - Z_0(s), f\rangle^2 \mathrm{d}s \geq n\right\}.$$

Then $\tau_n \to \infty$ as $n \to \infty$. By (12.26) and Schwarz's inequality,

$$\left\{\mathbf{E}\left[\int_0^{t\wedge\tau_n} \langle |l_k(s,\cdot)|, Y_k(s)\rangle \mathrm{d}s\right]\right\}^2$$
$$\leq C_k(t)\mathbf{E}\left[\int_0^{t\wedge\tau_n} \langle Z_k(s) - Z_0(s), f\rangle^2 \mathrm{d}s\right], \qquad (12.27)$$

where

$$C_k(t) = C^2 \int_0^t \mathbf{E}[\langle Y_k(s), f^2\rangle^2]\mathrm{d}s.$$

By (12.5) it is easy to show $\sup_{k\geq 1} C_k(t) < \infty$. Now (12.27) implies

$$\lim_{k\to\infty} \mathbf{E}\left[\int_0^{t\wedge\tau_n} \langle |l_k(s,\cdot)|, Y_k(s)\rangle \mathrm{d}s\right] = 0.$$

From (12.5) and (12.23) it is easy to show that the sequences

$$\{\langle Z_k(s), Af\rangle\}, \ \{\langle Y_k(s), cf^2\rangle\}, \ \{\langle Y_k(s), l(\cdot, Z_0(s))\rangle\}$$

are all uniformly integrable on $\Omega \times [0,t]$ relative to the product measure $\mathbf{P}(\mathrm{d}\omega)\mathrm{d}s$. Then letting $k \to \infty$ in (12.25) we obtain

$$G(\langle Z_0(t), f\rangle) = \int_0^t \left[G'(\langle Z_0(s), f\rangle)\langle Z_0(s), Af\rangle + G''(\langle Z_0(s), f\rangle)\langle \zeta, cf^2\rangle\right.$$

$$+ \int_{\mathbb{R}^d} l(x, Z_0(s))\zeta(\mathrm{d}x)\Big] \mathrm{d}s + \text{local mart.}$$

Here the local martingale is clearly a square-integrable martingale. □

Proposition 12.15 *For every* $\mu \in \mathscr{S}'(\mathbb{R}^d)$ *there is a process* $\{Z(t) : t \geq 0\}$ *with sample paths in* $D([0, \infty), \mathscr{S}'(\mathbb{R}^d))$ *so that for* $G \in C^\infty(\mathbb{R})$ *and* $f \in \mathscr{S}(\mathbb{R}^d)$ *we have*

$$G(\langle Z(t), f\rangle) = G(\langle \mu, f\rangle) + \int_0^t G'(\langle Z(s), f\rangle)\langle Z(s), Af\rangle \mathrm{d}s$$
$$+ \int_0^t G''(\langle Z(s), f\rangle)\langle \zeta, cf^2\rangle \mathrm{d}s$$
$$+ \int_0^t \mathrm{d}s \int_{\mathbb{R}^d} l(x, Z(s))\zeta(\mathrm{d}x) + \textit{mart.} \qquad (12.28)$$

Proof. Let $\{Z_0(t) : t \geq 0\}$ be as in Lemma 12.14 and let $Z(t) = P_t^b\mu + Z_0(t)$. Then (12.28) clearly holds. □

Proposition 12.16 *Suppose that* $\{Z(t) : t \geq 0\}$ *is a process that has sample paths in* $D([0, \infty), \mathscr{S}'(\mathbb{R}^d))$ *and solves the martingale problem given by* (12.28). *Then* $\{Z(t) : t \geq 0\}$ *is a Markov process with transition semigroup* $(Q_t^\zeta)_{t\geq 0}$ *defined by*

$$\int_{\mathscr{S}'(\mathbb{R}^d)} e^{i\langle \nu, f\rangle} Q_t^\zeta(\mu, \mathrm{d}\nu)$$
$$= \exp\left\{ i\langle \mu, P_t^b f\rangle + \int_0^t \langle \zeta, \phi(-iP_s^b f)\rangle \mathrm{d}s \right\}, \qquad (12.29)$$

where $f \in \mathscr{S}(\mathbb{R}^d)$.

Proof. Let $(t, z) \mapsto G(t, z)$ be a function on $[0, \infty) \times \mathbb{R}$ such that $z \mapsto G(t, z)$ belongs to $C^\infty(\mathbb{R})$ for every $t \geq 0$ and $t \mapsto G(t, z)$ is continuously differentiable for every $z \in \mathbb{R}$. Let $(t, x) \mapsto f_t(x)$ be a function on $[0, \infty) \times \mathbb{R}^d$ such that $x \mapsto f_t(x)$ belongs to $\mathscr{S}(\mathbb{R}^d)$ for every $t \geq 0$ and $t \mapsto f_t(x)$ is continuously differentiable for every $x \in \mathbb{R}^d$. Using Proposition 12.15 one can show as in the proof of Theorem 7.13 that

$$G(t, \langle Z(t), f_t\rangle) = G(0, \langle \mu, f_0\rangle) + \int_0^t G'_z(s, \langle Z(s), f_s\rangle)\langle Z(s), Af_s\rangle \mathrm{d}s$$
$$+ \int_0^t G''_{zz}(s, \langle Z(s), f_s\rangle)\langle \zeta, cf_s^2\rangle \mathrm{d}s$$
$$+ \int_0^t \Big[G'_z(s, \langle Z(s), f_s\rangle)\langle Z(s), f'_s\rangle + G'_s(s, \langle Z(s), f_s\rangle)\Big] \mathrm{d}s$$
$$+ \int_0^t \mathrm{d}s \int_E l_s(x, Z(s))\zeta(\mathrm{d}x) + \text{mart.},$$

where $f'_s(x) = (\mathrm{d}/\mathrm{d}s)f_s(x)$ and $l_s(x, \mu)$ is defined as in Lemma 12.10 with f and G replaced by f_s and $G(s, \cdot)$, respectively. Clearly, the equality above remains valid when $(t, z) \mapsto G(t, z)$ is a complex function. By applying this to $f_t = P^b_{T-t}f$ and

$$G(t, z) = \exp\left\{ iz + \int_0^{T-t} \langle \zeta, \phi(-iP^b_s f)\rangle ds \right\}$$

one sees that

$$t \mapsto \exp\left\{ i\langle Z(t), P^b_{T-t}f\rangle + \int_0^{T-t} \langle \zeta, \phi(-iP^b_s f)\rangle ds \right\}$$

is a complex martingale on $[0, T]$. Then $\{Z(t) : t \geq 0\}$ is a Markov process with transition semigroup $(Q^\zeta_t)_{t\geq 0}$ defined by (12.29). $\qquad\square$

By Propositions 12.15 and 12.16 there is a unique solution to the martingale problem given by (12.28) and the solution is a Markov process with transition semigroup $(Q^\zeta_t)_{t\geq 0}$. This process gives a description of the asymptotic fluctuations of the immigration superprocesses as the branching mechanisms are small. More precisely, we have the following:

Theorem 12.17 As $k \to \infty$, the process $\{Z_k(t) : t \geq 0\}$ converges weakly in $D([0, \infty), \mathscr{S}'(\mathbb{R}^d))$ to the unique solution $\{Z(t) : t \geq 0\}$ of the martingale problem given by (12.28) with $Z(0) = 0$.

Proof. By Lemma 12.13 the sequence $\{Z_k(t) : t \geq 0; k \geq 1\}$ is tight in the space $D([0, \infty), \mathscr{S}'(\mathbb{R}^d))$. Then we get the result by Lemma 12.14 and Proposition 12.16. $\qquad\square$

Example 12.1 Suppose that $\eta \in M(\mathbb{R}^d)$ is a finite measure with $\eta(\mathbb{R}^d) = b$ and $\phi(z)$ is a local branching mechanism given by (12.1) with (c, m) independent of $x \in \mathbb{R}^d$. Let $z_k(t) = k[\langle Y_k(t), 1\rangle - 1]$ for $t \geq 0$. A modification of the arguments in this section shows $\{z_k(t) : t \geq 0\}$ converges weakly in $D([0, \infty), \mathbb{R})$ to a Markov process $\{z(t) : t \geq 0\}$ with transition semigroup $(Q^b_t)_{t\geq 0}$ defined by

$$\int_{\mathbb{R}} e^{iuy} Q^b_t(x, dy) = \exp\left\{ e^{-bt}iux + \int_0^t \phi(-e^{-bt}iu)ds \right\}, \qquad u \in \mathbb{R}.$$

This is a one-dimensional *OU-type process*; see, e.g., Sato (1999, pp.106–108).

12.4 Fluctuation Limits in Sobolev Spaces

In this section, we show the fluctuation limit theorem actually holds in a suitable weighted Sobolev space. Recall that the weighted Sobolev space $H_n(\mathbb{R}^d)$ with index $n \geq 0$ is the completion of $\mathscr{S}(\mathbb{R}^d)$ with respect to the norm $\| \cdot \|_n$ defined in (12.16) and $H_{-n}(\mathbb{R}^d)$ denotes the dual space of $H_n(\mathbb{R}^d)$ with duality $\langle \cdot, \cdot \rangle$. Let $\{Z_k(t) : t \geq 0\}$ and $\{Z(t) : t \geq 0\}$ be defined as in Section 12.3.

Theorem 12.18 *For any integer $n > 2 + d/2$ the processes $\{Z_k(t) : t \geq 0\}$ and $\{Z(t) : t \geq 0\}$ live in the weighted Sobolev space $H_{-n}(\mathbb{R}^d)$ and $\{Z_k(t) : t \geq 0\}$ converges as $k \to \infty$ to $\{Z(t) : t \geq 0\}$ weakly in $D([0, \infty), H_{-n}(\mathbb{R}^d))$.*

Proof. For any $f \in \mathscr{S}(\mathbb{R}^d)$ we have $\langle \lambda, f^2 \rangle = \|f\|_0^2 \leq \|f\|_2^2$. By Proposition 12.7 there is a constant $C > 0$ such that

$$\langle \lambda, (Af)^2 \rangle \leq \langle \lambda, (\Delta f)^2 \rangle + 2b^2 \langle \lambda, f^2 \rangle \leq (1 + 2b^2) q_2(f)^2 \leq C\|f\|_2^2.$$

Then Lemma 12.12 implies

$$\sup_{k \geq 1} \mathbf{E}\left[\sup_{0 \leq s \leq t} \langle Z_k(s), f \rangle^2 \right] \leq C(t)\|f\|_2^2$$

for a locally bounded function $t \mapsto C(t)$. Thus for $t \geq 0$ and $\rho > 0$ we have

$$\sup_{k \geq 1} \mathbf{P}\left[\sup_{0 \leq s \leq t} |\langle Z_k(s), f \rangle| \geq \rho \right] \leq C(t)\|f\|_2^2/\rho^2.$$

By Theorem 12.5 and Proposition 12.8 the sequence $\{Z_k(t) : t \geq 0; k \geq 1\}$ is tight in $D([0, \infty), H_{-n}(\mathbb{R}^d))$. Then the result follows by Theorem 12.17. $\qquad\square$

Example 12.2 Let us consider the case $d = 1$. Suppose that $\{W(\mathrm{d}s, \mathrm{d}x)\}$ is a σ-finite orthogonal martingale measure on $[0, \infty) \times \mathbb{R}$ with covariance measure $2c(x)\mathrm{d}s\zeta(\mathrm{d}x)$ and $\{N(\mathrm{d}s, \mathrm{d}x, \mathrm{d}u)\}$ is a Poisson random measure on $(0, \infty) \times \mathbb{R} \times (0, \infty)$ with intensity $\mathrm{d}s\zeta(\mathrm{d}x)m(x, \mathrm{d}u)$. We assume $\{N(\mathrm{d}s, \mathrm{d}x, \mathrm{d}u)\}$ and $\{W(\mathrm{d}s, \mathrm{d}x)\}$ are defined on some filtered probability space $(\Omega, \mathscr{F}, \mathscr{F}_t, \mathbf{P})$ and are independent of each other. Let $p_t(x, y)$ denote the transition density of the killed Brownian motion generated by $A = \Delta/2 - b$. Given $X_0 \in H_0(\mathbb{R})$ we can define an $H_0(\mathbb{R})$-valued process $\{X_t : t \geq 0\}$ by

$$X_t(y) := P_t^b X_0(y) + \int_0^t \int_{\mathbb{R}} p_{t-s}(x, y)W(\mathrm{d}s, \mathrm{d}x)$$
$$+ \int_0^t \int_{\mathbb{R}} \int_0^\infty u p_{t-s}(x, y)\tilde{N}(\mathrm{d}s, \mathrm{d}x, \mathrm{d}u), \qquad (12.30)$$

where $\tilde{N}(\mathrm{d}s, \mathrm{d}x, \mathrm{d}u) = N(\mathrm{d}s, \mathrm{d}x, \mathrm{d}u) - \mathrm{d}s\zeta(\mathrm{d}x)m(x, \mathrm{d}u)$. In fact, it is easily seen that

$$\mathbf{E}[\|X_t\|_0^2] \leq 3\|X_0\|_0^2 + 3\int_{\mathbb{R}} \mathrm{d}y \int_0^t \mathrm{d}s \int_{\mathbb{R}} p_{t-s}(x, y)^2 \phi''(x, 0)\zeta(\mathrm{d}x) < \infty.$$

For $f \in \mathscr{S}(\mathbb{R})$ we have

$$\mathbf{E}\left[\exp\{i\langle X_t, f \rangle\} \right]$$
$$= \mathbf{E}\left[\exp\left\{ i\langle X_0, P_t^b f \rangle + \int_0^t \int_{\mathbb{R}} i P_{t-s}^b f(x)W(\mathrm{d}s, \mathrm{d}x) \right. \right.$$

$$+ \int_0^t \int_{\mathbb{R}} \int_0^\infty iuP_{t-s}^b f(x) \tilde{N}(\mathrm{d}s, \mathrm{d}x, \mathrm{d}u) \bigg\} \bigg]$$

$$= \exp\bigg\{ i\langle X_0, P_t^b f\rangle - \int_0^t \mathrm{d}s \int_{\mathbb{R}} [P_{t-s}^b f(x)]^2 c(x)\zeta(\mathrm{d}x)$$

$$+ \int_0^t \mathrm{d}s \int_{\mathbb{R}} \zeta(\mathrm{d}x) \int_0^\infty \Big(e^{iuP_{t-s}^b f(x)} - 1 - iuP_{t-s}^b f(x) \Big) m(x, \mathrm{d}u) \bigg\}$$

$$= \exp\bigg\{ i\langle X_0, P_t^b f\rangle + \int_0^t \langle \zeta, \phi(-iP_{t-s}^b f)\rangle \mathrm{d}s \bigg\}.$$

A comparison of this equality with (12.29) shows that for any $\mu \in H_0(\mathbb{R})$ the probability measure $Q_t^\zeta(\mu, \cdot)$ is actually supported by $H_0(\mathbb{R})$ and

$$\int_{H_0(\mathbb{R})} e^{i\langle \nu, f\rangle} Q_t^\zeta(\mu, \mathrm{d}\nu) = \exp\bigg\{ i\langle \mu, P_t^b f\rangle + \int_0^t \langle \zeta, \phi(-iP_s^b f)\rangle \mathrm{d}s \bigg\}. \quad (12.31)$$

By considering an approximating sequence from $\mathscr{S}(\mathbb{R})$ we see that the above formula holds for every $f \in H_0(\mathbb{R})$. That gives a special case of the transition semigroup defined by (11.7) and (11.31). A similar calculation based on the property of independent increments of $\{W(\mathrm{d}s, \mathrm{d}x)\}$ and $\{N(\mathrm{d}s, \mathrm{d}x, \mathrm{d}u)\}$ shows that $\{X_t : t \geq 0\}$ is a Markov process with transition semigroup $(Q_t^\zeta)_{t\geq 0}$ given by (12.31). Therefore $\{X_t : t \geq 0\}$ have identical finite-dimensional distributions with the limit process $\{Z(t) : t \geq 0\}$ in Theorems 12.17 and 12.18. In other words, the limiting fluctuation process is a generalized OU-process with state space $H_0(\mathbb{R})$ in the terminology of the last chapter.

12.5 Notes and Comments

A general reference for nuclear spaces is Treves (1967). For the theory of classical Sobolev spaces see Adams and Fournier (2003). There are several references for stochastic processes in nuclear spaces; see, e.g., Kallianpur and Xiong (1995) and Walsh (1986).

The equilibrium distributions of super-stable processes without immigration were characterized in Dawson (1977). A simplified approach to the asymptotic behavior of superprocesses was given in Wang (1997b, 1998b). Fluctuation limits of branching particle systems and superprocesses, which usually give rise to time-inhomogeneous OU-processes, have been studied extensively; see, e.g., Bojdecki and Gorostiza (1986, 1991, 2002), Dawson et al. (1989a) and the references therein. Engländer and Winter (2006) proved a law of large numbers for super-diffusions which improves an earlier result of Engländer and Turaev (2002). Chen et al. (2008) proved an almost sure scaling limit theorem for Dawson–Watanabe superprocesses. In Méléard (1996) fluctuation limits of McKean–Vlasov interacting particle systems were studied, where the limiting OU-process was characterized as the unique solution of a Langevin type equation in a weighted Sobolev space.

The fluctuation limit theorems given in this chapter are modifications of those in Gorostiza and Li (1998) and Li and Zhang (2006); see also Dawson et al. (2004b). Three different kinds of fluctuation limits (high-density fluctuation, small-branching fluctuation and large-scale fluctuation) of immigration superprocess with binary branching were studied in Li (1999), which led to generalized OU-diffusions. Some Gaussian processes with long-range dependence arising from occupation time fluctuations of immigration particle systems with or without branching were studied in Gorostiza et al. (2005).

A construction of the two-dimensional regular affine process in $D = \mathbb{R}_+ \times \mathbb{R}$ was given in Dawson and Li (2003) as the strong solution of a system of stochastic equations. Let $\{(x(t), z(t)) : t \geq 0\}$ be a realization of the affine process. Then the first coordinator $\{x(t) : t \geq 0\}$ is a one-dimensional CBI-process. In fact, Dawson and Li (2003) showed that the second coordinator $\{z(t) : t \geq 0\}$ may arise as the fluctuation limit of a generalized CBI-process with branching rate depending on the first one. A similar limit theorem for discrete-state branching processes with immigration was proved in Li and Ma (2008).

Appendix A
Markov Processes

For the convenience of the reader, in this appendix we give a summary of some of the concepts and results for general stochastic processes and Markov processes that are used in the main text. Many of them can be found in Sharpe (1988); see also Ethier and Kurtz (1986) and Getoor (1975).

A.1 Measurable Spaces

Given a class $\mathscr{F}$ of functions on a non-empty set E, we define $\mathrm{b}\mathscr{F} = \{f \in \mathscr{F} : f$ is bounded$\}$ and $\mathrm{p}\mathscr{F} = \{f \in \mathscr{F} : f$ is positive$\}$. We say $\mathscr{F}$ *separates points* if for every $x \neq y \in E$ there exists $f \in \mathscr{F}$ so that $f(x) \neq f(y)$. For a class $\mathscr{G}$ of functions on or subsets of E, we use $\sigma(\mathscr{G})$ to denote the σ-algebra on E generated by $\mathscr{G}$, that is, $\sigma(\mathscr{G}) = \cap\{\mathscr{F} : \mathscr{F}$ is a σ-algebra on E and all elements of $\mathscr{G}$ are $\mathscr{F}$-measurable$\}$. If $(E, \mathscr{E})$ is a measurable space, we also use $\mathscr{E}$ to denote the class of real $\mathscr{E}$-measurable functions on E. We write $\mu(f)$ for the integral of a function $f \in \mathscr{E}$ with respect to a measure μ on $(E, \mathscr{E})$ if the integral exists. Let $\mathbb{R}$ denote the one-dimensional Euclidean space.

Let $\|\cdot\|$ denote the supremum/uniform norm of functions. We say a sequence $\{f_n\}$ of functions on E *converges uniformly* to a function f on E if $\|f_n - f\| \to 0$ as $n \to \infty$. We say $\{f_n\}$ *converges boundedly and pointwise* to f if there is a constant $C \geq 0$ such that $\|f_n\| \leq C$ for all $n \geq 1$ and $f_n(x) \to f(x)$ as $n \to \infty$ for all $x \in E$.

A *monotone vector space* $\mathscr{L}$ on the set E is defined to be a collection of bounded real functions on E satisfying the conditions: (i) $\mathscr{L}$ is a vector space over $\mathbb{R}$; (ii) $\mathscr{L}$ contains the constant function 1_E; (iii) if $\{f_n\} \subset \mathrm{p}\mathscr{L}$ and $f_n \to f$ increasingly for a bounded function f, then $f \in \mathscr{L}$.

Proposition A.1 (Monotone Class Theorem; Sharpe, 1988, p.364) *Let $\mathscr{K}$ be a collection of bounded real functions on the set E which is closed under multiplication. If $\mathscr{L}$ is a monotone vector space containing $\mathscr{K}$, then $\mathscr{L} \supset \mathrm{b}\sigma(\mathscr{K})$.*

Z. Li, *Measure-Valued Branching Markov Processes*,
Probability and Its Applications, DOI 10.1007/978-3-642-15004-3,
© Springer-Verlag Berlin Heidelberg 2011

Proposition A.2 (Modified Monotone Class Theorem) *Let $\mathcal{K}$ be a vector space of bounded real functions on the set E which contains 1_E and is closed under multiplication. If another collection of bounded real functions $\mathcal{G}$ contains $\mathcal{K}$ and is closed under bounded pointwise convergence, then $\mathcal{G} \supset b\sigma(\mathcal{K})$.*

Proof. Let $\mathcal{L}$ be the intersection of all classes of bounded real functions that contain $\mathcal{K}$ and are closed under bounded pointwise convergence. Then $\mathcal{L}$ is closed under bounded pointwise convergence and $\mathcal{K} \subset \mathcal{L} \subset \mathcal{G}$. For $f \in \mathcal{L}$ let

$$\mathcal{L}_f = \{g \in \mathcal{L} : af + bg \in \mathcal{L} \text{ for all } a, b \in \mathbb{R}\}.$$

It is easy to see that $\mathcal{L}_f$ is closed under bounded pointwise convergence. For $f \in \mathcal{K}$ we have $\mathcal{K} \subset \mathcal{L}_f$ and so $\mathcal{L}_f = \mathcal{L}$. If $f \in \mathcal{L}$, for every $g \in \mathcal{K}$ we have $f \in \mathcal{L}_g$ and so $g \in \mathcal{L}_f$. It follows that $\mathcal{K} \subset \mathcal{L}_f$, yielding $\mathcal{L}_f = \mathcal{L}$. Therefore $\mathcal{L}$ is a vector space. By the monotone class theorem we have $\mathcal{L} \supset b\sigma(\mathcal{K})$, which implies the desired result. □

Let us consider a measurable space $(E, \mathcal{E})$. Suppose that μ is a σ-finite measure on $(E, \mathcal{E})$. A set $N \subset E$ is called a μ-*null set* if there is $N_0 \in \mathcal{E}$ so that $N \subset N_0$ and $\mu(N_0) = 0$. For $A, B \subset E$ we define the *symmetric difference*

$$A \triangle B := (A \setminus B) \cup (B \setminus A). \tag{A.1}$$

It is easy to show that

$$\mathcal{E}^\mu := \{A \subset E : A \triangle B \text{ is a } \mu\text{-null set for some } B \in \mathcal{E}\} \tag{A.2}$$

is a σ-algebra, which is called the μ-*completion* of $\mathcal{E}$. We can let $\mu(A) = \mu(B)$ for $B \in \mathcal{E}$ such that $A \triangle B$ is a μ-null set to extend μ uniquely to a σ-finite measure on $(E, \mathcal{E}^\mu)$. The measure space $(E, \mathcal{E}, \mu)$ is said to be *complete* if $\mathcal{E} = \mathcal{E}^\mu$. The *universal completion* of $\mathcal{E}$ is the σ-algebra $\mathcal{E}^u$ defined to be the intersection of the μ-completions of $\mathcal{E}$ as μ runs over all finite measures on $(E, \mathcal{E})$.

Proposition A.3 *If $\mathcal{E}_1$ and $\mathcal{E}_2$ are σ-algebras on the set E such that $\mathcal{E}_1 \subset \mathcal{E}_2 \subset \mathcal{E}_1^u$, then $\mathcal{E}_2^u = \mathcal{E}_1^u$.*

Proof. Let $A \in \mathcal{E}_1^u$ and let μ be a finite measure on $\mathcal{E}_2$. Since $A \in \mathcal{E}_1^\mu$, it is easy to find $A_1, A_2 \in \mathcal{E}_1 \subset \mathcal{E}_2$ so that $A_1 \subset A \subset A_2$ and $\mu(A_1) = \mu(A_2)$. Then $A \in \mathcal{E}_2^\mu$, implying $\mathcal{E}_1^u \subset \mathcal{E}_2^u$. To show the reverse inclusion, let $A \in \mathcal{E}_2^u$ and let μ be a finite measure on $\mathcal{E}_1$. Then μ extends uniquely to $\mathcal{E}_2 \subset \mathcal{E}_1^u$ and $A \in \mathcal{E}_2^\mu$. Consequently, there are $A_1, A_2 \in \mathcal{E}_2 \subset \mathcal{E}_1^u \subset \mathcal{E}_1^\mu$ such that $A_1 \subset A \subset A_2$ and $\mu(A_1) = \mu(A_2)$. This yields the existence of $B_1, B_2 \in \mathcal{E}_1$ such that $B_1 \subset A_1$, $A_2 \subset B_2$ and $\mu(B_1) = \mu(B_2)$. Then $A \in \mathcal{E}_1^\mu$, which implies $\mathcal{E}_2^u \subset \mathcal{E}_1^u$. □

Let $(E, \mathcal{E})$ be a measurable space. The *trace* or *restriction* of $\mathcal{E}$ on a subset $A \subset E$ is defined to be the σ-algebra $\mathcal{E}_A := \{B \cap A : B \in \mathcal{E}\}$. For a measure μ on $(E, \mathcal{E})$, the (outer) *trace* or *restriction* μ_A of μ on $(A, \mathcal{E}_A)$ is defined by $\mu_A(C) = \inf\{\mu(B) : C = B \cap A, B \in \mathcal{E}\}$. The trace μ_A can be realized as follows. Choose

$A_0 \in \mathcal{E}$ with $A_0 \supset A$ having minimal μ-measure. Then for $C \in \mathcal{E}_A$ of the form $C = B \cap A$ with $B \in \mathcal{E}$ we have $\mu_A(C) = \mu(B \cap A_0)$; see Sharpe (1988, p.367).

Proposition A.4 (Sharpe, 1988, p.368) *Let $A \subset E$ and let $\mathcal{E}_A$ be the trace of $\mathcal{E}$ on A. Then we have:*

(1) *given a finite measure μ on $(A, \mathcal{E}_A)$, the formula $\bar{\mu}(B) := \mu(B \cap A)$ for $B \in \mathcal{E}$ defines a finite measure $\bar{\mu}$ on $(E, \mathcal{E})$ whose trace on A is μ;*
(2) *$(\mathcal{E}^u)_A \subset (\mathcal{E}_A)^u$ and these two coincide if $A \in \mathcal{E}^u$.*

Suppose that $(E, \mathcal{E})$ and $(F, \mathcal{F})$ are measurable spaces. A σ-finite *kernel* from $(E, \mathcal{E})$ to $(F, \mathcal{F})$ is a function $K = K(\cdot, \cdot)$ on $E \times \mathcal{F}$ having values in $[0, \infty]$ such that:

(1) for each $A \in \mathcal{F}$ the mapping $x \mapsto K(x, A)$ is $\mathcal{E}$-measurable;
(2) for each $x \in E$ the mapping $A \mapsto K(x, A)$ is a σ-finite measure on $(F, \mathcal{F})$.

A kernel K is said to be *finite* or *bounded* if $x \mapsto K(x, F)$ is a finite or bounded, respectively, function on E. The kernel K is called *Markov* or *sub-Markov* if $K(x, F) = 1$ or $K(x, F) \leq 1$, respectively, for each $x \in E$. A kernel from $(E, \mathcal{E})$ to $(E, \mathcal{E})$ is simply called a kernel on $(E, \mathcal{E})$. Given a bounded kernel K from $(E, \mathcal{E})$ to $(F, \mathcal{F})$, for any $f \in b\mathcal{F}$ we can define $Kf \in b\mathcal{E}$ by

$$Kf(x) = K(x, f) = \int_F f(y) K(x, dy), \quad x \in E,$$

and for any finite measure μ on $(E, \mathcal{E})$ we can define a finite measure μK on $(F, \mathcal{F})$ by

$$\mu K(B) = \int_E K(x, B) \mu(dx), \qquad B \in \mathcal{F}.$$

Proposition A.5 (Sharpe, 1988, p.376) *A bounded kernel K from $(E, \mathcal{E})$ to $(F, \mathcal{F})$ extends in a unique way to a bounded kernel K from $(E, \mathcal{E}^u)$ to $(F, \mathcal{F}^u)$.*

For a metrizable topological space E with a metric d compatible with its topology, let $\mathcal{C}(E) := \mathcal{C}(E, d)$ denote the space of d-continuous real functions on (E, d) and let $\mathcal{C}_u(E) := \mathcal{C}_u(E, d)$ denote the space of d-uniformly continuous real functions on E. The advantage of $\mathcal{C}_u(E)$ is that if (E, d) is separable and totally bounded, then $b\mathcal{C}_u(E)$ with the supremum norm is separable, whereas $b\mathcal{C}(E)$ is not. The *Borel σ-algebra* $\mathcal{B}(E) = \mathcal{B}(E, d)$ on E is defined to be the σ-algebra generated by $b\mathcal{C}(E)$ or, equivalently, by all open subsets of E. If E is locally compact, we let $C_0(E)$ denote the space of continuous real functions on E vanishing at infinity. A topological space is called a *Radon topological space* or *Lusin topological space* if it is homeomorphic to a universally measurable subset or a Borel subset, respectively, of a compact metric space. A measurable space $(F, \mathcal{F})$ is called a *Radon measurable space* or *Lusin measurable space* if it is measurably isomorphic to $(E, \mathcal{B}(E))$ with E being a Radon or Lusin topological space, respectively.

A.2 Stochastic Processes

Let $(\Omega, \mathscr{G}, \mathbf{P})$ be a probability space. We shall use either $\mathbf{E}(X)$ or $\mathbf{P}(X)$ to denote the expectation of a random variable X defined on this space. A collection $(\mathscr{G}_t)_{t \in I}$ of sub-σ-algebras of $\mathscr{G}$ indexed by an interval $I \subset \mathbb{R}$ is called a *filtration* of $(\Omega, \mathscr{G})$ if $\mathscr{G}_r \subset \mathscr{G}_t$ for every $r \leq t \in I$. If a filtration $(\mathscr{G}_t)_{t \in I}$ is defined on $(\Omega, \mathscr{G}, \mathbf{P})$, we call $(\Omega, \mathscr{G}, \mathscr{G}_t, \mathbf{P})_{t \in I}$ a *filtered probability space*.

Suppose that $(\Omega, \mathscr{G}, \mathscr{G}_t, \mathbf{P})_{t \in I}$ is a filtered probability space. A random variable T taking values in $I \cup \{\infty\}$ is called a *stopping time* or an *optional time* over the filtration $(\mathscr{G}_t)_{t \in I}$ in case $\{\omega \in \Omega : T(\omega) \leq t\} \in \mathscr{G}_t$ for all $t \in I$. Given a stopping time T over $(\mathscr{G}_t)_{t \in I}$, we can define a σ-algebra

$$\mathscr{G}_T := \{A \in \mathscr{G}(I) : A \cap \{T \leq t\} \in \mathscr{G}_t \text{ for every } t \in I\}, \qquad (A.3)$$

where $\mathscr{G}(I) = \sigma(\cup_{t \in I} \mathscr{G}_t)$. Let $\tau = \sup(I)$ and let $\mathscr{G}_{t+} = \cap\{\mathscr{G}_s : t < s \in I\}$ for $t \in I \setminus \{\tau\}$. We say $(\mathscr{G}_t)_{t \in I}$ is *right continuous* if $\mathscr{G}_{t+} = \mathscr{G}_t$ for every $t \in I \setminus \{\tau\}$. Let $\mathscr{G}_{\tau+} = \mathscr{G}_\tau$ in case $\tau \in I$. If T is a stopping time over $(\mathscr{G}_{t+})_{t \in I}$, we define $\mathscr{G}_{T+}$ by (A.3) with $\mathscr{G}_t$ replaced by $\mathscr{G}_{t+}$.

The special case $I = [0, \infty)$ is often considered. Suppose that $(\Omega, \mathscr{G}, \mathscr{G}_t, \mathbf{P})_{t \geq 0}$ is a filtered probability space. Let $\bar{\mathscr{G}}$ be the $\mathbf{P}$-completion of $\mathscr{G}$ and let $\mathscr{N} = \{A \in \bar{\mathscr{G}} : \mathbf{P}(A) = 0\}$. Let $\bar{\mathscr{G}}_t = \sigma(\mathscr{G}_t \cup \mathscr{N})$ for $t \geq 0$. We call $(\bar{\mathscr{G}}, \bar{\mathscr{G}}_t)_{t \geq 0}$ the *augmentation* of $(\mathscr{G}, \mathscr{G}_t)_{t \geq 0}$ by the probability $\mathbf{P}$. If $\mathscr{G} = \bar{\mathscr{G}}$ and $\mathscr{G}_t = \bar{\mathscr{G}}_t$ for every $t \geq 0$, we say $(\mathscr{G}, \mathscr{G}_t)_{t \geq 0}$ are *augmented*. We say a filtered probability space $(\Omega, \mathscr{G}, \mathscr{G}_t, \mathbf{P})_{t \geq 0}$ satisfies the *usual hypotheses* if $(\mathscr{G}, \mathscr{G}_t)_{t \geq 0}$ are augmented and $(\mathscr{G}_t)_{t \geq 0}$ is right continuous.

Proposition A.6 *Suppose that $(\mathscr{G}, \mathscr{G}_t)_{t \geq 0}$ are augmented. If S and T are stopping times over $(\mathscr{G}_t)_{t \geq 0}$ such that $\mathbf{P}\{S \neq T\} = 0$, then $\mathscr{G}_S = \mathscr{G}_T$.*

Proof. For any $A \in \mathscr{G}_S$ we have $A \in \mathscr{G}_\infty$ and $A \cap \{S \leq t\} \in \mathscr{G}_t$ for $t \geq 0$. Since $(\mathscr{G}, \mathscr{G}_t)_{t \geq 0}$ are augmented and $\mathbf{P}\{S \neq T\} = 0$, we have $A \cap \{T \leq t\} \in \mathscr{G}_t$ for $t \geq 0$. Then $A \in \mathscr{G}_T$. That proves $\mathscr{G}_S \subset \mathscr{G}_T$. Similarly we have $\mathscr{G}_T \subset \mathscr{G}_S$.   □

We say the filtration $(\mathscr{G}_t)_{t \geq 0}$ is *quasi-left continuous* if for every increasing sequence of stopping times $\{T_n\}$ with limit T we have $\mathscr{G}_T = \sigma(\cup_{n=1}^\infty \mathscr{G}_{T_n})$. A stopping time T is called a *predictable time* if there is an *announcing sequence* of stopping times $\{T_n\}$ such that $\lim_{n \to \infty} T_n = T$ and $T_n < T$ on $\{T < \infty\}$ for each $n \geq 1$. A stopping time T is said to be *totally inaccessible* if for every predictable time S we have $S \neq T$ a.s. on $\{T < \infty\}$.

Suppose that E is a metrizable topological space. For clarity we sometimes write $\mathscr{B}^0(E)$ for the Borel σ-algebra $\mathscr{B}(E)$. Let $\mathscr{B}^u(E)$ denote the universal completion of $\mathscr{B}^0(E)$. Let $\mathscr{B}^\bullet(E)$ be a σ-algebra on E such that $\mathscr{B}^0(E) \subset \mathscr{B}^\bullet(E) \subset \mathscr{B}^u(E)$. Then Proposition A.3 implies that $\mathscr{B}^u(E)$ is also the universal completion of $\mathscr{B}^\bullet(E)$. Let $I \subset \mathbb{R}$ be an interval and let $(\Omega, \mathscr{G}, \mathbf{P})$ be a probability space. A collection $(X_t)_{t \in I}$ of measurable maps of $(\Omega, \mathscr{G})$ into $(E, \mathscr{B}^\bullet(E))$ is called a *stochastic process*. For fixed $\omega \in \Omega$, the map $t \mapsto X_t(\omega)$ from I to E is called a *sample path* of $(X_t)_{t \in I}$. The *natural σ-algebras* $\mathscr{F}^\bullet$ and $(\mathscr{F}_t^\bullet)_{t \in I}$ of $(X_t)_{t \in I}$ are defined by

$$\mathscr{F}^{\bullet} = \sigma(\{f(X_s) : s \in I, f \in b\mathscr{B}^{\bullet}(E)\})$$

and

$$\mathscr{F}_t^{\bullet} = \sigma(\{f(X_s) : s \in I_t, f \in b\mathscr{B}^{\bullet}(E)\}),$$

where $I_t = (-\infty, t] \cap I$. The process $(X_t)_{t \in I}$ is $\mathscr{B}^{\bullet}(E)$-*adapted* relative to a filtration $(\mathscr{G}_t)_{t \in I}$ in case $\mathscr{F}_t^{\bullet} \subset \mathscr{G}_t$ for every $t \in I$. It is $\mathscr{B}^{\bullet}(E)$-*progressive* relative to $(\mathscr{G}_t)_{t \in I}$ if the mapping $(\omega, s) \mapsto X_s(\omega)$ restricted to $\Omega \times I_t$ is $(\mathscr{G}_t \times \mathscr{B}(I_t))$-measurable for every $t \in I$. Clearly, a $\mathscr{B}^{\bullet}(E)$-progressive process is $\mathscr{B}^{\bullet}(E)$-adapted. We simply say $(X_t)_{t \in I}$ is *adapted* or *progressive* if it is $\mathscr{B}^0(E)$-adapted or $\mathscr{B}^0(E)$-progressive, respectively.

Let $(X_t)_{t \in I}$ be a stochastic process taking values in $(E, \mathscr{B}^{\bullet}(E))$. For any $t_1 < \cdots < t_n \in I$ let $P_{t_1, \ldots, t_n}$ be the probability measure on $(E^n, \mathscr{B}^{\bullet}(E)^n)$ induced by the mapping $\omega \mapsto (X_{t_1}(\omega), \ldots, X_{t_n}(\omega))$. We call

$$\{P_{t_1, \ldots, t_n} : t_1 < \cdots < t_n \in I, n = 1, 2, \ldots\}$$

the family of *finite-dimensional distributions* of $(X_t)_{t \in I}$. If another process $(Y_t)_{t \in I}$ has identical finite-dimensional distributions as $(X_t)_{t \in I}$, we say it is a *realization* of $(X_t)_{t \in I}$. If the processes $(X_t)_{t \in I}$ and $(Y_t)_{t \in I}$ are defined on the same probability space and if $\mathbf{P}\{X_t = Y_t\} = 1$ for every $t \in I$, we say $(Y_t)_{t \in I}$ is a *modification* of $(X_t)_{t \in I}$. We say a process $(X_t)_{t \in I}$ is *continuous* or *right continuous* if all its sample paths $t \mapsto X_t(\omega)$ are continuous or right continuous on I, respectively. A path or process $(X_t)_{t \geq 0}$ is said to be *càdlàg* (*continu à droite avec limites à gauche*) if it is right continuous at every $t \geq 0$ and possesses left limit at every $t > 0$.

Suppose that F is a non-empty set and $(t, x) \mapsto f(t, x)$ is a real or complex function defined on the product space $[0, \infty) \times F$. We say $(t, x) \mapsto f(t, x)$ is *locally bounded* provided

$$\sup_{0 \leq s \leq t} \sup_{x \in F} |f(s, x)| < \infty, \qquad t \geq 0.$$

A real or complex stochastic process $(X_t)_{t \geq 0}$ is said to be *locally bounded* if $(t, \omega) \mapsto X_t(\omega)$ is a locally bounded function on $[0, \infty) \times \Omega$.

Now let us consider a metric space (E, d). Suppose that T is a subset of $[0, \infty)$ such that $0 \in T$ and $t \mapsto x(t)$ is a path from T to E. For any $\varepsilon > 0$ the *number of ε-oscillations* of $t \mapsto x(t)$ on T is defined as

$$m(\varepsilon) := \sup\{n \geq 0 : \text{there are } 0 = t_0 < t_1 < \cdots < t_n \in T$$
$$\text{so that } d(x(t_{i-1}), x(t_i)) \geq \varepsilon \text{ for all } 1 \leq i \leq n\}.$$

An earlier version of the proof of the following proposition was suggested to the author by Tom Kurtz.

Proposition A.7 *Let (E, d) be a complete separable metric space and let $(X_t)_{t \geq 0}$ be a stochastic process in $(E, \mathscr{B}^0(E))$. If $(X_t)_{t \geq 0}$ has a càdlàg realization, then it has a càdlàg modification.*

Proof. Suppose that $(\xi_t)_{t \geq 0}$ is a càdlàg realization of $(X_t)_{t \geq 0}$. Let $(\mathscr{F}_t)_{t \geq 0}$ be the natural filtration of $(\xi_t)_{t \geq 0}$. Take a countable dense subset $T = \{0, r_1, r_2, \ldots\}$ of $[0, \infty)$ and let $T_n = \{0, r_1, \ldots, r_n\}$. For $\varepsilon > 0$ and $a > 0$ let $m^a(\varepsilon)$ and $m_n^a(\varepsilon)$ denote the numbers of ε-oscillations of $t \mapsto X_t$ on $T \cap [0, a]$ and $T_n \cap [0, a]$, respectively. Let $\mu^a(\varepsilon)$ and $\mu_n^a(\varepsilon)$ denote respectively those numbers of $t \mapsto \xi_t$. Then $m_n^a(\varepsilon) \to m^a(\varepsilon)$ and $\mu_n^a(\varepsilon) \to \mu^a(\varepsilon)$ increasingly as $n \to \infty$. Let $\tau_n^\varepsilon(0) = 0$ and for $k \geq 0$ define

$$\tau_n^\varepsilon(k+1) = \min\{t > \tau_n^\varepsilon(k) : t \in T_n, d(\xi_{\tau_n^\varepsilon(k)}, \xi_t) \geq \varepsilon\}$$

if $\tau_n^\varepsilon(k) < \infty$ and $\tau_n^\varepsilon(k+1) = \infty$ if $\tau_n^\varepsilon(k) = \infty$. Since T_n is discrete, for any $u \geq 0$ we have

$$\{\tau_n^\varepsilon(k+1) \leq u\} = \bigcup_{s < t \in T_n \cap [0,u]} (\{\tau_n^\varepsilon(k) = s\} \cap \{d(\xi_s, \xi_t) \geq \varepsilon\}). \qquad (A.4)$$

By the separability of (E, d) we have $\mathscr{B}^0(E \times E) = \mathscr{B}^0(E) \times \mathscr{B}^0(E)$. Then $\{d(\xi_s, \xi_t) \geq \varepsilon\} \in \mathscr{F}_u$ for $s < t \leq u$. Using (A.4) one can show inductively that each $\tau_n^\varepsilon(k)$ is a stopping time over $(\mathscr{F}_t)$. Since $\{\mu_n^a(\varepsilon) \geq k\} = \{\tau_n^\varepsilon(k) \leq a\}$, each $\mu_n^a(\varepsilon)$ is a random variable and hence so is $\mu^a(\varepsilon) = \lim_{n \to \infty} \mu_n^a(\varepsilon)$. Similarly, $m^a(\varepsilon) = \lim_{n \to \infty} m_n^a(\varepsilon)$ is a random variable. Since $(\xi_t)_{t \geq 0}$ is a càdlàg realization of $(X_t)_{t \geq 0}$, we get

$$\mathbf{P}\{m^a(\varepsilon) < \infty\} = \mathbf{P}\{\mu^a(\varepsilon) < \infty\} = 1.$$

Let $\Omega_1 = \cap_{j=1}^\infty \{m^j(1/j) < \infty\}$. Then $\mathbf{P}(\Omega_1) = 1$. It is simple to show that for $\omega \in \Omega_1$ the limit $Y_t(\omega) := \lim_{T \ni s \to t+} X_s(\omega)$ exists at $t \geq 0$ and $Z_t(\omega) := \lim_{T \ni s \to t-} X_s(\omega)$ exists at $t > 0$. Fix $x_0 \in E$ and let $Y_t(\omega) = x_0$ for all $t \geq 0$ and $\omega \in \Omega \setminus \Omega_1$. Then $(Y_t)_{t \geq 0}$ is a càdlàg process. Since $(X_t)_{t \geq 0}$ is clearly right continuous in probability, we have $Y_t = X_t$ a.s. for every $t \geq 0$. Therefore $(Y_t)_{t \geq 0}$ is a càdlàg modification of $(X_t)_{t \geq 0}$. $\qquad \square$

A.3 Right Markov Processes

Let E be a Radon topological space and let $\mathscr{B}^\bullet(E)$ be a σ-algebra such that $\mathscr{B}^0(E) \subset \mathscr{B}^\bullet(E) \subset \mathscr{B}^u(E)$. A family of Markov or sub-Markov kernels $(P_t)_{t \geq 0}$ on $(E, \mathscr{B}^\bullet(E))$ is called a *transition semigroup* if it satisfies the following *Chapman–Kolmogorov equation*:

$$P_{r+t}(x, B) = \int_E P_r(x, dy) P_t(y, B) \qquad (A.5)$$

for all $r, t \geq 0$, $x \in E$ and $B \in \mathscr{B}^{\bullet}(E)$. By Proposition A.5, we can always regard $(P_t)_{t \geq 0}$ as kernels on $(E, \mathscr{B}^u(E))$. A *Borel transition semigroup* $(P_t)_{t \geq 0}$ is a transition semigroup on a Lusin topological space E such that $P_t f \in b\mathscr{B}^0(E)$ for each $t \geq 0$ and $f \in b\mathscr{B}^0(E)$. We say the transition semigroup $(P_t)_{t \geq 0}$ is *Markov* or *conservative* if each P_t is a Markov kernel. We say $(P_t)_{t \geq 0}$ is *normal* if $P_0(x, \cdot) = \delta_x$ for every $x \in E$.

Let us consider a transition semigroup $(P_t)_{t \geq 0}$ on $(E, \mathscr{B}^{\bullet}(E))$. A family $(\mu_t)_{t \in \mathbb{R}}$ of σ-finite measures on $(E, \mathscr{B}^{\bullet}(E))$ is called an *entrance rule* for $(P_t)_{t \geq 0}$ if $\mu_s P_{t-s} \rightarrow \mu_t$ increasingly as $s \rightarrow t \in \mathbb{R}$. By an *entrance law* at $\alpha \in [-\infty, \infty)$ for $(P_t)_{t \geq 0}$ we mean a family of σ-finite measures $(\mu_t)_{t > \alpha}$ such that $\mu_s P_{t-s} = \mu_t$ for $t \geq s > \alpha$. We say $(\mu_t)_{t > \alpha}$ is *bounded* if $t \mapsto \mu_t(E)$ is a bounded function on (α, ∞). A *probability entrance law* is an entrance law $(\mu_t)_{t > \alpha}$ where each μ_t is a probability measure. If there is a σ-finite measure μ_α such that $\mu_t = \mu_\alpha P_{t-\alpha}$ for all $t > \alpha$, we say the entrance law $(\mu_t)_{t > \alpha}$ is *closable* and call $(\mu_t)_{t \geq \alpha}$ a *closed entrance law*. We say an entrance law $(\mu_t)_{t > \alpha}$ is *minimal* or *extremal* if every entrance law dominated by $(\mu_t)_{t > \alpha}$ is proportional to it. Note that an entrance law $(\mu_t)_{t > \alpha}$ at $\alpha \in [-\infty, \infty)$ may be extended to an entrance rule $(\mu_t)_{t \in \mathbb{R}}$ by setting $\mu_t = 0$ for $t \leq \alpha$. In this sense, we can regard the entrance law as a special case of the entrance rule. The concepts of entrance rules and entrance laws can obviously be extended to semigroups of bounded kernels. We sometimes make use of those extensions.

Let $\mathscr{K}^1(P)$ denote the set of all probability entrance laws $(\mu_t)_{t > 0}$ at zero for $(P_t)_{t \geq 0}$ endowed with the σ-algebra generated by all mappings $\{\mu \mapsto \mu_t(f) : t > 0, f \in b\mathscr{B}^{\bullet}(E)\}$. Let $\mathscr{K}^1_m(P)$ be the set of minimal probability entrance laws in $\mathscr{K}^1(P)$. From Dynkin (1978, Theorems 3.1 and 9.1) we know $\mathscr{K}^1(P)$ is a *simplex*, that is, $\mathscr{K}^1_m(P)$ is a measurable subset of $\mathscr{K}^1(P)$ and for each $\mu \in \mathscr{K}^1(P)$ there is a unique probability measure Q_μ on $\mathscr{K}^1_m(P)$ such that

$$\mu_t(\cdot) = \int_{\mathscr{K}^1_m(P)} \nu_t(\cdot) Q_\mu(d\nu), \qquad t > 0.$$

A σ-finite measure m on $(E, \mathscr{B}^{\bullet}(E))$ is called an *excessive measure* for $(P_t)_{t \geq 0}$ if $m P_t \leq m$ for every $t \geq 0$. The measure m is called a *purely excessive measure* if $m P_t \leq m$ for every $t \geq 0$ and $m P_t \rightarrow 0$ as $t \rightarrow \infty$, and it is called an *invariant measure* if $m P_t = m$ for every $t \geq 0$. For $\alpha \geq 0$ we say a function $f \in p\mathscr{B}^u(E)$ is α-*super-mean-valued* for $(P_t)_{t \geq 0}$ if $e^{-\alpha t} P_t f \leq f$ for all $t \geq 0$, and it is called an α-*excessive function* for $(P_t)_{t \geq 0}$ if $e^{-\alpha t} P_t f \rightarrow f$ increasingly as $t \rightarrow 0$. In the special case with $\alpha = 0$, we simply say f is *super-mean-valued* or *excessive*, respectively. Let $\mathscr{S}^\alpha$ denote the set of α-excessive functions for $(P_t)_{t \geq 0}$.

A family of bounded kernels $(U^\alpha)_{\alpha > 0}$ on $(E, \mathscr{B}^{\bullet}(E))$ is called a *resolvent* in case the *resolvent equation*

$$U^\alpha f(x) - U^\beta f(x) = (\beta - \alpha) U^\alpha U^\beta f(x) \qquad (A.6)$$

is satisfied for all $\alpha, \beta > 0$, $x \in E$ and $f \in b\mathscr{B}^{\bullet}(E)$. A resolvent $(U^\alpha)_{\alpha > 0}$ is called *Markov* or *conservative* if αU^α is a Markov kernel for all $\alpha > 0$. A function

$f \in p\mathscr{B}^u(E)$ is called α-*supermedian* for the resolvent $(U^\alpha)_{\alpha>0}$ if $\beta U^{\alpha+\beta} f \le f$ for all $\beta > 0$. Let $\tilde{\mathscr{S}}^\alpha$ denote the class of all α-supermedian functions for $(U^\alpha)_{\alpha>0}$.

If $(P_t)_{t\ge 0}$ is a transition semigroup on $(E, \mathscr{B}^\bullet(E))$ such that $(t, x) \mapsto P_t f(x)$ is measurable with respect to $\mathscr{B}([0, \infty)) \times \mathscr{B}^\bullet(E)$ for every $f \in b\mathscr{B}^\bullet(E)$, the operators $(U^\alpha)_{\alpha>0}$ defined by

$$U^\alpha f(x) = \int_0^\infty e^{-\alpha t} P_t f(x) \mathrm{d}t, \quad f \in b\mathscr{B}^\bullet(E), \tag{A.7}$$

constitute a resolvent, which is called the *resolvent of* $(P_t)_{t\ge 0}$. We also call U^α the α-*potential operator* of $(P_t)_{t\ge 0}$. The *potential operator* U of $(P_t)_{t\ge 0}$ is defined by

$$U f(x) = \int_0^\infty P_t f(x) \mathrm{d}t, \quad f \in p\mathscr{B}^\bullet(E). \tag{A.8}$$

However, this kernel may not be σ-finite. It is easy to show that if f is α-super-mean-valued for $(P_t)_{t\ge 0}$, it is α-supermedian for $(U^\alpha)_{\alpha>0}$.

A particularly important special case is where E is a locally compact separable metric space. In this case, its one-point compactification is metrizable. A normal and conservative transition semigroup $(P_t)_{t\ge 0}$ on a locally compact separable metric space E is called a *Feller semigroup* provided:

(1) $P_t(C_0(E)) \subset C_0(E)$ for all $t \ge 0$;
(2) $P_t f \to f$ pointwise as $t \to 0$ for all $f \in C_0(E)$.

If $(P_t)_{t\ge 0}$ is a Feller semigroup, then $P_t f \to f$ uniformly as $t \to 0$ for all $f \in C_0(E)$; see Sharpe (1988, p.50).

Suppose that $(P_t)_{t\ge 0}$ is a Markov transition semigroup on $(E, \mathscr{B}^\bullet(E))$ and $(\xi_t)_{t\in I}$ is a stochastic process in $(E, \mathscr{B}^\bullet(E))$ indexed by an interval $I \subset \mathbb{R}$. We assume that $(\xi_t)_{t\in I}$ is defined on $(\Omega, \mathscr{G}, \mathbf{P})$ and is $\mathscr{B}^\bullet(E)$-adapted to a filtration $(\mathscr{G}_t)_{t\in I}$ of $(\Omega, \mathscr{G})$. We say $\{(\xi_t, \mathscr{G}_t) : t \in I\}$ has the *simple $\mathscr{B}^\bullet(E)$-Markov prop-erty* with transition semigroup $(P_t)_{t\ge 0}$ if

$$\mathbf{P}\big[f(\xi_t)|\mathscr{G}_r\big] = P_{t-r} f(\xi_r), \quad r \le t \in I, f \in b\mathscr{B}^\bullet(E). \tag{A.9}$$

If $\{(\xi_t, \mathscr{G}_t) : t \ge 0\}$ satisfies the simple $\mathscr{B}^\bullet(E)$-Markov property with transition semigroup $(P_t)_{t\ge 0}$, the distribution μ_0 of ξ_0 is called the *initial law* of $(\xi_t)_{t\ge 0}$. In this case, we necessarily have

$$\begin{aligned}
\mathbf{P}\big[f_1(\xi_{t_1}) & f_2(\xi_{t_2}) \cdots f_n(\xi_{t_n})\big] \\
&= \mu_0\big(P_{t_1}(f_1 \cdots P_{t_{n-1}-t_{n-2}}(f_{n-1} P_{t_n-t_{n-1}} f_n)))
\end{aligned} \tag{A.10}$$

for $0 \le t_1 \le t_2 \le \cdots \le t_n$ and $f_1, f_2, \ldots, f_n \in b\mathscr{B}^\bullet(E)$, which is a simple consequence of (A.9) by an induction argument. Consequently, the restriction of $\mathbf{P}$ on the natural σ-algebra $\mathscr{F}^\bullet$ is determined uniquely by (A.10).

Proposition A.8 *Suppose that* $\{(\xi_t, \mathscr{G}_t) : t \geq 0\}$ *satisfies the simple* $\mathscr{B}^\bullet(E)$-*Markov property* (A.9). *Let* $(\mathscr{G}^*, \mathscr{G}_t^*)$ *denote the augmentations of* $(\mathscr{G}, \mathscr{G}_t)$ *with respect to* **P**. *Then* $\{(\xi_t, \mathscr{G}_t^*) : t \geq 0\}$ *satisfies the simple* $\mathscr{B}^u(E)$-*Markov property.*

Proof. Let μ_t denote the distribution of ξ_t on $(E, \mathscr{B}^\bullet(E))$. For $f \in b\mathscr{B}^u(E)$ we can choose $f_1, f_2 \in b\mathscr{B}^\bullet(E)$ so that $f_1 \leq f \leq f_2$ and $\mu_t(f_2 - f_1) = 0$. Then $f_1(\xi_t), f_2(\xi_t) \in b\mathscr{G}_t$ and

$$\mathbf{P}\big[f_2(\xi_t) - f_1(\xi_t)\big] = \mu_t(f_2 - f_1) = 0. \tag{A.11}$$

It follows that $f(\xi_t) \in b\mathscr{G}_t^*$, and so $(\xi_t)_{t \geq 0}$ is $\mathscr{B}^u(E)$-adapted relative to $(\mathscr{G}_t^*)_{t \geq 0}$. Then to get the desired result it suffices to show

$$\mathbf{P}\big[f(\xi_t)1_A\big] = \mathbf{P}\big[P_{t-r}f(\xi_r)1_A\big] \tag{A.12}$$

for $t \geq r \geq 0$, $A \in \mathscr{G}_r^*$ and $f \in b\mathscr{B}^u(E)$. Let $\mathscr{N} = \{N \in \mathscr{G}^* : \mathbf{P}(N) = 0\}$. Then there is $A_0 \in \mathscr{G}_r$ so that $A \triangle A_0 \in \mathscr{N}$. By (A.9) we have (A.12) for $f \in b\mathscr{B}^\bullet(E)$. For $f \in b\mathscr{B}^u(E)$ we can take $f_1, f_2 \in b\mathscr{B}^\bullet(E)$ so that $f_1 \leq f \leq f_2$ and (A.11) holds. Since (A.12) holds for both f_1 and f_2, it also holds for f. □

Corollary A.9 (Sharpe, 1988, p.6) *Suppose that* $(P_t)_{t \geq 0}$ *preserves* $\mathscr{B}^\bullet(E)$ *and* $\mathscr{B}^\diamond(E)$ *with* $\mathscr{B}^0(E) \subset \mathscr{B}^\bullet(E) \subset \mathscr{B}^\diamond(E) \subset \mathscr{B}^u(E)$. *Let* $\{(\xi_t, \mathscr{G}_t) : t \geq 0\}$ *satisfy the simple* $\mathscr{B}^\bullet(E)$-*Markov property* (A.9). *If* $(\xi_t)_{t \geq 0}$ *is* $\mathscr{B}^\diamond(E)$-*adapted to* $(\mathscr{G}_t)_{t \geq 0}$, *then* $\{(\xi_t, \mathscr{G}_t) : t \geq 0\}$ *satisfies the simple* $\mathscr{B}^\diamond(E)$-*Markov property.*

Proof. By Proposition A.8 we infer $\{(\xi_t, \mathscr{G}_t^*) : t \geq 0\}$ satisfies the simple $\mathscr{B}^u(E)$-Markov property. Then $\{(\xi_t, \mathscr{G}_t) : t \geq 0\}$ satisfies the simple $\mathscr{B}^\diamond(E)$-Markov property. □

Definition A.10 (Sharpe, 1988, p.7) *Suppose that* $(P_t)_{t \geq 0}$ *is a normal Markov transition semigroup on* $(E, \mathscr{B}^\bullet(E))$. *The collection* $\xi = (\Omega, \mathscr{G}, \mathscr{G}_t, \xi_t, \theta_t, \mathbf{P}_x)$ *is called a* $\mathscr{B}^\bullet(E)$-*Markov process with transition semigroup* $(P_t)_{t \geq 0}$ *in case* ξ *satisfies the following conditions:*

(1) $(\Omega, \mathscr{G}, \mathscr{G}_t)_{t \geq 0}$ *is a filtered measurable space, and* $(\xi_t)_{t \geq 0}$ *is an* E-*valued process* $\mathscr{B}^\bullet(E)$-*adapted to* $(\mathscr{G}_t)_{t \geq 0}$.
(2) $(\theta_t)_{t \geq 0}$ *is a collection of shift operators for* ξ, *that is, maps of* Ω *into itself satisfying* $\theta_s \circ \theta_t = \theta_{s+t}$ *and* $\xi_s \circ \theta_t = \xi_{s+t}$ *identically for* $t, s \geq 0$.
(3) *For every* $x \in E$, $\mathbf{P}_x$ *is a probability measure on* $(\Omega, \mathscr{G})$ *and* $x \mapsto \mathbf{P}_x(H)$ *is* $\mathscr{B}^\bullet(E)$-*measurable for each* $H \in b\mathscr{G}$.
(4) *For every* $x \in E$, *we have* $\mathbf{P}_x\{\xi_0 = x\} = 1$ *and the process* $(\xi_t)_{t \geq 0}$ *has the simple Markov property* (A.9) *relative to* $(\mathscr{G}_t, \mathbf{P}_x)$ *with transition semigroup* $(P_t)_{t \geq 0}$.

We say ξ *is right continuous if* $t \mapsto \xi_t(\omega)$ *is right continuous for every* $\omega \in \Omega$.

If the above conditions (1)–(4) are satisfied, we also say that ξ is a *realization* of the semigroup $(P_t)_{t \geq 0}$. In this case, for any finite measure μ on $(E, \mathscr{B}^\bullet(E))$ we may define the finite measure $\mathbf{P}_\mu$ on $(\Omega, \mathscr{G})$ by

$$\mathbf{P}_\mu(H) = \int_E \mathbf{P}_x(H)\mu(\mathrm{d}x), \quad H \in b\mathscr{G}. \tag{A.13}$$

In the sequel, we always assume μ is a probability measure unless stated otherwise. It is easy to verify that $(\xi_t)_{t\geq 0}$ has the simple $\mathscr{B}^\bullet(E)$-Markov property relative to $(\mathscr{G}_t, \mathbf{P}_\mu)$ with initial law μ. We mention that the measurability of $x \mapsto \mathbf{P}_x(H)$ is used in the definition (A.13) of the measure $\mathbf{P}_\mu$ on $(\Omega, \mathscr{G})$. Of course, this measurability follows automatically if $(\mathscr{G}, \mathscr{G}_t)$ are the natural σ-algebras of $\{\xi_t : t \geq 0\}$.

Consider a right continuous $\mathscr{B}^\bullet(E)$-Markov process $\xi = (\Omega, \mathscr{G}, \mathscr{G}_t, \xi_t, \theta_t, \mathbf{P}_x)$ with transition semigroup $(P_t)_{t\geq 0}$ and resolvent $(U^\alpha)_{\alpha>0}$ on $(E, \mathscr{B}^\bullet(E))$. Let $\mathscr{G}^\mu$ denote the $\mathbf{P}_\mu$-completion of $\mathscr{G}$ and let $\mathscr{N}^\mu(\mathscr{G})$ denote the family of $\mathbf{P}_\mu$-null sets in $\mathscr{G}^\mu$. Then define:

$$\bar{\mathscr{G}} = \cap\{\mathscr{G}^\mu : \mu \text{ is an initial law on } E\};$$
$$\mathscr{N}(\mathscr{G}) = \cap\{\mathscr{N}^\mu(\mathscr{G}) : \mu \text{ is an initial law on } E\};$$
$$\mathscr{G}_t^\mu = \sigma(\mathscr{G}_t \cup \mathscr{N}^\mu(\mathscr{G}));$$
$$\bar{\mathscr{G}}_t = \cap\{\mathscr{G}_t^\mu : \mu \text{ is an initial law on } E\}.$$

Therefore $(\mathscr{G}^\mu, \mathscr{G}_t^\mu)$ is the augmentation of $(\mathscr{G}, \mathscr{G}_t)$ by the probability $\mathbf{P}_\mu$. We call $(\bar{\mathscr{G}}, \bar{\mathscr{G}}_t)$ the *augmentation* of $(\mathscr{G}, \mathscr{G}_t)$ by the system of probabilities $\{\mathbf{P}_\mu : \mu \text{ is a probability on } E\}$. It is easy to see that $(\xi_t)_{t\geq 0}$ is $\mathscr{B}^u(E)$-adapted relative to $(\bar{\mathscr{G}}_t)_{t\geq 0}$ and each $\mathbf{P}_\mu$ extends uniquely to $\bar{\mathscr{G}}$. Moreover, for any $H \in b\bar{\mathscr{G}}$ the mapping $x \mapsto \mathbf{P}_x(H)$ is $\mathscr{B}^u(E)$-measurable and the equality in (A.13) remains true. Using Proposition A.8 and Corollary A.9 one can see $(\xi_t)_{t\geq 0}$ has the simple $\mathscr{B}^u(E)$-Markov property relative to $(\mathscr{G}_t^\mu, \mathbf{P}_\mu)$ and $(\bar{\mathscr{G}}_t, \mathbf{P}_\mu)$.

We say $(\mathscr{G}, \mathscr{G}_t)_{t\geq 0}$ are *augmented* with respect to the system $\{\mathbf{P}_\mu : \mu \text{ is a probability on } E\}$ provided $\mathscr{G} = \bar{\mathscr{G}}$ and $\mathscr{G}_t = \bar{\mathscr{G}}_t$ for all $t \geq 0$. As observed in Sharpe (1988, p.25), further application of augmentation procedure to $(\bar{\mathscr{G}}_t)_{t\geq 0}$ is fruitless in the sense that $[\bar{\mathscr{G}}_t]^\mu = \mathscr{G}_t^\mu$ and $[\bar{\mathscr{G}}_t]^- = \bar{\mathscr{G}}_t$. Similarly, letting $\mathscr{G}_{t+} = \cap_{s>t}\mathscr{G}_s$ we have $[\mathscr{G}_t^\mu]_+ = [\mathscr{G}_{t+}]^\mu$, which will be denoted simply by $\mathscr{G}_{t+}^\mu$. It is easy to see that for any initial law μ on E the filtered space $(\Omega, \mathscr{G}^\mu, \mathscr{G}_{t+}^\mu, \mathbf{P}_\mu)$ satisfies the usual hypotheses. Let $\mathscr{F}^u$ be the $\mathscr{B}^u(E)$-natural σ-algebra of $\{\xi_t : t \geq 0\}$. Let $\mathscr{F}^\mu$ denote the $\mathbf{P}_\mu$-completion of $\mathscr{F}^u$ and let $\mathscr{F} = \cap\{\mathscr{F}^\mu : \mu \text{ is an initial law on } E\}$. It was proved in Sharpe (1988, p.25) that

$$\mathbf{P}_\mu\big[F \circ \theta_t | \bar{\mathscr{G}}_t\big] = \mathbf{P}_\mu\big[F \circ \theta_t | \mathscr{G}_t^\mu\big] = \mathbf{P}_{\xi_t}(F) \tag{A.14}$$

for every $t \geq 0$, $F \in b\mathscr{F}$ and initial law μ.

Proposition A.11 *Suppose that T is a stopping time over $(\mathscr{G}_{t+})$. Then $\mathscr{G}_{T+}^\mu = \sigma(\mathscr{G}_{T+} \cup \mathscr{N}^\mu(\mathscr{G}))$.*

Proof. Since $\mathscr{G}_{T+}^\mu \supset \sigma(\mathscr{G}_{T+} \cup \mathscr{N}^\mu(\mathscr{G}))$ is obvious, we only need to verify the inclusion $\mathscr{G}_{T+}^\mu \subset \sigma(\mathscr{G}_{T+} \cup \mathscr{N}^\mu(\mathscr{G}))$. It suffices to show for every $A \in \mathscr{G}_{T+}^\mu$ there is $B \in \mathscr{G}_{T+}$ such that $A \triangle B \in \mathscr{N}^\mu(\mathscr{G})$, where "$\triangle$" denotes the symmetric difference defined by (A.1). For each $n \geq 1$ define the stopping time T_n over $(\mathscr{G}_t)$ by

$$T_n(\omega) = \begin{cases} k/2^n & \text{if } (k-1)/2^n \leq T(\omega) < k/2^n, \\ \infty & \text{if } T(\omega) = \infty. \end{cases} \tag{A.15}$$

Then $T_n \to T$ decreasingly as $n \to \infty$. In view of (A.15) we have $A \cap \{T_n = k/2^n\} \in \mathscr{G}^{\mu}_{k/2^n}$ for $1 \leq n < \infty$ and $1 \leq k = \infty$. Then there exists $A_{n,k} \in \mathscr{G}_{k/2^n}$ so that

$$(A \cap \{T_n = k/2^n\}) \triangle A_{n,k} \in \mathscr{N}^{\mu}(\mathscr{G}).$$

Since $\{T_n = k/2^n\} \in \mathscr{G}_{k/2^n}$, we have

$$B_{n,k} := A_{n,k} \cap \{T_n = k/2^n\} \in \mathscr{G}_{k/2^n}.$$

Observe also that

$$(A \cap \{T_n = k/2^n\}) \triangle B_{n,k} \in \mathscr{N}^{\mu}(\mathscr{G}).$$

Let $B_n = (\cup_{k=1}^{\infty} B_{n,k}) \cup B_{n,\infty}$. Then $A \triangle B_n \in \mathscr{N}^{\mu}(\mathscr{G})$ and

$$B_n \cap \{T_n = k/2^n\} = B_{n,k} \in \mathscr{G}_{k/2^n}.$$

It follows that $B_n \in \mathscr{G}_{T_n} \subset \mathscr{G}_{T_k}$ for $n \geq k$. By the right continuity of $(\mathscr{G}_{t+})$,

$$B := \bigcap_{k=1}^{\infty} \bigcup_{n=k}^{\infty} B_n \in \bigcap_{k=1}^{\infty} \mathscr{G}_{T_k+} = \mathscr{G}_{T+}.$$

Moreover, we have $A \triangle B \in \mathscr{N}^{\mu}(\mathscr{G})$. That gives the desired result. □

Corollary A.12 *For any initial law μ on E and any stopping time T for $(\mathscr{G}^{\mu}_{t+})$, there is a stopping time S for $(\mathscr{G}_{t+})$ so that $\{T \neq S\} \in \mathscr{N}^{\mu}(\mathscr{G})$. In this case, we have*

$$\mathscr{G}^{\mu}_{T+} = \mathscr{G}^{\mu}_{S+} = \sigma(\mathscr{G}_{S+} \cup \mathscr{N}^{\mu}(\mathscr{G})). \tag{A.16}$$

Proof. The first assertion was proved in Sharpe (1988, p.25). Then (A.16) follows by Propositions A.6 and A.11. □

Proposition A.13 (Sharpe, 1988, p.26) *Let $f \in \mathscr{B}^u(E)$ and let μ be an initial law on E. Then we have:*

(1) *If T is an stopping time over $(\mathscr{G}^{\mu}_{t+})$, then $f(\xi_T)1_{\{T<\infty\}} \in \mathscr{G}^{\mu}_{T+}$.*
(2) *If T is an stopping time over $(\bar{\mathscr{G}}_{t+})$, then $f(\xi_T)1_{\{T<\infty\}} \in \bar{\mathscr{G}}_{T+}$.*

Definition A.14 (Sharpe, 1988, p.26) *Suppose that $\xi = (\Omega, \mathscr{G}, \mathscr{G}_t, \xi_t, \theta_t, \mathbf{P}_x)$ is a right continuous $\mathscr{B}^{\bullet}(E)$-Markov process with transition semigroup $(P_t)_{t \geq 0}$ and $(\mathscr{G}, \mathscr{G}_t)$ have been augmented by $\{\mathbf{P}_{\mu} : \mu$ is an initial law on $E\}$. We say $(\xi_t)_{t \geq 0}$ satisfies the strong Markov property relative to $(\mathscr{G}_{t+})$ provided*

$$\mathbf{P}_{\mu}\big[f(\xi_t) \circ \theta_T 1_{\{T<\infty\}} | \mathscr{G}_{T+}\big] = P_t f(\xi_T) 1_{\{T<\infty\}} \tag{A.17}$$

for every $t \geq 0$, stopping time T over $(\mathscr{G}_{t+})$, initial law μ and $f \in b\mathscr{B}^u(E)$.

Theorem A.15 *Suppose that* $\xi = (\Omega, \mathscr{G}^\bullet, \mathscr{G}_t^\bullet, \xi_t, \theta_t, \mathbf{P}_x)$ *is a right continuous* $\mathscr{B}^\bullet(E)$-*Markov process with transition semigroup* $(P_t)_{t\geq 0}$. *Let* $(\mathscr{G}, \mathscr{G}_t)$ *be the augmentations of* $(\mathscr{G}^\bullet, \mathscr{G}_t^\bullet)$ *by* $\{\mathbf{P}_\mu : \mu$ *is an initial law on* $E\}$. *Then* $(\xi_t)_{t\geq 0}$ *has the strong Markov property relative to* $(\mathscr{G}_{t+})$ *if and only if*

$$\mathbf{P}_\mu\big[f(\xi_t) \circ \theta_{T_0} 1_{\{T_0 < \infty\}} | \mathscr{G}_{T_0+}^\bullet\big] = P_t f(\xi_{T_0}) 1_{\{T_0 < \infty\}} \tag{A.18}$$

for every $t \geq 0$, *stopping time* T_0 *over* $(\mathscr{G}_{t+}^\bullet)$, *initial law* μ *and* $f \in \mathrm{b}\mathscr{B}^\bullet(E)$.

Proof. Suppose that $(\xi_t)_{t\geq 0}$ has the strong Markov property relative to $(\mathscr{G}_{t+})$ and T_0 is a stopping time over $(\mathscr{G}_{t+}^\bullet)$. Since $t \mapsto \xi_t$ is clearly $\mathscr{B}^\bullet(E)$-progressive over $(\mathscr{G}_{t+}^\bullet)$, for any $f \in \mathrm{b}\mathscr{B}^\bullet(E)$ the process $t \mapsto f(\xi_t)$ is progressive over $(\mathscr{G}_{t+}^\bullet)$. Then $f(\xi_{T_0}) 1_{\{T_0 < \infty\}} \in \mathrm{b}\mathscr{G}_{T_0+}^\bullet$; see, e.g., Dellacherie and Meyer (1978, p.122) or Sharpe (1988, p.22). Consequently, we have $P_t f(\xi_{T_0}) 1_{\{T_0 < \infty\}} \in \mathrm{b}\mathscr{G}_{T_0+}^0$ for $t \geq 0$. By letting $T = T_0$ in (A.17) and taking the conditional expectation relative to $\mathscr{G}_{T_0+}^\bullet$ we obtain (A.18). For the converse, suppose that (A.18) holds for every stopping time T_0 over $(\mathscr{G}_{t+}^\bullet)$ and every $f \in \mathrm{b}\mathscr{B}^\bullet(E)$. Let T be a stopping time over $(\mathscr{G}_{t+})$ and let $A \in \mathscr{G}_{T+}$. By Corollary A.12 there is a stopping time T_0 over $(\mathscr{G}_{t+}^\bullet)$ and an event $A_0 \in \mathscr{G}_{T_0+}^\bullet$ so that $\{T \neq T_0\} \in \mathscr{N}^\mu(\mathscr{G}^\bullet)$ and $A \triangle A_0 \in \mathscr{N}^\mu(\mathscr{G}^\bullet)$. By (A.18) for every $f \in \mathrm{b}\mathscr{B}^\bullet(E)$ we have

$$\mathbf{P}_\mu\big[1_A f(\xi_t) \circ \theta_T 1_{\{T < \infty\}}\big] = \mathbf{P}_\mu\big[1_A P_t f(\xi_T) 1_{\{T < \infty\}}\big]. \tag{A.19}$$

As in the proof of Proposition A.8 it is easy to see the above equality also holds for $f \in \mathrm{b}\mathscr{B}^u(E)$. That gives (A.17) for $f \in \mathrm{b}\mathscr{B}^u(E)$. $\square$

If (A.18) is satisfied for every $t \geq 0$, stopping time T_0 over $(\mathscr{G}_{t+}^\bullet)$, initial law μ and $f \in \mathrm{b}\mathscr{B}^\bullet(E)$, we say the process $(\xi_t)_{t\geq 0}$ satisfies the *strong Markov property* relative to $(\mathscr{G}_{t+}^\bullet)$. Note that the condition that $(P_t)_{t\geq 0}$ sends $\mathrm{b}\mathscr{B}^\bullet(E)$ into itself is used to guarantee the measurability of the right-hand side relative to $\mathscr{G}_{T_0+}^\bullet$.

A real-valued process $(Z_t)_{t\geq 0}$ is called $\mathbf{P}_\mu(\mathscr{G})$-*evanescent* in case $\{\omega \in \Omega : Z_t(\omega) \neq 0$ for some $t \geq 0\} \in \mathscr{N}^\mu(\mathscr{G})$. Let $\mathscr{I}^\mu(\mathscr{G})$ denote the class of $\mathbf{P}_\mu(\mathscr{G})$-evanescent processes and let $\mathscr{I}(\mathscr{G}) = \cap\{\mathscr{I}^\mu(\mathscr{G}) : \mu$ is an initial law on $E\}$. Let $\mathscr{D}(\mathscr{G}_t)$ denote the class of bounded right continuous real processes adapted to $(\mathscr{G}_t)_{t\geq 0}$. Let $\mathscr{O}^\mu(\mathscr{G}_t)$ be the σ-algebra on $\Omega \times [0, \infty)$ generated by $\mathscr{D}(\mathscr{G}_t) \cup \mathscr{I}^\mu(\mathscr{G})$ and let $\tilde{\mathscr{O}}(\mathscr{G}_t) = \cap\{\mathscr{O}^\mu(\mathscr{G}_t) : \mu$ is an initial law on $E\}$. We say an extended real function f on E is *nearly optional* relative to ξ provided $(\omega, t) \mapsto f(\xi_t(\omega))$ is $\tilde{\mathscr{O}}(\mathscr{G}_t)$-measurable. Clearly, a continuous function on E is nearly optional. By the monotone class theorem it is easy to see that a Borel function on E is also nearly optional. The function f is said to be *nearly Borel* relative to ξ if for every initial law μ there are Borel functions g and h on E so that $g \leq f \leq h$ and $\mathbf{P}_\mu\{g(\xi_t) = h(\xi_t)$ for all $t \geq 0\} = 1$. Let d be a metric on E compatible with its topology. Recall that $\mathscr{C}_u(E) = \mathscr{C}_u(E, d)$ denotes the set of real d-uniformly continuous functions on (E, d) and $\mathscr{S}^\alpha$ is the class of all α-excessive functions for $(P_t)_{t\geq 0}$.

Theorem A.16 (Sharpe, 1988, p.31) *Suppose that* $\xi = (\Omega, \mathscr{G}, \mathscr{G}_t, \xi_t, \theta_t, \mathbf{P}_x)$ *is a right continuous* $\mathscr{B}^\bullet(E)$-*Markov process with transition semigroup* $(P_t)_{t\geq 0}$ *and*

$(\mathscr{G}, \mathscr{G}_t)$ *have been augmented by* $\{\mathbf{P}_\mu : \mu$ *is an initial law on* $E\}$. *Then the following conditions are equivalent:*

(1) $\{t \mapsto f(\xi_t)$ *is not right continuous*$\} \in \mathscr{N}(\mathscr{G})$ *for every* $\alpha > 0$ *and every* $f \in \mathscr{S}^\alpha$;

(2) $\{t \mapsto U^\alpha f(\xi_t)$ *is not right continuous*$\} \in \mathscr{N}(\mathscr{G})$ *for every* $\alpha > 0$ *and every* $f \in b\mathscr{C}_u(E)$;

(3) $(\xi_t)_{t \geq 0}$ *satisfies the strong Markov property relative to* $(\mathscr{G}_{t+})$, *and* $U^\alpha f$ *is nearly optional relative to* $(\xi_t, \mathscr{G}_{t+})$ *for every* $\alpha > 0$ *and every* $f \in b\mathscr{C}_u(E)$;

(4) $(\xi_t)_{t \geq 0}$ *satisfies the strong Markov property relative to* $(\mathscr{G}_{t+})$, *and* $P_s f$ *is nearly optional relative to* $(\xi_t, \mathscr{G}_{t+})$ *for every* $s \geq 0$ *and every* $f \in b\mathscr{C}_u(E)$;

(5) $\{t \mapsto P_s f(\xi_t)$ *is not right continuous*$\} \in \mathscr{N}(\mathscr{G})$ *for every* $s \geq 0$ *and every* $f \in b\mathscr{C}_u(E)$;

(6) $P_s f$ *is nearly optional relative to* $(\xi_t, \mathscr{G}_{t+})$ *for every* $s \geq 0$ *and every* $f \in b\mathscr{C}_u(E)$, *and*

$$\mathbf{P}_\mu\{f(\xi_t)1_{\{T<t\}}|\mathscr{G}_{T+}^\mu\} = P_{t-T}f(\xi_T)1_{\{T<t\}}, \quad t \geq 0, f \in b\mathscr{C}_u(E),$$

for every optional time T *over* $(\mathscr{G}_{t+})$ *and every initial law* μ *on* E;

(7) $\{s \mapsto P_{t-s}f(\xi_s)1_{[0,t)}(s)$ *is not right continuous*$\} \in \mathscr{N}(\mathscr{G})$ *for every* $t \geq 0$ *and every* $f \in b\mathscr{C}_u(E)$;

(8) $\{s \mapsto P_{t-s}f(\xi_s)1_{[0,t)}(s)$ *is not right continuous*$\} \in \mathscr{N}(\mathscr{G})$ *for every* $t \geq 0$ *and every* $f \in b\mathscr{B}^u(E)$.

Corollary A.17 (Sharpe, 1988, p.36) *Let* $(\mathscr{F}^\bullet, \mathscr{F}_t^\bullet)$ *be the natural* σ-*algebras of* ξ *generated by* $\{\xi_t : t \geq 0\}$ *and let* $(\mathscr{F}, \mathscr{F}_t)$ *be their augmentations. If* ξ *satisfies one of the conditions in Theorem A.16 relative to* $(\mathscr{F}_t)$, *then* $(\mathscr{F}_t^\mu)$ *and* $(\mathscr{F}_t)$ *are right continuous.*

The properties in Theorem A.16 depend not only on the transition semigroup $(P_t)_{t \geq 0}$, but also on the particular realization ξ. In particular, when $(P_t)_{t \geq 0}$ is a Borel semigroup, for every $\alpha > 0$ and every $f \in \mathscr{C}_u(E)$ the function $U^\alpha f$ is nearly optional relative to $(\xi_t, \mathscr{G}_{t+})$, so the properties hold if and only if $(\xi_t)_{t \geq 0}$ satisfies the strong Markov property relative to $(\mathscr{G}_{t+})_{t \geq 0}$.

Definition A.18 (Sharpe, 1988, p.38) *The system* $\xi = (\Omega, \mathscr{G}, \mathscr{G}_t, \xi_t, \theta_t, \mathbf{P}_x)$ *is called a right Markov process or simply a right process with transition semigroup* $(P_t)_{t \geq 0}$ *provided:*

(1) ξ *is a right continuous realization of* $(P_t)_{t \geq 0}$;

(2) ξ *satisfies the conditions in Theorem A.16;*

(3) $(\mathscr{G}, \mathscr{G}_t)_{t \geq 0}$ *are augmented and* $(\mathscr{G}_t)_{t \geq 0}$ *is right continuous.*

We call ξ *a Borel right process if it is a right process with Borel transition semigroup. A Markov transition semigroup* $(P_t)_{t \geq 0}$ *is called a right transition semigroup if it is the transition semigroup of a right process.*

Proposition A.19 (Sharpe, 1988, p.39) *The minimum of two* α-*excessive functions of a right semigroup is also* α-*excessive.*

The *fine topology* of a right process ξ is the smallest topology on E rendering continuous all functions in $\cup_{\alpha \geq 0} \mathscr{S}^{\alpha}$ as maps from E to $[0, \infty]$; see Sharpe (1988, p.53 and p.232). A function $f \in b\mathscr{B}^0(E)$ is finely continuous relative to ξ if and only if $t \mapsto f(\xi_t)$ is a.s. right continuous on $[0, \infty)$. More generally, we have:

Theorem A.20 (Sharpe, 1988, p.53 and p.55) *Let $f \in \mathscr{B}^u(E)$. If $t \mapsto f(\xi_t)$ is a.s. right continuous at $t = 0$, then f is finely continuous relative to ξ. Conversely, if f is finely continuous and nearly optional relative to ξ, then $t \mapsto f(\xi_t)$ is a.s. right continuous on $[0, \infty)$.*

A right process ξ is called a *Hunt process* if it is *quasi-left continuous*, that is, for every increasing sequence of stopping times $\{T_n\}$ with limit T we have $\xi_{T_n} \to \xi_T$ a.s. on $\{T < \infty\}$. If ξ is a Hunt process, then $t \mapsto \xi_t$ is a.s. càdlàg on $[0, \infty)$; see Sharpe (1988, p.221).

Let us consider two Radon topological spaces E and F. Suppose that $\xi = (\Omega, \mathscr{G}, \mathscr{G}_t, \xi_t, \theta_t, \mathbf{P}_x)$ is a right process in E with transition semigroup $(P_t)_{t \geq 0}$ and ψ is a map of E to F. In addition, we assume:

(1) ψ is surjective and measurable relative to the σ-algebras $\mathscr{B}^u(E)$ and $\mathscr{B}^u(F)$;
(2) for every $t \geq 0$ and every $f \in \mathscr{B}^u(F)$ there exists a function $Q_t f \in \mathscr{B}^u(F)$ so that $P_t(f \circ \psi) = (Q_t f) \circ \psi$;
(3) the path $t \mapsto X_t := \psi(\xi_t)$ is a.s. right continuous in F.

Under the above conditions, the operator $f \mapsto Q_t f$ determines a probability kernel on $(F, \mathscr{B}^u(F))$ and $(Q_t)_{t \geq 0}$ form a Markov transition semigroup. Let $\Omega_1 = \{\omega \in \Omega : t \mapsto X_t(\omega) \text{ is right continuous}\}$. The above property (3) implies $\mathbf{P}_x(\Omega_1) = 1$ for every $x \in E$, so we can replace Ω by Ω_1 in the definition of ξ. Let $(\mathscr{F}^u, \mathscr{F}_t^u)$ be the $\mathscr{B}^u(F)$-natural σ-algebras of $\{X_t : t \geq 0\}$ on Ω_1. A simple calculation shows that $\mathbf{P}_{x_1}$ and $\mathbf{P}_{x_2}$ coincide on $\mathscr{F}^u$ if $\psi(x_1) = \psi(x_2) = x$. We denote their common restriction on $\mathscr{F}^u$ by $\mathbf{Q}_x$. Let $(\mathscr{F}, \mathscr{F}_t)$ be the augmentations of $(\mathscr{F}^u, \mathscr{F}_t^u)$ relative to the family of probability measures $\{\mathbf{Q}_x : x \in F\}$.

Theorem A.21 (Sharpe, 1988, p.75) *The system $X = (\Omega_1, \mathscr{F}, \mathscr{F}_t, X_t, \theta_t, \mathbf{Q}_x)$ is a right process in F with transition semigroup $(Q_t)_{t \geq 0}$.*

A general transition semigroup $(P_t)_{t \geq 0}$ on $(E, \mathscr{B}^{\bullet}(E))$ may be extended to a conservative transition semigroup on a larger space. Simply take an abstract point $\partial \notin E$ and let $\bar{E} = E \cup \{\partial\}$ be the Radon topological space obtained by adjoining ∂ to E as an isolated point. Let $\mathscr{B}^{\bullet}(\bar{E}) = \sigma(\mathscr{B}^{\bullet}(E) \cup \{\partial\})$ and define $(\tilde{P}_t)_{t \geq 0}$ on $(\bar{E}, \mathscr{B}^{\bullet}(\bar{E}))$ by

$$\tilde{P}_t(x, B) = \begin{cases} P_t(x, B) & \text{if } x \in E \text{ and } B \in \mathscr{B}^{\bullet}(E), \\ 1 - P_t(x, E) & \text{if } x \in E \text{ and } B = \{\partial\}, \\ 1_B(\partial) & \text{if } x = \partial. \end{cases} \qquad (A.20)$$

It is trivial to see that $(\tilde{P}_t)_{t \geq 0}$ is a conservative transition semigroup on $(\bar{E}, \mathscr{B}^{\bullet}(\bar{E}))$. We call $(P_t)_{t \geq 0}$ a *right transition semigroup* if $(\tilde{P}_t)_{t \geq 0}$ is a right semigroup in the sense of Definition A.18. In this case, suppose that $\tilde{\xi} = (\Omega, \mathscr{G}, \mathscr{G}_t, \xi_t, \theta_t, \tilde{\mathbf{P}}_x)$ is a right process on $\tilde{E}$ realizing $(\tilde{P}_t)_{t \geq 0}$. It is obvious from (A.20) that ∂ is a trap for the process. That is, $\tilde{\mathbf{P}}_\partial \{\xi_t = \partial$ for all $t \geq 0\} = 1$. Let $\zeta = \inf\{t \geq 0 : \xi_t = \partial\}$. By the strong Markov property, we have $\tilde{\mathbf{P}}_x \{\xi_t = \partial$ for all $t \geq \zeta\} = 1$ and all $x \in \tilde{E}$. In many respects, the process $\tilde{\xi}$ is interesting only when it is in E. Indeed, if $(P_t)_{t \geq 0}$ is the object of interest, the adjunction of ∂ is quite artificial. In this situation, one may simplify the notation by making the convention that every function f on E is automatically extended to $\tilde{E}$ by setting $f(\partial) = 0$. Then $P_t f$ means exactly the same thing as $\tilde{P}_t f$. Let $\mathbf{P}_x = \tilde{\mathbf{P}}_x$ for $x \in E$. The system $\xi = (\Omega, \mathscr{G}, \mathscr{G}_t, \xi_t, \theta_t, \mathbf{P}_x)$ is called a *right process* on E with lifetime ζ and transition semigroup $(P_t)_{t \geq 0}$. In the special case $\mathbf{P}_x \{\zeta = \infty\} = 1$ for all $x \in E$, the process ξ is said to be *conservative*. We call ξ a *Hunt process* if $\tilde{\xi}$ is a Hunt process in $\tilde{E}$.

Theorem A.22 (Fitzsimmons, 1988, p.349 and p.350) *Suppose that ξ is a Borel right process with bounded potential operator U. Let $M(E)$ denote the space of finite Borel measures on E endowed with the topology of weak convergence. Let $f \in \mathrm{b}\mathscr{B}^0(E)$. Then we have:*

(1) *f is finely continuous relative to ξ if and only if $\lim_{n \to \infty} \nu_n(f) = \nu(f)$ for all $\nu_n, \nu \in M(E)$ satisfying $\uparrow \lim_{n \to \infty} \nu_n U = \nu U$;*
(2) *$t \mapsto f(\xi_t)$ has left limits on $(0, \infty)$ a.s. if and only if $\lim_{n \to \infty} \nu_n(f)$ exists for all $\nu_n \in M(E)$ such that $\downarrow \lim_{n \to \infty} \nu_n U$ exists in $M(E)$;*
(3) *if $t \mapsto f(\xi_t)$ is quasi-left continuous, then $\lim_{n \to \infty} \nu_n(f) = \nu(f)$ for all $\nu_n, \nu \in M(E)$ satisfying $\downarrow \lim_{n \to \infty} \nu_n U = \nu U$.*

A.4 Ray–Knight Completion

We first assume that E is a compact metrizable space. A *Ray resolvent* $(U^\alpha)_{\alpha > 0}$ on $(E, \mathscr{B}(E))$ is by definition a Markov resolvent such that $U^\alpha \mathscr{C}(E) \subset \mathscr{C}(E)$ and $\cup_{\alpha > 0} \mathscr{S}^\alpha \cap \mathscr{C}(E)$ separates the points of E, where $\mathscr{S}^\alpha$ is the class of α-supermedian functions for $(U^\alpha)_{\alpha > 0}$. It was proved in Getoor (1975, p.9) that to every Ray resolvent $(U^\alpha)_{\alpha > 0}$ there corresponds a unique Markov transition semigroup $(P_t)_{t \geq 0}$ on $(E, \mathscr{B}(E))$ such that $t \mapsto P_t f(x)$ is right continuous for $x \in E$ and $f \in \mathscr{C}(E)$, and

$$U^\alpha f(x) = \int_0^\infty e^{-\alpha t} P_t f(x) dt, \quad \alpha > 0, x \in E, f \in \mathscr{C}(E). \quad (A.21)$$

The Markov transition semigroup $(P_t)_{t \geq 0}$ defined by (A.21) is called the *Ray semigroup* associated with $(U^\alpha)_{\alpha > 0}$. A Ray semigroup is not necessarily normal. The set of *branch points* for $(P_t)_{t \geq 0}$ is $B := \{x \in E : P_0(x, \cdot) \neq \delta_x(\cdot)\}$ and the set of *non-branch points* for $(P_t)_{t \geq 0}$ is $D := \{x \in E : P_0(x, \cdot) = \delta_x(\cdot)\} = E \setminus B$.

Proposition A.23 (Sharpe, 1988, p.44) *Let $(P_t)_{t \geq 0}$ be a Ray semigroup on E and let B and D be defined as above. Then:*

(1) *for any $\{g_n\}$ uniformly dense in $\mathscr{C}(E) \cap \mathscr{S}^1$ we have $B = \cup_n \{P_0 g_n < g_n\}$;*

(2) B is an F_σ set in E and hence $B \in \mathscr{B}(E)$;

(3) for any $t \geq 0$ and $x \in E$ the probability measure $P_t(x, \cdot)$ is carried by D.

Theorem A.24 (Sharpe, 1988, p.46) *The restriction of* $(P_t)_{t \geq 0}$ *to* D *is a right semigroup which may be realized on the space* Ω *of right continuous maps of* $[0, \infty)$ *into* D *having left limits in* E.

Theorem A.25 (Sharpe, 1988, p.49) *Suppose that* E *is a locally compact, noncompact separable metric space and* $(U^\alpha)_{\alpha > 0}$ *is a Markov resolvent on* $(E, \mathscr{B}(E))$ *such that* $U^\alpha(C_0(E)) \subset C_0(E)$ *for all* $\alpha > 0$ *and* $\alpha U^\alpha f \to f$ *pointwise as* $\alpha \to \infty$ *for all* $f \in C_0(E)$. *Then there is a right process* ξ *with state space* E *having resolvent* $(U^\alpha)_{\alpha > 0}$ *such that:*

(1) ξ *is quasi-left continuous;*

(2) *for all* $t > 0$ *the set* $\{\xi_s(\omega) : 0 \leq s \leq t\}$ *a.s. has compact closure in* E;

(3) *a.s. the left limit* $\xi_{t-} := \lim_{s \to t-} \xi_s$ *exists in* E *for all* $t > 0$.

The conditions in Theorem A.25 are satisfied if $(U^\alpha)_{\alpha > 0}$ is the resolvent generated by a Feller semigroup. Then a Feller semigroup has a Hunt realization.

Now suppose we are given a general Radon topological space E with a totally bounded metric d for its topology. Let $(U^\alpha)_{\alpha > 0}$ be a Markov resolvent on $(E, \mathscr{B}^u(E))$ satisfying

$$\mathscr{B}(E) \subset \sigma(\{U^\alpha f : \alpha > 0, f \in \mathscr{C}_u(E, d)\}). \qquad (A.22)$$

A set $\mathscr{Y} \subset \mathrm{pb}\mathscr{B}^u(E)$ is called a *rational cone* if it is closed under positive rational linear combinations. For $\mathscr{D} \subset \mathrm{pb}\mathscr{B}^u(E)$, we denote by $q(\mathscr{D})$ the rational cone generated by $\mathscr{D}$, that is, the smallest rational cone containing $\mathscr{D}$. For a rational cone $\mathscr{Y} \subset \mathrm{pb}\mathscr{B}^u(E)$, set $\lambda(\mathscr{Y}) = \{f_1 \wedge \cdots \wedge f_n : n \geq 1, f_i \in \mathscr{Y}\}$ and $u(\mathscr{Y}) = \{U^{\alpha_1} f_1 + \cdots + U^{\alpha_n} f_n : n \geq 1, f_i \in \mathscr{Y}$ and strictly positive rationals $\alpha_i\}$. It is obvious that $u(\mathscr{Y})$ is a rational cone contained in the cone $\cup_{\alpha > 0} \mathrm{b}\mathscr{S}^\alpha$. That $\lambda(\mathscr{Y})$ is also a rational cone comes from the trivial identities $(\wedge_i a_i) + b = \wedge_i (a_i + b)$ and $(\wedge_i a_i) + (\wedge_j b_j) = \wedge_{i,j}(a_i + b_j)$.

For a given function class $\mathscr{D} \subset \mathrm{pb}\mathscr{B}^u(E)$, we set $\mathscr{R}_0 = u(q(\mathscr{D}))$ and set $\mathscr{R}_n = \lambda(\mathscr{R}_{n-1} + u(\mathscr{R}_{n-1}))$ for $n \geq 1$ inductively, where $\mathscr{R}_{n-1} + u(\mathscr{R}_{n-1}) = \{f + g : f \in \mathscr{R}_{n-1}$ and $g \in u(\mathscr{R}_{n-1})\}$. The set $\mathscr{R}(\mathscr{D}) := \cup_{n \geq 0} \mathscr{R}_n$ is called the *rational Ray cone* generated by $(U^\alpha)_{\alpha > 0}$ and $\mathscr{D}$; see Getoor (1975, p.58) and Sharpe (1988, p.90). The rational Ray cone $\mathscr{R} = \mathscr{R}(\mathscr{D})$ generated by $(U^\alpha)_{\alpha > 0}$ and $\mathscr{D} \subset \mathrm{pb}\mathscr{B}^u(E)$ is the smallest rational cone contained in $\mathrm{pb}\mathscr{B}^u(E)$ such that:

(1) $U^\alpha(\mathscr{R}) \subset \mathscr{R}$ for all rationals $\alpha > 0$;

(2) $f, g \in \mathscr{R}$ implies $f \wedge g \in \mathscr{R}$;

(3) $\mathscr{R}$ contains $u(q(\mathscr{D}))$.

Clearly, for each $f \in \mathscr{R}(\mathscr{D})$ there is a constant $\beta = \beta(f) > 0$ so that f is an β-supermedian function for $(U^\alpha)_{\alpha > 0}$. Furthermore, if $(U^\alpha)_{\alpha > 0}$ is the resolvent associated with a conservative right semigroup $(P_t)_{t \geq 0}$ on (E, d), for each $f \in \mathscr{R}(\mathscr{D})$ there exists $\beta = \beta(f) > 0$ so that f is β-excessive for $(P_t)_{t \geq 0}$.

Proposition A.26 (Sharpe, 1988, p.90) *If $\mathscr{D}$ is a countable uniformly dense subset of $p\mathscr{C}_u(E, d)$ and contains the constant function 1_E, then the rational Ray cone $\mathscr{R} = \mathscr{R}(\mathscr{D})$ is countable, contains the positive rational constant functions, and separates the points of E.*

In the remainder of this section, we assume $\mathscr{D} \subset p\mathscr{C}_u(E, d)$ satisfies the conditions of Proposition A.26. Recall that $\|\cdot\|$ denotes the supremum norm. We give the rational Ray cone $\mathscr{R} = \mathscr{R}(\mathscr{D})$ an enumeration $\{g_0, g_1, g_2, \dots\}$ with $g_0 = 0$. Clearly,

$$\rho(x, y) = \sum_{n=1}^{\infty} \frac{|g_n(x) - g_n(y)|}{2^n \|g_n\|}, \qquad x, y \in E, \tag{A.23}$$

defines a metric ρ on E, and each g_n is ρ-uniformly continuous. Let $(\bar{E}, \bar{\rho})$ denote the completion of (E, ρ). Observe that the map $x \mapsto (g_n(x))_{n \geq 1}$ of E into $K := \prod_{n=1}^{\infty}[0, \|g_n\|]$ with the metric q defined by

$$q(a, b) = \sum_{n=1}^{\infty} \frac{|a_n - b_n|}{2^n \|g_n\|}, \qquad a, b \in K,$$

is an isometry. It follows that the completion $(\bar{E}, \bar{\rho})$ is compact. The topology on E induced by the metric ρ is called the *Ray topology* of $(U^\alpha)_{\alpha > 0}$.

Proposition A.27 (Sharpe, 1988, p.91) *Each function $f \in \mathscr{C}_u(E, \rho)$ extends to a unique $\bar{f} \in \mathscr{C}_u(\bar{E}, \bar{\rho})$. For each $\alpha > 0$, we have $U^\alpha(\mathscr{C}_u(E, \rho)) \subset \mathscr{C}_u(E, \rho)$ and $U^\alpha(\mathscr{C}_u(E, d)) \subset \mathscr{C}_u(E, \rho)$, and $\mathscr{C}_u(E, \rho)$ is the uniform closure of $\mathscr{R} - \mathscr{R} := \{f - g : f, g \in \mathscr{R}\}$.*

Proposition A.28 (Sharpe, 1988, p.91) *If $U^\alpha(\mathscr{C}_u(E, d)) \subset \mathscr{C}_u(E, d)$ for all $\alpha > 0$, then the Ray topology is coarser than the original topology.*

Proposition A.29 (Sharpe, 1988, p.92) *Let $\mathscr{B}^r(E)$ denote the σ-algebra on E generated by the Ray topology. Then $\mathscr{B}(E) \subset \mathscr{B}^r(E) \subset \mathscr{B}^u(E)$ and $U^\alpha \mathscr{B}^r(E) \subset \mathscr{B}^r(E)$ for every $\alpha > 0$.*

Proposition A.30 (Sharpe, 1988, pp.92–93) *We have $E \in \mathscr{B}^u(\bar{E})$, so (E, ρ) is a Radon space. If (E, d) is Lusin and if $(U^\alpha)_{\alpha > 0}$ maps $b\mathscr{B}(E)$ into itself, then $\mathscr{B}(E) = \mathscr{B}^r(E)$ and $E \in \mathscr{B}(\bar{E})$, so (E, ρ) is a Lusin space.*

For every $\bar{f} \in \mathscr{C}_u(\bar{E}, \bar{\rho})$ we clearly have $f := \bar{f}|_E \in \mathscr{C}_u(E, \rho)$. Then $U^\alpha f \in \mathscr{C}_u(E, \rho)$ by Proposition A.27 and so $U^\alpha f$ extends continuously to some $(U^\alpha f)^- \in \mathscr{C}_u(\bar{E}, \bar{\rho})$. Define the operators $(\bar{U}^\alpha)_{\alpha > 0}$ on $\mathscr{C}_u(\bar{E}, \bar{\rho})$ by

$$\bar{U}^\alpha \bar{f} = (U^\alpha f)^-, \qquad \alpha > 0, \bar{f} \in \mathscr{C}_u(\bar{E}, \bar{\rho}).$$

Theorem A.31 (Sharpe, 1988, p.93) *For $\alpha > 0$ and $x \in E$ the measure $\bar{U}^\alpha(x, \cdot)$ is carried by $E \in \mathscr{B}^u(\bar{E})$ and its restriction to E is $U^\alpha(x, \cdot)$. Moreover, the family $(\bar{U}^\alpha)_{\alpha > 0}$ is a Ray resolvent on the space $\bar{E}$.*

We call $(\bar{U}^\alpha)_{\alpha>0}$ the *Ray extension* of $(U^\alpha)_{\alpha>0}$. The space $(\bar{E}, \bar{\rho})$ constructed above is called the *Ray–Knight completion* of (E, ρ) with respect to $(U^\alpha)_{\alpha>0}$. It depends not only on E, d and $(U^\alpha)_{\alpha>0}$ but also on the choice of the family $\mathscr{D} \subset \mathrm{p}\mathscr{C}_u(E, d)$. If $(U^\alpha)_{\alpha>0}$ is the resolvent associated with a conservative right semigroup $(P_t)_{t\geq 0}$, we also call $(\bar{E}, \bar{\rho})$ the *Ray–Knight completion* of (E, ρ) with respect to $(P_t)_{t\geq 0}$. In this case, the Ray semigroup $(\bar{P}_t)_{t\geq 0}$ associated with $(\bar{U}^\alpha)_{\alpha\geq 0}$ is called the *Ray extension* of $(P_t)_{t\geq 0}$.

Theorem A.32 (Sharpe, 1988, p.94) *Let $(P_t)_{t\geq 0}$ be a conservative right semigroup on E. Then there is a realization $\xi = (\Omega, \mathscr{G}, \mathscr{G}_t, \xi_t, \theta_t, \mathbf{P}_x)$ of $(P_t)_{t\geq 0}$ which is a right process in both (E, d) and (E, ρ) and the left limit $\xi_{t-} := \lim_{s\to t-} \xi_s$ taken in the Ray topology exists in $\bar{E}$ for all $t > 0$.*

Theorem A.33 *Suppose that $(P_t)_{t\geq 0}$ is a conservative Borel right semigroup on a Lusin topological space E. Then every right continuous realization of $(P_t)_{t\geq 0}$ with the augmented natural σ-algebras is a right process. In particular, the semigroup can be realized canonically on the space of right continuous paths from $[0, \infty)$ to E.*

Proof. Let ξ be a right process with semigroup $(P_t)_{t\geq 0}$. Then each $f \in \mathscr{S}^\alpha$ is a nearly Borel function of ξ relative to the Ray topology; see Sharpe (1988, p.95). By Proposition A.30, we have $\mathscr{B}(E) = \mathscr{B}^r(E)$, so each $f \in \mathscr{S}^\alpha$ is nearly Borel in the original topology. Then the result follows by Sharpe (1988, p.98). $\square$

If $(P_t)_{t\geq 0}$ is a right semigroup on (E, d) not necessarily conservative, the associated resolvent $(U^\alpha)_{\alpha>0}$ is not necessarily Markov. In this case, we let $\tilde{E} = E \cup \{\partial\}$ for an abstract point $\partial \notin E$. Let $(\tilde{E}, \tilde{d})$ be a topological extension of (E, d) with ∂ being an isolated point and let $(\tilde{P}_t)_{t\geq 0}$ denote the conservative extension of $(P_t)_{t\geq 0}$ on $\tilde{E}$ with ∂ being a cemetery. Let $(\tilde{U}^\alpha)_{\alpha>0}$ denote the resolvent associated with $(\tilde{P}_t)_{t\geq 0}$. Let $\tilde{\mathscr{R}}$ be the countable Ray cone for $(\tilde{U}^\alpha)_{\alpha>0}$ constructed from $\tilde{\mathscr{D}}$, which is a countable uniformly dense subset of $\mathrm{p}\mathscr{C}_u(\tilde{E}, \tilde{d})$ and contains the constant function $1_{\tilde{E}}$. Let $(\bar{E}, \bar{\rho}, \bar{U}^\alpha, \bar{P}_t)$ be the corresponding Ray–Knight completion of $(\tilde{E}, \tilde{d}, \tilde{U}^\alpha, \tilde{P}_t)$.

Proposition A.34 *In the situation described above, if there are constants $\alpha > 0$ and $\varepsilon > 0$ such that $U^\alpha 1_E(x) \geq \varepsilon$ for all $x \in E$, then ∂ is an isolated point of $\bar{E}$.*

Proof. Since $\tilde{\mathscr{D}}$ is uniformly dense in $\mathrm{p}\mathscr{C}_u(\tilde{E}, \tilde{d})$, there is a function $\tilde{g} \in \tilde{\mathscr{D}}$ such that $\tilde{g}(\partial) < \alpha\varepsilon/2$ and $\tilde{g}(x) \geq 1$ for every $x \in E$. Fix a rational $\beta \in (\alpha/2, \alpha)$. Then $\tilde{f} := \tilde{U}^\beta \tilde{g} \in \tilde{\mathscr{R}}$ by the construction of $\tilde{\mathscr{R}}$. Since ∂ is a cemetery for $(\tilde{P}_t)_{t\geq 0}$, we have $\tilde{f}(\partial) = \beta^{-1}\tilde{g}(\partial) < \varepsilon$. However, for every $x \in E$ we have $\tilde{f}(x) = \tilde{U}^\beta \tilde{g}(x) \geq U^\alpha 1_E(x) \geq \varepsilon$. Let $\bar{f}$ be the unique continuous extension of $\tilde{f}$ to $\bar{E}$. It follows that $\bar{f}(x) \geq \varepsilon$ for every $x \in \bar{E} \setminus \{\partial\}$. Then the point ∂ must be isolated in $\bar{E}$. $\square$

By Proposition A.34, if $(P_t)_{t\geq 0}$ is a conservative right semigroup, then ∂ is an isolated point of $\bar{E}$. In that case, the topology of E inherited from $\bar{E}$ coincides with its Ray topology defined directly by $(P_t)_{t\geq 0}$. In the general case, we also call the inherited topology of E the *Ray topology* of $(P_t)_{t\geq 0}$.

A.5 Entrance Space and Entrance Laws

Let $\xi = (\Omega, \mathscr{G}, \mathscr{G}_t, \xi_t, \theta_t, \mathbf{P}_x)$ be a right process on the Radon topological space E with transition semigroup $(P_t)_{t\geq 0}$ and resolvent $(U^\alpha)_{\alpha>0}$. We first assume $(P_t)_{t\geq 0}$ is conservative. Let d be a metric for the topology of E such that the d-completion of E is compact. Let $(\bar{E}, \bar{\rho}, \bar{U}^\alpha)$ be a Ray–Knight completion of (E, d, U^α) and let $(\bar{P}_t)_{t\geq 0}$ be the Ray extension of $(P_t)_{t\geq 0}$. Set $E_R = \{x \in \bar{E} : \bar{U}^1(x, \cdot)$ is carried by $E\}$, which is called the *Ray space* for ξ or $(P_t)_{t\geq 0}$. It was proved in Sharpe (1988, p.191) that $E_R \in \mathscr{B}^u(\bar{E})$ is a Radon topological space and $E \subset E_R$. By the resolvent equation we have $E_R = \{x \in \bar{E} : \bar{U}^\alpha(x, \cdot)$ is carried by $E\}$ for each $\alpha > 0$.

Theorem A.35 (Sharpe, 1988, p.191) *Let $(\bar{E}_1, \bar{\rho}_1, \bar{U}_1^\alpha)$ and $(\bar{E}_2, \bar{\rho}_2, \bar{U}_2^\alpha)$ be Ray–Knight completions of (E, d_1, U^α) and (E, d_2, U^α) respectively, where d_1 and d_2 are totally bounded metrics for the original topology of E. Then the corresponding Ray spaces E_R^1 and E_R^2 are homeomorphic under a mapping $\psi : E_R^1 \to E_R^2$ satisfying $\bar{U}_1^\alpha(x, B) = \bar{U}_2^\alpha(\psi(x), B)$ for all $\alpha > 0$ and $B \in \mathscr{B}(E)$.*

Therefore, the Ray space E_R together with the resolvent $(\bar{U}^\alpha)_{\alpha>0}$ restricted to E_R is uniquely determined, up to homeomorphism, by the original topology on E and $(U^\alpha)_{\alpha>0}$. This makes the Ray space a natural object. Let D denote the set of non-branch points of $(\bar{P}_t)_{t\geq 0}$ on $\bar{E}$, and let $E_D = D \cap E_R = \{x \in E_R : \bar{P}_0(x, \cdot) = \delta_x(\cdot)\}$ which is called the *entrance space* for $(P_t)_{t\geq 0}$. Since $D \in \mathscr{B}(\bar{E})$, we have $E_D \in \mathscr{B}(E_R, \rho)$.

Proposition A.36 (Sharpe, 1988, pp.192–193) *We have:*

(1) *For $x \in E_R$ and $t > 0$, $\bar{P}_t(x, \cdot)$ is carried by E.*
(2) *For $x \in E_R$, $\bar{P}_0(x, \cdot)$ is carried by E_D.*
(3) *For $x \in B$, $\bar{P}_0(x, \cdot)$ is not concentrated at any point of $\bar{E}$.*

The restriction $(Q_t)_{t\geq 0}$ of $(\bar{P}_t)_{t\geq 0}$ to (E_D, ρ) is a right semigroup, and $E_D \setminus E$ is quasi-polar for any realization Y of $(Q_t)_{t\geq 0}$ as a right process, that is, for every initial law μ on (E_D, ρ) the $\mathbf{Q}_\mu$-outer measure of $\{\omega : Y_t(\omega) \in E$ for all $t > 0\}$ is equal to one; see Sharpe (1988, p.193). The following theorem gives a complete characterization of probability entrance laws for a conservative right semigroup.

Theorem A.37 (Sharpe, 1988, p.196) *For every probability entrance law $(\eta_t)_{t>0}$ for $(P_t)_{t\geq 0}$ on E, there is a unique probability measure η_0 on $\mathscr{B}^u(E_D, \rho)$ such that $\eta_t = \eta_0 \bar{P}_t$ for every $t > 0$.*

Corollary A.38 *If E is a locally compact separable metric space and $(P_t)_{t\geq 0}$ is a Feller semigroup on E, then all probability entrance laws for $(P_t)_{t\geq 0}$ are closable.*

Proof. We assume E is not compact, for otherwise the proof is easier. Let $\bar{E} = E \cup \{\partial\}$ be a one-point compactification of E. Then $\bar{E}$ is compact and separable, so it is metrizable. Let $\bar{d}$ be a metric on $\bar{E}$ compatible with its topology and let d be the restriction of $\bar{d}$ to E. It is easy to see that the Ray–Knight completion of E given by d and $(P_t)_{t\geq 0}$ coincides with $\bar{E}$ and the entrance space is just E. Then the result follows from Theorem A.37. $\square$

In the remainder of this section, let E be a Lusin topological space and consider a Borel right semigroup $(P_t)_{t \geq 0}$ on E which is not necessarily conservative. Let $E^\partial = E \cup \{\partial\}$ be a topological extension of E with ∂ being an isolated point. Let Ω denote the space of right continuous paths w from $\mathbb{R}$ to E^∂ such that there are constants $\alpha(w) < \beta(w)$ so that $w_t \in E$ for $t \in (\alpha(w), \beta(w))$ and $w_t = \partial$ for $t \in (\alpha(w), \beta(w))^c$. Let $(\mathscr{F}^0, \mathscr{F}_t^0)_{t \in \mathbb{R}}$ be the natural σ-algebras on Ω generated by the coordinate process. For $r \in [-\infty, \infty)$ let Ω_r denote the space of right continuous paths $w \in \Omega$ satisfying $\alpha(w) = r$.

Theorem A.39 (Getoor and Glover, 1987, pp.57–58) *Let $(\eta_t)_{t \in \mathbb{R}}$ be an entrance rule for $(P_t)_{t \geq 0}$. Then there exists a Radon measure $\rho(\mathrm{d}s)$ on $\mathbb{R}$ and an entrance law $(\nu_t^r)_{t > r}$ for every $r \in [-\infty, \infty)$ so that*

$$\eta_t = \nu_t^{-\infty} + \int_{-\infty}^{\infty} \nu_t^s \rho(\mathrm{d}s), \qquad t \in \mathbb{R}, \tag{A.24}$$

where $\nu_t^s = 0$ for $t \leq s$ by convention.

Theorem A.40 (Getoor and Glover, 1987, p.63) *To each entrance rule $(\eta_t)_{t \in \mathbb{R}}$ for $(P_t)_{t \geq 0}$ there corresponds a unique σ-finite measure $\mathbf{Q}_\eta$ on $(\Omega, \mathscr{F}^0)$ so that*

$$\begin{aligned}
&\mathbf{Q}_\eta \{w_{t_1} \in \mathrm{d}x_1, w_{t_2} \in \mathrm{d}x_2, \ldots, w_{t_n} \in \mathrm{d}x_n\} \\
&= \eta_{t_1}(\mathrm{d}x_1) P_{t_2 - t_1}(x_1, \mathrm{d}x_2) \cdots P_{t_n - t_{n-1}}(x_{n-1}, \mathrm{d}x_n)
\end{aligned} \tag{A.25}$$

for all $\{t_1 < \cdots < t_n\} \subset \mathbb{R}$ and $\{x_1, \ldots, x_n\} \subset E$. Moreover, if $(\eta_t)_{t \in \mathbb{R}}$ is an entrance law at $r \in \mathbb{R}$, then $\mathbf{Q}_\eta$ is carried by Ω_r.

The measure $\mathbf{Q}_\eta$ defined by (A.25) is called the *Kuznetsov measure* corresponding to the entrance rule $(\eta_t)_{t \in \mathbb{R}}$; see Getoor and Glover (1987). This property roughly means that $\{w_t : t \in \mathbb{R}\}$ is a Markov process with transition semigroup $(P_t)_{t \geq 0}$ and one-dimensional distributions $(\eta_t)_{t \in \mathbb{R}}$. For the entrance rule $(\eta_t)_{t \in \mathbb{R}}$ given by (A.24), the measure $\mathbf{Q}_\eta$ can be represented as

$$\mathbf{Q}_\eta(\mathrm{d}w) = \mathbf{Q}^{-\infty}(\mathrm{d}w) + \int_{-\infty}^{\infty} \mathbf{Q}^s(\mathrm{d}w) \rho(\mathrm{d}s), \tag{A.26}$$

where $\mathbf{Q}^s$ is the Kuznetsov measure corresponding to the entrance law $(\nu_t^s)_{t > s}$; see Getoor and Glover (1987, p.66). The theory of Kuznetsov measures was developed systematically in Dellacherie et al. (1992) and Getoor (1990).

Example A.1 Let $(P_t)_{t \geq 0}$ be the transition semigroup of the absorbing-barrier Brownian motion in $(0, \infty)$. For any $t > 0$ the kernel $P_t(x, \mathrm{d}y)$ has density

$$p_t(x, y) = g_t(x - y) - g_t(x + y), \quad x, y > 0, \tag{A.27}$$

where

$$g_t(z) = \frac{1}{\sqrt{2\pi t}} \exp\{-z^2/2t\}, \quad t > 0, z \in \mathbb{R}. \tag{A.28}$$

We can define an entrance law $(\kappa_t)_{t>0}$ for $(P_t)_{t\geq 0}$ by

$$\kappa_t(f) = \frac{2}{t}\int_0^\infty yg_t(y)f(y)\mathrm{d}y = \frac{\mathrm{d}}{\mathrm{d}x}P_tf(0+), \quad f \in b\mathscr{B}(0,\infty). \quad \text{(A.29)}$$

The corresponding Kuznetsov measure $n(\mathrm{d}w)$ is called *Itô's excursion law*, which is carried by the set of positive continuous paths $\{w_t : t > 0\}$ such that $w_{0+} = w_t = 0$ for every $t \geq \tau_0(w) := \inf\{s > 0 : w_s = 0\}$; see, e.g., Ikeda and Watanabe (1989, p.124).

A.6 Concatenations and Weak Generators

Suppose that E is a Lusin topological space and $(P_t)_{t\geq 0}$ is a Borel right semigroup on this space. We consider a right process $\xi = (\Omega, \mathscr{G}, \mathscr{G}_t, \xi_t, \theta_t, \mathbf{P}_x)$ with transition semigroup $(P_t)_{t\geq 0}$. Let $(\mathscr{F}, \mathscr{F}_t)$ be the augmentations of the $\mathscr{B}(E)$-natural σ-algebras $(\mathscr{F}^0, \mathscr{F}_t^0)$ generated by $\{\xi_t : t \geq 0\}$. A right continuous $(\mathscr{F}_t)$-adapted increasing process $\{K(t) : t \geq 0\}$ is called an *additive functional* of ξ if $K_0 = 0$ and for every bounded stopping time T we have a.s.

$$K_{T+t} = K_T + K_t \circ \theta_T, \qquad t \geq 0. \quad \text{(A.30)}$$

Clearly, an additive functional $\{K(t) : t \geq 0\}$ defines a σ-finite random measure $K(\mathrm{d}s)$ on $[0,\infty)$. For any $\beta \in b\mathscr{B}(E)$ write

$$K_t(\beta) = \int_{[0,t]} \beta(\xi_s)K(\mathrm{d}s), \qquad t \geq 0.$$

Let $b\mathscr{E}(K)$ denote the set of functions $\beta \in b\mathscr{B}(E)$ so that $t \mapsto \mathrm{e}^{-K_t(\beta)}$ is a locally bounded stochastic process. Note that $b\mathscr{E}(K) \supset \mathrm{pb}\mathscr{B}(E)$. We say an additive functional $\{K(t) : t \geq 0\}$ is *admissible* if each $\omega \mapsto K_t(\omega)$ is measurable with respect to the natural σ-algebra $\mathscr{F}^0$ and

$$k(t) := \sup_{x\in E} \mathbf{P}_x[K(t)] \to 0, \qquad t \to 0. \quad \text{(A.31)}$$

In the sequel, we assume $\{K(t) : t \geq 0\}$ is a continuous admissible additive functional of ξ. Let $b \in b\mathscr{E}(K)$ and let $\gamma(x, \mathrm{d}y)$ be a bounded Borel kernel on E. For $f \in b\mathscr{B}(E)$ we consider the linear evolution equation

$$q_t(x) = \mathbf{P}_x[\mathrm{e}^{-K_t(b)}f(\xi_t)] + \mathbf{P}_x\left[\int_0^t \mathrm{e}^{-K_s(b)}\gamma(\xi_s, q_{t-s})K(\mathrm{d}s)\right], \quad \text{(A.32)}$$

where $t \geq 0$ and $x \in E$. Recall that $\|\cdot\|$ denotes the supremum norm.

Proposition A.41 *For every* $f \in b\mathscr{B}(E)$ *there is a unique locally bounded Borel function* $(t, x) \mapsto q_t(x)$ *on* $[0, \infty) \times E$ *solving (A.32), which is given by*

$$q_t(x) = \mathbf{P}_x\big[e^{-K_t(b)}f(\xi_t)\big] + \mathbf{P}_x\bigg\{\int_0^t e^{-K_{s_1}(b)}K(ds_1)\mathbf{P}_{\mu_{s_1}}f(\xi_{t-s_1})\bigg\}$$

$$+\sum_{i=2}^\infty \mathbf{P}_x\bigg\{\int_0^t e^{-K_{s_1}(b)}K(ds_1)\mathbf{P}_{\mu_{s_1}}\bigg\{\int_0^{t-\sigma_1}e^{-K_{s_2}(b)}K(ds_2)\cdots$$

$$\mathbf{P}_{\mu_{s_{i-1}}}\bigg\{\int_0^{t-\sigma_{i-1}}e^{-K_{s_i}(b)}K(ds_i)\mathbf{P}_{\mu_{s_i}}\big[e^{-K_{t-\sigma_i}(b)}f(\xi_{t-\sigma_i})\big]\bigg\}\cdots\bigg\}\bigg\},$$

where $\sigma_i = \sum_{j=1}^i s_j$ and $\mu_s = \gamma(\xi_s, \cdot)$. Moreover, the operators $\pi_t : f \mapsto q_t$ form a locally bounded semigroup $(\pi_t)_{t\geq 0}$.

Proof. For $r \geq 0$ it is not hard to see that $(t, x) \mapsto q_t(x)$ satisfies (A.32) for $t \geq 0$ if and only if it satisfies the equation for $0 \leq t \leq r$ and $(t, x) \mapsto q_{r+t}(x)$ satisfies

$$q_{r+t}(x) = \mathbf{P}_x\big[e^{-K_t(b)}q_r(\xi_t)\big] + \mathbf{P}_x\bigg[\int_0^t e^{-K_s(b)}\gamma(\xi_s, q_{r+t-s})K(ds)\bigg] \quad (A.33)$$

for $t \geq 0$. Let $t \mapsto l(t)$ be an increasing deterministic function so that $e^{-K_t(b)} \leq l(t)$ for all $t \geq 0$. Fix a constant $\delta > 0$ so that $k(\delta)l(\delta)\|\gamma(\cdot, 1)\| < 1$. Observe that the i-th term of the series in the definition of $q_t(x)$ is bounded by $k(t)^i l(t)^i\|\gamma(\cdot, 1)\|^i\|f\|$. Then the series converges uniformly on $[0, \delta] \times E$. Since each $\omega \mapsto K_t(\omega)$ is measurable with respect to the natural σ-algebra, it is easy to see that $(t, x) \mapsto q_t(x)$ is jointly measurable and satisfies (A.32) on $[0, \delta] \times E$. By the relation of (A.32) and (A.33) we can extend $(t, x) \mapsto q_t(x)$ to a solution of (A.32) on $[0, \infty) \times E$. Moreover, the operator $f \mapsto q_t$ determines a bounded Borel kernel $\pi_t(x, dy)$ on E and $(\pi_t)_{t\geq 0}$ form a locally bounded semigroup. To show the uniqueness of the solution of (A.32), suppose that $(t, x) \mapsto v_t(x)$ is a locally bounded solution of (A.32) with $v_0(x) \equiv 0$. It is easily seen that

$$\|v_t\| \leq l(t)\|\gamma(\cdot, 1)\| \sup_{x\in E} \mathbf{P}_x\bigg[\int_0^t \|v_{t-s}\|K(ds)\bigg],$$

and hence

$$\sup_{0\leq s\leq t}\|v_s\| \leq k(t)l(t)\|\gamma(\cdot, 1)\| \sup_{0\leq s\leq t}\|v_s\|$$

for every $t \geq 0$. Then we must have $\|v_t\| = 0$ for $0 \leq t \leq \delta$. Using the above procedure and the relation of (A.32) and (A.33) successively we get $\|v_t\| = 0$ for all $t \geq 0$. Since (A.32) is a linear equation, that gives the uniqueness of the solution. $\square$

Proposition A.42 *Let $f \in b\mathscr{B}(E)$ and let $(t, x) \mapsto \pi_t f(x)$ be defined by (A.32). Then $t \mapsto \pi_t f(x)$ is right continuous pointwise on E if and only if so is $t \mapsto P_t f(x)$.*

Proof. Clearly, the second term on the right-hand side of (A.32) tends to zero as $t \to 0$. Moreover, by (A.31) we have

$$\lim_{t\to 0}\big|\mathbf{P}_x[(1 - e^{-K_t(b)})f(\xi_t)]\big| \leq \lim_{t\to 0}\|f\|\mathbf{P}_x\big[|1 - e^{-K_t(b)}|\big] = 0.$$

It follows that

$$\lim_{t\to 0} \pi_t f(x) = \lim_{t\to 0} \mathbf{P}_x\big[e^{-K_t(b)} f(\xi_t)\big] = \lim_{t\to 0} P_t f(x),$$

which means if one of the limits exists, so do the other two and the equalities hold. Then we get the result by the semigroup properties of $(P_t)_{t\geq 0}$ and $(\pi_t)_{t\geq 0}$. $\qquad\square$

Now suppose that $b(x) \geq \gamma(x,1)$ for every $x \in E$. Let $(\pi_t)_{t\geq 0}$ be defined by (A.32). Since $(P_t)_{t\geq 0}$ is not conservative in general, we can only understand $\xi = (\Omega, \mathscr{G}, \mathscr{G}_t, \xi_t, \mathbf{P}_x)$ as a right process in the extended state space $E\cup\{\partial\}$ with ∂ being an isolated cemetery. Let $\bar{E}$ be a Ray–Knight completion of $E \cup \{\partial\}$ relative to ξ. Then Proposition A.30 implies $E \in \mathscr{B}(\bar{E})$. By Theorem A.32 we have $\mathbf{P}_x\{$the left limit $\xi_{t-} := \lim_{s\to t-} \xi_s$ taken in the Ray topology exists in $\bar{E}$ for all $t > 0\} = 1$. Let $\hat{\gamma}(x, dy)$ be a sub-Markov kernel on E satisfying $\gamma(x, dy) = b(x)\hat{\gamma}(x, dy)$. We extend $\hat{\gamma}(x, dy)$ to a Markov kernel from E to $\bar{E}$ by setting $\hat{\gamma}(x, \{\partial\}) = 1-\hat{\gamma}(x, E)$. Fix $x_0 \in E$ and let $b(x) = b(x_0)$ and $\hat{\gamma}(x, \cdot) = \hat{\gamma}(x_0, \cdot)$ for $x \in \bar{E} \setminus E$. Let $\hat{\xi} = (\Omega, \mathscr{G}, \hat{\mathscr{G}}_t, \hat{\xi}_t, \hat{\mathbf{P}}_x)$ be the subprocess with lifetime ζ constructed from ξ and the strictly positive multiplicative functional $t \mapsto \exp\{-K_t(b)\}$. Then $\hat{\xi}$ is also a right process; see Sharpe (1988, p.287). Let $\tilde{\xi} = (\tilde{\Omega}, \tilde{\mathscr{G}}, \tilde{\mathscr{G}}_t, \tilde{\xi}_t, \tilde{\mathbf{P}}_x)$ be the concatenation defined from an infinite sequence of copies of $\hat{\xi}$ and the transfer kernel $\eta(\omega, dy) := \hat{\gamma}(\xi_{\zeta(\omega)-}(\omega), dy)$ as in Sharpe (1988, p.82). The intuitive idea of this concatenation is described as follows. The process $\tilde{\xi}$ evolves as ξ until time ζ, it is then revived by means of the kernel η, and evolves again as ξ and so on. It is known that $\tilde{\xi}$ is also a right process; see Sharpe (1988, p.79 and p.82). Suppose that every $f \in b\mathscr{B}(E)$ is extended trivially to $\bar{E} \setminus E$. Then we have the renewal equation

$$\tilde{\mathbf{P}}_x[f(\tilde{\xi}_t)] = \mathbf{P}_x[1_{\{t<\zeta(\omega)\}}f(\xi_t(\omega))] + \mathbf{P}_x[1_{\{\zeta(\omega)\leq t\}}\tilde{\mathbf{P}}_{\eta(\omega,\cdot)}[f(\tilde{\xi}_{t-\zeta(\omega)}(\tilde{\omega}))]],$$

where the expectations of $\tilde{\xi}_t$ or $\tilde{\omega}$ are taken with respect to $\tilde{\mathbf{P}}_x$ or $\tilde{\mathbf{P}}_{\eta(\omega,\cdot)}$ and those of ω are taken with respect to $\mathbf{P}_x$. By Sharpe (1988, p.210), we have $\mathbf{P}_x\{$the path $t \mapsto \xi_t$ has at most countably many jumps$\} = 1$. Let $(\tilde{P}_t)_{t\geq 0}$ denote the transition semigroup of $\tilde{\xi}$. The above equation can be rewritten as

$$\tilde{P}_t f(x) = \mathbf{P}_x\big[f(\xi_t)e^{-K_t(b)}\big] + \mathbf{P}_x\bigg[\int_0^t \hat{\gamma}\tilde{P}_{t-s}f(\xi_s)e^{-K_s(b)}b(\xi_s)K(ds)\bigg].$$

Then $(t, x) \mapsto \tilde{P}_t f(x)$ is a solution of (A.32). Since the processes ξ and $\tilde{\xi}$ coincide during the time interval $[0, \zeta)$, they induce identical fine topologies on E.

Theorem A.43 *If $b(x) \geq \gamma(x,1)$ for every $x \in E$, then the semigroup $(\pi_t)_{t\geq 0}$ defined by (A.32) is a right semigroup which induces the same fine topology on E as $(P_t)_{t\geq 0}$. Moreover, if $(P_t)_{t\geq 0}$ has a Hunt realization, so does the semigroup $(\pi_t)_{t\geq 0}$.*

Proof. The first assertion follows from the arguments given above and the uniqueness of the solution of (A.32). Since $t \mapsto \exp\{-K_t(b)\}$ is continuous and strictly positive, the lifetime ζ of the subprocess $\hat{\xi}$ is totally inaccessible. Then $\tilde{\xi}$ is a Hunt process if so is ξ. That proves the second assertion.   □

Let $b\mathscr{C}_\xi(E)$ be the set of functions $f \in b\mathscr{B}(E)$ that are finely continuous relative to ξ. Theorem A.20 implies that $t \mapsto P_t f(x)$ is right continuous pointwise for every $f \in b\mathscr{C}_\xi(E)$. Let $(U^\alpha)_{\alpha>0}$ denote the resolvent of ξ.

Lemma A.44 *The set of functions $U^\beta b\mathscr{C}_\xi(E)$ is independent of $\beta > 0$. Moreover, if $g_1, g_2 \in b\mathscr{C}_\xi(E)$ and $U^\beta g_1 = U^\beta g_2$ for some $\beta > 0$, then $g_1 = g_2$.*

Proof. Let us consider two constants $\alpha, \beta > 0$. If $f \in U^\beta b\mathscr{C}_\xi(E)$, we have $f = U^\beta g$ for some $g \in b\mathscr{C}_\xi(E)$. Then the resolvent equation implies that

$$f = U^\alpha g - (\beta - \alpha)U^\alpha U^\beta g = U^\alpha h,$$

where $h = g - (\beta - \alpha)U^\beta g \in b\mathscr{C}_\xi(E)$. It follows that $U^\beta b\mathscr{C}_\xi(E) \subset U^\alpha b\mathscr{C}_\xi(E)$. By symmetry we have $U^\alpha b\mathscr{C}_\xi(E) \subset U^\beta b\mathscr{C}_\xi(E)$. That proves the first assertion. Suppose that $g_1, g_2 \in b\mathscr{C}_\xi(E)$ and $U^\beta g_1 = U^\beta g_2$ for some $\beta > 0$. By the resolvent equation we have $U^\alpha g_1 = U^\alpha g_2$ for every $\alpha > 0$. Since $t \mapsto P_t g_1(x)$ and $t \mapsto P_t g_2(x)$ are right continuous, we have $g_1 = g_2$ by the uniqueness of Laplace transforms.   □

Fix $\beta > 0$ and let $\mathscr{D}(A) = U^\beta b\mathscr{C}_\xi(E)$. For $f = U^\beta g \in \mathscr{D}(A)$ with $g \in b\mathscr{C}_\xi(E)$ set $Af = \beta f - g$, which is well-defined by Lemma A.44. We call $(A, \mathscr{D}(A))$ the *weak generator* of $(P_t)_{t\geq0}$. By the resolvent equation of $(U^\alpha)_{\alpha>0}$ it is easy to show that $(A, \mathscr{D}(A))$ is independent of the choice of $\beta > 0$. We can also define a multi-valued version of the weak generator following Ethier and Kurtz (1986). Let $\mathscr{D}(\tilde{A}) = U^\beta b\mathscr{B}(E)$ and for any $f \in \mathscr{D}(\tilde{A})$ let $\tilde{A}f = \{\beta f - g : g \in b\mathscr{B}(E)$ and $U^\beta g = f\}$. It is easy to show that $(\tilde{A}, \mathscr{D}(\tilde{A}))$ is also independent of the choice of $\beta > 0$. In particular, for any $f \in \mathscr{D}(A)$ we have $Af \in \tilde{A}f$.

Proposition A.45 *Let $\alpha > 0$. Then $U^\alpha(\alpha - A)f = f$ for every $f \in \mathscr{D}(A)$ and $(\alpha - A)U^\alpha f = f$ for every $f \in b\mathscr{C}_\xi(E)$.*

Proof. For any $f \in \mathscr{D}(A)$ there is $g \in b\mathscr{C}_\xi(E)$ so that $f = U^\beta g$. Then the definition of Af and the resolvent equation yields

$$U^\alpha(\alpha - A)f = U^\alpha(\alpha f - \beta f + g) = (\alpha - \beta)U^\alpha U^\beta g + U^\alpha g = U^\beta g = f,$$

giving the first assertion. For any $f \in b\mathscr{C}_\xi(E)$ we first use the resolvent equation to see

$$U^\alpha f = U^\beta f + (\beta - \alpha)U^\beta U^\alpha f = U^\beta h,$$

where $h = f + (\beta - \alpha)U^\alpha f$. Therefore

$$(\alpha - A)U^\alpha f = \alpha U^\alpha f - AU^\beta h = \alpha U^\alpha f - \beta U^\beta h + h = f.$$

That gives the second assertion.   □

Theorem A.46 *Let* $(A, \mathcal{D}(A))$ *be the weak generator of* $(P_t)_{t\geq 0}$. *Then for* $f \in \mathcal{D}(A)$ *we have*

$$P_t f(x) = f(x) + \int_0^t P_s A f(x) \mathrm{d}s, \qquad t \geq 0, x \in E. \tag{A.34}$$

Proof. Suppose that $f = U^\beta g$ for $g \in b\mathscr{C}_\xi(E)$. Then $U^\alpha A f = \alpha U^\alpha f - f$ for $\alpha > 0$ by Proposition A.45. Using this relation it is easy to show

$$\int_0^\infty e^{-\alpha t} \mathrm{d}t \int_0^t P_s A f(x) \mathrm{d}s = \int_0^\infty e^{-\alpha t} (P_t f - f)(x) \mathrm{d}t.$$

Since f is finely continuous relative to ξ, the function $t \mapsto P_t f(x)$ is right continuous for every $x \in E$. Then (A.34) follows by the uniqueness of Laplace transforms. $\qquad \square$

Corollary A.47 *Let* $(A, \mathcal{D}(A))$ *be the weak generator of* $(P_t)_{t\geq 0}$. *Then for* $f \in \mathcal{D}(A)$ *we have*

$$A f(x) = \lim_{t\to 0} \frac{1}{t} \big[P_t f(x) - f(x) \big], \qquad x \in E. \tag{A.35}$$

Proof. Since $Af \in b\mathscr{C}_\xi(E)$, we have (A.35) from (A.34). $\qquad \square$

In the remainder of this section we consider the semigroup $(\pi_t)_{t\geq 0}$ defined by (A.32) in the special case with $K(\mathrm{d}s) = \mathrm{d}s$ being the Lebesgue measure. Given a function $b \in b\mathscr{B}(E)$, we define a locally bounded semigroup of Borel kernels $(P_t^b)_{t\geq 0}$ on E by the following *Feynman–Kac formula*:

$$P_t^b f(x) = \mathbf{P}_x \Big[e^{-\int_0^t b(\xi_s) \mathrm{d}s} f(\xi_t) \Big], \qquad x \in E, f \in b\mathscr{B}(E). \tag{A.36}$$

Then we can rewrite (A.32) into

$$\pi_t f(x) = P_t^b f(x) + \int_0^t P_{t-s}^b \gamma \pi_s f(x) \mathrm{d}s, \qquad t \geq 0, x \in E. \tag{A.37}$$

Lemma A.48 (Gronwall's inequality) *Suppose that* $t \mapsto g(t) \geq 0$ *and* $t \mapsto h(t)$ *are integrable functions on the interval* $[0, T]$. *If there is a constant* $C > 0$ *such that*

$$g(t) \leq h(t) + C \int_0^t g(s) \mathrm{d}s, \qquad 0 \leq t \leq T, \tag{A.38}$$

then

$$g(t) \leq h(t) + C \int_0^t e^{C(t-s)} h(s) \mathrm{d}s, \qquad 0 \leq t \leq T. \tag{A.39}$$

Proof. Let $f(t)$ denote the right-hand side of (A.39). By integration by parts,

$$\int_0^t f(s)ds = \int_0^t h(s)ds + C \int_0^t \left[e^{Cu} \int_0^u e^{-Cs}h(s)ds \right] du$$

$$= \int_0^t h(s)ds + e^{Ct} \int_0^t e^{-Cs}h(s)ds - \int_0^t h(s)ds$$

$$= \int_0^t e^{C(t-s)}h(s)ds.$$

It follows that

$$f(t) = h(t) + C \int_0^t f(s)ds, \quad 0 \le t \le T. \tag{A.40}$$

Let $\Delta(t) = f(t) - g(t)$. From (A.38) and (A.40) we have

$$\Delta(t) \ge C \int_0^t \Delta(s)ds \ge C^2 \int_0^t ds \int_0^s \Delta(r)dr = C^2 \int_0^t (t-r)\Delta(r)dr$$

$$\ge C^3 \int_0^t (t-r)dr \int_0^r \Delta(s)ds = \frac{C^3}{2} \int_0^t (t-s)^2 \Delta(s)ds$$

$$\ge \cdots$$

$$\ge \frac{C^n}{(n-1)!} \int_0^t (t-s)^{n-1} \Delta(s)ds.$$

The right-hand side goes to zero as $n \to \infty$. Then $\Delta(t) \ge 0$ and (A.39) follows. $\square$

Proposition A.49 *For any $f \in b\mathscr{B}(E)$ the solution to (A.37) is also the unique locally bounded solution to*

$$\pi_t f(x) = P_t f(x) + \int_0^t P_{t-s}(\gamma - b)\pi_s f(x)ds, \quad t \ge 0, x \in E. \tag{A.41}$$

Moreover, we have $\|\pi_t f\| \le \|f\|e^{c_0 t}$ for all $t \ge 0$, where $c_0 = \|b^-\| + \|\gamma(\cdot, 1)\|$ and $b^- = 0 \vee (-b)$.

Proof. Let $(t, x) \mapsto \pi_t f(x)$ be the unique locally bounded solution of (A.37). We can use the Markov property of ξ and Fubini's theorem to write

$$\int_0^t ds \int_E b(y) P_s^b f(y) P_{t-s}(x, dy)$$

$$= \int_0^t \mathbf{P}_x \left\{ b(\xi_{t-s}) \mathbf{P}_{\xi_{t-s}} \left[e^{-\int_0^s b(\xi_u)du} f(\xi_s) \right] \right\} ds$$

$$= \int_0^t \mathbf{P}_x \left[b(\xi_{t-s})e^{-\int_{t-s}^t b(\xi_u)du} f(\xi_t) \right] ds$$

$$= \mathbf{P}_x \left[f(\xi_t) \int_0^t b(\xi_{t-s})e^{-\int_{t-s}^t b(\xi_u)du} ds \right]$$

$$= \mathbf{P}_x \left\{ \left[1 - e^{-\int_0^t b(\xi_u)du} \right] f(\xi_t) \right\}$$

$$= P_t f(x) - P_t^b f(x).$$

By similar calculations,

$$
\int_0^t ds \int_E \left[b(y) \int_0^s P_{s-r}^b \gamma \pi_r f(y) dr \right] P_{t-s}(x, dy)
$$

$$
= \int_0^t ds \int_0^s \mathbf{P}_x \left\{ b(\xi_{t-s}) \mathbf{P}_{\xi_{t-s}} \left[e^{-\int_0^{s-r} b(\xi_u) du} \gamma \pi_r f(\xi_{s-r}) \right] \right\} dr
$$

$$
= \int_0^t ds \int_0^s \mathbf{P}_x \left[b(\xi_{t-s}) e^{-\int_{t-s}^{t-r} b(\xi_u) du} \gamma \pi_r f(\xi_{t-r}) \right] dr
$$

$$
= \mathbf{P}_x \left[\int_0^t \gamma \pi_r f(\xi_{t-r}) dr \int_r^t b(\xi_{t-s}) e^{-\int_{t-s}^{t-r} b(\xi_u) du} ds \right]
$$

$$
= \mathbf{P}_x \left\{ \int_0^t \gamma \pi_r f(\xi_{t-r}) \left[1 - e^{-\int_0^{t-r} b(\xi_u) du} \right] dr \right\}
$$

$$
= \int_0^t P_{t-r} \gamma \pi_r f(x) dr - \int_0^t P_{t-r}^b \gamma \pi_r f(x) dr.
$$

Then we can add up the two equations and use (A.37) to get (A.41). For any solution $(t, x) \mapsto \pi_t f(x)$ of (A.41) we have

$$
\| \pi_t f \| \le \| f \| + c_0 \int_0^t \| \pi_{t-s} f \| ds = \| f \| + c_0 \int_0^t \| \pi_s f \| ds.
$$

Then Gronwall's inequality implies $\| \pi_t f \| \le \| f \| e^{c_0 t}$. That gives the uniqueness of the solution since (A.41) is a linear equation. $\qquad\square$

Now we prove some analytic properties of the semigroup $(\pi_t)_{t \ge 0}$ defined by (A.37) or (A.41). By Proposition A.49 we have $\| \pi_t f \| \le \| f \| e^{c_0 t}$ for $t \ge 0$ and $f \in b\mathscr{B}(E)$. Then we can define the operators $(R^\alpha)_{\alpha > c_0}$ on $b\mathscr{B}(E)$ by

$$
R^\alpha f(x) = \int_0^\infty e^{-\alpha t} \pi_t f(x) dt, \quad x \in E, f \in b\mathscr{B}(E). \tag{A.42}
$$

Proposition A.50 *For every $\alpha > c_0$ and $f \in b\mathscr{B}(E)$ we have*

$$
R^\alpha f(x) = U^\alpha f(x) + U^\alpha (\gamma - b) R^\alpha f(x), \quad x \in E.
$$

Proof. By taking the Laplace transforms of both sides of (A.41) we have

$$
R^\alpha f(x) = U^\alpha f(x) + \int_0^\infty e^{-\alpha t} dt \int_0^t P_{t-s}(\gamma - b) \pi_s f(x) ds
$$

$$
= U^\alpha f(x) + \int_0^\infty ds \int_s^\infty e^{-\alpha t} P_{t-s}(\gamma - b) \pi_s f(x) dt
$$

$$
= U^\alpha f(x) + \int_0^\infty e^{-\alpha s} U^\alpha (\gamma - b) \pi_s f(x) ds
$$

$$
= U^\alpha f(x) + U^\alpha (\gamma - b) R^\alpha f(x).
$$

That proves the desired equation. $\qquad\square$

Proposition A.51 *Let* $f \in b\mathscr{B}(E)$. *Then we have* $f \in (\alpha - \tilde{A})U^{\alpha}f$ *for* $\alpha > 0$ *and* $f \in (\alpha - \tilde{A} - \gamma + b)R^{\alpha}f$ *for* $\alpha > c_0$.

Proof. Let $h = f + (\beta - \alpha)U^{\alpha}f$. By the resolvent equation we have $U^{\alpha}f = U^{\beta}h$. Then the definition of $\tilde{A}$ implies $f = (\alpha - \beta)U^{\alpha}f + h \in \{\alpha U^{\alpha}f - \beta U^{\alpha}f + g : g \in b\mathscr{B}(E)$ and $U^{\beta}g = U^{\alpha}f\} = (\alpha - \tilde{A})U^{\alpha}f$, which gives the first assertion. For $\alpha > c_0$ we get from Proposition A.50 that

$$(\alpha - \tilde{A} - \gamma + b)R^{\alpha}f = (\alpha - \tilde{A})R^{\alpha}f - (\gamma - b)R^{\alpha}f$$
$$= (\alpha - \tilde{A})U^{\alpha}[f + (\gamma - b)R^{\alpha}f] - (\gamma - b)R^{\alpha}f.$$

By the first assertion, the set represented by the first term on the right-hand side includes $f + (\gamma - b)R^{\alpha}f$. Then $(\alpha - \tilde{A} - \gamma + b)R^{\alpha}f$ includes f, proving the second assertion. $\square$

Lemma A.52 *Let* $\alpha > c_0$ *and* $f \in \mathscr{D}(\tilde{A})$. *Then for any* $h \in (\alpha - \tilde{A} - \gamma + b)f$ *we have* $\|h\| \geq (\alpha - c_0)\|f\|$.

Proof. For $h \in (\alpha - \tilde{A} - \gamma + b)f$ we have $\alpha f - \gamma f + bf - h \in \tilde{A}f$. By the definition of $\tilde{A}$, there exist $g \in b\mathscr{B}(E)$ so that $U^{\alpha}g = f$ and $\alpha f - \gamma f + bf - h = \alpha f - g$. Therefore $\|f\| \leq \alpha^{-1}\|g\|$ and $h = g - \gamma f + bf$. It follows that $\|h\| \geq \|g\| - \|(\gamma + b^-)f\| \geq (\alpha - c_0)\|f\|$. $\square$

Lemma A.53 *If* f_1 *and* f_2 *are distinct functions from* $\mathscr{D}(\tilde{A})$, *then for any* $\alpha > c_0$ *the intersection* $(\alpha - \tilde{A} - \gamma + b)f_1 \cap (\alpha - \tilde{A} - \gamma + b)f_2$ *is empty.*

Proof. Suppose that $h \in (\alpha - \tilde{A} - \gamma + b)f_1 \cap (\alpha - \tilde{A} - \gamma + b)f_2$. Then there exist $h_1 \in \tilde{A}f_1$ and $h_2 \in \tilde{A}f_2$ so that $h = (\alpha - \gamma + b)f_1 - h_1 = (\alpha - \gamma + b)f_2 - h_2$. By the definition of $\tilde{A}$, there exist $g_1, g_2 \in b\mathscr{B}(E)$ so that $f_i = U^{\beta}g_i$ and $h_i = \beta f_i - g_i$ for $i = 1$ and 2. It follows that $(f_2 - f_1) = U^{\beta}(g_2 - g_1)$ and $h_2 - h_1 = \beta(f_2 - f_1) - (g_2 - g_1)$. Those imply $h_2 - h_1 \in \tilde{A}(f_2 - f_1)$, and so

$$0 = (\alpha - \gamma + b)(f_2 - f_1) - (h_2 - h_1) \in (\alpha - \gamma + b - \tilde{A})(f_2 - f_1),$$

which is in contradiction to Lemma A.52. $\square$

Lemma A.54 *For any* $\alpha > c_0$ *and* $f \in \mathscr{D}(A)$ *we have* $f = R^{\alpha}(\alpha - A - \gamma + b)f$.

Proof. Clearly, for $f \in \mathscr{D}(A)$ the set $(\alpha - \tilde{A} - \gamma + b)f$ contains the function $h := (\alpha - A - \gamma + b)f$. By Proposition A.51 we have $h \in (\alpha - \tilde{A} - \gamma + b)R^{\alpha}h$. Then the sets $(\alpha - \tilde{A} - \gamma + b)f$ and $(\alpha - \tilde{A} - \gamma + b)R^{\alpha}h$ have a non-empty intersection. Thus Lemma A.53 implies that $f = R^{\alpha}h$. $\square$

Theorem A.55 *Let* $f \in \mathscr{D}(A)$ *and let* $(t, x) \mapsto \pi_t f(x)$ *be defined by* (A.37) *or* (A.41). *Then we have*

$$\pi_t f(x) = f(x) + \int_0^t \pi_s (A + \gamma - b)f(x)\mathrm{d}s, \qquad t \geq 0, x \in E.$$

Proof. By Theorems A.16 and A.20, any $f \in \mathscr{D}(A)$ is finely continuous relative to ξ, so $t \mapsto P_t f(x)$ is right continuous pointwise. Then Proposition A.42 implies $t \mapsto \pi_t f(x)$ is right continuous pointwise. By integration by parts it is easy to show

$$\int_0^\infty e^{-\alpha t} dt \int_0^t \pi_s(A + \gamma - b) f ds = \frac{1}{\alpha} R^\alpha (A + \gamma - b) f.$$

Using Lemma A.54 one can see the above value is equal to

$$R^\alpha f - \frac{1}{\alpha} f = \int_0^\infty e^{-\alpha t} (\pi_t f - f) dt.$$

Then the desired equation follows by the uniqueness of the Laplace transform. □

A.7 Time–Space Processes

In this section we discuss briefly time–space processes associated with inhomogeneous Markov processes. For simplicity we only consider those processes with Borel transition semigroups. Suppose that $I \subset \mathbb{R}$ is an interval and F is a Lusin topological space. Let $\tilde{E}$ be a Borel subset of $I \times F$. For $t \in I$ let $E_t = \{x \in F : (t, x) \in \tilde{E}\}$. Then each E_t is a Lusin topological space. We fix an abstract point $\partial \notin I \times F$ and assume all functions on $\tilde{E} \subset I \times F$ have been extended trivially to $\tilde{E}^c \cup \{\partial\}$.

Suppose that for each pair $r \leq t \in I$ there is a Markov kernel $P_{r,t}$ from $(E_r, \mathscr{B}(E_r))$ to $(E_t, \mathscr{B}(E_t))$. The family $(P_{r,t} : r \leq t \in I)$ is called an *inhomogeneous transition semigroup* with *global state space* $\tilde{E}$ if it satisfies the *Chapman–Kolmogorov equation*

$$P_{r,t}(x, B) = \int_{E_s} P_{r,s}(x, dy) P_{s,t}(y, B) \tag{A.43}$$

for all $r \leq s \leq t \in I$, $x \in E_r$ and $B \in \mathscr{B}(E_t)$. In this work, we assume for every $f \in b\mathscr{B}(\tilde{E})$ the function

$$(r, x, t) \mapsto 1_{\{r \leq t\}} \int_{E_t} f(t, y) P_{r,t}(x, dy)$$

is measurable with respect to the σ-algebra $\mathscr{B}(\tilde{E} \times I)$.

Definition A.56 *The collection* $\xi = (\Omega, \mathscr{G}, \mathscr{G}_{r,t}, \xi_t, \mathbf{P}_{r,x})$ *is called an inhomogeneous Markov process with global state space* $\tilde{E}$ *and transition semigroup* $(P_{r,t} : r \leq t \in I)$ *if the following conditions are satisfied:*

(1) *For every* $r \in I$, $(\Omega, \mathscr{G}, \mathscr{G}_{r,t} : t \in I \cap [r, \infty))$ *is a filtered measurable space so that* $\mathscr{G}_{s,t} \subset \mathscr{G}_{r,u}$ *for* $r \leq s \leq t \leq u \in I$.

(2) *For every $r \leq t \in I$, $\omega \mapsto \xi_t(\omega)$ is a measurable mapping from $(\Omega, \mathcal{G}_{r,t})$ to $(E_t, \mathcal{B}(E_t))$.*
(3) *For every $(r, x) \in \tilde{E}$, $\mathbf{P}_{r,x}$ is a probability measure on $(\Omega, \mathcal{G})$ such that for every $H \in b\mathcal{G}$ the function $(r, x) \mapsto \mathbf{P}_{r,x}(H)$ is $\mathcal{B}(\tilde{E})$-measurable.*
(4) *For every $(r, x) \in \tilde{E}$ we have $\mathbf{P}_{r,x}\{\xi_r = x\} = 1$ and the following simple Markov property holds:*

$$\mathbf{P}_{r,x}\big[f(\xi_t)|\mathcal{G}_{r,s}\big] = P_{s,t}f(\xi_s), \quad r \leq s \leq t \in I, f \in b\mathcal{B}(E_t).$$

We say ξ is right continuous if $t \mapsto \xi_t(\omega)$ is right continuous for every $\omega \in \Omega$.

Given an inhomogeneous transition semigroup $(P_{r,t} : r \leq t \in I)$ with global state space $\tilde{E}$, we can define a homogeneous Borel transition semigroup $(\tilde{P}_t)_{t\geq 0}$ on $\tilde{E}$ by

$$\tilde{P}_t f(r, x) = 1_I(r + t) \int_{E_{r+t}} f(r + t, y) P_{r,r+t}(x, \mathrm{d}y), \qquad (A.44)$$

where $t \geq 0$, $(r, x) \in \tilde{E}$ and $f \in b\mathcal{B}(\tilde{E})$. We call $(\tilde{P}_t)_{t\geq 0}$ the *time–space semigroup* associated with $(P_{r,t} : r \leq t \in I)$. Suppose that $\xi = (\Omega, \mathcal{G}, \mathcal{G}_{r,t}, \xi_t, \mathbf{P}_{r,x})$ is a right continuous inhomogeneous Markov process with transition semigroup $(P_{r,t} : r \leq t \in I)$. Let $\tilde{\Omega} = I \times \Omega$. For $(v, \omega) \in \tilde{\Omega}$ define

$$\tilde{\xi}_t(v, \omega) = \begin{cases} (v + t, \xi_{v+t}(\omega)) & \text{if } t \geq 0 \text{ and } v + t \in I, \\ \partial & \text{if } t \geq 0 \text{ and } v + t \notin I. \end{cases} \qquad (A.45)$$

Let $(\tilde{\mathscr{F}}, \tilde{\mathscr{F}}_t)$ be the $\mathcal{B}(\tilde{E})$-natural σ-algebras generated by $\{\tilde{\xi}_t : t \geq 0\}$. For $(r, x) \in \tilde{E}$ let $\tilde{\mathbf{P}}_{r,x}$ be the probability measure on $(\tilde{\Omega}, \tilde{\mathscr{F}})$ induced by $\mathbf{P}_{r,x}$ via the mapping $\omega \mapsto (r, \omega)$.

Theorem A.57 *The system $\tilde{\xi} = (\tilde{\Omega}, \tilde{\mathscr{F}}, \tilde{\mathscr{F}}_t, \tilde{\xi}_t, \tilde{\mathbf{P}}_{r,x})$ is a right continuous Markov process in $\tilde{E}$ with transition semigroup $(\tilde{P}_t)_{t\geq 0}$.*

Proof. Let $(r, x) \in \tilde{E}$ and $t \geq s \geq 0$. Let $f = f_v(x) = f(v, x)$ be a bounded Borel function on $\tilde{E}$. Since $\tilde{\mathbf{P}}_{r,x}$ is concentrated on $\{r\} \times \Omega$, if $r + t \in I$, we have

$$\begin{aligned} \tilde{\mathbf{P}}_{r,x}\big[f(\tilde{\xi}_t)|\tilde{\mathscr{F}}_s\big] &= \tilde{\mathbf{P}}_{r,x}\big[f(r + t, \xi_{r+t})|\sigma(\{\xi_{r+u} : 0 \leq u \leq s\})\big] \\ &= \mathbf{P}_{r,x}\big[f(r + t, \xi_{r+t})|\mathcal{G}_{r,r+s}\big] = P_{r+s,r+t}f_{r+t}(\xi_{r+s}) \\ &= \tilde{P}_{t-s}f(r + s, \xi_{r+s}) = \tilde{P}_{t-s}f(\tilde{\xi}_s). \end{aligned}$$

If $r + t \notin I$, both sides of the above equality are equal to zero. Then $\tilde{\xi}$ is a Markov process with transition semigroup $(\tilde{P}_t)_{t\geq 0}$. $\qquad \square$

We call $\tilde{\xi} = (\tilde{\Omega}, \tilde{\mathscr{F}}, \tilde{\mathscr{F}}_t, \tilde{\xi}_t, \tilde{\mathbf{P}}_{r,x})$ the *time–space process* of ξ. By Theorem A.57, the study of the inhomogeneous process ξ can be reduced to that of the homogeneous time–space process $\tilde{\xi}$. If $\tilde{\xi}$ has a right realization, we call $(P_{r,t} : r \leq t \in I)$

an *inhomogeneous right transition semigroup*. The following theorem shows that the terminology is consistent with that in the homogeneous case.

Theorem A.58 *Suppose that* $\tilde{E} = [0, \infty) \times E$ *for a Lusin topological space E and there is a homogeneous Borel transition semigroup $(P_t)_{t \geq 0}$ on E so that $P_{r,t} = P_{t-r}$ for $t \geq r \geq 0$. Then $(\tilde{P}_t)_{t \geq 0}$ is a right semigroup if and only if $(P_t)_{t \geq 0}$ is a right semigroup.*

Proof. Suppose that $(\tilde{P}_t)_{t \geq 0}$ is a right semigroup. Let $\tilde{\xi} = (\tilde{\Omega}, \tilde{\mathcal{G}}, \tilde{\mathcal{G}}_t, (\alpha_t, \xi_t), \tilde{\mathbf{P}}_{r,x})$ be a right realization of $(\tilde{P}_t)_{t \geq 0}$. One can use Theorem A.16 to see that $\xi = (\tilde{\Omega}, \tilde{\mathcal{G}}, \tilde{\mathcal{G}}_t, \xi_t, \tilde{\mathbf{P}}_{0,x})$ is a right process with transition semigroup $(P_t)_{t \geq 0}$. Then $(P_t)_{t \geq 0}$ is a Borel right semigroup. The converse was obtained in Sharpe (1988, p.86). $\square$

Starting from a realization of the corresponding time–space process we can also reconstruct the inhomogeneous process with transition semigroup $(P_{r,t} : t \geq r \in I)$. For this purpose, let us consider a right continuous realization $\tilde{\xi} = (\Omega, \tilde{\mathcal{G}}, \tilde{\mathcal{G}}_t, (\alpha_t, y_t), \mathbf{P}_{r,x})$ of $(\tilde{P}_t)_{t \geq 0}$. In view of (A.44) we have $\alpha_t = \alpha_0 + t$ for $t \geq 0$. For $\omega \in \Omega$ define

$$\xi_t(\omega) = \begin{cases} y_{t-\alpha_0(\omega)}(\omega) & \text{if } t \in I \cap [\alpha_0(\omega), \infty), \\ \partial & \text{if } t \in I \cap (-\infty, \alpha_0(\omega)). \end{cases} \tag{A.46}$$

Let $\mathcal{F} = \sigma(\{\xi_t : t \in I\})$ and let $\mathcal{F}_{r,t} = \sigma(\{\xi_s : r \leq s \leq t\})$ for $t \geq r \in I$.

Theorem A.59 *The system $\xi = (\Omega, \mathcal{F}, \mathcal{F}_{r,t}, \xi_t, \mathbf{P}_{r,x})$ is a right continuous inhomogeneous Markov process with transition semigroup $(P_{r,t} : t \geq r \in I)$.*

Proof. Let $(r, x) \in \tilde{E}$ and $r \leq s \leq t \in I$. Since $\mathbf{P}_{r,x}\{\alpha_0 = r\} = 1$, for any $f \in b\mathcal{B}(E_t)$ we have

$$\begin{aligned} \mathbf{P}_{r,x}[f(\xi_t)|\mathcal{F}_{r,s}] &= \mathbf{P}_{r,x}[f(\xi_t)|\sigma(\{\xi_u : r \leq u \leq s\})] \\ &= \mathbf{P}_{r,x}[f(y_{t-r})|\sigma(\{y_{u-r} : r \leq u \leq s\})] \\ &= \mathbf{P}_{r,x}[f(y_{t-r})|\tilde{\mathcal{G}}_{s-r}] = \tilde{P}_{t-s}f(\alpha_{s-r}, y_{s-r}) \\ &= \tilde{P}_{t-s}f(s, \xi_s) = P_{s,t}f(\xi_s). \end{aligned}$$

That gives the desired Markov property of ξ. $\square$

Example A.2 Suppose that E is a complete separable metric space. Let $D_E := D([0, \infty), E)$ be the space of càdlàg paths from $[0, \infty)$ to E equipped with the usual Skorokhod metric. Then D_E is also a complete separable metric space. Suppose that $(P_t)_{t \geq 0}$ is a Borel right semigroup on E with a càdlàg realization. For simplicity we consider the canonical realization $\xi = (D_E, \mathcal{F}^0, \mathcal{F}^0_t, \xi_t, \mathbf{P}_x)$, where $(\mathcal{F}^0, \mathcal{F}^0_t)$ are the natural σ-algebras of D_E and $\xi_t(\omega) = \omega(t)$ for $t \geq 0$ and $\omega \in D_E$. Let $y^s(t) = y(t \wedge s)$ for $s, t \geq 0$ and $y \in D_E$. Let $S = \{(t, y) \in [0, \infty) \times D_E : y = y^t\}$. For $t \geq 0$ let

$$D_E^t = \{y \in D_E : y = y^t\} = \{y \in D_E : (t, y) \in S\}.$$

Then we have $D_E^s \subset D_E^t$ for $t \geq s \geq 0$. Given $r \geq 0$ and $y_1, y_2 \in D_E$ we define $y_1/r/y_2 \in D_E$ by

$$(y_1/r/y_2)(t) = \begin{cases} y_1(t) & \text{if } 0 \leq t < r, \\ y_2(t-r) & \text{if } r \leq t < \infty. \end{cases}$$

The operators $y \mapsto y^s$ and $(r, y_1, y_2) \mapsto y_1/r/y_2$ are Borel measurable; see Dellacherie and Meyer (1978, p.146). We can define an inhomogeneous Borel transition semigroup $(\bar{P}_{r,t} : t \geq r \geq 0)$ with global state space S by

$$\bar{P}_{r,t} f(y) = \mathbf{P}_{y(r)}[f(y/r/\xi^{t-r})], \quad y \in D_E^r, f \in \mathrm{b}\mathscr{B}(D_E^t). \qquad (A.47)$$

From Proposition 2.1.2 of Dawson and Perkins (1991, p.14) it follows that $(\bar{P}_{r,t} : t \geq r \geq 0)$ is a right transition semigroup. For $\omega \in D_E$ and $t \geq 0$ let $\bar{\xi}_t(\omega) = \omega^t \in D_E^t$. It is easy to see $\mathscr{F}_{r,t}^0 := \sigma(\{\bar{\xi}_s : r \leq s \leq t\}) = \mathscr{F}_t^0$ for $t \geq r \geq 0$. For $r \geq 0$ and $y \in D_E^r$ define the probability measure $\bar{\mathbf{P}}_{r,y}$ on $(D_E, \mathscr{F}^0)$ by

$$\bar{\mathbf{P}}_{r,y}(A) = \mathbf{P}_{y(r)}(\{\omega \in D_E : y/r/\omega \in A\}), \quad A \in \mathscr{F}^0. \qquad (A.48)$$

Then $\bar{\xi} = (D_E, \mathscr{F}^0, \mathscr{F}_{r,t}^0, \bar{\xi}_t, \bar{\mathbf{P}}_{r,y})$ is a càdlàg realization of $(\bar{P}_{r,t} : t \geq r \geq 0)$. This process is called the *path process* of ξ; see Dawson and Perkins (1991). Clearly, the path process records all the information of the history of the sample path of ξ.

References

Abraham, R. and Delmas, J.-F. (2008): Fragmentation associated with Lévy processes using snake. *Probab. Theory Related Fields* **141**, 113–154.

Abraham, R. and Delmas, J.-F. (2009): Changing the branching mechanism of a continuous state branching process using immigration. *Ann. Inst. H. Poincaré Probab. Statist.* **45**, 226–238.

de Acosta, A., Araujo, A. and Giné, E. (1978): On Poisson measures, Gaussian measures and the central limit theorem in Banach spaces. In: *Probability on Banach Spaces*, 1–68. Adv. Probab. Related Topics **4**. Dekker, New York.

Adams, R.A. and Fournier, J.J.F. (2003): *Sobolev Spaces*. 2nd Ed. Elsevier and Academic Press, Amsterdam.

Aldous, D. (1991): The continuum random tree I. *Ann. Probab.* **19**, 1–28.

Aldous, D. (1993): The continuum random tree III. *Ann. Probab.* **21**, 248–289.

Aliev S.A. (1985): A limit theorem for the Galton–Watson branching processes with immigration. *Ukrainian Math. J.* **37**, 535–438.

Aliev, S.A. and Shchurenkov, V.M. (1982): Transitional phenomena and the convergence of Galton–Watson processes to Jiřina processes. *Theory Probab. Appl.* **27**, 472–485.

Applebaum, D. (2007): On the infinitesimal generators of Ornstein–Uhlenbeck processes with jumps in Hilbert space. *Potential Anal.* **26**, 79–100.

Athreya, K.B. and Ney, P.E. (1972): *Branching Processes*. Springer, Berlin.

Athreya, S.R., Barlow, M.T., Bass, R.F. and Perkins, E.A. (2002): Degenerate stochastic differential equations and super-Markov chains. *Probab. Theory Related Fields* **123**, 484–520.

Athreya, S.R. and Swart, J.M. (2005): Branching–coalescing particle systems. *Probab. Theory Related Fields* **131**, 376–414.

Athreya, S.R. and Swart, J.M. (2009): Erratum: Branching–coalescing particle systems. *Probab. Theory Related Fields* **145**, 639–640.

Berg, C., Christensen, J.P.R. and Ressel, P. (1984): *Harmonic Analysis on Semigroups*. Springer, Berlin.

Bertoin, J. (1996): *Lévy Processes*. Cambridge Univ. Press, Cambridge.

Bertoin, J. (2006): *Random Fragmentation and Coagulation Processes*. Cambridge Univ. Press, Cambridge.

Bertoin, J. and Le Gall, J.-F. (2000): The Bolthausen-Sznitman coalescent and the genealogy of continuous-state branching processes. *Probab. Theory Related Fields* **117**, 249–266.

Bertoin, J. and Le Gall, J.-F. (2003): Stochastic flows associated to coalescent processes. *Probab. Theory Related Fields* **126**, 261–288.

Bertoin, J. and Le Gall, J.-F. (2005): Stochastic flows associated to coalescent processes II: Stochastic differential equations. *Ann. Inst. H. Poincaré Probab. Statist.* **41**, 307–333.

Bertoin, J. and Le Gall, J.-F. (2006): Stochastic flows associated to coalescent processes III: Infinite population limits. *Illinois J. Math.* **50**, 147–181.

Bertoin, J., Le Gall, J.-F. and Le Jan, Y. (1997): Spatial branching processes and subordination. *Canad. J. Math.* **49**, 24–54.

Beznea, L. (2010): Potential theoretical methods in the construction of measure-valued Markov branching processes. *J. Europ. Math. Soc.* To appear.

Billingsley, P. (1999): *Convergence of Probability Measures.* 2nd Ed. Wiley, New York.

Bogachev, V.I. and Röckner, M. (1995): Mehler formula and capacities for infinite-dimensional Ornstein–Uhlenbeck processes with general linear drift. *Osaka J. Math.* **32**, 237–274.

Bogachev, V.I., Röckner, M. and Schmuland, B. (1996): Generalized Mehler semigroups and applications. *Probab. Theory Related Fields* **105**, 193–225.

Bojdecki, T. and Gorostiza, L.G. (1986): Langevin equation for $\mathscr{S}'$-valued Gaussian processes and fluctuation limits of infinite particle system. *Probab. Theory Related Fields* **73**, 227–244.

Bojdecki, T. and Gorostiza, L.G. (1991): Gaussian and non-Gaussian distribution-valued Ornstein–Uhlenbeck processes. *Canad. J. Math.* **43**, 1136–1149.

Bojdecki, T. and Gorostiza, L.G. (2002): Self-intersection local time for $\mathscr{S}'(\mathbb{R}^d)$-Ornstein–Uhlenbeck processes arising from immigration systems. *Math. Nachr.* **238**, 37–61.

Borodin, A.N. and Salminen, P. (1996): *Handbook of Brownian Motion — Facts and Formulae.* Birkhäuser, Basel.

Bose, A. and Kaj, I. (2000): A scaling limit process for the age-reproduction structure in a Markov population. *Markov Process. Related Fields* **6**, 397–428.

Burdzy, K. and Le Gall, J.-F. (2001): Super-Brownian motion with reflecting historical paths. *Probab. Theory Related Fields* **121**, 447–491.

Burdzy, K. and Mytnik, L. (2005): Super-Brownian motion with reflecting historical paths II: Convergence of approximations. *Probab. Theory Related Fields* **133**, 145–174.

Caballero, M.E., Lambert, A. and Uribe Bravo G., (2009): Proof(s) of the Lamperti representation of continuous-state branching processes. *Probab. Surv.* **6**, 62–89.

Ceci, C. and Gerardi, A. (2006): Modelling a multitype branching brownian motion: Filtering of a measure-valued process. *Acta Appl. Math.* **91**, 39–66.

Cerrai, S. (1994): A Hille–Yosida theorem for weakly continuous semigroups. *Semigroup Forum* **49**, 349–367.

Champagnat, N. and Roelly, S. (2008): Limit theorems for conditioned multitype Dawson–Watanabe processes and Feller diffusions. *Elect. J. Probab.* **13**, 777–810.

Chen, M.F. (2004): *From Markov Chains to Non-Equilibrium Particle Systems.* 2nd Ed. World Sci., River Edge, NJ.

Chen, Z.Q., Ren, Y.X. and Wang, H. (2008): An almost sure scaling limit theorem for Dawson–Watanabe superprocesses. *J. Funct. Anal.* **159**, 267–294.

Chojnowska-Michalik, A. (1987): On processes of Ornstein–Uhlenbeck type in Hilbert space. *Stochastics* **21**, 251–286.

Conway, J.B. (1990): *A Course in Functional Analysis.* Springer, Berlin.

Cox, J.T., Durrett, R. and Perkins, E.A. (2000): Rescaled voter models converge to super-Brownian motion. *Ann. Probab.* **28**, 185–234.

Cox, J.T. and Klenke, A. (2003): Rescaled interacting diffusions converge to super Brownian motion. *Ann. Appl. Probab.* **13**, 501–514.

Cox, J.T. and Perkins, E.A. (2004): An application of the voter model–super-Brownian motion invariance principle. *Ann. Inst. H. Poincaré Probab. Statist.* **40**, 25–32.

Cox, J.T. and Perkins, E.A. (2005): Rescaled Lotka–Volterra models converge to super-Brownian motion. *Ann. Probab.* **33**, 904–947.

Cox, J.T. and Perkins, E.A. (2007): Survival and coexistence in stochastic spatial Lotka–Volterra models. *Probab. Theory Related Fields* **139**, 89–142.

Cox, J.T. and Perkins, E.A. (2008): Renormalization of the two-dimensional Lotka–Volterra model. *Ann. Appl. Probab.* **18**, 747–812.

Da Prato, G. and Zabczyk, J. (1992): *Stochastic Equations in Infinite Dimensions.* Cambridge Univ. Press, Cambridge.

Dawson, D. (1975): Stochastic evolution equations and related measure processes. *J. Multivariate Anal.* **5**, 1–52.

Dawson, D.A. (1977): The critical measure diffusion process. *Z. Wahrsch. verw. Geb.* **40**, 125–145.

Dawson, D.A. (1978): Geostochastic calculus. *Canad. J. Statist.* **6**, 143–168.

Dawson, D.A. (1992): Infinitely divisible random measures and superprocesses. In: *Stochastic Analysis and Related Topics* (Silivri, 1990), 1–129. Progr. Probab. **31**. Birkhäuser, Boston, MA.

Dawson, D.A. (1993): Measure-valued Markov processes. In: *Ecole d'Eté de Probabilités de Saint-Flour, XXI-1991* 1–260. Lecture Notes Math. **1541**. Springer, Berlin.

Dawson, D.A., Etheridge, A.M., Fleischmann, K., Mytnik, L., Perkins, E.A. and Xiong, J. (2002a): Mutually catalytic branching in the plane: Finite measure states. *Ann. Probab.* **30**, 1681–1762.

Dawson, D.A., Etheridge, A.M., Fleischmann, K., Mytnik, L., Perkins, E.A. and Xiong, J. (2002b): Mutually catalytic branching in the plane: Infinite measure states. *Elect. J. Probab.* **7**, Paper 15, 1–61.

Dawson, D.A. and Fleischmann, K. (1991): Critical branching in a highly fluctuating random medium, *Probab. Theory Related Fields* **90**, 241–274.

Dawson, D.A. and Fleischmann, K. (1992): Diffusion and reaction caused by a point catalysts. *SIAM J. Appl. Math.* **52**, 163–180.

Dawson, D.A. and Fleischmann, K. (1997a): A continuous super-Brownian motion in a super-Brownian medium. *J. Theoret. Probab.* **10**, 213–276.

Dawson, D.A. and Fleischmann, K. (1997b): Longtime behavior of a branching process controlled by branching catalysts. *Stochastic Process. Appl.* **71**, 241–257.

Dawson, D.A. and Fleischmann, K. (2002): Catalytic and mutually catalytic super-Brownian motions. In: *Seminar on Stochastic Analysis, Random Fields and Applications III* (Ascona, 1999) 89–110. Progr. Probab. **52**. Birkhäuser, Basel.

Dawson, D.A., Fleischmann, K. and Gorostiza, L.G. (1989a): Stable hydrodynamic limit fluctuations of a critical branching particle system in a random medium. *Ann. Probab.* **17**, 1083–1117.

Dawson, D.A., Fleischmann, K. and Leduc, G. (1998): Continuous dependence of a class of super-processes on branching parameters and applications. *Ann. Probab.* **26**, 562–601.

Dawson, D.A., Fleischmann, K., Mytnik, L., Perkins, E.A. and Xiong, J. (2003): Mutually catalytic branching in the plane: Uniqueness. *Ann. Inst. H. Poincaré Probab. Statist.* **39**, 135–191.

Dawson, D.A., Gorostiza, L.G. and Li, Z.H. (2002c): Non-local branching superprocesses and some related models. *Acta Appl. Math.* **74**, 93–112.

Dawson, D.A., Gorostiza, L.G. and Wakolbinger, A. (2004a): Hierarchical equilibria of branching populations. *Elect. J. Probab.* **9**, 316–381.

Dawson, D.A., Iscoe, I. and Perkins, E.A. (1989b): Super-Brownish motion: Path properties and hitting probabilities. *Probab. Theory Related Fields* **83**, 135–205.

Dawson, D.A. and Ivanoff, D. (1978): Branching diffusions and random measures. In: *Branching Processes* (Conf., Saint Hippolyte, Que., 1976), 61–103. Adv. Probab. Related Topics **5**. Dekker, New York.

Dawson, D.A. and Li, Z.H. (2003): Construction of immigration superprocesses with dependent spatial motion from one-dimensional excursions. *Probab. Theory Related Fields* **127**, 37–61.

Dawson, D.A. and Li, Z.H. (2004): Non-differentiable skew convolution semigroups and related Ornstein–Uhlenbeck processes. *Potent. Anal.* **20**, 285–302.

Dawson, D.A. and Li, Z.H. (2006): Skew convolution semigroups and affine Markov processes. *Ann. Probab.* **34**, 1103–1142.

Dawson, D.A. and Li, Z.H., Schmuland, B. and Sun, W. (2004b): Generalized Mehler semigroups and catalytic branching processes with immigration. *Potent. Anal.* **21**, 75–97.

Dawson, D.A., Li, Z.H. and Wang, H. (2001): Superprocesses with dependent spatial motion and general branching densities. *Elect. J. Probab.* **6**, Paper 25, 1–33.

Dawson, D.A., Li, Z.H. and Zhou, X.W. (2004c): Superprocesses with coalescing Brownian spatial motion as large scale limits. *J. Theoret. Probab.* **17**, 673–692.

Dawson, D.A. and Perkins, E.A. (1991): *Historical Processes*. Memoirs Amer. Math. Soc. **93**, no. 454. Providence, RI.

Dawson, D.A. and Perkins E.A. (1998): Long-time behavior and coexistence in a mutually catalytic branching model. *Ann. Probab.* **26**, 1088–1138.

Dellacherie, C., Maisonneuve, B. and Meyer, P.A. (1992): *Probabilités et Potential*. Chapitres XVII–XXIV. Hermann, Paris.

Dellacherie, C. and Meyer, P.A. (1978): *Probabilities and Potential*. Chapters I–IV. North-Holland, Amsterdam.

Dellacherie, C. and Meyer, P.A. (1982): *Probabilities and Potential*. Chapters V–VIII. North-Holland, Amsterdam.

Delmas, J.-F. and Dhersin, J.-S. (2003): Super Brownian motion with interactions. *Stochastic Process. Appl.* **107**, 301–325.

Donnelly, P. and Kurtz, T.G. (1996): A countable representation of the Fleming–Viot measure-valued diffusion. *Ann. Probab.* **24**, 698–742.

Donnelly, P. and Kurtz, T.G. (1999a): Particle representations for measure-valued population models. *Ann. Probab.* **27**, 166–205.

Donnelly, P. and Kurtz, T.G. (1999b): Genealogical processes for Fleming–Viot models with selection and recombination. *Ann. Appl. Probab.* **9**, 1091–1148.

Duffie, D., Filipović, D. and Schachermayer, W. (2003): Affine processes and applications in finance. *Ann. Appl. Probab.* **13**, 984–1053.

Duquesne, T. and Le Gall, J.-F. (2002): *Random Trees, Lévy Processes and Spatial Branching Processes*. Astérisque **281**.

Durrett, R. (1995): Ten lectures on particle systems. In: *Ecole d'Eté de Probabilités de Saint-Flour XXIII-1993*, 97–201. Lecture Notes Math. **1608**. Springer, Berlin.

Durrett, R. (2008): *Probability Models for DNA Sequence Evolution*. 2nd Ed. Springer, Berlin.

Durrett, R. and Perkins, E.A. (1999): Rescaled contact processes converge to super-Brownian motion in two or more dimensions. *Probab. Theory Related Fields* **114**, 309–399.

Dynkin, E.B. (1978): Sufficient statistics and extreme points. *Ann. Probab.* **6**, 705–730.

Dynkin, E.B. (1989a): Superprocesses and their linear additive functionals. *Trans. Amer. Math. Soc.* **314**, 255–282.

Dynkin, E.B. (1989b): Three classes of infinite dimensional diffusion processes. *J. Funct. Anal.* **86**, 75–110.

Dynkin, E.B. (1991a): Branching particle systems and superprocesses. *Ann. Probab.* **19**, 1157–1194.

Dynkin, E.B. (1991b): A probabilistic approach to one class of nonlinear differential equations. *Probab. Theory Related Fields* **89**, 89–115.

Dynkin, E.B. (1991c): Path processes and historical superprocesses. *Probab. Theory Related Fields* **90**, 1–36.

Dynkin, E.B. (1993a): Superprocesses and partial differential equations. *Ann. Probab.* **21**, 1185–1262.

Dynkin, E.B. (1993b): On regularity of superprocesses. *Probab. Theory Related Fields* **95**, 263–281.

Dynkin, E.B. (1994): *An Introduction to Branching Measure-valued Processes*. Amer. Math. Soc., Providence, RI.

Dynkin, E.B. (2002): *Diffusions, Superdiffusions and Partial Differential Equations*. Amer. Math. Soc., Providence, RI.

Dynkin, E.B. (2004): *Superdiffusions and Positive Solutions of Nonlinear Partial Differential Equations*. Amer. Math. Soc., Providence, RI.

Dynkin, E.B., Kuznetsov, S.E. and Skorokhod, A.V. (1994): Branching measure-valued processes. *Probab. Theory Related Fields* **99**, 55–96.

El Karoui, N. and Méléard, S. (1990): Martingale measures and stochastic calculus. *Probab. Theory Related Fields* **84**, 83–101.

El Karoui, N. and Roelly, S. (1991): Propriétés de martingales, explosion et representation de Lévy–Khintchine d'une classe de processus de branchement à valeurs mesures. *Stochastic Process. Appl.* **38**, 239–266.

Engländer, J. (2007): Branching diffusions, superdiffusions and random media. *Probab. Surv.* **4**, 303–364.

Engländer, J. and Kyprianou, A.E. (2004): Local extinction versus local exponential growth for spatial branching processes. *Ann. Probab.* **32**, 78–99.

Engländer, J. and Turaev, D. (2002): A scaling limit theorem for a class of superdiffusions. *Ann. Probab.* **30**, 683–722.

Engländer, J. and Winter, A. (2006): Law of large numbers for a class of superdiffusions. *Ann. Inst. H. Poincaré Probab. Statist.* **42**, 171–185.

Etheridge, A.M. (2000): *An Introduction to Superprocesses.* Amer. Math. Soc., Providence, RI.

Etheridge, A.M. (2004): Survival and extinction in a locally regulated population. *Ann. Appl. Probab.* **14**, 188–214.

Etheridge, A.M. and Fleischmann, K. (1998): Persistence of a two-dimensional super-Brownian motion in a catalytic medium. *Probab. Theory Related Fields* **110**, 1–12.

Etheridge, A.M. and March, P. (1991): A note on superprocesses. *Probab. Theory Related Fields* **89**, 141–147.

Etheridge, A.M. and Williams, D.R.E. (2003): A decomposition of the $(1 + \beta)$-superprocess conditioned on survival. *Proc. Roy. Soc. Edinburgh Sect. A* **133**, 829–847.

Ethier, S.N. and Kurtz, T.G. (1986): *Markov Processes: Characterization and Convergence.* Wiley, New York.

Ethier, S.N. and Kurtz, T.G. (1993): Fleming–Viot processes in population genetics. *SIAM J. Control Optim.* **31**, 345–386.

Evans, S.N. (1991): Trapping a measure-valued Markov branching process conditioned on nonextinction. *Ann. Inst. H. Poincaré Probab. Statist.* **27**, 215–220.

Evans, S.N. (1992): The entrance space of a measure-valued Markov branching process conditioned on nonextinction. *Canad. Math. Bull.* **35**, 70–74.

Evans, S.N. (1993): Two representations of conditioned superprocess. *Proc. Roy. Soc. Edinburgh Sect. A* **123**, 959–971.

Evans, S.N. (2008): Probability and real trees. In: *Ecole d'Eté de Probabilités de Saint-Flour XXXV-2005.* Lecture Notes Math. **1920**. Springer, Berlin.

Evans, S.N. and Perkins, E. (1990): Measure-valued Markov branching processes conditioned on nonextinction. *Israel J. Math.* **71**, 329–337.

Evans, S.N., Pitman, J. and Winter, A. (2006): Rayleigh processes, real trees, and root growth with re-grafting. *Probab. Theory Related Fields* **134**, 81–126.

Evans, S.N. and Winter, A. (2006): Subtree prune and regraft: A reversible real tree-valued Markov process. *Ann. Probab.* **34**, 918–961.

Ewens, W.J. (2004): *Mathematical Population Genetics.* Vol. 1. Springer, Berlin.

Feller, W. (1951): Diffusion processes in genetics. In: *Proceedings 2nd Berkeley Symp. Math. Statist. Probab., 1950,* 227–246. Univ. of California Press, Berkeley and Los Angeles.

Feller, W. (1971): *An Introduction to Probability Theory and its Applications.* Vol. 2. 2nd Ed. Wiley, New York.

Feng, S. (2010): *The Poisson–Dirichlet Distribution and Related Topics.* Springer, Berlin.

Fitzsimmons, P.J. (1988): Construction and regularity of measure-valued Markov branching processes. *Israel J. Math.* **64**, 337–361.

Fitzsimmons, P.J. (1992): On the martingale problem for measure-valued Markov branching processes. In: *Seminar on Stochastic Processes, 1991* (Los Angeles, CA, 1991), 39–51. Progr. Probab. **29**. Birkhäuser, Boston, MA.

Fleischmann, K., Gärtner, J. and Kaj, I. (1996): A Schilder type theorem for super-Brownian motion. *Canad. J. Math.* **48**, 542–568.

Fleischmann, K. and Mueller, C. (1997): A super-Brownian motion with a locally infinite catalytic mass. *Probab. Theory Related Fields* **107**, 325–357.

Fleischmann, K. and Sturm, A. (2004): A super-stable motion with infinite mean branching. *Ann. Inst. H. Poincaré Probab. Statist.* **40**, 513–537.

Fournier, N. and Méléard, S. (2004): A microscopic probabilistic description of a locally regulated population and macroscopic approximations. *Ann. Appl. Probab.* **14**, 1880–1919.

Friedman, A. (1964): *Partial Differential Equations of Parabolic Type.* Prentice-Hall, Englewood Cliffs, NJ.

Fu, Z.F. and Li, Z.H. (2004): Measure-valued diffusions and stochastic equations with Poisson process. *Osaka J. Math.* **41**, 727–744.

Fu, Z.F. and Li, Z.H. (2010): Stochastic equations of non-negative processes with jumps. *Stochastic Process. Appl.* **120**, 306–330.

Fuhrman, M. and Röckner, M. (2000): Generalized Mehler semigroups: The non-Gaussian case. *Potent. Anal.* **12**, 1–47.

Getoor, R.K. (1975): *Markov Processes: Ray Processes and Right Processes.* Lecture Notes Math. **440**. Springer, Berlin.

Getoor, R.K. (1990): *Excessive Measures.* Birkhäuser, Boston, MA.

Getoor, R.K. and Glover, J. (1987): Constructing Markov processes with random times of birth and death. In: *Seminar on Stochastic Processes, 1986* (Charlottesville, Va., 1986), 35–69. Progr. Probab. Statist. **13**. Birkhäuser, Boston, MA.

Gill, H.S. (2009): Superprocesses with spatial interactions in a random medium. *Stochastic Process. Appl.* **119**, 3981–4003.

Goldys, B. and Kocan, M. (2001): Diffusion semigroups in spaces of continuous functions with mixed topology. *J. Differential Equations* **173**, 17–39.

Goldys, B. and van Neerven, J.M.A.M. (2003): Transition semigroups of Banach space valued Ornstein–Uhlenbeck processes. *Acta Appl. Math.* **76**, 283–330.

Gorostiza, L.G. and Li, Z.H. (1998): Fluctuation limits of measure-valued immigration processes with small branching. In: *Stochastic Models* (Guanajuato, 1998), 261–268. Aportaciones Mat. Investig. **14**. Soc. Mat. Mexicana, México.

Gorostiza, L.G. and Lopez-Mimbela, J.A. (1990): The multitype measure branching process. *Adv. Appl. Probab.* **22**, 49–67.

Gorostiza, L.G., Navarro, R. and Rodrigues, E.R. (2005): Some long-range dependence processes arising from fluctuations of particle systems. *Acta Appl. Math.* **86**, 285–308.

Gorostiza, L.G. and Roelly, S. (1991): Some properties of the multitype measure branching process. *Stochastic Process. Appl.* **37**, 259–274.

Gorostiza, L.G., Roelly, S. and Wakolbinger, A. (1992): Persistence of critical multitype particle and measure branching process. *Probab. Theory Related Fields* **92**, 313–335.

Greven, A., Klenke, A. and Wakolbinger, A. (1999): The longtime behavior of branching random walk in a catalytic medium. *Elect. J. Probab.* **4**, Paper 12, 1–80.

Greven, A., Popovic, L. and Winter, A. (2009): Genealogy of catalytic branching models. *Ann. Appl. Probab.* **19**, 1232–1272.

Grey, D.R. (1974): Asymptotic behavior of continuous time, continuous state-space branching processes. *J. Appl. Probab.* **11**, 669–677.

Handa, K. (2002): Quasi-invariance and reversibility in the Fleming–Viot process. *Probab. Theory Related Fields* **122**, 545–566.

Hara, T. and Slade, G. (2000): The scaling limit of the incipient infinite cluster in high-dimensional percolation II: Integrated super-Brownian excursion. *J. Math. Phys.* **41**, 1244–1293.

Harris, T.E. (1963): *The Theory of Branching Processes.* Springer, Berlin.

He, H. (2009): Discontinuous superprocesses with dependent spatial motion. *Stochastic Process. Appl.* **119**, 130–166.

He, H. (2010): Rescaled Lotka–Volterra models converge to super-stable processes. *J. Theoret. Probab.* In press.

Hewitt, E. and Stromberg, K. (1965): *Real and Abstract Analysis.* Springer, Berlin.

van der Hofstad, R. (2006): Infinite canonical super-Brownian motion and scaling limits. *Commun. Math. Phys.* **265**, 547–583.

van der Hofstad, R. and Slade, G. (2003): Convergence of critical oriented percolation to super-Brownian motion above $4+1$ dimensions. *Ann. Inst. H. Poincaré Probab. Statist.* **39**, 413–485.

Hong, W.M. (2002): Moderate deviation for super-Brownian motion with super-Brownian immigration. *J. Appl. Probab.* **39**, 829–838.

Hong, W.M. (2003): Large deviations for the super-Brownian motion with super-Brownian immigration. *J. Theoret. Probab.* **16**, 899–922.

Hong, W.M. (2005): Quenched mean limit theorems for the super-Brownian motion with super-Brownian immigration. *Infin. Dimens. Anal. Quantum Probab. Relat. Top.* **8**, 383–396.

Hong, W.M. and Li, Z.H. (1999): A central limit theorem for super Brownian motion with super Brownian immigration. *J. Appl. Probab.* **36**, 1218–1224.

Hong, W.M. and Zeitouni, O. (2007): A quenched CLT for super-Brownian motion with random immigration. *J. Theoret. Probab.* **20**, 807–820.

Hsu, P. (1986): Brownian exit distribution of a ball. In: *Seminar on Stochastic Processes, 1985* (Gainesville, Fla., 1985), 108–116. Progr. Probab. Statist. **12**. Birkhäuser, Boston, MA.

Ikeda, N., Nagazawa, M. and Watanabe, S. (1968a): Branching Markov processes (I). *J. Math. Kyoto Univ.* **8**, 233–278.

Ikeda, N., Nagazawa, M. and Watanabe, S. (1968b): Branching Markov processes (II). *J. Math. Kyoto Univ.* **8**, 365–410.

Ikeda, N., Nagazawa, M. and Watanabe, S. (1969): Branching Markov processes (III). *J. Math. Kyoto Univ.* **9**, 95–160.

Ikeda, N. and Watanabe, S. (1989): *Stochastic Differential Equations and Diffusion Processes.* 2nd Ed. North-Holland, Amsterdam; Kodansha, Tokyo.

Iscoe, I. (1986): A weighted occupation time for a class of measure-valued branching processes. *Probab. Theory Related Fields* **71**, 85–116.

Iscoe, I. (1988): On the supports of measure-valued critical branching Brownian motion. *Ann. Probab.* **16**, 200–221.

Jacod, J. and Shiryaev, A.N. (2003): *Limit Theorems for Stochastic Processes.* 2nd Ed. Springer, Berlin.

Jagers, P. (1975): *Branching Processes with Biological Applications.* Wiley, New York.

Jagers, P. (1995): Branching processes as population dynamics. *Bernoulli* **1**, 191–200.

Jiřina, M. (1958): Stochastic branching processes with continuous state space. *Czech. Math. J.* **8**, 292–313.

Jiřina, M. (1964): Branching processes with measure-valued states. In: *1964 Trans. 3rd Prague Conf. Information Theory, Statist. Decision Functions, Random Processes* (Liblice, 1962), 333–357. Publ. House Czech. Acad. Sci., Prague.

Kaj, I. and Sagitov, S. (1998): Limit processes for age-dependent branching particle systems. *J. Theoret. Probab.* **11**, 225–257.

Kallenberg, O. (1975): *Random measures.* Academic Press, New York.

Kallenberg, O. (2008): Some local approximations of Dawson–Watanabe superprocesses. *Ann. Probab.* **36**, 2176–2214.

Kallianpur, G. and Xiong, J. (1995): *Stochastic differential equations in infinite-dimensional spaces.* Inst. Math. Statist., Hayward, CA.

Kawazu, K. and Watanabe, S. (1971): Branching processes with immigration and related limit theorems. *Theory Probab. Appl.* **16**, 36–54.

Keller-Ressel, M., Schachermayer, W. and Teichmann, J. (2010): Affine processes are regular. *Probab. Theory Related Fields.* In press.

Kelley, J.L. (1955): *General Topology.* Springer, Berlin.

Klenke, A. (2000): A review on spatial catalytic branching. In: *Stochastic Models* (Ottawa, ON, 1998), 245–263. CMS Conf. Proc. **26**. Amer. Math. Soc., Providence, RI.

Klenke, A. (2003): Catalytic branching and the Brownian snake. *Stochastic Process. Appl.* **103**, 211–235.

Knight, F. (1963): Random walks and a sojourn density process of Brownian motion. *Trans. Amer. Math. Soc.* **107**, 56–86.

Konno, N. and Shiga, T. (1988): Stochastic partial differential equations for some measure-valued diffusions. *Probab. Theory Related Fields* **79**, 201–225.

Krylov, N.V. (1996): On L_p-theory of stochastic partial differential equations in the whole space. *SIAM J. Math Anal.* **27**, 313–340.

Krylov, N.V. (1997): SPDE's and superdiffusions. *Ann. Probab.* **25**, 1789–1809.

Kurtz, T.G. (1998): Martingale problems for conditional distributions of Markov processes. *Elect. J. Probab.* **3**, Paper 9, 1–29.

Kurtz, T.G. and Ocone, D.L. (1988): Unique characterization of conditional distributions in non-linear filtering. *Ann. Probab.* **16**, 80–107.

Kurtz, T.G. and Xiong, J. (1999): Particle representations for a class of SPEDs. *Stochastic Process. Appl.* **83**, 103–126.

Kuznetsov, S.E. (1994): Regularity properties of a supercritical superprocess. In: *The Dynkin Festschrift*, 221–235. Progr. Probab. **34**. Birkhäuser, Boston, MA.

Kyprianou, A.E. and Pardo, J.C. (2008): Continuous-state branching processes and self-similarity. *J. Appl. Probab.* **45**, 1140–1160.

Lalley, S.P. (2009): Spatial epidemics: Critical behavior in one dimension. *Probab. Theory Related Fields* **144**, 429–469.

Lambert, A. (2007): Quasi-stationary distributions and the continuous-state branching process conditioned to be never extinct. *Elect. J. Probab.* **12**, 420–446.

Lamperti, J. (1967a): The limit of a sequence of branching processes. *Z. Wahrsch. verw. Geb.* **7**, 271–288.

Lamperti, J. (1967b): Continuous state branching processes. *Bull. Amer. Math. Soc.* **73**, 382–386.

Le Gall, J.-F. (1991): Brownian excursions, trees and measure-valued branching processes. *Ann. Probab.* **19**, 1399–1439.

Le Gall, J.-F. (1993): A class of path-valued Markov processes and its applications to superprocesses. *Probab. Theory Related Fields* **95**, 25–46.

Le Gall, J.-F. (1995): The Brownian snake and solution of $\Delta u = u^2$ in a domain. *Probab. Theory Related Fields* **102**, 393–432.

Le Gall, J.-F. (1999): *Spatial Branching Processes, Random Snakes and Partial Differential Equations*. Lectures in Mathematics ETH Zürich, Birkhäuser, Basel.

Le Gall, J.-F. (2005): Random trees and applications. *Probab. Surv.* **2**, 245–311.

Le Gall, J.-F. (2007): The topological structure of scaling limits of large planar maps. *Invent. Math.* **169**, 621–670.

Le Gall, J.-F. and Le Jan, Y. (1998a): Branching processes in Lévy processes: The exploration process. *Ann. Probab.* **26**, 213–252.

Le Gall, J.-F. and Le Jan, Y. (1998b): Branching processes in Lévy processes: Laplace functionals of snakes and superprocesses. *Ann. Probab.* **26**, 1407–1432.

Le Gall, J.-F. and Paulin, F. (2008): Scaling limits of bipartite planar maps are homeomorphic to the 2-sphere. *Geomet. Funct. Anal.* **18**, 893–918.

Leduc, G. (2000): The complete characterization of a general class of superprocesses. *Probab. Theory Related Fields* **116**, 317–358.

Leduc, G. (2006): Martingale problem for superprocesses with non-classical branching functional. *Stochastic Process. Appl.* **116**, 1468–1495.

Lescot, P. and Röckner, M. (2002): Generators of Mehler-type semigroups as pseudo-differential operators. *Infin. Dimens. Anal. Quantum Probab. Relat. Top.* **5**, 297–315.

Lescot, P. and Röckner, M. (2004): Perturbations of generalized Mehler semigroups and applications to stochastic heat equations with Lévy noise and singular drift. *Potent. Anal.* **20**, 317–344.

Li, Z.H. (1991): Integral representations of continuous functions. *Chinese Sci. Bull. Chinese Ed.* **36**, 81–84. *English Ed.* **36**, 979–983.

Li, Z.H. (1992a): A note on the multitype measure branching process. *Adv. Appl. Probab.* **24**, 496–498.

Li, Z.H. (1992b): Measure-valued branching processes with immigration. *Stochastic Process. Appl.* **43**, 249–264.

Li, Z.H. (1995/6): Convolution semigroups associated with measure-valued branching processes. *Chinese Sci. Bull. Chinese Ed.* **40**, 2018–2021. *English Ed.* **41**, 276–280.

Li, Z.H. (1996): Immigration structures associated with Dawson–Watanabe superprocesses. *Stochastic Process. Appl.* **62**, 73–86.

Li, Z.H. (1998a): Immigration processes associated with branching particle systems. *Adv. Appl. Probab.* **30**, 657–675.

Li, Z.H. (1998b): Entrance laws for Dawson–Watanabe superprocesses with non-local branching. *Acta Math. Sci. English Ed.* **18**, 449–456.

Li, Z.H. (1999): Measure-valued immigration diffusions and generalized Ornstein–Uhlenbeck diffusions. *Acta Math. Appl. Sinica English Ser.* **15**, 310–320.

Li, Z.H. (2000): Asymptotic behavior of continuous time and state branching processes. *J. Austral. Math. Soc. Ser. A* **68**, 68–84.

Li, Z.H. (2002): Skew convolution semigroups and related immigration processes. *Theory Probab. Appl.* **46**, 274–296.

Li, Z.H. (2006): A limit theorem for discrete Galton–Watson branching processes with immigration. *J. Appl. Probab.* **43**, 289–295.

Li, Z.H. and Ma, C.H. (2008): Catalytic discrete state branching models and related limit theorems. *J. Theoret. Probab.* **21**, 936–965.

Li, Z.H. and Shiga, T. (1995): Measure-valued branching diffusions: Immigrations, excursions and limit theorems. *J. Math. Kyoto Univ.* **35**, 233–274.

Li, Z.H., Shiga, T. and Yao, L.H. (1999): A reversibility problem of Fleming–Viot processes. *Elect. Comm. Probab.* **4**, 65–76.

Li, Z.H., Wang, H. and Xiong, J. (2004): A degenerate stochastic partial differential equation for superprocesses with singular interaction. *Probab. Theory Related Fields* **130**, 1–17.

Li, Z.H., Wang, H. and Xiong, J. (2005): Conditional log-Laplace functionals of immigration superprocesses with dependent spatial motion. *Acta Appl. Math.* **88**, 143–175.

Li, Z.H. and Zhang, M. (2006): Fluctuation limit theorems of immigration superprocesses with small branching. *Statist. Probab. Letters* **76**, 401–411.

Liggett, T.M. (1985): *Interacting Particle Systems.* Springer, Berlin.

Liggett, T.M. (1999): *Stochastic Interacting Systems: Contact, Voter and Exclusion Processes.* Springer, Berlin.

Limic, V. and Sturm, A. (2006): The spatial Λ-coalescent. *Elect. J. Probab.* **11**, 363–393.

Linde W. (1986): *Probability in Banach Spaces: Stable and Infinitely Divisible Distributions.* Wiley, New York.

Liu, L. and Mueller, C. (1989): On the extinction of measure-valued critical branching Brownian motion. *Ann. Probab.* **17**, 1463–1465.

Liu, R.L. and Ren, Y.X. (2009): Some properties of superprocesses conditioned on non-extinction. *Sci. China Ser. A* **52**, 771–784.

Liu, R.L., Ren, Y.X. and Song, R. (2009): $L \log L$ criterion for a class of superdiffusions. *J. Appl. Probab.* **46**, 479–496.

Ma, C.H. (2009): A limit theorem of two-type Galton–Watson branching processes with immigration. *Statist. Probab. Letters* **79**, 1710–1716.

Ma, Z.M. and Xiang, K.N. (2001): Superprocesses of stochastic flows. *Ann. Probab.* **29**, 317–343.

Malliavin, P. (1997): *Stochastic Analysis.* Springer, Berlin.

Méléard, M. (1996): Asymptotic behaviour of some interacting particle systems: McKean–Vlasov and Boltzmann models. In: *Probabilistic Models for Nonlinear Partial Differential Equations* (Montecatini Terme, 1995), 42–95. Lecture Notes Math. **1627**. Springer, Berlin.

Méléard, M. and Roelly, S. (1993): Interacting measure branching processes: some bounds for the support. *Stochastics Stochastics Rep.* **44**, 103–121.

Mitoma, I. (1983): Tightness of probabilities on $C([0, 1], \mathscr{S})$ and $D([0, 1], \mathscr{S}')$. *Ann. Probab.* **11**, 989–999.

Mörters, P. and Vogt, P. (2005): A construction of catalytic super-Brownian motion via collision local time. *Stochastic Process. Appl.* **115**, 77–90.

Mselati, B. (2004): *Classification and Probabilistic Representation of the Positive Solutions of a Semilinear Elliptic Equation.* Memoirs Amer. Math. Soc. **168**, no. 798. Providence, RI.

Mueller, C. (1998): The heat equation with Lévy noise. *Stochastic Process. Appl.* **74**, 67–82.

Mueller, C. (2009): Some tools and results for parabolic stochastic partial differential equations. In: *A Minicourse on Stochastic Partial Differential Equations*, 111–144. Lecture Notes Math. **1962**. Springer, Berlin.

Mytnik, L. (1998a): Weak uniqueness for the heat equation with noise. *Ann. Probab.* **26**, 968–984.

Mytnik, L. (1998b): Uniqueness for a mutually catalytic branching model. *Ann. Probab.* **112**, 245–253.

Mytnik, L. (2002): Stochastic partial differential driven by stable noise. *Probab. Theory Related Fields* **123**, 157–201.

Overbeck, L. (1993): Conditioned super-Brownian motion. *Probab. Theory Related Fields* **96**, 545–570.

Pakes, A.G. (1988): Some limit theorems for continuous-state branching processes. *J. Austral. Math. Soc. (Ser. A).* **44**, 71–87.

Pakes, A.G. (1999): Revisiting conditional limit theorems for mortal simple branching processes. *Bernoulli* **5**, 969–998.

Pakes, A.G. and Trajstman, A.C. (1985): Some properties of continuous-state branching processes, with applications to Bartoszynski's virus model. *Adv. Appl. Probab.* **17**, 23–41.

Parthasarathy, K.R. (1967): *Probability Measures on Metric Spaces*. Academic Press, New York.

Patie, P. (2009): Exponential functional of a new family of Lévy processes and self-similar continuous state branching processes with immigration. *Bull. Sci. Math.* **133**, 355–382.

Pazy, A. (1983): *Semigroups of Linear Operators and Applications to Partial Differential Equations*. Springer, Berlin.

Perkins, E.A. (1988): A space-time property of a class of measure-valued branching diffusions. *Trans. Amer. Math. Soc.* **305**, 743–795.

Perkins, E.A. (1992): Conditional Dawson–Watanabe processes and Fleming–Viot processes. In: *Seminar on Stochastic Processes, 1991* (Los Angeles, CA, 1991), 143–156. Progr. Probab. **29**. Birkhäuser, Boston, MA.

Perkins, E.A. (1995): *On the Martingale Problem for Interacting Measure-valued Branching Diffusions*. Memoirs Amer. Math. Soc. **115**, no. 549. Providence, RI.

Perkins, E.A. (2002): Dawson–Watanabe superprocesses and measure-valued diffusions. In: *Ecole d'Eté de Probabilités de Saint-Flour XXIX-1999*, 125–324. Lecture Notes Math. **1781**. Springer, Berlin.

Pinsky, M.A. (1972): Limit theorems for continuous state branching processes with immigration. *Bull. Amer. Math. Soc.* **78**, 242–244.

Pinsky, R.G. (1995): On the large time growth rate of the support of supercritical super-Brownian motion. *Ann. Probab.* **23**, 1748–1754.

Pinsky, R.G. (1996): Transience, recurrence and local extinction properties of the support for supercritical finite measure-valued diffusions. *Ann. Probab.* **24**, 237–267.

Pitman, J. (1999): Coalescents with multiple collisions. *Ann. Probab.* **27**, 1870–1902.

Pitman, J. (2006): Combinatorial stochastic processes. In: *Ecole d'Eté de Probabilités de Saint-Flour XXXII-2002*. Lecture Notes Math. **1875**. Springer, Berlin.

Priola, E. and Zabczyk, J. (2010): Structural properties of semilinear SPDEs driven by cylindrical stable processes. *Probab. Theory Related Fields*. In press.

Ray, D.B. (1963): Sojourn times of a diffusion process. *Illinois J. Math.* **7**, 615–630.

Reimers, M. (1989): One dimensional stochastic differential equations and the branching measure diffusion. *Probab. Theory Related Fields* **81**, 319–340.

Ren, Y.X. (2001): Construction of super-Brownian motions. *Stochastic Anal. Appl.* **19**, 103–114.

Ren, Y.X., Song R. and Wang H. (2009): A class of stochastic partial differential equations for interacting superprocesses on a bounded domain. *Osaka J. Math.* **46**, 373–401.

Revuz, D. and Yor, M. (1999): *Continuous Martingales and Brownian Motion*. 3rd Ed. 3rd Pr. Springer, Berlin.

Rhyzhov, Y.M. and Skorokhod, A.V. (1970): Homogeneous branching processes with a finite number of types and continuous varying mass. *Theory Probab. Appl.* **15**, 704–707.

Röckner, M. and Wang, F.Y. (2003): Harnack and functional inequalities for generalized Mehler semigroups. *J. Funct. Anal.* **203**, 237–261.

Roelly, S. (1986): A criterion of convergence of measure-valued processes: Application to measure branching processes. *Stochastics* **17**, 43–65.

Roelly, S. and Rouault, A. (1989): Processus de Dawson–Watanabe conditionné par le futur lointain. *C. R. Acad. Sci. Paris Ser. I Math.* **309**, 867–872.

Sagitov, S. (1999): The general coalescent with asynchronous mergers of ancestral lines. *J. Appl. Probab.* **36**, 1116–1125.

Salisbury, T.S. and Verzani, J. (1999): On the conditioned exit measures of super Brownian motion. *Probab. Theory Related Fields* **115**, 237–285.

Salisbury, T.S. and Verzani, J. (2000): Non-degenerate conditionings of the exit measures of super Brownian motion. *Stochastic Process. Appl.* **87**, 25–52.

Sato, K. (1999): *Lévy Processes and Infinitely Divisible Distributions.* Cambridge Univ. Press, Cambridge.

Schied, A. (1996): Sample large deviations for super-Brownian motion. *Probab. Theory Related Fields* **104**, 319–347.

Schied, A. (1999): Existence and regularity for a class of infinite-measure (ξ, ψ, K)-superprocesses. *J. Theoret. Probab.* **12**, 1011–1035.

Schmuland, B. and Sun W. (2001): On the equation $\mu_{t+s} = \mu_s * T_s \mu_t$. *Statist. Probab. Letters* **52**, 183–188.

Schmuland, B. and Sun, W. (2002): A cocycle proof that reversible Fleming–Viot processes have uniform mutation. *C. R. Math. Acad. Sci., Soc. Royale Canad.* **24**, 124–128.

Schweinsberg, J. (2000): A necessary and sufficient condition for the Λ-coalescent to come down from infinity. *Elect. Comm. Probab.* **5**, 1–11.

Schweinsberg, J. (2003): Coalescent processes obtained from supercritical Galton–Watson processes. *Stochastic Process. Appl.* **106**, 107–139.

Sharpe, M. (1988): *General Theory of Markov Processes.* Academic Press, New York.

Shiga, T. (1990): A stochastic equation based on a Poisson system for a class of measure-valued diffusion processes. *J. Math. Kyoto Univ.* **30**, 245–279.

Silverstein, M.L. (1967/8): A new approach to local time. *J. Math. Mech.* **17**, 1023–1054.

Silverstein, M.L. (1968): Markov processes with creation of particles. *Z. Wahrsch. verw. Geb.* **9**, 235–257.

Silverstein, M.L. (1969): Continuous state branching semigroups. *Z. Wahrsch. verw. Geb.* **14**, 96–112.

Skoulakis, G. and Adler, R.J. (2001): Superprocess over a stochastic flow. *Ann. Appl. Probab.* **11**, 488–543.

Slade, G. (2002): Scaling limits and super-Brownian motion. *Notices Amer. Math. Soc.* **49**, 1056–1067.

Stannat, W. (2003): Spectral properties for a class of continuous state branching processes with immigration. *J. Funct. Anal.* **201**, 185–227.

Stroock, D.W. and Varadhan, S.R.S. (1979): *Multidimensional Diffusion Processes.* Springer, Berlin.

Treves, F. (1967): *Topological Vector Spaces, Distributions and Kernels.* Academic Press, New York.

Tribe, R. (1992): The behavior of superprocesses near extinction. *Ann. Probab.* **20**, 286–311.

Tribe, R. (1994): A representation for super Brownian motion. *Stochastic Process. Appl.* **51**, 207–219.

Walsh, J.B. (1986): An introduction to stochastic partial differential equations. In: *Ecole d'Eté de Probabilités de Saint-Flour XIV-1984*, 265–439. Lecture Notes Math. **1180**. Springer, Berlin.

Wang, F.Y. (2005): *Functional Inequalities, Markov Semigroups and Spectral Theory.* Science Press, Beijing.

Wang, H. (1997a): State classification for a class of measure-valued branching diffusions in a Brownian medium. *Probab. Theory Related Fields* **109**, 39–55.

Wang, H. (1998a): A class of measure-valued branching diffusions in a random medium. *Stochastic Anal. Appl.* **16**, 753–786.

Wang, Y.J. (1997b): A proof of the persistence criterion for a class of superprocesses. *J. Appl. Probab.* **34**, 559–563.

Wang, Y.J. (1998b): Criterion on the limits of superprocesses. *Sci. China Ser. A* **41**, 849–858.

Wang, Y.J. (2002): An alternative approach to super-Brownian motion with a locally infinite branching mass. *Stochastic Process. Appl.* **102**, 221–233.

Watanabe, S. (1968): A limit theorem of branching processes and continuous state branching processes. *J. Math. Kyoto Univ.* **8**, 141–167.

Watanabe, S. (1969): On two dimensional Markov processes with branching property. *Trans. Amer. Math. Soc.* **136**, 447–466.

Wentzell, A.D. (1981): *A Course in the Theory of Stochastic Processes*. McGraw-Hill, New York.

Winter, A. (2007): *Tree-valued Markov limit dynamics*. Habilitationsschrift. Mathematisches Institut, Universität Erlangen-Nürnberg, Erlangen.

Xiang, K.N. (2009): Measure-valued flows given consistent exchangeable families. *Acta Appl. Math.* **105**, 1–44.

Xiang, K.N. (2010): Explicit Schilder type theorem for super-Brownian motions. *Comm. Pure Appl. Math.* **63**, 1381–1431.

Xiong, J. (2004): A stochastic log-Laplace equation. *Ann. Probab.* **32**, 2362–2388.

Zhang, M. (2004a): Large deviations for super-Brownian motion with immigration. *J. Appl. Probab.* **41**, 187–201.

Zhang, M. (2004b): Moderate deviation for super-Brownian motion with immigration. *Sci. China Ser. A* **47**, 440–452.

Zhang, M. (2005a): Functional central limit theorem for the super-Brownian motion with super-Brownian immigration. *J. Theoret. Probab.* **18**, 665–685.

Zhang, M. (2005b): A large deviation principle for a Brownian immigration particle system. *J. Appl. Probab.* **42**, 1120–1133.

Zhang, M. (2008): Some scaled limit theorems for an immigration super-Brownian motion. *Sci. China Ser. A* **51**, 203–214.

Zhang, Q. (2002): The boundary behavior of heat kernels of Dirichlet Laplacians. *J. Differential Equations* **182**, 416–430.

Zhao, X.L. (1994): Excessive functions of a class of superprocesses. *Acta Math. Sci. English Ser.* **14**, 393–399.

Zhao, X.L. (1996): Harmonic functions of superprocesses and conditioned superprocesses. *Sci. China Ser. A* **39**, 1268–1279.

Zhou, X. (2007): A superprocess involving both branching and coalescing. *Ann. Inst. H. Poincaré Probab. Statist.* **43**, 599–618.

Zhou, X. (2008): A zero-one law of almost sure local extinction for $(1 + \beta)$-super-Brownian motion. *Stochastic Process. Appl.* **118**, 1982–1996.

Subject Index

A

absolutely continuous linear part 263
adapted process 305
adapted process, $\mathscr{B}^{\bullet}(E)$- 305
additive functional 321
admissible additive functional 33, 321
affine process 85
affine semigroup 85
age-structured superprocess 133
announcing sequence 304
augmentation by a probability 304
augmentation of a filtration 310
augmented filtration 304, 310

B

Bernstein polynomials 21
binary local branching mechanism 46
Borel σ-algebra 1, 303
Borel function 1
Borel measure 2
Borel right process 313
Borel transition semigroup 307
bounded entrance law 307
bounded kernel 303
bounded pointwise convergence 301
bounded set in a nuclear space 284
branch point 315
branching mechanism 42, 66
branching particle system 88
branching property 29
Brownian motion with drift 81
Brownian snake, ξ- 96

C

càdlàg path 305

càdlàg process 305
Campbell measure 54
canonical entrance rule 212
catalyst measure 46
catalytic super-Brownian motion 46
CB-process 45
CBI-process 66
Chapman–Kolmogorov equation 306, 329
characteristic function 27
characteristic functional 256
classical Mehler formula 259
classical OU-process 258
closable entrance law 307
closable entrance path 266
closed entrance law 307
cluster representation 18
complete measure space 302
completely monotone function 21
completion, μ- 302
compound Poisson random measure 15
conditioned superprocess 138
conservative process 315
conservative resolvent 307
conservative transition semigroup 307
continuous process 305
continuous-state branching process 45
continuous-state branching process with immigration 66
convolution 15
countably Hilbert nuclear space 285
covariance measure 161
critical CB-process 58
critical superprocess 48
cumulant semigroup 31, 58

Z. Li, *Measure-Valued Branching Markov Processes*,
Probability and Its Applications, DOI 10.1007/978-3-642-15004-3,
© Springer-Verlag Berlin Heidelberg 2011

Symbol Index